Cases on Green Energy and Sustainable Development

Peter Yang
Case Western Reserve University, USA

A volume in the Practice, Progress, and Proficiency in Sustainability (PPPS) Book Series

IGI Global
DISSEMINATOR OF KNOWLEDGE

Published in the United States of America by
IGI Global
Engineering Science Reference (an imprint of IGI Global)
701 E. Chocolate Avenue
Hershey PA, USA 17033
Tel: 717-533-8845
Fax: 717-533-8661
E-mail: cust@igi-global.com
Web site: http://www.igi-global.com

Library of Congress Cataloging-in-Publication Data

Names: Yang, Peter, editor.
Title: Cases on green energy and sustainable development / Peter Yang, editor.
Description: Hershey, PA : Engineering Science Reference, [2019] | Includes bibliographical references.
Identifiers: LCCN 2018055435| ISBN 9781522585596 (hardcover) | ISBN 9781522585602 (softcover) | ISBN 9781522585619 (ebook)
Subjects: LCSH: Clean energy. | Renewable energy sources. | Sustainable development.
Classification: LCC TJ808 .C36 2019 | DDC 333.79/4--dc23 LC record available at https://lccn.loc.gov/2018055435

This book is published in the IGI Global book series Practice, Progress, and Proficiency in Sustainability (PPPS) (ISSN: 2330-3271; eISSN: 2330-328X)

British Cataloguing in Publication Data
A Cataloguing in Publication record for this book is available from the British Library.

Practice, Progress, and Proficiency in Sustainability (PPPS) Book Series

ISSN:2330-3271
EISSN:2330-328X

Editor-in-Chief: Ayman Batisha, International Sustainability Institute, Egypt

MISSION

In a world where traditional business practices are reconsidered and economic activity is performed in a global context, new areas of economic developments are recognized as the key enablers of wealth and income production. This knowledge of information technologies provides infrastructures, systems, and services towards sustainable development.

The **Practices, Progress, and Proficiency in Sustainability (PPPS) Book Series** focuses on the local and global challenges, business opportunities, and societal needs surrounding international collaboration and sustainable development of technology. This series brings together academics, researchers, entrepreneurs, policy makers and government officers aiming to contribute to the progress and proficiency in sustainability.

COVERAGE

- Eco-Innovation
- Strategic Management of IT
- Socio-Economic
- E-Development
- Intellectual Capital
- Innovation Networks
- Outsourcing
- ICT and knowledge for development
- Global Business
- Sustainable Development

IGI Global is currently accepting manuscripts for publication within this series. To submit a proposal for a volume in this series, please contact our Acquisition Editors at Acquisitions@igi-global.com or visit: http://www.igi-global.com/publish/.

The Practice, Progress, and Proficiency in Sustainability (PPPS) Book Series (ISSN 2330-3271) is published by IGI Global, 701 E. Chocolate Avenue, Hershey, PA 17033-1240, USA, www.igi-global.com. This series is composed of titles available for purchase individually; each title is edited to be contextually exclusive from any other title within the series. For pricing and ordering information please visit http://www.igi-global.com/book-series/practice-progress-proficiency-sustainability/73810. Postmaster: Send all address changes to above address.

Titles in this Series

For a list of additional titles in this series, please visit:
http://www.igi-global.com/book-series/practice-progress-proficiency-sustainability/73810

Driving the Development, Management, and Sustainability of Cognitive Cities
Kiran Ahuja (DAV Institute of Engineering and Technology, India) and Arun Khosla (Dr. B.R. Ambedkar National Institute of Technoloy, India)
Engineering Science Reference • ©2019 • 337pp • H/C (ISBN: 9781522580850) • US $225.00

Technology-Driven Innovation in Gulf Cooperation Council (GCC) Countries ...
Anna Visvizi (Deree – The American College of Greece, Greece & Effat University, Saudi Arabia) Saad Bakry (King Saud University, Saudi Arabia) Miltiadis D. Lytras (Deree – The American College of Greece, Greece & Effat University, Saudi Arabia) and Wadee Alhalabi (Effat University, Saui Arabia)
Business Science Reference • ©2019 • 180pp • H/C (ISBN: 9781522590125) • US $195.00

Novel Technologies and Systems for Food Preservation
Pedro Dinis Gaspar (University of Beira Interior, Portugal) and Pedro Dinho da Silva (University of Beira Interior, Portugal)
Engineering Science Reference • ©2019 • 416pp • H/C (ISBN: 9781522578949) • US $345.00

Bioeconomical Solutions and Investments in Sustainable City Development
José G. Vargas-Hernández (University of Guadalajara, Mexico) and Justyna Anna Zdunek-Wielgołaska (Warsaw University of Technology, Poland)
Engineering Science Reference • ©2019 • 306pp • H/C (ISBN: 9781522579588) • US $195.00

Optimizing the Use of Farm Waste and Non-Farm Waste to Increase Productivity and...
Leighton Naraine (Clarence Fitzroy Bryant College, St. Kitts-Nevis)
Engineering Science Reference • ©2019 • 223pp • H/C (ISBN: 9781522579342) • US $155.00

Handbook of Research on International Collaboration, Economic Development, and ...
Vasilii Erokhin (Harbin Engineering University, China) Tianming Gao (Harbin Engineering University, China) and Xiuhua Zhang (Harbin Engineering University, China)
Business Science Reference • ©2019 • 703pp • H/C (ISBN: 9781522569541) • US $235.00

For an entire list of titles in this series, please visit:
http://www.igi-global.com/book-series/practice-progress-proficiency-sustainability/73810

IGI Global
DISSEMINATOR OF KNOWLEDGE

701 East Chocolate Avenue, Hershey, PA 17033, USA
Tel: 717-533-8845 x100 • Fax: 717-533-8661
E-Mail: cust@igi-global.com • www.igi-global.com

Editorial Advisory Board

Table of Contents

Detailed Table of Contents

This case study examines the progress being made by 12 least developed countries (LDCs) in their effort to achieve Sustainable Development Goal 7 (SDG7) – access to clean and sustainable energy for all. Focusing on solar photovoltaics (PV), the authors look at what can be done to further the spread of renewable energy, and the role various actors have to playing in helping these countries to meet SDG7. Furthermore, with countries on the cusp of submitting their revised contributions under the Paris Agreement, they look at the role solar PV can play in helping LDCs to participate in taking action against climate change. After outlining the current policy landscape, and efforts being made within these countries, they look at the obstacles, opportunities, and the role of solar PV going forward. They also look at the steps that policymakers, both national and international, can take to encourage the rapid uptake of renewable energy in developing nations.

As a member of the European Union (EU), Bulgaria has been implementing the EU's policy targets designed to increase the share of renewable energy (RE) use in gross final energy consumption by 2020. The target for Bulgaria, set at 16%, was accomplished eight years earlier than mandated, in 2012. The result of rapid but poorly regulated growth in renewables—seemingly a success story—illustrates the potential pitfalls of RE policy implementation. Having met its target, Bulgaria undertook a series of restrictive policy measures that undermined short-term RE growth, increased regulatory uncertainty and market stagnation. The objective of this chapter is to understand the factors that shaped these unintended policy measures and outcomes. Drawing on key informant interviews, the chapter presents a case study of renewable energy policy implementation in a multi-level governance system and illustrates the boomerang effects associated with top-down policy implementation.

Renewable Energy Sources-Based Electricity (RES-E) plays a key role in sustainable development – of meeting current energy demands, without adding to global warming concerns. However, as of 2017, only 8.5% of the total electricity generation came from RES-E. To boost this contribution, countries rely on strong legislative and policy support/tools. This case focuses on studying the legislative or regulatory frameworks put in place by the top three developed countries, and compares it with three developing countries, each of which are forerunners in RES-E, as of 2017. The comparative study suggests that while no single policy can be credited with the success behind rising RES-E in these countries, two key incentives are most important – namely feed-in-tariffs and renewable purchase obligations. Feed-in-tariffs act as floor price guarantee to the generator and renewable purchase obligations assures the generator of quantum of sale of the RES-E generated. When combined, these two incentives remain the most trusted policy tools even today for countries starting their journey in increasing their RES-E footprints.

Renewable energy generation is a fundamental component of the transition to a low-carbon economy. The world needs to invest up to USD 600 million annually to meet the electricity demand in a sustainable way whereas the current investment level stands at USD 280 billion. Scaling-up the current level of investment requires a larger implication of the private sectors and a different role for the public sector. The challenge lays in the fact that different investors are motivated by a different risk and return profiles. The current chapter presents the trends in renewable energy financial flows and investment vehicles. It looks at the risks associated with the investment in renewable energy and the relevant risk mitigation instruments. Finally, it applies these concepts to the case of the Lake Turkana wind farm in Kenya, a project that faced many challenges and involved more than 15 investors.

Sub-Saharan Africa (SSA) is one of the least electrified regions in the world and also a region that is characterized by poverty and inequality due to high levels of climate change vulnerability. In order to reduce greenhouse gas emissions and facilitate the attainment of the Sustainable Development Goals, SSA policymakers are compelled to devise new innovative strategies and policies to enhance investments in renewable energy technologies (RETs). Accordingly, this chapter provides an assessment of some strategies to accelerate RET deployment and the potential of polycentric governance systems to improve RET deployment. The assessment concluded that even though renewable energy investments through climate finance and microfinance modalities are not at a level sufficient to ensure that universal energy access can be attained in the region, SSA can still accelerate its progress on RET deployment by utilizing nationally determined contributions as instruments to direct South-South aid, trade, and investments into priority renewable energy sub-sectors.

The City of Boulder, Colorado has for 10 years attempted to break up with its electric utility, Xcel Energy, in favor of forming its own municipal utility. Environmental proponents of the separation argue that a democratically accountable, local utility would be better suited to achieve Boulder's ambitious environmental and climate action goals. However, other environmentalists disagree and instead argue that Xcel Energy is a willing and capable environmental partner. This case examines this conflict in order to illustrate a divide in Boulder's environmental community, which mirrors a divide in the larger environmental movement, between structural environmentalists on the one hand and neoliberal environmentalists on the other. The case offers a review of the theoretical work that informs these conflicting perspectives. Finally, it analyzes structural and neoliberal sentiments expressed in the opinion pages of the city's newspaper in order to demonstrate how they intervene and shape Colorado electricity politics.

This multi-faceted case study investigates sustainable land development using permaculture as the design tool. Permaculture, coined by Bill Mollison and David Holmgren, is a sustainable design theory that builds off three ethical principles used to produce a set of guidelines to follow in order to create an ecologically focused project. Permaculture, a contraction of perma-nent and initially agri-culture, has evolved to perma-nent and culture, understanding that without agriculture, culture is impossible. This chapter begins with an overview of the environmental issues followed by a description and brief history of sustainable development, with emphasis placed on the United Nations (UN) Sustainable Development Goals (SDGs). The focus will be a three-part case study examining different scales (urban, suburban, and rural) of permaculture land development in the midwestern United States (U.S.). These permaculture designs will illustrate how SDGs can be achieved to forge a sustainable future.

This chapter identifies the content of airlines' operations strategy and reports the strategy patterns adopted by the airlines of each region. A detailed configuration of the airlines' green operational strategy is developed, using the content analysis of the sustainability reports from 23 airlines in five regions (North America, Europe, Asia, the Middle East, and South America). The green operational strategy adopted by each region is identified; each region adopted a green pattern that was unlike those of any other region. The indicative models for each region and across regions are developed by using a simple and special tailored quantitative analytical technique. The results of this chapter raise a set of questions about the impact of contextual factors on whichever green strategy pattern is adopted, indicating the need to conduct more in-depth analysis of green actions. This is one of a few studies to have developed a comprehensive definition of airlines' green operations strategy and explore the green strategy patterns adopted by airlines from different regions.

This chapter outlines the relevance of sustainable development as a key for family firm success and its ability to guarantee long-term survival and spread positive effects in social, economic, and natural environments. By particularly analyzing a single case study of a Sardinian family business, this work explores the intertwined relationships among sustainability, owner innovativeness, and firm success. Moreover, the importance of family businesses and the scarcity of the study conducted to date have suggested a focus on how these companies tackle sustainability challenges.

The Ottawa Renewable Energy Cooperative is considering installing solar modules on the roofs of two buildings while they stay connected to the public electricity grid. Solar power produced over their own needs would be sent to the public electricity grid for a credit on their electricity bill. When they need more power than they are generating, these buildings would purchase electricity from the grid. In addition to paying for the electricity they purchase, they would be subject to a "demand charge" that applies each month to the hour during which their consumption is at a peak for that month. Any electricity consumed during that peak hour would be charged at a rate about 100 times the rate for other hours. The case addresses three questions: (1) Is it profitable for these organizations to install solar on their roofs? (2) Can profitability be increased by adding a battery? and (3) How sensitive is profitability to uncertainty in future electricity prices? The case shows how the answers to these questions depend on the profile of hourly electricity consumption during the day, which is very different from one building to the other.

To reduce reliance on fossil fuels and increase demands for clean energy technology worldwide, there is currently a growing interest in the use of fuel cells as energy-efficient and environmentally-friendly power generators. With this inevitable depletion, fossil fuels will not be able to respond to energy demand for future. Among all major types of fuel cells, hydrogen fuel cells (HFCs) are in the forefront stage and have gained substantial attention for vehicle and portable applications, which is composed of a cathode, an anode, and a PEM. The heart of the fuel cells is membrane electrode assembly (MEA). An electro-deposition technique for preparing the nano-catalyst layer in PEMFCs has been designed, which may enable an increase in the level of Pt utilization currently achieved in these systems. Functionalization process has been done using a mixture of concentrated nitric acid and sulfuric acid in refluxing condition. The hydrocarbon-based polymer membrane has been used as electrolyte part.

The electric grid that has the tendency to communicate two-way and can sense various parameters in the transmission line is termed as smart grid. This chapter deals about the overview of smart grid evolution, characteristics, and operation. There are various benefits in smart grid like improvement in efficiency, adaptive, self-healing, and optimized than conventional grid. The smart grid composition is complex and defined based on standards adaption, technical components perspective, technical perspective, and conceptual reference model perspective. In the architecture of smart grid, the role of advanced metering infrastructure (AMI) plays a vital role to sense, measure, record, and communicate the data from load centre to data centre. AMI consists of smart meter, communication network, data reception, and management system. This chapter also covers the IEEE and IEC standards defined for smart grid operation. It also envisages the barriers in the implementation of smart grids.

This chapter describes technology that ensures reliable pumping of drainage and sewage water during electromechanical and hydro-mechanical transients from blocks of the hydroelectric power stations and pumping stations. As an alternative source of energy, it is proposed to use the energy of the liquid column of the pressure penstock of the stations, and as an auxiliary, to use jet pumps. Transmission of energy to the suction stream is carried out without direct usage of electrical and mechanical energy. During total shutdown of electric power, reliable evacuation of drainage and seepage water and reduction the influence of electromechanical and hydro-mechanical transients on power equipment and pipelines can be ensured with the use of self-regulating jet pumps over a period of several days; this cannot be accomplished by any other pump. The scientific results of the research are recommended to allow efficient use of water and energy resources and to ensure reliable operation of the power equipment of stations, especially in the events of sudden power outages.

With the rapid growth of distributed generation currently, DC microgrids energy system structure is being deployed in parallel with, or independently from, the main power grid network. The DC microgrids energy system structure is designed to provide an effective coordination with the aggregating distributed generators, energy storage, and connected loads. In this sense, the DC microgrids energy system structure can be connected to the grid network or can be off-grid network. In the mode of grid network connected, DC microgrids energy system structure is presented as a controllable entity. When it is necessary, DC microgrids energy system is connected in islanded mode to deliver reliable power to the grid network during the interrupted power supply from the grid network system. Having said that, the DC microgrids energy system structure is encompassed of renewable energy sources, energy storages and loads, and not excluding the grid network transmission. Hence, this chapter proposes to focus on designing and modelling a self-assistive controller using voltage droop method for DC distributed generators and storages which is a part of the DC microgrids energy system structure.

This chapter concerns energy storage technologies. It firstly outlines two popular storage technologies, batteries and supercapacitors, while their working principles are revealed. The key issues of these two technologies, such as costs, key types, capacities, etc., are also discussed. Afterwards, a hybrid electrical energy storage (HEES) system consisting of both technologies are demonstrated where the electrical circuit is illustrated. The design of the system aims to demonstrate different characteristics of these two technologies via their charging and discharging process. A test rig is explained in detail while other components, including a load bank, an inverter, a data acquisition subsystem (both the hardware and the software) are also clarified. The experimental results are illustrated and analyzed thereafter. Also, this chapter presents several other promising technologies where their key features, pros and cons, and core applications are pointedly reviewed. The concerned storage technologies include photovoltaic (PV) systems, pumped hydro-energy storage (PHES), superconducting magnetic energy storage (SMES), gas, and other alternatives sources. The authors provide the readers with a brief insight of various energy storage technologies and the inspiration of developing a low-cost, accessible energy storage system for the reader's own purposes.

The main drivers of the third industrial revolution era were the internet technologies and rise of renewable and distributed energy technologies. Transition to green and decentralized energy resources and digital transformation of the existing industrial

Preface

OVERVIEW OF THE SUBJECT MATTER

Fossil fuels have helped mankind experience and expand industrial and technological revolution, generate and consume unforeseen quantities of goods and services, and facilitate the economic and social development. However, extracting and consuming these carbon-based energy sources have also caused unprecedented environmental and ecological impacts. In addition to releasing various pollutants, such as nitrogen oxides, sulfur dioxide, volatile organic compounds and heavy metals, harmful to air, water and human health, combustion of fossil fuels also caused rapidly increasing mass amount of CO_2 emissions, causing unprecedented atmospheric temperature rise since 1950 and climate change. In addition, massive extraction of finite fossil fuels will ultimately lead to depletion of these energy sources. Both anthropogenic CO_2 emissions and fossil fuel depletion endanger the sustainability of the ecosystem that humankind depends on.

To substantially reduce the irreversible risks and effects of climate change caused by fossil fuel dependent economic activities, the Paris Climate Agreement, signed by 195 countries, set the goal of holding the increase in global average temperature to well below 2°C above pre-industrial levels and to limit the increase to 1.5°C. To reach this goal, the Paris Agreement calls for undertaking economy-wide absolute emission reduction targets in the developed world and enhancing mitigation efforts and are moving over time towards economy-wide emission reduction or limitation targets in the developing countries (UN, 2015).

Energy transformation, or transition from fossil fuels to renewable energy, and energy efficiency, is the primary path towards meeting these needs. UN Sustainable Development Goal 7 demands to substantially increase the share of renewable energy in the global energy mix and the renewable energy share in the total final energy consumption by 2030 (UN, 2018). The REN21 *Renewables 2018 Global Status Report* (GSR) reveals that on the one hand, a revolution in the power sector is driving rapid change towards a renewable energy future, and one the other, the overall transition is not advancing with the speed needed. At the same time, the report points out the urgent need for an accelerated energy transformation: While renewables are progressing positively in the power generation sector, the heating, cooling and transport sectors, which combined account for about 80% of global total final energy demand, are lagging behind (REN21, 2018).

Despite the urgent need for action, there is an astonishing lack of understanding of the benefits and needs of using renewable energy sources and energy sustainability not only to reduce carbon emissions and climate change, but also to grow a sustainable economy and society. On the other hand, there is also a significant lack of public knowledge of the technological and financial challenges of renewable energy technologies that is necessary for the strong public support in increased R&D, rapid improvement and cost reduction, and deployment of renewable energy technologies.

To achieve the emissions reductions envisioned by the Paris Agreement or the UN Sustainable Development Goal 7, it is extremely important to educate citizens, especially students, train professionals, raise public awareness, and increase public involvement in and public access to information and cooperation at all levels on energy transformation and sustainable development.

OBJECTIVE

The objective of *Cases on Green Energy and Sustainable Development* is to contribute to meeting these needs of informing, educating and training students and young professionals about renewable energy and energy efficiency technologies and sustainable development. Since students and young professionals are future stakeholders of the world and will face increased sustainability and energy security issues, they need to be better educated, trained and prepared for the energy transformation and a sustainable future.

This book will provide relevant theoretical frameworks and the latest empirical research findings on the important role the renewable energy and energy efficiency play in the energy transition and sustainable development; covering economics and promotion policies of major renewable energy and energy efficiency technologies; reflecting the latest status of related knowledge, R&D, and market penetration of these technologies; and incorporating a comprehensive, connecting and active learning approach. It is intended to nurture our students and young professionals to help them grow toward comprehensively educated citizens prepared for contributing to energy transition and sustainable development in the world's combat against global warming and climate change and for the universal renewable energy access, especially in the least developed countries (LDCs).

TARGET AUDIENCE

This case book is designed to be used as a textbook for the senior high school social science or science courses, the college general education course, or the introductory to intermediate-level college social science, science or engineering courses. It can be either used for high school courses or introductory level college liberal arts courses on a selective basis, or be used for introductory- or intermediate-level college social science, science or engineering courses on a more complete basis. In addition, it can be used as a reference book for anyone who is interested to become an informed professional, politician, or citizen on energy transition, energy efficiency, and sustainable development.

ORGANIZATION OF THE BOOK

The book is organized into two sections and 17 chapters. Section 1 covers the first nine chapters that are related to strategies, policies, and management of deployment of renewable energy and energy efficiency technologies and sustainable development. Section 2 includes the remaining eight chapters that present case studies on the deployment of renewable energy and energy efficiency technologies. A brief description of each chapter follows:

Chapter 1 provides an overview of countries' progress in solar PV deployment in support of the implementation of nationally determined contributions (NDCs) as the core of the Paris Agreement, and the contribution to achieving the long-term Sustainable Development Goal 7 (SDG 7). By pointing out that one of the world's challenges of today that over one billion people still lack basic access to power, the author's study of the implementation of NDCs and the achievement of SDG 7 – access to affordable, reliable, sustainable and modern energy for all – is of great importance for tackling of the identified development challenges.

Chapter 2 presents a case study on Bulgaria's renewable energy policy implementation in a multi-level governance system, and illustrates the boomerang effects associated with top-down policy implementation. It recognizes this EU member's achievement in having accomplished its EU policy target of the share of renewable energy consumption in gross final energy mix (16%) by 2020 in 2012, eight years earlier. However, the study also identifies the challenges that sprang from the rapid but poorly regulated growth in renewables. For example, a series of restrictive policy measures were implemented since 2012 that undermined short-term RE growth, increased regulatory uncertainty and market stagnation. With this case study, the authors attempt to help the reader understand the factors that shaped the unintended policy measures and outcomes.

Chapter 3 examines promotional policies and legislative support for grid connected renewable energy projects. In grappling with increasing environmental and climate issues caused by increased carbon emissions from increased fossil fuel consumption, countries have promoted investments into the renewable energy generation sector, using various policy and legislative tools. Offered incentives took various forms, such as providing price based incentives to reduce the price of renewable power or quantity based incentives through promising purchases of part or all of the electricity generated from renewable energy sources. Literature suggests that no country has relied upon only one form of incentive, but they devised a "policy package". This chapter provides an overview of some of the widely used legislative levers, which have fostered the growth in RES-E generation in select developed and developing countries. It also includes a brief discussion on problems associated with increased deployment of renewable energy technologies, which indicates that increased green power can also have unintended consequences. Cases from select countries studied in this chapter provide an understanding of best practices in RES-E legislative support mechanisms.

Chapter 4 presents the trends in renewable energy financial flows and investment vehicles, and identifies the related challenges, such as different investors are motivated by different risks and return profiles. Considering that the renewable energy generation plays a fundamental role in the transition to a low-carbon economy, the world needs to scale the renewable energy investment from the current level of USD 280 billion up to the level of USD 600 million annually to meet the electricity demand in a sustainable way. The author suggests that the key of scaling up the current level of investment lies in a larger role of the private sectors and a different role for the public sector. The study examines the risks associated with the investment in renewable energy and the relevant risk mitigation instruments, and applies the findings of this examination to the case of the Lake Turkana wind farm in Kenya, a project that faces many challenges and involves more than fifteen investors.

Chapter 5 reviews the financing and governance strategies for renewable energy deployment in urban Sub-Saharan Africa (SSA), one of the world's least electrified regions and a region that is most stricken by poverty, inequality, and high levels of climate change vulnerability. The author assesses the new innovative strategies and policies to enhance investments in Renewable Energy Technologies (RETs) devised by the SSA policymakers. Accordingly, this chapter provides an assessment of some strategies to accelerate RET deployment and the potential of polycentric governance systems to improve RET deployment in order to reduce greenhouse gas emissions and achieve the Sustainable Development Goals. The assessment is focused on renewable energy microfinance and polycentric governance. The study finds that renewable energy investments through climate finance and microfinance modalities are not at a sufficient level to ensure that universal energy access could be attained in the region. However, SSA is still found to be able to use Nationally Determined Contributions as instruments to direct South-South aid, trade and investments into priority renewable energy sub-sectors to accelerate its progress on RET deployment.

Chapter 6 presents a case study on the ten-year attempt of the City of Boulder, Colorado to break up with its electric utility, Xcel Energy, in order to form its own utility. This study examines the conflict between two environmental groups. One of the two that is in favor of the separation argues that a democratically accountable, local utility would be better suited to achieve Boulder's ambitious environmental and climate action goals. However, the other that is contending the separation considers Xcel Energy a willing and capable environmental partner. Through this study that illustrates a divide in Boulder's environmental community, the authors argue that it mirrors the divide in the larger environmental movement, between structural environmentalists on the one hand and neoliberal environmentalists on the other. They offer a review of the theoretical work that informs these conflicting perspectives, and analyze structural and neoliberal sentiments expressed in the opinion pages of the city's newspaper in order to demonstrate how they intervene and shape Colorado electricity politics.

Chapter 7 is a multi-faceted case study that investigates sustainable land development using permaculture as the design tool. The author defines Permaculture, coined by Bill Mollison and David Holmgren, as a sustainable design theory that builds on three ethical principles used to produce a set of guidelines to follow in order to create an ecologically focused project. Permaculture, a contraction of perma-nent and initially agri-culture, has evolved to perma-nent and culture; understanding that without agriculture, culture is impossible. This chapter begins with an overview of the environmental issues followed by a description and brief history of sustainable development, with emphasis placed on the UN Sustainable Development Goals (SDGs). It then focuses on a three-part case study on different scales (urban, suburban, and rural) of permaculture land development in the Midwestern United States. Through these permaculture designs, the study illustrates how permaculture can help achieve SDGs to forge a sustainable future.

Chapter 8 identifies airlines' green operational strategies and reports the strategy patterns adopted by the airlines of each region. Using the analysis of the sustainability reports from 23 airlines in five regions (North America, Europe, Asia, the Middle East and South America), the author develops a detailed configuration of the airlines' green operational strategies. In the identification of the green operational strategy adopted by each region and the green pattern of each region that differed from those of any other regions, the study develops indicative models for each regions and across regions by using a simple and special tailored quantitative analytical technique. The results of this chapter raise a set of questions about the impact of contextual factors on whichever green strategy pattern is adopted, indicating the need to conduct more in-depth analysis of green actions. This study can be seen as one of a few studies that have developed a comprehensive definition of airlines' green operational strategies and explore the green strategy patterns adopted by airlines from different regions.

Chapter 9 outlines the relevance of sustainable development as a key for family firm success and its ability to guarantee long-term survival and spread positive effects in social, economic and natural environments. By particularly analyzing a single case study of a Sardinian family business, this case study explores the intertwined relationships among sustainability, owner innovativeness and firm success. Moreover, the importance of family businesses and the scarcity of the study conducted to date have suggested a focus on how these companies tackle sustainability challenges.

Chapter 10 examines the issues related to economically optimal solar power generation. As the subject of their case study, the authors use the Ottawa Renewable Energy Cooperative, which is considering installing solar modules on the roofs of two buildings while they stay connected to the public electricity grid. Solar power over their own needs is sent to the public electricity grid for a credit on their electricity bill. When they need more power than they generate, they purchase from the grid. In addition to paying for the electricity they purchase, they are subject to a "demand charge" that applies to their grid power consumption during the peak hours each month. Any grid electricity they consume during peak hours is charged at a rate about 100 times the rate for off-peak hours. The case addresses three questions: (1) Is it profitable for these organizations to install solar on their roofs? (2) Can profitability be increased by adding a battery? (3) How sensitive is profitability to uncertainty in future electricity prices? The case shows how the answers to these questions depend on the profile of hourly electricity consumption during the day, which differs significantly from one building to the other.

Chapter 11 reviews hydrogen fuel cells for green power generation. To reduce reliance on fossil fuels and increase demands for clean energy technology worldwide, there is currently a growing interest in the use of fuel cells as energy-efficient and environmentally friendly power generators. Among all major types of fuel cell, hydrogen fuel cells (HFCs) are in the forefront stage and have gained substantial attention for vehicle and portable applications, which is composed of a cathode, an anode, and a polymer electrolyte membrane (PEM). The heart of the fuel cells is membrane electrode assembly (MEA). An electro-deposition technique for preparing the nano catalyst layer in PEM fuel cells has been designed, which may enable an increase in the level of Pt utilization currently achieved in these systems. Functionalization process has been done using a mixture of concentrated nitric acid and sulfuric acid in refluxing condition. The hydrocarbon based polymer membrane has been used as electrolyte part.

Chapter 12 provides an overview of smart grid, which is fundamentally important for the increased and efficient deployment of renewable energy. This chapter reviews the importance of smart grid technologies, differences between the conventional and smart grid technologies, smart grid architecture, major components of the smart grid, and advanced metering infrastructure (AMI). The word grid refers to the electric grid which includes network of transmission lines, substations and transformers. The term "smart grid" means the grid based utilization of digital technology that allows for two-way communication between the utility and the customers and also the sensing along the transmission lines. The smart grid includes controls, computers, automation and new technologies and equipment working together. The current grid consists of more than 9,200 power generating units with more than one million megawatts generating capacity connected to more than 3,00,000 miles of transmission lines. It is necessary to substantially enhance the existing grid by incorporating digital and computerized equipment and technologies into the power grid. The application of smart grid technology will greatly improve the performance of the grid for increased renewable energy deployment. It can automate and manage the complexity and needs of electricity in this era.

Chapter 13 presents the auxiliary system technology that ensures reliable pumping of drainage and sewage water during electromechanical and hydro-mechanical transients from blocks of the hydroelectric power stations and pumping stations. As an alternative source of energy, it is proposed to use the energy of the liquid column of the pressure penstock of the stations, and as an auxiliary, to use jet pumps. Transmission of energy to the suction stream is carried out without direct usage of electrical and mechanical energy. During total shutdown of electric power, reliable evacuation of drainage and seepage water and reduction the influence of electromechanical and hydro-mechanical transients on power equipment and pipelines can be ensured with the use of self-regulating jet pumps over a period of several days, which cannot be accomplished by any other pump. The authors recommend the results of their research to allow efficient use of water and energy resources and to ensure reliable operations of the power equipment of stations, especially in the events of sudden power outages.

Chapter 14 proposes to design and model a self-assistive controller using voltage droop method for DC distributed generators and storages as part of the renewable energy micro grids. The renewable energy micro grids are being deployed in parallel with, or independently from the main power grid network. Renewable energy micro grids are designed to provide an effective coordination with the aggregating distributed generators, energy storages and connected loads. In this regard, renewable energy micro grids can be connected to the grid network or can be off-grid network. In the mode of grid network connected, renewable energy micro grids are presented as a controllable entity. When it is necessary, renewable energy micro grids are connected in islanded mode to deliver reliable power to the grid network during the interrupted power supply from the grid network system. The renewable energy micro grids are encompassed of renewable energy sources, energy storages and loads, and do not exclude the grid network transmission.

Chapter 15 presents a case study on hybrid electric energy storage and its dynamic performance. The authors conduct tests with the application of a hybrid electrical energy system which consists of super-capacitors and batteries. They review a series of studies in terms of electrical energy storage devices being used in applications such as electrical vehicle and smart grid. Then the authors classify the advantage and disadvantages of these technologies. Subsequently, they demonstrate trial test process and explained the potential for the energy storage device with respect to other energy resources for future perspective. Finally, the study shows trial tests and their outcomes.

Chapter 16 reviews energy informatics using the distributed ledger technology and advanced data analytics. The authors view internet technologies and rise of renewable and distributed energy technologies as the main drivers of the third industrial revolution era and the transition to green and decentralized energy resources and digital transformation of the existing industrial infrastructure as the biggest achievements of this industrial revolution. In comparison, they identify artificial intelligence (AI), quantum computing, advanced biotechnology, internet of things, additive manufacturing, and most importantly distributed ledger technology (DLT) as the main drivers of the fourth era, and categorized energy forecasting such as wind and solar power forecasting models as the most common energy AI based informatics applications in the energy sector. In addition, they expect the use of DLT to be an industrial standard in various industrial sectors including energy business in the coming decade. Their study emphasizes description of energy forecasting using AI and energy DLT and future developments and solutions to overcome challenges especially which are associated with standardization of the energy DLT applications.

Chapter 17 reviews green building technologies. The author first provides an overview of buildings' significant responsibility in the global energy use (40 percent) and global greenhouse gas emissions (one third). As these impacts are being acknowledged and addressed in specialist building design techniques and technologies that aim to reduce the environmental impacts of buildings, the author refers these techniques and technologies as green building technologies. This chapter describes green building technologies and shows why they are vital in addressing climate change and reducing the negative environmental impacts associated with built environments. A structured approach is presented which can be applied to identify and integrate green building technologies into new and existing buildings. By combining global implications with technical detail, the author provides a valuable guide to green building technologies and their role in supporting a transition to a more sustainable future.

Peter Yang
Case Western Reserve University, USA

REFERENCES

Renewable Energy Policy Network for the 21st Century (REN21). (2018). *Renewables 2018 Global Status Report (GSR)*. Retrieved from http://www.ren21.net/wp-content/uploads/2018/06/17-8652_GSR2018_FullReport_web_-1.pdf

United Nations (UN). (2015). *Paris Agreement*. Retrieved from https://unfccc.int/sites/default/files/english_paris_agreement.pdf

United Nations (UN). (2018, January 1). *Sustainable Development Goals (SDGs)*. Retrieved from https://www.un.org/sustainabledevelopment/news/communications-material/

Acknowledgment

I would like to sincerely thank everyone who submitted chapter proposals, every chapter author, and every member of Editorial Advisory Board for their genuine interest, enthusiasm, energy, and cooperation, as well as contributions and insights in this book project, without which this publication of *Cases on Green Energy and Sustainable Development* would be impossible.

I also greatly appreciate the IGI Global editorial support team, especially Ms. Jordan Tepper and Ms. Jan Travers, for their prompt responses and solutions, which have been instrumental for the smooth progression of every phase of the entire publication process.

Last but not the least, I express my whole-hearted gratitude to my wife, Yuezeng Yang, for her love, understanding, and support in this publication.

Peter Yang
Case Western Reserve University, USA

Section 1
Strategies, Policies, and Management

Chapter 1
Countries Progress in Solar PV in Support of NDC Implementation and Contribution to Achieving SDG7

Dereje Azemraw Senshaw
Global Green Growth Institute (GGGI), South Korea

Alexander Edwards
Global Green Growth Institute (GGGI), South Korea

EXECUTIVE SUMMARY

This case study examines the progress being made by 12 least developed countries (LDCs) in their effort to achieve Sustainable Development Goal 7 (SDG7) – access to clean and sustainable energy for all. Focusing on solar photovoltaics (PV), the authors look at what can be done to further the spread of renewable energy, and the role various actors have to playing in helping these countries to meet SDG7. Furthermore, with countries on the cusp of submitting their revised contributions under the Paris Agreement, they look at the role solar PV can play in helping LDCs to participate in taking action against climate change. After outlining the current policy landscape, and efforts being made within these countries, they look at the obstacles, opportunities, and the role of solar PV going forward. They also look at the steps that policymakers, both national and international, can take to encourage the rapid uptake of renewable energy in developing nations.

DOI: 10.4018/978-1-5225-8559-6.ch001

OBJECTIVE

The objective of this study is to examine the role of solar PV in helping Least Developed Countries (LDCs) achieve Sustainable Development Goal 7 (SDG7) – access to clean and sustainable energy for all, while simultaneously keeping up with their climate change contributions under the Paris Agreement. With a recent Renewable Energy Policy Network for the 21st Century (REN21) report (2018) showing that renewables accounted for 70% of global power additions in 2018, and that a further 55% of that was from solar PV (coupled with the massive uptake of PV technology in China), it is clear that solar power is going to play a key role in our shared energy future. We look at the challenges developing countries face when dealing with PV, the policies currently in place, and what decisions we can make going forward to help encourage the uptake of PV in addition to a widening of access to energy.

BACKGROUND TO SOLAR PV PROLIFERATION IN DEVELOPING COUNTRIES

While as much as 85.3% (United Nations Economic and Social Council, 2017) of the world's population had access to electricity in 2014, that still leaves over 1 billion people lacking basic access to energy. This is highly problematic as without reliable access to energy, populations may experience limited economic opportunities, lack of access to quality education, comparatively worse health conditions and medical services, and inability to perform basic everyday tasks such as cooking or purifying water.

With this in mind, the international community devised a range of measures aimed at tackling such pressing issues such as Agenda 2030 and the Sustainable Development Goals, with SDG 7 in particular – access to affordable, reliable, sustainable and modern energy for all – aiming to make this problem a thing of the past.

For developing countries, however, this poses a significant challenge in that while committing to increasing energy access under the Agenda 2030 framework, they have also pledged to make this energy sustainable and modern all the while reducing their GHG emissions and curbing their environmental impact under the Paris Agreement, and simultaneously reducing poverty and hunger.

With the 2020 revision of countries' Nationally Determined Contributions [NDCs] (at their most basic level an outline of the plan each country has devised in order to help it meet its obligations under the Paris Agreement) rapidly approaching, it is more crucial than ever for countries to develop concrete strategies for how they intend to achieve a more sustainable future such as the one put forward under the global agreements listed above.

For many LDCs this naturally means turning towards renewable energy as a means of not only increasing energy access but also doing so in a way that is both modern, sustainable, and will not negatively impact on their Paris Agreement commitments. Despite many LDCs having access to a wealth of renewable resources including, in particular, high levels of solar irradiance (solar energy received per unit area), the proliferation of renewable technology within these countries has been far from straightforward.

This case describes how 12 LDC across Africa, South East Asia, and the Pacific, have progressed in terms of the deployment of solar PV over the recent years. Drawing on data regarding the increased installed capacity of solar PV, in addition to developments in the policy landscape and international efforts aimed at tackling this problem, this case looks at the challenges they face, the actions they are taking to overcome these challenges, and potential solutions to these often-complex issues.

SETTING THE STAGE

For this study, as noted above, 12 LDCs from Africa, South-East Asia, and the Pacific were analysed in terms of their progress regarding the installation of solar PV, the accompanying policy landscape, and capacity building that has taken place. These countries are Burkina Faso, Cambodia, Ethiopia, Kiribati, Lao PDR, Mozambique, Myanmar, Nepal, Rwanda, Senegal, Uganda, and Vanuatu.

These countries were selected, as they are the 12 LDC member countries of the Global Green Growth Institute (GGGI), an international organisation working in the field of sustainable development and green growth. As such, these 12 countries represent the LDCs which have made a conscious effort to acknowledge both the threats and opportunities posed by climate change and sustainable development. They thus present an excellent opportunity to examine less developed countries who are making an effort to develop sustainably, and to contribute to the Paris Agreement. Furthermore, they are also countries which are rich in terms of renewable resources, yet are at the beginning of their respective journey's with little existing infrastructure in place meaning they offer substantial potential improvements both in terms of renewable energy generation capacity and energy access.

As hinted above, despite all these countries having an abundance of solar energy, by the year 2015 not one of them saw solar represent more than a 0.85% share of total final energy consumption (International Energy Agency, International Renewable Energy Agency, United Nations Statistics Division, World Bank Group, & World Health Organisation (IEA et al.), 2018). Despite this, there are promising signs for the future, with significant infrastructure expansion plans in many of these countries including large scale PV projects designed to provide a substantial boost to renewable generation capabilities.

In order to subsequently allow a more thorough analysis of how these countries PV capacity has progressed and what can still be done to further improve its prospects going forward, this section aims to present a comprehensive picture of solar PV, as well as the countries in question, and the sustainable development project landscape. Data will be presented outlining the progress of each country in terms of their installed solar PV capacity, their plans for the future, and the policy landscape in place relevant to this issue.

The Players

Trying to effectively narrow down all the parties involved when it comes to the expansion of renewable energy in developing countries can be difficult given the wide reach of the sector across governments, the private sector, and international organisations coupled with the fact the precise nature of the relationship between all the parties can be increasingly blurred relative to developed country equivalents.

At the most basic level, however, we can think of the key players in the development of solar PV within LDCs as a) the national government, b) the contractor undertaking the work, and c) the financier(s). This process typically involves national governments forming an agreement with a solar firm, or putting a contract out to tender (essentially an auction), with a financing organisation providing/helping to secure funding.

The national government also plays a key role in defining the landscape in which this occurs, with some governments choosing policies designed to encourage or lower the price of such an endeavour such as tax breaks, or priority grid connection.

Furthermore, the government is crucial in that in many cases, it signs a deal with the supplier guaranteeing to purchase the energy for a set period of time and at a specified price (referred to as a Power Purchase Agreement [PPA]) in order to provide financial security for potential contractors, and in certain cases will also assume ownership of the energy generating assets at the end of this period such as the Scatec Solar Project in Mozambique (Scatec Solar, 2018).

International organisations and the international community at large can also play a key role in this process in terms of helping to shape the policy landscape, and in working to encourage private sector investment, and public-private partnerships.

Where the real complexity lies is the source of the finance, with different cases showing a multitude of different mechanisms in place. The United Nationals Report of the

Intergovernmental Committee of Experts on Sustainable Development Financing (2014) noted four key sources of finance for such projects. These are national public sources, national private sources, international public sources, and international private sources.

To complicate the matter, in addition to these sources, we see a host of intermediaries such as national and regional development banks, bi/multilateral aid agencies, finance institutions, blended institutions (e.g. global health funds, public-private infrastructure funds), and investors with long/short-medium term liabilities such as pension funds, banks, and hedge funds.

Furthermore, modern finance instruments are becoming increasingly complex with what is termed blended finance (Mawdsley, 2018) becoming more commonplace. What this essentially means is using development finance and philanthropic funds to mobilise private sector investment in order to increase the money available for sustainable development (Organisation for Economic Cooperation and Development & World Economic Forum [OECD & WEF], 2015).

This can lead to the deployment of a range of various instruments as required for each specific situation. Therefore, when attempting to assess the key parties for a specific PV project, it is necessary to do so on a case by case basis given the potential complexities.

Energy Access

The following table outlines the extent of the issue these countries face regarding energy access, one of the core tenets of SDG7 (See Table 1).

Table 1. Access to electricity by country

Access to electricity (% of population)							
	Total					Urban	Rural
LDC	1990	2000	2010	2014	2016	2016	2016
Burkina Faso	3	9	13	19	19	61	1
Cambodia	-	17	31	56	50	100	36
Ethiopia	-	13	25	27	43	85	27
Kiribati	32	52	63	81	85	88	82
Lao PDR	15	43	70	81	87	97	80
Mozambique	-	7	17	22	24	64	5
Myanmar	36	44	49	52	57	89	40
Nepal	-	28	67	85	91	95	85
Rwanda	-	6	10	20	29	80	18
Senegal	25	62	54	61	65	88	38
Uganda	1	8	15	20	27	58	18
Vanuatu	7	22	37	43	58	91	46

(Source: The World Bank, 2018)

As we can see, there is a significant level of variance between countries, with some fairing considerably better than others (e.g. Lao PDR at 87%, Nepal at 91%). What is also immediately noticeable is the gap between urban and rural electrification, with the rural population experiencing far higher levels of disconnection that their urban counterparts. We also see distinctly different levels of progress, with the data from the year 2000 indicating a far smaller gap between the best and worst performers relatively to 2016. With increasing energy access being one of the core objectives for LDCs regarding Agenda 2030, finding a way to solve this issue is of paramount importance.

It is important to note that we are not advocating for solar PV to fill in all the gaps in energy access. Certain countries, such as Nepal for example, are rich in hydro with an estimated commercially exploitable potential of as much as 42,000 MW (despite only having around 800 MW installed (Asian Development Bank, 2014). Solar PV, however, has the added benefit of being relatively geographically unrestricted (depending on the scale) and as such, is an excellent option when looking to tackle the ongoing issues of poor rural access to electricity.

Solar PV Development

The rate of decline in solar PV pricing over the last five to ten years has been nothing short of spectacular with the International Renewable Energy Agency (2016a) predicting the continuation of this trend in the years to come. Recent years have seen record lows in terms of solar PV prices. While various sources may choose to show this fact using slightly different metrics such as $ per Watt (W), per kilowatt hour (kWh), or per Watt Peak (Wp), at a functional level, these are all very similar, and they all reflect the price per unit of energy. What is most relevant here is the overall decrease we are seeing.

Auctions in May 2016 for example, saw prices in Dubai fall as low as USD 0.0299/kWh while in 2015, prices in Europe, China, India, South Africa and the US where regularly in the range of USD 0.06-0.1/kWh (International Renewable Energy Agency, 2016b). Meanwhile, PV modules prices in Europe fell by as much as 83% between Q1 2010 and 2017 (International Renewable Energy Agency, 2018).

This is particularly important as it is now driving the levelized cost of electricity [LCOE] towards being increasingly competitive with traditional fossil fuel generation. This is significant as the levelized cost of electricity is one of the most complete ways of comparing the price of energy. By taking into account the total lifetime cost of a generation point, and diving by its lifetime output, the levelized cost of electricity allows us to really compare the cost of generating one unit of energy with another method of generation and is therefore crucial information from a policy perspective.

On top of this, the International Renewable Energy Agency (2016a) is estimating that balance of system costs (costs for every component of a PV system other than the panels) could fall by as much as 69% in less efficient markets such as those in LDCs and that the global average total installed cost of utility-scale PV systems could fall even further from roughly USD 1.8/W in 2015 to USD 0.8/W in 2025.

This is in addition to the fact that PV module prices could foreseeably continue to fall by as much as one third during the next decade (International Renewable Energy Agency, 2016a), the combined effect of which would see solar PV becoming one of, if not the most cost-effective methods of energy generation. The following graph outlines the considerable drop in global PV prices since 2013 and up until 2023.

With large capital costs frequently inhibiting investments in solar PV projects in LDCs (Nygaard, Handsen, Mackenzie, & Pederson, 2017), this decrease in price is crucial in order for solar PV to flourish as an alternative to fossil fuel generation. This is not to say that other methods of generation do not also experience high upfront costs. In fact, a recent report by the U.S. Energy Information Administration [U.S. EIA] (2019) stated that solar PV is significantly cheaper (around 60-70% cheaper than coal, offshore wind, or nuclear) in terms of $/kW than many other established methods of generation in developed countries, but the unique situations of developing countries was key regarding solar.

Figure 1. Global average price for Chinese tier 1 crystalline silicon pv modules (2018 onwards estimated)
Source: GTM Solar Summit

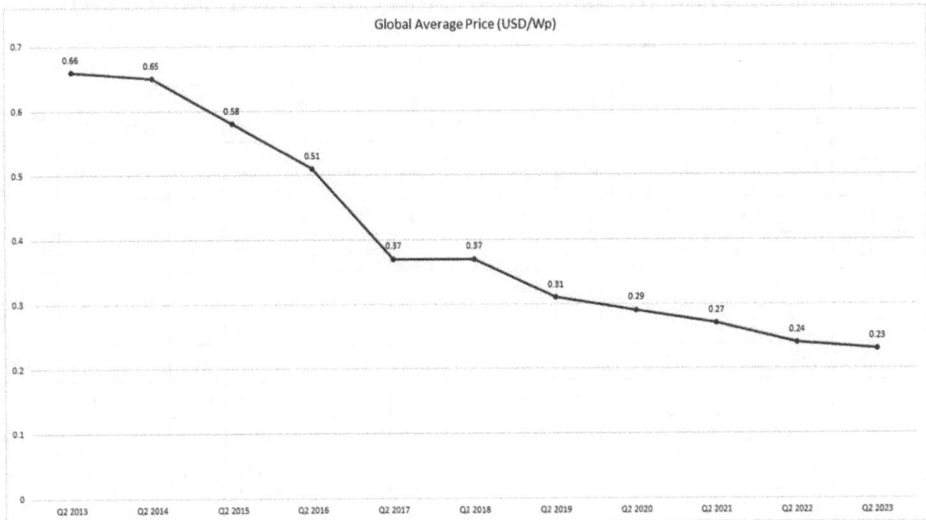

While we will go into this in more depth later, Nygaard et al. (2017) noted multiple relevant factors including few suppliers and low competition, high transport and importation costs, a lack of standards, the intermittent nature of renewable generation, minimal access to finance, and poor levels of R&D. This is offset somewhat when talking about more traditional methods of generation such as coal or biomass, and as such raised the price comparatively.

Installed Solar PV Capacity

Here we look at how solar PV has developed in these countries, both in absolute and relative terms, in order to paint a clearer picture of the progress being made to date.

One of the most notable trends is that the majority of countries are exhibiting a steady upwards trend in their total installed solar PV capacity. While Ethiopia, Nepal, Uganda and Senegal stand out as the leaders in terms of absolute capacity, it is promising to note that almost all countries are exhibiting not just an increase year on year, but a fairly consistent and steady increase at that.

Despite this, in the last year where concrete data exists regarding energy generation figures, it is clear to see that, with the exception of Kiribati that generates a massive 13% of its total energy from solar, that PV still represents a relatively minor portion of total energy generation. While solar PV is only one of many forms of renewable energy generation, it is clear that to achieve levels of around 29-36% (Sustainable Energy for All [SE4ALL], 2015) as is the ambition of SDG7, significant investment is required.

Table 2. Estimated installed solar PV capacity by country (Megawatt [MW])

	2012			2013			2014			2015			2016		
LDC	On Grid	Off Grid	Total	On Grid	Off Grid	Total	On Grid	Off Grid	Total	On Grid	Off Grid	Total	On Grid	Off Grid	Total
Burkina Faso	0.20	5.80	6.00	-	6.30	6.30	-	7.00	7.00	-	8.00	8.00	-	10.00	10.00
Cambodia	-	5.00	5.00	-	6.00	6.00	-	9.00	9.00	-	12.00	12.00	-	12.00	12.00
Ethiopia	0.40	10.60	11.00	-	23.40	23.40	0.30	32.70	33.00	0.10	57.90	58.00	-	70.00	70.00
Kiribati	0.41	1.60	2.00	0.40	1.60	2.00	0.40	1.60	2.00	0.40	1.60	2.00	1.36	1.64	3.00
Lao PDR	-	-	-	-	-	-	-	-	-	-	1.00	1.00	-	1.00	1.00
Mozambique	0.50	1.50	2.00	0.10	4.90	5.00	-	7.00	7.00	-	10.00	10.00	-	13.00	13.00
Myanmar	-	2.00	2.00	-	5.00	5.00	0.30	7.70	8.00	-	12.00	12.00	-	16.00	16.00
Nepal	-	20.00	20.00		25.00	25.00		32.00	32.00	-	32.00	32.00	-	32.00	32.00
Rwanda	-	-	-	-	-	-	8.80	0.20	9.00	8.60	0.40	9.00	8.40	0.60	0.9
Senegal	0.10	5.90	6.00	-	7.00	7.00	2.00	8.00	10.00	2.00	9.00	11.00	44.00	10.00	54.00
Uganda	-	17.30	17.30	0.50	18.50	19.00	-	22.00	20.00	-	22.00	22.00	10.00	24.00	34.00
Vanuatu	-	-	-	-	0.60	0.60	-	0.06	0.06	-	0.06	0.06	0.94	0.06	1.00

(Source: International Renewable Energy Agency Renewable Energy Statistics, 2017a)

Figure 2. Total installed PV capacity by year

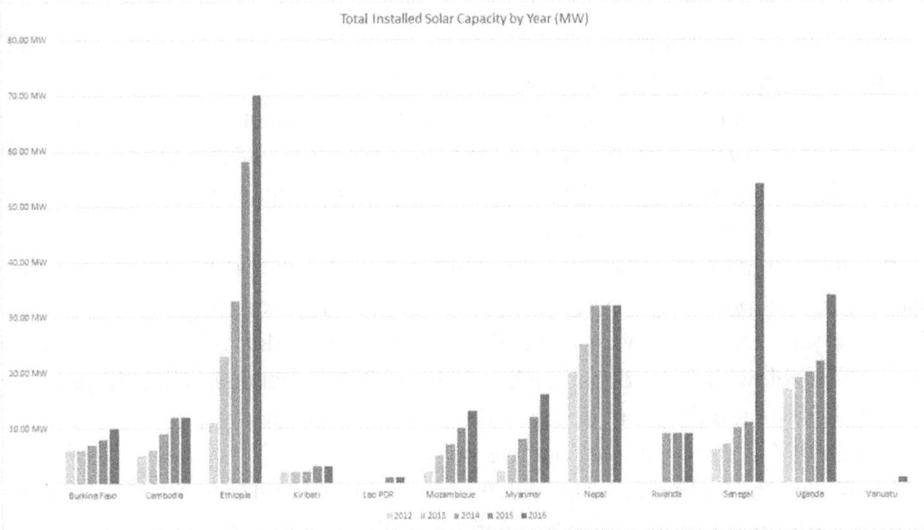

Table 3. Solar PV generation and total energy generation by country

LDC	Solar PV Generation 2015 (GWh)	Total Annual Energy Generation (GWh)	% from Solar PV
Burkina Faso	11.00	860.00	1.26
Cambodia	3.00	1,652.00	0.18
Ethiopia	95.00	8,692.00	1.09
Kiribati	3.00	23.00	13.00
Lao PDR	1.00	4,713.00	0.02
Mozambique	16.00	15,163.00	0.11
Myanmar	17.00	10,870.00	0.16
Nepal	48.00	2,266.00	2.12
Rwanda	14.00	533.00	2.63
Senegal	18.00	3,227.00	0.56
Uganda	35.00	3,464.40	1.01
Vanuatu	-	69.90	-

(Source: International Renewable Energy Agency Renewable Energy Statistics, 2017; Pacific Energy Conference, 2016; Regulatory Indicators for Sustainable Energy)

Future Solar PV Capacity

Another point worth noting, is that with a couple of exceptions such as Senegal and Rwanda, PV capacity up until 2016 is comprised almost entirely of off-grid PV with utility scale generation being noticeably absent. This is changing rapidly however with almost every country in this chapter set to invest heavily in utility scale PV projects in the coming years. The following figure outlines the potential increase in solar PV in each country over the coming years.

It is important to note that this figure takes into account feasibility studies being undertaken for potential projects, requests for qualification, and projects where PPAs are yet to be formally signed and as a result, the reality of future capacity is likely to be below the figures shown here (although by what degree it is impossible to say). This should therefore be taken more as a general measure of the interest in large scale PV as opposed to a concrete projection.

Even allowing for the earlier caveat, this demonstrates that there is a clear interest in the development of large-scale solar PV within these countries compared to previous years. It is also interesting to note that while the World Bank or similar organisations such as the African Development Bank or Asian Development Bank often play a role in implementing the tender process or providing risk reduction, a significant amount of these potential and upcoming projects are privately funded highlighting the increasingly competitive nature of solar PV.

Figure 3. Potential increase in PV capacity

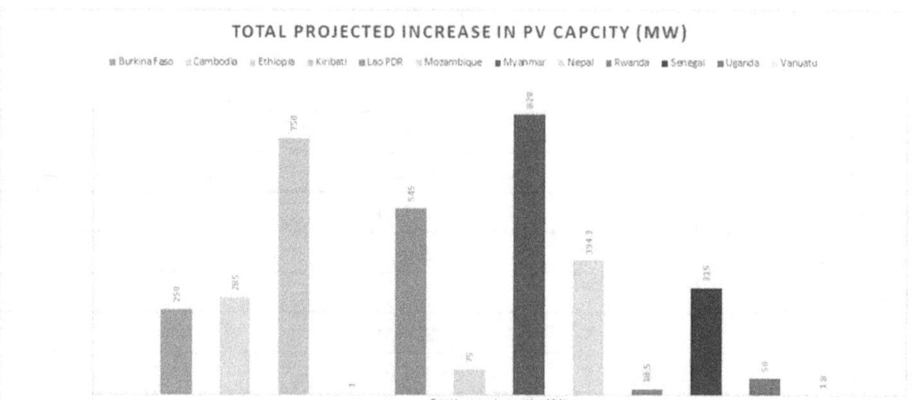

	Total increase in capacity (MW)
Burkina Faso	250
Cambodia	285
Ethiopia	750
Kiribati	1
Lao PDR	545
Mozambique	75
Myanmar	820
Nepal	394.3
Rwanda	18.5
Senegal	315
Uganda	50
Vanuatu	1.8

Around the years 2014 to 2015, there has been a significant increase in terms of planning and interest for the development of solar PV, and in particular, utility scale projects capable of making a significant dent in the share of total energy generation, quite possibly due to a combination of the fall in price coupled with the introduction of the NDCs under the Paris agreement.

Policy Landscape

The existence of political priorities and development policies may be seen as one of the key drivers of renewable energy integration (Ahlborg & Hammar, 2014). With the International Renewable Energy Agency (2017) noting how 59% of current Solar PV benefits from some form of Feed-in-Tariff [FiT] (a tariff where you get paid an agreed amount per unit of renewable energy you feed into the grid), it is crucial to examine the policy framework within the LDCs in this study in order to help determine what can be learned from them and to compare their experiences.

The Regulatory Indicators for Sustainable Energy [RISE] score from Table 4 is designed to provide an overall score outlining the state of the policy framework within a country regarding the suitability for meeting SDG7 objectives of energy access, energy efficiency, and renewable energy. The variance within these scores is relatively low with Uganda and Senegal leading the way and Mozambique and Vanuatu bringing up the rear.

Table 4. Policy framework by country: key policies and tools

LDC	RISE Score	Legal framework for renewable energy development	Existence of renewable energy strategy	Existence and monitoring of officially approved electrification plan	Existence of official renewable energy target	Existence of official solar PV targets	Existence of FIT covering solar PV	Existence of other generation support schemes	Existence of fiscal incentives for RE generation
Burkina Faso	31	X	X	X	X	X			X
Cambodia	42			X	X			X	X
Ethiopia	36	X	X		X	X			X
Kiribati	N/A		X		X				
Lao PDR	34	X	X	X	X	X			X
Mozambique	25						X	X	
Myanmar	38			X	X				X
Nepal	36	X	X		X				X
Rwanda	40	X	X	X	X			X	X
Senegal	48	X	X	X	X			X	
Uganda	55	X	X	X	X		X	X	X
Vanuatu	25		X	X	X				

(Source: RISE.esmap.org)

Every country with the exception of Mozambique has renewable energy targets in place, with the NDCs playing a key role in this. Surprisingly, only two countries, including Mozambique, have a Feed-in Tariff in place, however the majority of countries offer some form of fiscal incentive, often with some other forms of generation support, coupled with a renewable energy strategy.

CASE DESCRIPTION

When looking at the expansion and deployment of solar PV in pursuit of NDCs and SDG 7, there are a range of factors which may be seen as crucial to its successful diffusion. This section outlines the key areas of concern faced by the LDCs in this study when attempting to pursue a strategy of renewable energy deployment. These include the financial, fiscal/tax, legislative, political, and technological and environmental aspects of renewable energy integration (Abdmouleh et al., 2015).

Financial Concerns

One of the key characteristics of renewable energy generation is that it has relatively high capital costs (although as noted, these are falling), necessitating a large initial investment in addition to having long payback times (Nygaard et al., 2017). This is coupled with the fact that it can often be considered a riskier investment given technology and resource uncertainties (Abdmouleh et al., 2015) resulting in significantly higher interest rates. This in turn makes everything more expensive as the money you borrow to pay for it costs you more.

This is a particularly significant issue in LDCs, where a general lack of financing instruments as well as underdeveloped or fragmented financial sectors can make investment in such projects an undesirable prospect due to the high level of perceived risk (Karekezi & Kithyoma, 2003; Ohunakin et al., 2014). With national governments typically being unable to fund such projects, the majority of financing is coming from international financing organizations and the private sector, and it is therefore crucial for the government to collaborate with these organizations regarding PV projects (Abubakar Mas'ud et al., 2016).

In addition to this, the market for solar PV is often underdeveloped or exhibiting signs of market failure, thus further discouraging investment. This can be from a range of factors, including underinvestment in R&D programs, underpriced environmental impacts, and the existence of monopolies in the energy generation sector (Baumol & Oates, 1998; Baumol et al.,1982; Beck, 1995; Foxon & Pearson, 2008; Margolis & Kammen, 1999; Sen & Ganguly, 2017). In order to address these issues, a range of measures must be undertaken with the fundamental aim of increasing investor confidence and building the capacity of LDCs in terms of their ability to attract finance.

These include for example, increasing the availability of risk capital aimed at lowering the risk for private investors (essentially guaranteeing investments to varying degrees). A small amount of public funding for example, can be used to promote far larger private investment (Sen & Ganguly, 2017). In addition to this, countries must look to increase the flow of bankable projects (essentially projects which are financially self-sustaining in the long-run) and develop new planning stage risk mitigation mechanisms (Sen & Ganguly, 2017).

As Sen and Ganguly (2017) noted, comprehensive project preparation can serve multiple purposes, including setting up public-private partnerships, supporting individual projects and financing proposals in addition to creating enabling conditions to support local market development. Furthermore, "targeted public guarantees and new public-private risk sharing facilities" (Sen & Ganguly, 2017, p. 1176) can help enable more projects to reach the construction phase.

They went on to note how the majority of investment is made during the construction phase in the form of debt financing, therefore strategies should focus on increasing public-private co-lending stating (p.1176) " Development/Commercial Finance Institutions (DFIs/CFIs) and green banks can adapt public-private partnerships and finance mechanisms to bring in commercial banks and non-traditional private participants such as institutional investors" and that scaling up risk mitigation mechanisms would allow more projects to secure financing for construction.

What is important is that international partners bring certainty and stability to investors, allowing them to have confidence when investing, thus increasing availability and lowering the price of capital for these countries.

Fiscal and Tax Concerns

As noted, renewable energy is historically non-competitive given the lower generation cost for fossil fuel or nuclear power. This is exacerbated by the fact that most conventional power plants were not only built with large subsidies, but also that their initial capital costs have largely been paid off by this stage, in addition to the failure to properly account for the negative environmental and health impacts of these plants (Abdmouleh et al., 2015). In order to address this relative lack of competitiveness, measures must be taken to correct this.

One of the key ways to do this, are through fiscal measures such as providing tax exemptions or reductions to potential investors in renewable goods or services, and by imposing carbon or energy taxes on conventional methods of generation (Ecotec, 2001). Environmental taxes may serve to adjust total costs of generation and bring traditional methods in line with renewable generation (Menanteau et al.,

2003) although this is often difficult in the case of LDCs as power companies are often state owned, and furthermore, many people in developing nations rely on cheap fossil fuel to survive.

An alternative to this is to implement tax incentives for investment, thus lowering the cost for investing in renewable technologies. Sieminski (2013) noted how this can be as effective as a subsidy with certain tax cuts representing the equivalent of subsidizing as much as 35% of the overall project.

Legislative Concerns

With renewable energy, particularly in developing countries, suffering from a lack of investor confidence in certain cases, it is also crucial for a clear and precise legal framework to exist which promotes the uptake of renewable technologies into the generation mix. As outlined in Table 4 in the previous section, however, Uganda is the only country to achieve a RISE (Regulatory Indicators for Sustainable Energy) score of greater than 50 points implying that of the countries in this study, almost all still show room for improvement.

Through legislation though, countries can aim to guarantee a market for renewable energy, using two key techniques (Abdmouleh et al., 2015) which are legislation organizing power purchase, and making it easier to access the grid.

Figure 4. Market incentives and enablers experienced by current solar PV generation schemes
(Source: International Renewable Energy Agency, 2017c)

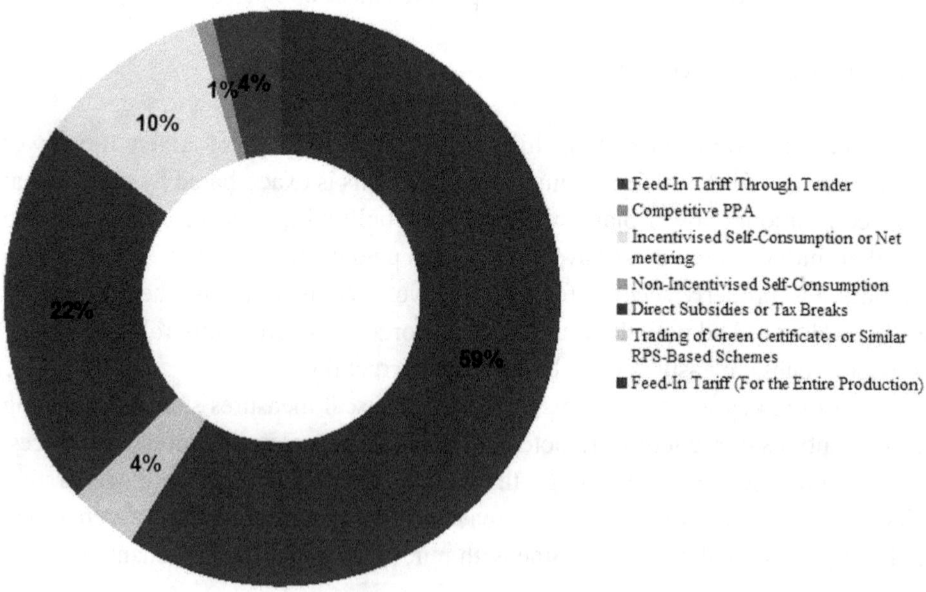

Power purchase mechanisms come in a range of shapes and sizes, but at their core, offer a guarantee that energy generated will bring in a minimum price, thus removing uncertainty. There are several main methods including Feed-in-Tariffs (FiT), renewable portfolio standards, and tendering agreements.

As noted in Figure 4, as much as 59% of current solar PV employs FiT schemes with as little as 1% coming through competitive PPA (International Renewable Energy Agency, 2017c). A strong legislative approach is key to the success of such projects for at least the immediate future, until generation costs fall to truly competitive levels (although this might not be far away).

What is interesting to note is that the only two countries highlighted in Table 4. as having a Feed-in-Tariff in place (Mozambique and Uganda), were not the highest in terms of current capacity, implying that while having some sort of generation reward scheme in place is certainly a strong incentive, there are clearly other factors at work regarding the uptake of PV (potentially including other economic instruments which might be better tailored to the situation of a specific country).

The other legislative approach is grid access legislation (Abdmouleh et al., 2015). This is because renewable projects often face problems in terms of grid access, potentially due to the fact that renewable projects are often located in decentralized locations given their need for large amounts of space and specific environmental conditions. Therefore, policies aimed to extending or prioritizing grid connectivity for solar PV or renewable projects can have a large impact. Furthermore, given the intermittent nature of solar PV, certain grids have policies in place which favour continuous and consistent generators (Ecotec, 2001)

Political Concerns

In a recent study looking at Mozambique and Tanzania (Ahlborg & Hammar, 2014), the authors found that the main drivers in both countries regarding renewable energy integration was the existence of political priorities and development policies. The political opportunities surrounding solar PV and renewable energy generation are not insignificant.

These include but are not limited to, job creation, reduction of CO2 footprint, increase in local added value, promotion of socio-economic stability and development, meeting national energy needs, poverty reduction and environmental protection (Abdmouleh et al., 2015; Ohunakin et al., 2014; Polack, 2010). Despite the myriad of opportunities, it is only recently, and in particular with the introduction of the NDCs, that we have started to see a coherent approach to renewable energy across the developing world.

Like the majority of other aspects, the crucial part of the politics surrounding solar PV, is to create a stable investment environment which encourages confidence and promotes financial capacity building. The political will driving solar PV integration is also heavily linked to a range of other factors including price support mechanisms, R&D funding and tax incentives (Abdmouleh et al., 2015). The most important aspect, however, is the adoption of specific national targets for levels of solar PV uptake which serves as a crucial message encouraging deployment (Abdmouleh et al., 2015). This will serve to increase the market scale leading to decreasing costs over time as it subsequently grows.

As noted, however, there can be significant political challenges, as a duty to deal with what are often perceived as more immediate problems such as poverty or ensuring access to medical care can sometimes conflict with a more sustainable approach to development in the short term. It is therefore crucial to consider the potentially hidden impacts of any decisions and their political implications when considering policy.

Technological and Environmental Concerns

Finally, there is the technological and environmental aspect. LDCs are in a unique position in this regard in that environmentally, they are blessed with some of the most abundant solar resources, especially in the case of Africa, yet technologically, much like in terms of financial capacity, they are lagging sorely behind the majority of developed nations. Issues surrounding this include, as noted, a lack of grid infrastructure, the need for energy storage due to non-constant generation (which can significantly raise the overall cost), insufficient technology transfer and a lack of understanding of how innovation spreads and becomes mainstream (Al Badi et al., 2009; Polack, 2010).

Furthermore, replacement parts are often made overseas and are difficult to obtain, particularly in the case of isolated sites (Mirza et al, 2009; Polack, 2010) and in cases where they are, it is often economically prohibitive (Polack, 2010; Roper, 2005).

In addition to the lack of access (at least for a reasonable price) to required materials, many of the countries in that we have analysed for this chapter experience a lack of access to skilled personnel (Ahlborg & Hammar, 2014). Many LDCs also lack a significant R&D presence in addition to lacking the highly skilled labour required for the construction and maintenance of utility scale solar PV systems in particular (Engelken et al., 2016). It is therefore crucial to focus on building institutional, technical and human capacity to support renewable energy deployment (Sen & Ganguly, 2017).

Efforts must be undertaken to not only share best practice within regions, but also to facilitate global frameworks of exchange. Therefore, capacity building projects must be put in place from "policy and regulatory design and management to project preparation, valuation, development, implementation and financing," and "a wide array of skills needs to be built up in government ministries, financing institutions, regulatory agencies and utilities (Sen & Ganguly, 2017, p.1178)."

The environmental challenges faced by solar PV are also significant, closely linked to the technological ones we saw above, and also go to show that falling prices do not guarantee the widespread diffusion of solar PV. It will also take proper management, both from governments and the citizens.

As noted, solar power is intermittent by nature, meaning it comes and goes and is therefore not constant. With regular blackouts being a crucial issue to solve for developing countries in terms of energy access, there are obvious issues with being overly reliant on a single source of energy. One solution to this is battery technology, however this is victim to many of the same issues that solar was previously, in that it is expensive to import, an emerging technology, and not necessarily readily available.

To add to this, you have the fact that batteries are most needed in areas which are off the grid, and therefore are unable to rely on traditional forms of generation when solar PV plants are not operating at full capacity (i.e. night-time). This exposes another issue in that often areas which lack energy access are geographically remote, and difficult to access. This can increase upfront costs, and as we saw, make it harder to connect to the grid.

Furthermore, other renewable sources such as wind are also intermittent, so solutions must be tailored to ensure the energy supply remains consistent without relying on conventional generation based on fossil fuels.

CONCLUSION AND RECOMMENDATIONS

There is a growing trend in adoption of policy instruments that supporting deployment of renewable energy, among which implementing of auctions – usually blended with other instruments – are becoming the common mechanism that helps RE deployment in the off-grid sector in both developed and developing countries (IRENA, 2018). Driven by rapid cost decline in renewables like solar and wind energy, renewables based mini-/off-grid systems plus energy storage systems are becoming more technically and financially viable and replacing fossil run diesel. Raising ambitious RE targets in the next NDC 2020 will be needed to achieve RE transformation and achieve long-term climate objectives. The raised RE targets provides important high-level signal to private sector, power utilities and investors and helps for achieving mitigation actions of countries.

- Delivering support on capacity building and technology transfer is needed to developing countries in order to effectively conduct technical and financial assessment of viable RE projects to attract financing. Fostering knowledge and sharing experiences of good practices in RE deployment needs to be strengthening to enhance RE project implementation.
- Building on good track from growing blended Purchasing Power Parity (PPP) finance instrument, countries should strengthen their legal and regulatory frameworks to attract and leverage private investments. Leveraging private investment in RE deployment through de-risking mechanism is very essential. Developing a national financing vehicle is essential in investment de-risking mechanisms to attract Public Private cooperation to implement RE projects.
- As an early energy planning exercise, conducting various stage extensive consultations with stakeholders are very critical to buy-in all actors and get political will and mandate to implement RE.
- Effective climate change actions will include decarbonization efforts, such as phasing out fossil fuel subsidies in all forms, introducing forms of carbon pricing, and using a share of the public revenues generated to facilitate a just transition.
- Increasing the uptake of RE in developing countries would also involve building institutional, technical and human capacity to support renewable energy research, development, deployment and maintenance.

REFERENCES

Abubakar Mas'ud, A., Vernyuy Wirba, A., Muhammad-Sukki, F., Albarracin, R., Abu-Bakar, S. H., Munir, A. B., & Aini Bani, N. (2016). A review on the recent progress made on solar photovoltaic in selected countries of sub-Saharan Africa. *Renewable & Sustainable Energy Reviews*, *62*, 441–452.

Adbmouleh, Z., Rashid A.M., & Gastli, A. (2015). Review of policies encouraging renewable energy integration & best practices. *Renewable & Sustainable Energy Reviews*, *45*, 249–262.

Ahlborg, H., & Hammar, L. (2014). Drivers and barriers to rural electrification in Tanzania and Mozambique – Grid-extension, off-grid, and renewable energy technologies. *Renewable Energy*, *62*, 117–124.

Al-Badi, A. H., Malik, A., & Gastli, A. (2009). Assessment of renewable energy resources potential in Oman and identification of barrier to their significant utilization. *Renewable & Sustainable Energy Reviews*, *13*, 2734–2739.

Asian Development Bank. (2014). *Technical Assistance for the South Asia Economic Integration Partnership – Power Trading in Bangladesh and Nepal (TA 8658-REG)*. Manilla: Author.

Baumol, W. J., & Oates, W. E. (1998). *The theory of environmental policy*. Cambridge, UK: Cambridge University Press.

Baumol, W. J., Panzar, J. C., & Willig, T. D. (1982). *Contestable markets and the theory of industry structure*. New York, NY: Harcourt Brace Jovanovich.

Beck, U. (1995). *Ecological politics in an age of risk*. Malden, MA: Blackwell Publishers Inc.

Ecotec Research Consulting Ltd. (2001). *Aphrodite Mourelatou European Environment Agency, "Renewable energies: success stories*. Copenhagen: European Environment Agency.

Engelken, M., Römer, B., Drescher, M., Welpe, I. M., & Picot, A. (2016). Comparing drivers, barriers, and opportunities of business models for renewable energies: A review. *Renewable & Sustainable Energy Reviews*, *60*, 795–809.

Foxon, T., & Pearson, P. (2008). Overcoming barriers to innovation and diffusion of cleaner technologies: Some features of a sustainable innovation policy regime. *Journal of Cleaner Production*, *16*(1), 148–161.

Government of the Republic of Kiribati. (2009). *Kiribati National Energy Policy*. Retrieved from http://www.mfed.gov.ki/sites/default/files/Kiribati%20National%20 Energy%20Policy.pdf

Government of the Republic of Kiribati. (2015). *Intended Nationally Determined Contribution*. Retrieved from http://www4.unfccc.int/Submissions/INDC/Published

International Energy Agency, International Renewable Energy Agency, United Nations Statistics Division, World Bank Group, & World Health Organisation. (2018). *Tracking SDG7: The Energy Progress Report*. Retrieved from http://www.irena.org/-/media/Files/IRENA/Agency/Publication/2018/May/SDG7_Tracking_report_executive_summary_2018.pdf

International Renewable Energy Agency. (2016a). *The Power to Change: Solar and Wind Cost Reduction Potential to 2025*. Abu Dhabi: The International Renewable Energy Agency.

International Renewable Energy Agency. (2016b). *Letting in the Light: How Solar Photovoltaics Will Revolutionise the Electricity System*. Abu Dhabi: The International Renewable Energy Agency.

International Renewable Energy Agency. (2017a). *Renewable Energy Statistics 2017*. Abu Dhabi: The International Renewable Energy Agency.

International Renewable Energy Agency. (2017b). *Kiribati Integrated Energy Roadmap: 2017-2025*. Abu Dhabi: The International Renewable Energy Agency.

International Renewable Energy Agency. (2017c). *Trends in Photovoltaic Applications 2017*. Abu Dhabi: The International Renewable Energy Agency.

International Renewable Energy Agency. (2018). *Renewable Power Generation Costs in 2017*. Abu Dhabi: The International Renewable Energy Agency.

Karakezi, S., & Kithyoma, W. (2003). *Renewable energy in Africa: prospects and limits*. Paper presented at Republic of Senegal and United Nations Workshop for African Energy Experts on Operationalizing the NEPAD Energy Initiative, Dakar, Senegal.

Margolis, R. M., & Kammen, D. M. (1999). Underinvestment: The energy technology and R&D policy challenge. *Science*, *285*, 690–692. PMID:10426983

Mawdsley, E. (2018). 'From billions to trillions': Financing the SDGs in a world 'beyond aid'. *Dialogues in Human Geography*, *8*(2), 191–195.

Menanteau, P., Finon, D., & Lamy, M. L. (2003). Prices quantities: Choosing policies for promoting the development of RE. *Energy Policy*, *31*(8), 799–812.

Mirza, U. K., Ahmad, N., Harijan, K., & Majeed, T. (2009). Identifying and addressing barriers to renewable energy development in Pakistan. *Renewable & Sustainable Energy Reviews*, *13*, 927–931.

New Zealand Government & European Union. (2016). *Pacific Energy Country Profiles*. New Zealand Ministry of Foreign Affairs and Trade.

Nygaard, I., Handsen, U. E., Mackenzie, G. A., & Pederson, M. b. (2017). Measures for diffusion of solar PV in selected Africa countries. *International Journal of Sustainable Energy*, *36*(7), 707–721.

Ohunakin, O. S., Adaramola, M. S., Oyewola, O. M., & Fagbenle, R. O. (2014). Solar energy applications and development in Nigeria: Drivers and barriers. *Renewable & Sustainable Energy Reviews*, *32*, 294–301.

Organisation for Economic Cooperation and Development & World Economic Forum. (2015). *A How-To Guide for Blended Finance*. Retrieved from http://www3.weforum.org/docs/WEF_Blended_Finance_How_To_Guide.pdf

Polack, A. R. (2010). *Drivers and Barriers of Renewable Energy in the Electrification of Vanuatu*. Murdoch, Australia: Murdoch University, School of Engineering and Energy.

Regulatory Indicators for Sustainable Energy. (2018). *Kiribati Country Energy Profile*. Retrieved from http://rise.esmap.org/countries%20Documents/Kiribati/1/INDC_KIRIBATI.pdf

GTM Research. (2018). *Trends in Solar Technology and System Prices*. Presented at GTM 2018 Solar Summit Breakfast Briefing.

Roper, T. (2005). Small Island States – Setting an Example on Green Energy Use. *Review of European Community & International Environmental Law*, *14*(2), 108–116.

Sen, S., & Ganguly, S. (2017). Opportunities, barriers and issues with renewable energy development – A discussion. *Renewable & Sustainable Energy Reviews*, *69*, 1170–1181.

Sieminksi, A. (2013). International energy outlook 2013. Washington, DC: Independent Statistics and Analysis, US Energy Information Administration.

Solar, S. (2018). *Scatec Solar closes financing for Mozambique's first large scale solar plant*. Retrieved from https://www.scatecsolar.com/Investor/Stock-exchange-notices/Scatec-Solar-closes-financing-for-Mozambique-s-first-large-scale-solar-plant

Sustainable Energy for All (SE4ALL). (2015). *Scaling Up Finance for Sustainable Energy Investments – Report of the SE4ALL Advisory Board's Finance Committee – 2015*. Retrieved from https://www.seforall.org/sites/default/files/SE4All-Advisory-Board-Finance-Committee-Report.pdf

United Nations Economic and Social Council (UN ESC). (2017, May). *Progress towards the Sustainable Development Goals*. Paper presented at the July 2017 session of the Economic and Social Council, New York, NY.

United Nations Intergovernmental Committee of Experts on Sustainable Development Financing. (2014). *Report of the Intergovernmental Committee of Experts on Sustainable Development Financing*. New York, NY: United Nations.

ADDITIONAL READING

Assmann, D., Laumanns, U., & Uh, D. (2006). *Renewable energy a global review of technologies, policies and market*. New York, NY: Earthscan.

Bacon, R., & Besant-Jones, J. (2001). Global electric power reform, privatization, and liberalization of the electric power industry in developing countries. *Annual Review of Energy and the Environment*, *26*, 331–359.

Barnes, D. F. (2007). The challenge of rural electrification, Washington: Resources for the Future Press (RFF Press); 2007.

Baurzhan, S., & Jenkins, G. P. (2016). Off-grid solar PV: Is it an affordable or appropriate solution for rural electrification in Sub-Saharan African countries? *Renewable & Sustainable Energy Reviews*, *60*, 1405–1418.

Bazilian, M., Nussbaumer, P., Rogner, H.-H., Brew-Hammond, A., Foster, V., & Pachauri, S. (2012). Energy access scenarios to 2030 for the power sector in sub-Saharan Africa. *Utilities Policy*, *12*, 1–16.

Besant-Jones, J. E. (2006). *Reforming Power Markets in Developing Countries: What Have We Learned?* Energy and Mining Sector Board Discussion Paper No. 19, September, The World Bank Group, Washington, DC.

Bugaje, I. M. (2006). Renewable energy for sustainable development in Africa: A review. *Renewable & Sustainable Energy Reviews*, *10*, 603–612.

Chowdhury, S. A., Aziz, S., Groh, S., Kirchhoff, H., & Leal Filho, W. (2015). Off-grid rural area electrification through solar–diesel hybrid minigrids in Bangladesh: Resource-efficient design principles in practice. *Journal of Cleaner Production*, *95*, 194–202.

Dornan, M. (2015). Renewable energy development in small island developing states of the Pacific. *Resources*, *4*(3), 490–506.

Foley, G. (1992). Rural electrification - The institutional dimension. *Utilities Policy*, *2*(4), 283–292.

Huang, Z., Yu, H., Peng, Z., & Zhao, M. (2015). Methods and tools for community energy planning: A review. *Renewable & Sustainable Energy Reviews*, *42*, 1335–1348.

Kankam, S., & Boon, E. (2009). Energy delivery and utilization for rural development: Lessons from Northern Ghana. *Energy for Sustainable Development*, *13*(3), 212–218.

Kaundinya, D. P., Balachandra, P., & Ravindranath, N. H. (2009). Grid-connected versus stand-alone energy systems for decentralized power, a review of literature. *Renewable & Sustainable Energy Reviews*, *13*, 2041–2050.

Kaygusuz, K. (2002). Energy for sustainable development: Key issues and challenges. *Energy Sources*, *2*, 73–83.

Keay, M., Rhys, J., & Robinson, D. (2013). *Decarbonization of the electricity industry – is there still a place for markets?* OIES Working Paper EL9, Oxford Institute for Energy Studies, November.

Liu, L. Q., Wang, Z. X., Zhang, H. Q., & Xue, Y. C. (2010). Solar energy development in China—A review. *Renewable & Sustainable Energy Reviews*, *14*, 301–311.

Martinot, E., Cabraal, A., & Mathur, S. (2001). World Bank/GEF solar home system projects experiences and lessons learned, 1993–2000. *Renewable & Sustainable Energy Reviews*, *5*, 39–57.

Mondal, MdAH., Kamp, LM., & Pachova, NI. (2010). Drivers, barriers, and strategies for implementation of renewable energy technologies in rural areas in Bangladesh—an innovative system analysis. *Energy Policy, 38*, 4626–4634.

Mulder, P., & Tembe, J. (2008). Rural electrification in an imperfect world: A case study from Mozambique. *Energy Policy*, *36*, 2785–2794.

Nepal, R., & Jamasb, T. (2012). Reforming small electricity systems under political instability: The case of Nepal. *Energy Policy*, *40*, 242–251.

Nepal, R., Jamasb, T., & Sen, A. (2018). Small systems, big targets: Power sector reforms and renewable energy in small systems. *Energy Policy*, *116*, 19–29.

Nguyen, K. Q. (2007). Alternatives to grid extension for rural electrification: Decentralized renewable energy technologies in Vietnam. *Energy Policy*, *35*, 2579–2589.

Ondraczek, J. (2013). The Sun Rises in the East (of Africa): A Comparison of the Development and Status of Solar Energy Markets in Kenya and Tanzania. *Energy Policy, 56*, 407–417.

Ondraczek, J. (2014). Are we There Yet? Improving solar PV Economics and Power Planning in Developing Countries: The Case of Kenya. *Renewable & Sustainable Energy Reviews*, *30*, 604–615.

Owen, A. (2006). Renewable energy: Externality costs as market barriers. *Energy Policy*, *34*, 632–642.

Painuly, J. (2001). Barriers to renewable energy penetration: A framework for analysis. *Renewable Energy*, *24*(1), 83–89.

Pollitt, M. (2004). Electricity reform in Chile: Lessons for developing countries. *Competition and Regulation in Network Industries*, *5*(3), 221–263.

Pollitt, M. (2012). The role of policy in energy transitions: Lessons from energy liberalisation era. *Energy Policy*, *50*, 128–137.

Quansah, D. A., Adaramola, M. S., & Mensah, L. D. (2016). Solar Photovoltaics in sub-Saharan Africa – Addressing Barriers, Unlocking Potential. *Energy Procedia, 106*, 97–110.

Sen, S., Ganguly, S., Das, A., Sen, J., & Dey, S. (2016). Renewable energy scenario in India: Opportunities and challenges. *Journal of African Earth Sciences, 122*, 25–31.

Turkenburg, W. C., Arent, D. J., Bertani, R., Faaij, A., Hand, M., & Krewitt, W. (2012). Chapter 11 – Renewable energy. Global energy assessment – toward a sustainable future, Cambridge UniversityPress, Cambridge, UK and NewYork, NY, USA and the International Institute for Applied Systems Analysis, Laxenburg, Austria, 761–900.

Turkson, J., & Wohlgemuth, N. (2001). Power sector reform and distributed generation in sub-Saharan Africa. *Energy Policy, 29*, 135–145.

Weisser, D. (2004). Power sector reform in small island developing states: What role for renewable energy technologies? *Renewable & Sustainable Energy Reviews, 8*, 101–127.

Zahnd, A., & McKay Kimber, H. (2009). Benefits from a renewable energy village electrification System. *Renewable Energy, 34*, 362–368.

KEY TERMS AND DEFINITIONS

Capacity Building: Improving the ability of governments, organizations, and individuals to solve problems and function effectively.

Development Bank: Financial Institutions which provide risk capital and finance for development projects. These are typically split into national and multilateral (internationally established and following international law).

Energy Access: The term used to measure what percentage of a group or population has access to electricity.

Feed-in-Tariff (FiT): FiTs are policy instruments which promote the use of renewable energy by paying individuals or organizations generating renewable energy a fixed price for each unit they generate.

Intermittent Generation: One of the core challenges of renewable energy, the fact that the supply is not constant (for example night and day/clouds with solar PV, the fact that certain days/seasons have stronger wind). As supply is not constant, it is vital to think of a way to ensure that this does not interfere with a constant supply of energy to the grid.

Kilowattpeak (kWp): In basic terms, a unit of measurement for the power of your PV installation/plant. The power a plant or panel would generate under standard conditions (a defined amount of incoming sunlight).

Levelized Cost of Electricity: The total cost to build and operate a power generation point over its lifetime divided by its lifetime output. Can be thought of as a way to compare the price of generating one unit of energy on a consistent basis. Useful for comparing various methods of generation.

Nationally Determined Contributions (NDC): Under the Paris Agreement, each party must prepare, communicate, and maintain successive nationally determined contributions. These essentially act as self-determined goals for how a country will contribute to achieving the goals of the Paris Agreement, and as noted, are updated and revised over time.

Off-Grid (PV/Energy): The grid may be thought of as a network of supply lines, able to carry energy to anywhere connected to it. Locations that are off-grid, are not reachable using traditional energy generation, and must rely on local generation for energy, or gain access to the grid.

Power Purchase Agreement (PPA): An agreement where a government consents to purchasing a set amount of power in advance. Used to reduce the risk of developing new power plants by guaranteeing revenue.

Project Tenders: Refers to the process where contracts for large projects are opened up to bids which must be submitted by a certain deadline.

Proliferation of Renewable Energy: This refers to the rapid uptake, and mainstream diffusion of renewable energy.

PV Capacity: The total amount of installed solar PV generation capability a country possesses.

PV Generation: Different from capacity in that it refers to the actual amount of power generated, rather than the potential amount implied by capacity.

PV Module: The term commonly used to refer to a solar panel, many of which are required to make a solar PV array, or plant.

Solar Irradiance: Refers to the energy per unit area received by the sun. On a more practical level, the higher the level of solar irradiance in an area, the greater the potential for solar power.

Solar PV: The most common form of solar energy today, solar PV or Photovoltaics, is when sunlight is converted directly into electricity, as opposed to other systems which focus the sun's energy to power traditional methods of generation such as steam turbines.

Sustainable Development Goals: Set of objectives developed by the United Nations aimed at outlining the key challenges to a better and sustainable future for all, including climate change, poverty, hunger, and many more. There are 17 total overarching goals, each with multiple sub-targets. These form part of the UN "Agenda 2030" which aims to transform our world by the year 2030.

ENDNOTE

1. RISE data unavailable for Kiribati; therefore, data was retrieved from NDC, International Renewable Energy Agency (2017) Integrated Energy Roadmap and Kiribati National Energy Policy, 2009.

APPENDIX: ANALYSIS AND DISCUSSION QUESTIONS

1. What are the different types of challenges/concerns regarding the uptake of renewable energy in developing countries?

   The core challenges faced by developing countries regarding the deployment of renewable energy can be broadly grouped into the following categories: Financial, fiscal, legislative, political, technological, and environmental. These deal with the prohibitive factors regarding the diffusion of solar PV such as high costs, undeveloped infrastructure, and a lack of finance.

2. Provide three reasons why the diffusion of solar PV could be more challenging in a developing context?

   There are many reasons for this. These include:
   - Higher finance costs – financing for such projects is harder to access, and often more expensive given higher levels of uncertainty
   - Higher capital costs – Most of the parts required are not made in these countries, and have to be imported, often at a high tax rate, with minimal competition.
   - Lack of skilled workers and research – There are a lack of workers available capable of building/maintaining solar PV technology, particularly at a large scale.
   - Lack of experience at a national level – Governments and industry have little experience with solar PV.
   - Geographic issues – Areas often lack infrastructure, are remote, or lack grid access.
   - Potential lack of developed policy framework

3. What role can the international community play in helping to develop solar PV in LDCs?

   The international community has several key roles to play. The first is helping to provide access to finance, be it through affordable loans, helping to lower the cost of credit through providing guarantees or assurances, and helping to build public-private partnerships. Another key aspect of their involvement is promoting knowledge sharing and helping to spread best practice between both the developed and developing world, and also between developing countries.

4. What problems are caused by the intermittent nature of renewable energy, and in particular solar PV?

   The big problem with relying on renewable energy, and in particular solar PV, is that it is a not a steady supply. Therefore, if you only have solar power, you will not have stable access to electricity and issues such as constant blackouts will continue to be a major problem. This is clearly problematic as for certain buildings or organisations such as hospitals, having constant access to power is critical. Furthermore, it can interrupt economic activity, weaken access to education, and make basic day-to-day functions such as water purification only possible at certain times of day.

5. What are the main solutions to this problem?

   The first solution would be energy storage, or battery technology. This allows the energy to be stored if not used, and then used at the time it is required. One issue with battery technology is that it can be expensive and inaccessible, and therefore it is not always suitable. The next solution is to connect the area to the grid, so that traditional power generation can pick up when renewable power is out. This of course is dependent upon the accessibility of the grid in the first place, which is not always the case due to their often-remote nature. One final option is to try and setup a mix of renewables capable of round-the clock generation, although doing this is largely dependent on if you are fortunate enough to have access to such a steady stream of renewable energy and is therefore highly unlikely to be simple in reality. In reality, each situation will most likely require a tailored solution, taking into account the unique characteristics of every scenario.

6. What can governments do to try and increase investor confidence in renewable energy projects?

   There are several key tactics that governments can employ to increase investor confidence. The first major role they can play, is by setting ambitious and comprehensive targets regarding renewable energy generation and usage, and then outlining a clear plan for how to get there including such things as legislative measures, training programmes, and financial incentives. This sends a clear signal to investors that there is a sustainable future in renewable projects within the country. They can also look to sign power purchase agreements in order to provide a more secure return on any investments, as well as forming public-private partnerships to help reduce risk and share knowledge.

7. What are the potential complications for developing countries regarding meeting the Sustainable Development Goals and the Nationally Determined Contributions at the same time in terms of renewable energy?

   The main complication from this, is that the NDCs are more specifically targeted at curbing the impact of climate change, whereas the SDGs are far more varied and range across issues such as poverty, the environment, health, and education just to name some. What is problematic is that the benefits of sustainability are not always immediate, whereas the consequences of many other issues are more apparent. Therefore, solar PV for example might be good for the environment, but the potentially higher cost could lead to increased poverty. It is therefore crucial to consider the hidden impacts of decisions and how they interact with each other.

8. What are the opportunities presented by renewable energy, and how can developing nations take advantage of them? Provide arguments in support of your answer.
   - ***Rapid and significant cost reductions of renewable energy***: Developing countries should take this advantage by raising ambition in RE targets and to improve policy and regulatory environment.
   - ***Commercially attractive investment opportunities***: Countries should take this advantage by phasing out fossil fuel subsides and increase on RE auctions.
   - ***Partnership and involvement of multiple stakeholders and actors***: This will help developing countries to take action by multiple stakeholders and actors at different levels of governance and to enhance the knowledge and capacity at national and subnational level. It can also raise awareness in public and private sector on renewable energy.
   - ***Availability of international climate finance***: Developing countries should take advantage to establish and set up the national financial vehicles and instruments in de-risking renewable energy projects to access to climate finance (e.g. the Green Climate Fund (GCF)). Also, to adopt blended concessional finance since it is a critical evolving tool to develop private sector markets, foster innovation, and crowd in private finance in some of the most challenging settings for deploying renewable energy.

Chapter 2
Implementing the European Union Renewable Energy Policy Targets in Bulgaria

Tatyana B. Ruseva
Appalachian State University, USA

Maria A. Petrova
University of Massachusetts – Boston, USA

EXECUTIVE SUMMARY

As a member of the European Union (EU), Bulgaria has been implementing the EU's policy targets designed to increase the share of renewable energy (RE) use in gross final energy consumption by 2020. The target for Bulgaria, set at 16%, was accomplished eight years earlier than mandated, in 2012. The result of rapid but poorly regulated growth in renewables—seemingly a success story—illustrates the potential pitfalls of RE policy implementation. Having met its target, Bulgaria undertook a series of restrictive policy measures that undermined short-term RE growth, increased regulatory uncertainty and market stagnation. The objective of this chapter is to understand the factors that shaped these unintended policy measures and outcomes. Drawing on key informant interviews, the chapter presents a case study of renewable energy policy implementation in a multi-level governance system and illustrates the boomerang effects associated with top-down policy implementation.

DOI: 10.4018/978-1-5225-8559-6.ch002

INTRODUCTION

Renewable energy is becoming a valuable source of widely available, cleaner, and relatively inexpensive electricity, yet the transition to non-fossil fuel-based economies is far from over. Since 2007 European Union (EU) member states have been working to increase the share of renewables in EU energy consumption, increase energy efficiency, and reduce greenhouse gas (GHG) emissions by 20% from 1990 levels by the year 2020. A cornerstone of the EU 2020 strategy is the Renewable Energy (RE) Directive (2009/28/EC), which set a 20% target for the overall share of renewables in EU gross final energy consumption.[1] Article 4 of the RE Directive mandates that member states develop National Renewable Energy Action Plans, which detail how member states envisage implementing the Directive and reaching the 2020 targets (European Parliament [EP], 2009).

To achieve the common 20% policy target for renewables, each EU member state has agreed to binding, country-specific targets for the overall share of renewable energy sources (RES) in three areas: 1) electricity (RES-E), 2) heating and cooling (RES-H&C), and 3) transport (RES-T). National renewable energy targets differ for each member state because they are calculated as the share of renewable consumption to gross final energy consumption, and take into consideration member states' different starting points, renewable energy potential, and economic performance (e.g., GDP per capita, economic growth forecasts). There is substantial variation in national RES targets, which range from 10% in Malta to 49% in Sweden, placing the overall EU countries' mean at 21% and the median at 18% (2009/28/EC)[2]. Bulgaria's national target for the share of RE consumption was set at 16%.

This chapter examines Bulgaria's experience with the implementation of the national renewable energy policy target under the EU 2009 Renewable Energy Directive. Bulgaria is a fairly recent EU member state (since 2007), whose size and resource endowments are similar to those of other Southeast European countries (e.g., Serbia, Croatia, Moldova). Given the strong trend for policy diffusion in Europe, Bulgaria's implementation of the RE Directive could provide lessons for other countries in the region with comparable political history and aspirations for EU membership. The case also illustrates the challenges of implementing EU policies in recent member states, and the potential for unintended consequences associated with top-down policy goals. Ambitious policy measures can backfire, i.e. they can create unintended consequences or contradictory responses, known as boomerang effects. Boomerang effects reduce the prevalence of desired policy goals and the effectiveness of measures promoting those goals (Kinzig et al. 2013; Brehm & Brehm 1981).

Our case reveals that there were serious boomerang effects in the implementation of the 2020 RE policy targets in Bulgaria. A rapid and poorly regulated growth in wind and solar energy production in the period 2007-2011 led to the achievement of the national target in 2012, eight years earlier than mandated. This outcome, itself a boomerang effect, was followed by a series of retroactive, restrictive policy measures, increased regulatory uncertainty, and market stagnation that undermined RE sector growth (Hiteva & Maltby 2017; Nikolaev & Konidari 2017). In 2013, Bulgaria's Ministry of Economy and Energy (MEE) introduced restrictions on new renewable generation capacity and previously contracted RES capacities were put on hold (MEE, 2014). The period 2012-2014 included a series of anti-RE interventions, restrictions of new solar and wind power installments, retroactive and discriminatory policy measures, and increasing uncertainty in the regulatory framework. These were combined with a growing public discontent and opposition to renewables, as well as spiraling debt and financial difficulties for the main utility and electricity distribution companies. As our case illustrates the boom in renewable energies did not contribute to a sustainable long-term RES development, but rather to a crisis in the energy sector and increased regulatory uncertainty that ultimately undermined RE growth in the country.

The objective of this book chapter is to describe this seeming policy paradox and to understand the factors generating boomerang effects (i.e. unintended consequences) in the implementation of renewable energy policy targets in the case of Bulgaria. To organize our analysis, we draw on the Institutional Analysis and Development (IAD) framework (Ostrom, 1990, 2005). The IAD framework helps us examine how Bulgaria's energy sector's attributes, socio-economic and political conditions, as well as local institutions, impact decision-making processes and subsequent policy outputs. Data came from 30 semi-structured interviews with representatives from government, industry, and nongovernmental organizations in Bulgaria's energy sector, supplemented by official documents, laws, and news reports from 2011 to 2015.

The rest of the chapter provides a description of Bulgaria's energy sector, the main actors, and institutions, as well as an analysis of the factors that help explain the boomerang effects of renewable energy policy implementation. While the case does not allow extrapolation of findings, it calls attention to the importance of policy implementation, the need for long-term renewable energy development strategy, regulatory stability, and administrative capacity in the promotion of renewable energies. The case and its findings also inform our understanding of RE promotion policies and regulations in new market economy contexts. It offers lessons about the importance of policy implementation and regulatory (un)certainty in a multi-level governance system.

BACKGROUND

In 2009, acknowledging the importance of renewable energy, the European Union passed the Renewable Energy Directive (2009/28/EC) which requires that by 2020 at least 20% of the overall energy consumed in the EU and 10% of the energy in the transportation sector come from renewable sources.[3] This policy target is calculated across all member states and requires each member to develop national renewable energy targets that contribute to the common 20% RE target.[4] It needs to be stressed that national RE targets refer to the portion of RES in gross final energy consumption.[5] Member states were directed to develop thorough National Renewable Energy Action Plans (NREAPs) and report these plans to the EU. Bulgaria's national target was to increase the share of RES in gross final energy consumption from 9.4% in 2005 to 16% by 2020. Sectorally, this target meant achieving 20.8% RES-E share in gross final consumption of electricity, 23.8% RES-H&C share in final consumption of heating and cooling energy, and 10.8% RES-T share in transportation by 2020 (Ministry of Economy, Energy, and Tourism [MEET], 2011). The Bulgarian NREAP envisioned a strong focus on hydroelectric energy for overall energy production and an emphasis on biofuels for the transportation sector.

In 2016, the European Commission proposed a framework for 2030 that included a goal of 27% share of RES in the total energy consumed in the Union (European Commission [EC], 2016). The 2030 proposal emphasizes a multi-level framework that specifies EU, national, and regional goals, thus giving significant discretion to individual member states. On the grounds of this proposal, in June 2018, the European Commission, Council, and Parliament struck an agreement that set the intended share of renewables to 32% by 2030, with a review clause for an upward revision of the EU-level target in 2023 (EC, 2018a).

The Role of Institutions as Central Pillars of Governance and the IAD Framework

EU renewable energy policy is part of a broader governance process that encompasses formulation, adoption, and implementation of policy objectives at multiple levels. The logic of governance integrates key concepts and approaches from the study of institutions and public administration (Frederickson, Smith, Larimer, & Licari, 2012; Lynn, Heinrich, & Hill, 2000; O'Toole & Meier, 1999). Lynn et al. (2000) define governance as "regimes of laws, administrative rules, judicial rulings, and practices that constrain, prescribe, and enable government activity, where such activity is broadly defined as the production and delivery of publicly supported goods and services" (p. 235). Renewable energy governance, therefore, is a useful approach for describing the multilevel, inter-jurisdictional relations among private and public organizations pertaining to the production, distribution, and consumption of electricity from renewable energy sources (Frederickson et al., 2012).

Institutions are central pillars in renewable energy governance. Institutions are understood as prescriptive statements (rules) that require, prohibit, or permit some action or outcome (Ostrom 1990, 2005). At their core, institutional rules structure social interactions by way of defining the choices and actions of actors (North, 1995; Ostrom, 1999). The Institutional Analysis and Development (IAD) framework is a useful conceptual approach for identifying *institutional rules* (regulations, norms, and shared strategies), *participants,* and *decision situations* in governance contexts (e.g., market exchanges, policy implementation) (Kiser & Ostrom, 1982; Ostrom, 2005). We employ the IAD framework as a way of organizing our inquiry of the factors leading to boomerang effects in the implementation of RE policy targets in Bulgaria. Specifically, these factors include *institutional rules* (examined under section "Institutional Rules for RES Promotion in Bulgaria"), *biophysical or resource sector conditions* (see section "Bulgaria's Energy Sector"), *and country-specific attributes* (see "Bulgaria's Political and Socio-Economic Challenges"). Together these three sets of factors impact decision-making processes and subsequent policy implementation outcomes in Bulgaria's RE sector in the 2011-2015 period (Figure 1).

The IAD framework has been effectively used over the past thirty years to study common-pool resource management (Ostrom, 1990), metropolitan governance (Parks & Oakerson, 2000), water and energy systems (Theesfeld, 2004; Villamayor-Tomas, 2017; Villamayor-Tomas, Grundmann, Epstein, Evans, & Kimmich, 2015), forestry (Agrawal, Wollenberg, & Persha, 2014), and renewable energy and climate adaptation (Iychettira, Hakvoort, & Linares, 2017; Roggero, Bisaro, & Villamayor-Tomas, 2018). We use the IAD framework here to organize our data collection and analysis of the implementation of the European RE policy target in Bulgaria.

Institutional Rules and Actors in the Bulgarian Energy Sector

The explicit focus of the IAD framework on multiple levels of rules is particularly useful in the context of EU multi-scale and multi-actor system of political decision-making. The implementation of the EU RE targets can be studied in terms of three levels of rules that cumulatively shape the decisions of participants in the RE sector. The three levels of rules are: 1) constitutional rules, such as the EU Renewable Energy Directive, 2) collective-choice rules, such as member states' national laws, including Bulgaria's Energy from Renewable Sources Act (ERSA), and 3) operational rules, such as decisions by state regulatory agencies (Ostrom, 2005).

At the national level, Bulgarian policies are carried out by public and private organizations (RE sector participants), who make decisions related to provision, production, and consumption of electricity from renewable energy sources, transmission and distribution, sanctioning, information distribution, and coordination. The latter comprise the range of activities and decisions occurring in the renewable energy action arena (Figure 1). Each of these decisions can be defined as a decision situation with an existing set of rules, participants, and external influences (McGinnis, 2011; Ostrom, 2005).

Two external factors shape the decisions of participants, and the resulting policy outputs, namely: 1) the *attributes of the community,* (i.e. the EU member state of Bulgaria) and 2) the *biophysical characteristics* of the resource system (i.e., the electric grid and resource endowments). As applied here, the IAD framework allows one to not only identify the set of institutional rules that affect outcomes in the RE sector, but also the participants, their positions, decisions, information, as well as perceived costs and benefits associated with potential outcomes.

Participants in the RE sector in Bulgaria include government agencies, regulators, industry associations, RE producers, distributors, and other market players. Participants adopt strategies in line with their interests, available information, and existing rules (Ostrom, 2005). Specifically, the National Electricity Company (NEK) – a parastatal company[6] – plays a central role in the electricity system, acting both as a single buyer on the high-voltage electricity grid and as a single supplier of electricity (i.e. public provider) to electricity distribution companies (EDCs). On the free market, NEK is a supplier of last resort to industrial consumers (EC, 2013). NEK is also a subsidiary of the Bulgarian Energy Holding, which owns about 45% of installed generation capacity.

Figure 1. The institutional analysis and development (IAD) framework
(Ostrom 2005; Hardy & Koontz 2009)

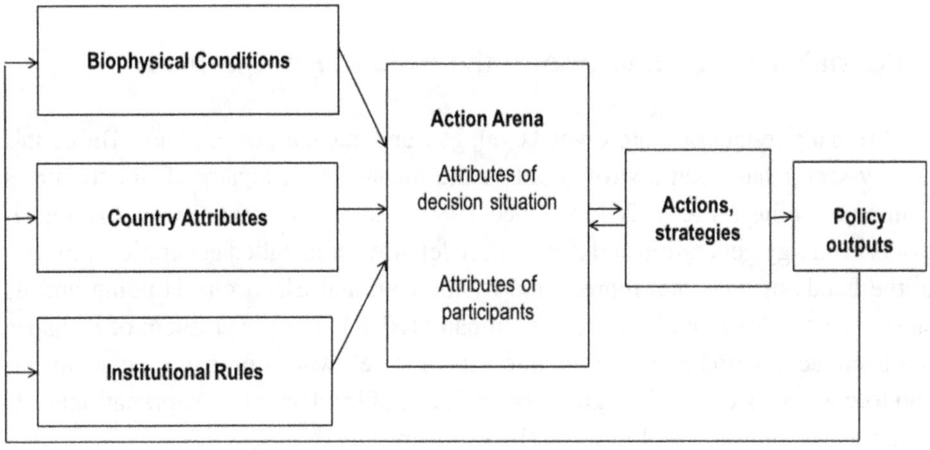

Most of the produced electricity is sold at regulated prices set by the Energy and Water Regulatory Commission (EWRC), and the rest is traded at negotiated prices. The EWRC is responsible for the regulation of activities related to production, transmission, and distribution of electricity, including electricity from renewables, as well as licensing of RES producers. The EDCs (EVN, EnergoPro, and CEZ[7]) act as end-user suppliers at regulated prices of electricity. The EDCs purchase electricity produced by RES power plants at preferential prices on behalf of NEK, and are subsequently compensated by NEK. Finally, the Electricity System Operator (ESO), a former subsidiary of NEK, operates the transmission grid.[8]

Case Study Sources and Data Collection

We draw on primary and secondary sources for this research, including 30 in-depth interviews with participants in the RES sector in Bulgaria, conducted between May 2014 and August 2015 (see Appendix), EU documents, government reports, print media, and news articles. Interview participants were selected using a snowball sampling technique, following an initial selection based on publications in the national media from 2013 onwards (e.g., news articles, documentaries, government and industry reports). Interviews were transcribed and data were subjected to a thematic analysis guided by the IAD framework's main constructs, namely the biophysical or resource conditions, the country's political and socio-economic conditions, and institutional rules.

POLICY IMPLEMENTATION IN A MULTI-LEVEL GOVERNANCE SYSTEM

Bulgaria's Energy Sector and the Role of Renewables

Traditionally reliant on state-owned coal, gas, and nuclear power plants, Bulgaria's energy sector has been a stronghold of the management legacy of the country's communist elite (Hiteva, 2013). Since 1989, a slow and partial privatization of assets in energy generation and distribution left 45% of installed generation capacity in the hands of the state, represented by the National Electricity Holding and its subsidiary the National Electricity Company (NEK). The government of Bulgaria has been faced with balancing the interests of state-owned companies, consumers, and foreign investors, and juggling the demands of implementing supranational EU policy goals and responsibilities at the same time.

About half of Bulgaria's energy still comes from coal, gas, and nuclear power plants, with over a third of energy supply coming from imported crude oil and gas (MEE, 2011; Nikolaev & Konidari 2017). Domestic sources of energy include coal (4.81 Mtoe lignite, 0.29 Mtoe brown coal as of 2016), natural gas (0.16 Mtoe), and crude oil (0.03 Mtoe) (National Statistical Institute [NSI], 2017). Hydroelectricity from predominantly state-owned hydro power plants (HPPs) has traditionally been the dominant source of power generation from renewables (MEE, 2013).

In its 2008 energy strategy, Bulgaria identified wind, biomass, and HPPs as the three main RES by means of which the 2020 RE target could be met (MEE, 2011). Solar photovoltaic (PV) power plants were not mentioned, even though Bulgaria has potential for energy from both solar thermal and solar PV systems. Wind potential is limited to the Black Sea coast and the mountain areas. The potential for hydroelectricity in the country has already been tapped through large HPPs which were built in the past, and only small-sized hydro power plants (< 500 kW capacity) remain feasible for future projects. Solid biomass (firewood) is traditionally used in household heating, while refined biomass (pellets and briquettes) and waste from wood logging and processing are considered future trends. Biogas from municipal landfills and animal manure was also considered for cogeneration of heat and power (CHP) (MEE, 2011).

Energy Consumption

The period 2007-2016 was marked by variability in the energy supply – with an increase in 2011 and a downturn in 2013 – and a fairly stable energy consumption (Nikolaev & Konidari, 2017). Over half of Bulgaria's energy consumption in 2016 came from fossil fuels (Figure 2). Transportation (3.5 Mtoe), industry (2.6 Mtoe), and residential (2.3 Mtoe) were the leading sectors in energy consumption, followed by the services (1.1 Mtoe) and agriculture (0.2 Mtoe) sectors (NSI, 2017). The share of Bulgaria's electricity from RES in gross electricity consumption in 1997 was 7%, with large-scale hydro power being the main source (Eurostat, 2018).

As Table 1 shows, the share of RES in gross final energy consumption doubled between 2007 (9.2%) and 2016 (18.8%), achieving the EU target of 16% in 2012, eight years ahead of the target date. Substantial reliance on solid biomass for household heating was the primary factor explaining the growth in RES-H&C (Table 1, Interviews 1, 2). The post-2012 growth in RES-T was largely driven by the country's transport fuel blending obligation under the 2009 EU RE Directive, which required companies to sell fuel mixed with bio-fuels on a schedule intended to meet the 2020 targets (MEE, 2013; Nikolaev & Konidari, 2017).

Figure 2. Bulgaria's 2016 final energy consumption by fuel type (Total=9.7 Mtoe; EC, 2018b)

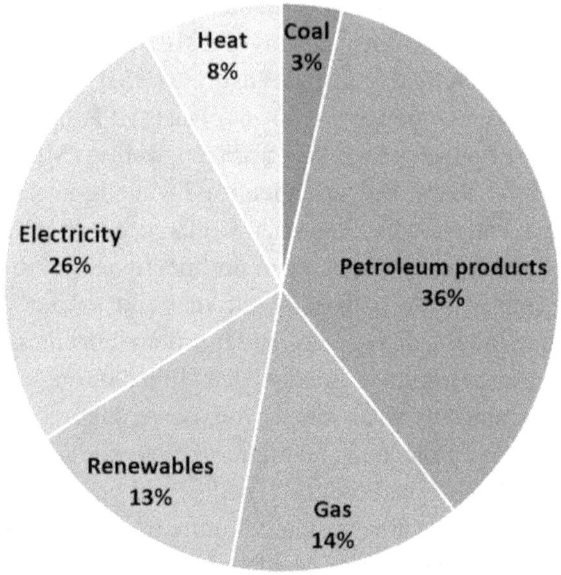

Table 1. Share of RE in Bulgaria's gross final energy consumption by sector

	2007	2008	2009	2010	2011	2012	2013	2014	2015	2016
Overall RES share (%)	9.2	10.5	12.1	14.1	14.3	16.0	19.0	18.0	18.2	18.8
RES-E (%)	9.4	10.0	11.3	12.7	12.9	16.1	18.9	18.9	19.1	19.2
RES-H&C (%)	13.9	17.3	21.7	24.4	24.9	27.5	29.2	28.3	28.6	30.0
RES-T (%)	0.9	0.9	1.0	1.4	0.8	0.6	6.0	5.8	6.5	7.3

(Eurostat, 2018)

Hydro power is the dominant source of RES-E in Bulgaria (Vladimirov, Georgiev, & Kolarova, 2018). However, the rapid growth of PV and wind capacity was the most notable development in RES technologies in the 2007-2016 period (Figure 3). There was a six-fold increase in Solar PV capacity in just one year (2011-2012), and an exponential growth in wind capacity between 2008 and 2012. This growth, however, was not predicted and was one of the drivers behind the wave of anti-RES measures in the post-2012 period (see section on Boomerang Effects) (Hiteva & Maltby, 2017; Interview 10).

Figure 3. Installed electricity capacity for renewables by type (2007-2016). Biomass includes solid biomass, wood, and wood waste
(NSI, 2018).

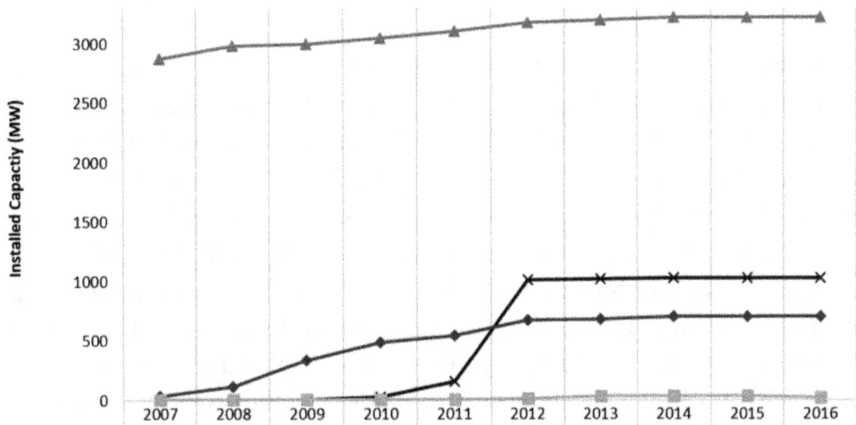

	2007	2008	2009	2010	2011	2012	2013	2014	2015	2016
Hydro	2876	2984	3001	3048	3108	3181	3203	3219	3219	3,223
Solar PV	0	0	2	25	154	1,013	1,020	1,026	1,029	1,028
Wind	30	114	333	488	541	677	683	699	699	699
Biomass	6	6	6	6	6	14	30	30	34	19

Electrical Grid System

According to an assessment by the European Commission, the installed RES capacity in the country was high (4,969 MW in 2016, compared to 2,912 MW in 2007) (Figure 3), but there was insufficient transmission capacity (EC, 2013). The overcapacity of electricity in the transmission system, exacerbated by the unusually warm 2012-13 winter and falling electricity exports (a decrease of 40% in 2013), led to introductions of scheduled power plant closures and unscheduled curtailment of PV and wind output by the ESO. Curtailment is a reduction in the amount of wind or solar energy generated and sold onto the power grid, relative to what can be produced with available resources (Jones 2017; Bird, Cochran, &Wang 2014). The Bulgarian Photovoltaic Association and the Bulgarian Wind Energy Association individually filed complaints with ESO and requested to get an explanation from NEK about the unscheduled curtailments of production capacities.

Interviews and document analyses suggest that the condition of the electrical grid system (a resource system factor under the IAD framework) was central to the boomerang effects experienced in Bulgaria's RE sector after 2012. The outdated condition of the grid was identified as one of the major reasons for restricting the production of RES-E by ESO: *The major challenges were first the grid infrastructure.*

...When our wind farm started operation, it had operated at 85 MWh, almost half of its production capability, which is 157 MWh, because of the bottlenecks of the grid in the northeastern part of Bulgaria. [It is] Because they could not evacuate all the electricity generated and there wasn't enough consumption at certain periods of time. (Interviews 11, 12)

In addition, the growing financial liabilities and budgetary imbalances prevented NEK and its subsidiary ESO from making much needed upgrades and investments in the electric grid. In September 2014, NEK reported a loss of BGN 425 million (USD 143 million) for the first nine months of 2014 (Novinite, 2014). The system of payments for electricity produced by wind power plants was based on forecasted production levels, which suffer from significant errors in prediction accuracy. Discrepancies between forecasted levels and actual production levels for wind power plants further complicated the financial situation of the EDCs, which are contractually obligated to purchase the electricity produced by RES installations. In short, both insufficient infrastructure and financial challenges for the major actors in the energy sector created an environment that stifled further growth in RES capacities, as witnessed in the leveling off of PV and wind RES capacities after 2012 (Figure 3).

Bulgaria's Political and Socio-Economic Challenges

In 1989 Bulgaria embarked on a transition from a planned to a market economy and from communism to democracy, but the transition process was very slow. The 1990s were marked by political instability, strikes, and growing economic inequality. After a series of economic and legal reforms in the early 2000s, the country signed an accession treaty with the EU in 2005, and formally joined the European Union on January 1, 2007. As a price for its EU membership, Bulgaria agreed to close down four of the six reactors of its Kozloduy nuclear power plant, a facility built in the communist period and supplying over a third of Bulgaria's electricity. This was a significant blow to the country, which traditionally had been a major exporter of electricity in Southeastern Europe, mainly because of electricity produced by the six nuclear reactors. In February 2013, rising residential utility bills forced the prime minister and his cabinet to step down, bringing to light the problems in Bulgaria's energy sector and pointing the finger to RES technologies and producers.

Political Unpredictability and Transparency

Political instability and lack of transparency were identified as key contextual factors by participants in the RE sector. Interviewees discussed uncertainty in the regulatory framework, frequent rule changes, unpredictability, and rapid turnover in agency leadership as significant challenges in RES promotion and investment.

In 2013 alone, there were three changes of the Chairman of the Energy and Water Regulatory Commission, with two of the three occurring within a span of 6 months (Energy and Water Regulatory Commission [EWRC], 2014; World Bank [WB], 2013).

Other factors shaping the decisions and actions of RE sector participants relate to lack of trust in government and perceptions of deep-rooted corruption (WB, 2013). According to data from Transparency International and reports by the World Bank, Bulgaria received one of the lowest scores on the Corruption Perception Index (CPI), which is a measure of the perceived level of public sector corruption. In 2012, of all EU countries, Bulgaria ranked only above Greece on the CPI (WB, 2013). In describing a recently completed new hydroelectric power plant, a government representative shared the following: *Because the dam is situated low in the valley and the conditions of the existing road were inadequate they had to build a new road. The cost was covered by the investor. It happened to be a very expensive road as the terrain was very rocky, raising the overall cost of the project. There has been a mention of some sort of corruptive practices, politicians' names were involved, some had ostensibly taken commission fees, consultant fees, etc.* (Interview 2).

Socioeconomic Inequality

Socioeconomic conditions in the country are also important contextual factors, particularly the divergence in living standards in Bulgaria with those in the EU (World Bank 2015). Low income levels, low living standards, corruption, and organized crime have remained serious issues for the country over the past decade. In 2010 the percentage of people living in poverty or in social exclusion was 49.1% (WB, 2015). The financial crisis in 2008 was particularly hard on the poorest households, whose real income decreased by nearly 13% in the period 2009-2010 (WB, 2013). These are also the households that spent the most of their earnings on electricity; when compared to their EU counterparts, Bulgarian households are in general more dependent on electricity (WB, 2013, 2015).

Based on a conventional measure of energy poverty, defined as spending more than 10% of household resources to cover energy needs, Bulgaria ranked second (61% of households) after Hungary (80%). In addition, the average number of energy poor households was higher in rural areas compared to urban areas (WB, 2013). Social assistance programs, such as guaranteed minimum income and heating allowances, have been ineffective at reaching those most in need, largely due to budget cutoffs since the early 2000 (WB, 2013). As explained: *Bulgaria is a relatively poor country and if you need to explain to the people - some of whom live in small villages and have very limited budget and earnings, and probably live on social security benefits – that they are going to pay for green energy in order for someone to carry out their [RE] investment, to get a fair return on capital, and etcetera, that would be absolutely rejected* (Interview 11).

A Culture of Distrust

Against the backdrop of the political and socioeconomic issues described above, participants in the RE sector operated in a culture of distrust in government institutions and the shadow of post-communist history. These mental models are pervasive even today and shape the behavior of the public, as well as that of RE sector participants. Interpersonal trust and trust in institutions is generally low among younger generations, a finding consistent with existing research (Coleman, 2015). As one interviewee put it: *Actually, the Bulgarian mentality is one against popular support, against popular trust in government measures, and of course very disrupted consumer trust. So everybody has a feeling that somebody is lying to them, somebody is stealing from them their resources and that actually you get much less than what you pay for. This is the general mentality* (Interview 12).

Consistent with this, we found that public distrust against renewables and discontent with high electricity prices were barriers to needed political will and liberalization in the energy sector. Political decision-makers appeared to be highly sensitive to increases in electricity prices, as political consequences can be significant during elections. Anecdotal evidence speaks to occasions when electricity arrears of whole neighborhoods had been written off in pre-election periods (Hiteva, 2009). Our interviews suggest that most people in the country are not willing to pay higher electricity prices. Bulgaria's household electricity prices are the lowest among EU member states (EUR 0.07 per kWh), with an EU-28 average of EUR 0.13 per kWh and a high of EUR 0.20 per kWh in Ireland (Eurostat, 2014). However, measured in terms of purchasing power parity, Bulgarians pay one of the highest electricity prices in the EU. A representative of a non-profit group explained this as follows: *A dominant view among the public emerged that renewables are the cause for the persistent growth in the cost of electricity, which to a large degree is untrue. There are other much more serious market and nonmarket global forces. But it is a fact that such a viewpoint is shared among the general population* (Interview 8). In sum, lack of political predictability and transparency, rising socioeconomic inequality, as well as growing public distrust and discontent with rising electricity prices created an unfavorable context for promoting and sustaining RES growth in Bulgaria.

Institutional Rules for RES Promotion in Bulgaria

The institutional rules and policy tools for RES promotion in Bulgaria were one of the key determinants of the boom in RES-E in the early 2000s. Our findings point to the high level of feed-in-tariffs and statutory obligation for purchase as initial drivers of the high investor interest and RE project development in Bulgaria in the 2007-2012 period. Unfortunately, this period was followed by a series of restrictive policy measures and growing regulatory uncertainty, which curtailed and set back RES development in the country.

The main policy tool for RES promotion in Bulgaria is a system of feed-in-tariffs (FITs) combined with a purchase obligation, introduced first in 2007 with the Renewable and Alternative Energy Sources and Biofuels Act (RAESBA), and maintained later in the 2011 Energy from Renewable Sources Act (ERSA). The FITs and purchase obligations apply to power purchase agreements (PPAs) for projects implemented prior to meeting the policy target embedded in the National Renewable Energy Action Plan and in line with the 2009 EU Renewable Energy Directive. The system of FITs with PPAs applies to all energy for which certificates of origin have been issued.

All RES technologies are eligible for FITs, with minimum payment rates specified in ERSA (art. 18 par. 1 item 6 ERSA) and actual rates set annually by EWRC (art. 32 par. 1 ERSA). The focus of ERSA was largely on RES-E, with limited attention to promotion strategies for RES-T and RES-H&C. The 2020 Energy Strategy for Bulgaria published in 2011 acknowledged that: *Beside the system of feed-in tariffs, practically no other mechanisms for promotion of RES development are applied. We also acknowledge that in the case of energy generation by renewable sources for heating and cooling purposes the support mechanisms are not sufficiently developed* (MEE 2011, p. 19).

A summary list of the policy tools for RES promotion in Bulgaria included:

- **Mandated purchase of RES:** Electricity distribution companies (EDCs) are required to purchase all RES-E. The purchase requirement is implemented through long-term PPAs.
- **Feed-in tariffs:** A set price for RES-E generated and exported to the public grid. As in many other countries (e.g., Germany, USA, Spain), payments are awarded as long-term contracts for eligible RES technologies, typically for a period of 15 to 20 years. The regulatory agency (EWRC) sets the FITs based on the typical production cost for the specific technology, plus an addition, and as such the FITs are fixed for the entire PPA period.
- **Priority grid connection:** EDCs are required to provide preferential connection to the electric grid to all new RES installations; RES-E producers cover the cost for grid connection.
- **Quotas**: Quotas for grid connection for new RES-E installations. Annual RES-E quotas are set by EWRC annually in accordance with the progress towards the RE target of the NREAP. Quotas were introduced in 2012 and

set to zero, with the exclusion of small biomass plants (Nikolaev & Konidari 2017).

- **Certificate of origin:** A document certifying the producer, the amount, the period of electricity generated, the production unit, and the unit's capacity.
- **Transport fuel blending obligation:** Conventional fuels must be mixed with bio-fuels following a schedule for meeting the bio-fuel share under the 2020 RE target.

Compared to other policy tools, *a FIT is a performance-based rather than an investment-based incentive*, and in that respect, it is more similar to production tax credits and renewable energy credits than to investment tax credits or other subsidies. To be effective in attracting investments, FIT rates are set above the retail cost of electricity. However, without additional controls, generous FIT levels can lead to more investment than intended, as witnessed in the case of Bulgaria's PV and solar technologies (Figure 3). Another example is the Spanish experience in which the government significantly reduced the tariff a year after its start, and suspended the FIT altogether in 2012 to contain costs to the government and other utility customers (Couture & Cory, 2013).

The replacement of RAESBA with ERSA in 2011 was necessitated because of the significant volume of statutory amendments required to transpose the 2009 EU Renewable Energy Directive in Bulgaria's legislation (Interview 1). ERSA stipulated that adjustments to the FIT rates would no longer be regulated by law and could be set (or reduced) by the regulator at any time. In June 2012, a year after ERSA came into effect, the level of FITs for wind and solar PV decreased by 20%. ERSA also reduced the length of PPAs from 25 to 20 years for solar PV and biomass, and from 15 to 12 years for wind and small-scale (<10MW) hydropower plants. These measures, however, were only the first in a series of policy changes undertaken by

Table 2. Restrictive Renewable Energy Policy Measures in Bulgaria, 2011-14

Year	Policy measures	Technologies
2011	ERSA readjusts FIT levels; length of PPA amended	Wind, PV, biomass
2012	Grid access tariffs for wind and solar (EWRC Decision) FIT levels reduced for newly installed capacities, 22% for wind and 50% for PV plants (EWRC Decision) Delays for adding new RES capacity (MEE Decision)	Wind, PV
2013	Temporary curtailment of RES power plants; Violations of the priority distribution and obligatory purchase of RES-E (ESO Decision)	Wind, PV
2014	20% tax on revenues from wind and PV energy production (Voted by the National Assembly of Bulgaria)	Wind, PV

(Source: Personal interviews and document analysis)

key veto players (e.g., EWRC, NEK, ESO) in an effort to "tame" the boom in RES-E (Table 2). From the perspective of the IAD framework, these policy changes shaped the behavior of RE sector participants leading to a set of outcomes (boomerang effects) undermining RES promotion.

Boomerang Effects

Starting in 2012, the state regulator, EWRC, and the ESO undertook a series of measures aimed at curbing RES-E production levels. These measures included: putting contracted new capacities on hold, reducing FIT rates, and introducing retroactive measures and regulations considered discriminatory in the context of EU fair competition rules. In principle, retroactive measures can send negative signals to RES investors, increase market risk, and stifle RES development (Hiteva & Maltby, 2017). The factors contributing to these measures in that period in Bulgaria were largely institutional and political: *The main thing was the unstable regulatory framework with the frequent legal act amendments. This creates unpredictability in the regulatory environment, but also uncertainty among investors. It drives them away. There is also lack of transparency in agency decision-making, as [seen] with the approval of trading rules [rules for purchase of electricity from renewables] that are not based on any economic indicators but rather on political factors.* (Interview 1).

In September 2012, the EWRC introduced a grid access fee of BGN 2.95 (USD 1.72) for all wind and PV power plants built in the period 2010-2012. The fee reduced tariff rates for affected RES power plants, retroactively, by as much as 39% (EC, 2013). The following explains the rationale for the grid access fee: *This [fee] came as an urgent measure. In mid-2012 it was one year after the new renewable law [ERSA] had come into force. It was a critical milestone for projects for generation of electricity from renewables. After that period the FIT rates decreased drastically and everybody urged to secure their commercial operations earlier - although they weren't really ready – so that they could get their FITs fixed for the rest of their off-take period at the higher FIT rate. This created a big crash. It was explained that a lot of small photovoltaic power plants just bought [the PPA] for one night before that critical deadline, and after that the cost of electricity, especially for photovoltaics -- which has the highest price on the grid - exploded. So the costs rose to a level that was not initially taken into account. This led to deficits in the respective off-takers and they had to take urgent measures to take out some of the renewable producers, for example by imposing the access fee.* (Interviews 11, 12).

The grid access fee was challenged by a coalition of wind and PV producers, the Bulgarian Photovoltaic Association (BPVA), and Wind Energy Association (BWEA) in a lawsuit filed in the Supreme Administrative Court. The court overruled EWRC's decision and ordered NEK/ESO to reimburse the parties affected by the fee. However, this ruling came at a time of rising financial difficulties for NEK and the EDCs. According to one RES producer: *It [the grid access fee] was repealed by Bulgaria's Supreme Administrative Court for being illegal and contradictory to the legal regulations, however, we have not received the money back from the electricity operator because of the fact that the energy system is totally exhausted in terms of [financial] resources* (Interviews 11, 12).

In March 2013, the EWRC issued a new set of rules for the purchase of electricity produced from renewables – one of a series of changes to the so called 'trading rules'. As explained, the cost of RES-E generation was significantly higher than the generation cost of state-owned power plants, which arguably put pressure on electricity prices (WB, 2013). The reason was that the RES-E cost was an aggregated cost including cost of CHP (cogeneration) in addition to the cost of RES power plants. The wind and PV industry associations issued statements arguing that the new rules disregarded any economic cost considerations, because they were largely driven by political factors.

In spring 2013, the Electricity System Operator ordered the three distribution companies (CES, EVN, EnergoPro) to limit the output of RES installations. The procedures used by ESO were described by BPVA and BWEA as non-transparent and illegal – in violation of the statutory requirement (ERSA) for preferential transmission and distribution of electricity from renewables (Interviews 1, 5, 6). The explanation offered by ESO for the temporary limits on solar and wind energy production was the potential overload and risk from connecting excessive renewable capacity to the aging electric grid system. Specifically, ESO cited the need to ensure the stability of the power grid (Novinite, 2014; Interview 6). Talking about this episode, one government representative shared this: *The lack of fully developed competitive market creates these problems. There is no way the conventional system [thermal power producers] can operate at 100% capacity. So they will be forced to go on the market. But we cannot reduce their potential either just because we now have renewable energy producers. They [the thermal power plants] also have power purchase agreements – we cannot violate their agreements just because RES producers have appeared* (Interview 2).

In the fall of 2013, in a public statement, the Bulgarian Deputy Prime Minister in charge of economic issues admitted that "Bulgaria's energy sector had yet again reached a critical phase due to serious political mistakes and poor calculations" (Novinite, 2014). The head of the Bulgarian Energy Holding followed with a proposal for a 30% tax on revenues of PV producers (Novinite, 2014). In December 2013, the National Assembly of Bulgaria passed a 20% tax on revenues from wind and solar power plants through a rider in the budget bill. The decision came into effect in January 2014 and was in force till August 8, 2014 when Bulgaria's Constitutional Court overturned the 20% tax on revenues. This high-level legal battle was championed by the President of Bulgaria who filed the case on behalf of RES producers affected by the 20% tax on revenues. In short, the anti-RES measures described above illustrate the limitations in RES promotion policies when implemented without regard to local resource conditions, socio-economic, political, and institutional factors.

RECOMMENDATIONS AND SOLUTIONS

The implementation of the EU Renewable Energy Directive in Bulgaria underscores an important policy lesson. Ambitious policy goals set and carried out without consideration of local conditions, institutional capacity, and grid infrastructure can undermine both policy goals and future RES promotion. The poorly regulated growth in RES-E in Bulgaria, coupled with political and economic legacies, aging and insufficient grid system, as well as frequently changing rules created the conditions for a crisis in the energy sector, and a spill-over suite of restrictive RES measures after 2012.

There are several policy-relevant recommendations from this case. First, it is important to develop a long-term national strategy for RES deployment in line with socio-economic, political, and cultural realities. Such a strategy should rely on institutional fit – a correspondence between RES promotion tools and institutional context. A coordinated long-term energy plan should also account for needed administrative capacity and robust institutions in order to sustain transitions to a RES-based economy. Other important and necessary factors include: regulatory expertise and independence, transparency of and trust in government, regulatory stability, as well as market liberalization and fair competition. Notably, administrative capacity, a free market, and political independence of the regulatory agencies are cornerstones of effective policy implementation (Hiteva & Maltby, 2017).

Second, there is a need for a long-term vision for implementation of RES policy tools. The initial rate of FITs is a critical decision, along with considerations about future adjustments. Evolving technology costs and market prices may require policy-makers to make periodic adjustments in FIT policies. However, changing tariff rates too frequently can backfire (as this case shows), because such changes increase regulatory risk and investor uncertainty (U.S. Energy Information Agency [US-EIA], 2013). One possible solution is a tariff digression scheme, whereby FIT rates decrease gradually by a pre-set percentage every year. Such an incremental decrease in FITs can be combined with FIT policy adjustments every 3-4 years based on a systematic methodology reflective of market changes, technological innovation, and progress toward policy targets. This approach effectively integrates flexibility in policy implementation and investor risk safeguards (US-EIA, 2013).

Third, investments and upgrades in electricity transmission infrastructure should not be delayed, and should not be undermined in the process of transitioning to a low fossil fuel-based economy. The case of Bulgaria's RES development showcases the limitations of an aging electric grid system at the backdrop of a restructuring energy system, decreasing energy exports, and consumption. A long-term national energy strategy should incorporate considerations of electric grid capacity because in the absence of quality infrastructure (the biophysical or resource attributes under the IAD framework), policy incentives and institutional rules are insufficient for RES growth and promotion.

FUTURE RESEARCH DIRECTIONS

Bulgaria has a little over 10 years of experience with the Europeanization process, which involves not only legal synchronization, but also a balance between EU targets and national interests (Hiteva & Maltby, 2017). As a small and fairly recent member, Bulgaria is "an EU energy policy-taker rather than policy-shaper" (Hiteva & Maltby, 2017, p. 239). In its desire to quickly reach mandated EU targets for RES, the government together with the utilities and EDCs, was unable to effectively conduct a strategic evaluation of what policy instruments and energy targets would have been the most feasible and beneficial in the long run. Therefore, a future comparative analysis of both short-term and long-term strategic plans for RES development is needed, specifically in the context of multi-level governance systems where RES targets are reflective of top-down rather than bottom-up policy prescription.

In a comparison of top-down Europeanization dynamics with regard to the 2009 RE Directive, Solorio and Jorgens (2017) find that Bulgaria went through a "fast, but uncoordinated implementation of EU directives", while Poland had an incomplete implementation, and Romania, a reluctant compliance with the RE Directive (p. 295). The case of Romania, in particular, bears similarities to our case study in that it illustrates the political, intuitional, and biophysical limits to implementing top-down policy goals (Davidescu, 2017). Romania's implementation of the 2009 EU RE Directive relied on the use of green certificates and a set of generous government incentives that attracted significant foreign investment in a short period of time. This led to Romania reaching its national RE target in 2014 despite delays in policy implementation driven by an economic crisis, structural constraints, and opposition by key political players (Davidescu, 2017).

It is important to note that there is still uncertainty regarding the future of RES in Bulgaria, and that there is a "passive resistance to developing RES-E beyond the minimum required" (Hiteva & Maltby, 2017, p. 238). Thus, new policy frameworks that rely on transparency and regulatory stability would be necessary to aid the implementation of the EU 2030 goals. In view of this, comparative case studies of policy implementation in multi-level governance systems can be fruitful future research endeavors.

Further, the public costs and revenues associated with RES policy implementation require future research in order to be accurately estimated and documented. The following questions may need to be addressed: What are the costs of RES promotion measures from a government perspective, but also from a societal perspective? What can be done to ensure the 'right' pace of transition to a RES-based energy system, particularly in countries where the social and financial burden of energy bills is a salient public issue? What mechanisms are effective at building broad social support for RES development?

With regard to long-term strategic planning it is important to account for new and phased-out electricity capacities beyond 2020, as well as forecasted energy demand and consumption. Questions that may need to be addressed include: What is a realistic, yet ambitious RES target for 2030? For 2050? A recent study estimated that the most feasible scenario for Bulgaria, representing both a cost-effective and moderate RES growth, includes 27.6% overall share of RES for 2030 (Nikolaev & Konidari, 2017). What types of institutional rules or policies are needed to meet this goal? How does the implementation of such policies ensure transparency, predictability for investors, and regulatory stability?

As the rest of the chapters in this book reveal, the transition towards a renewable energy future could be slow and painful. We need a better understanding of RES promotion strategies in a regional or multi-level governance system where some states act as policy takers and others as policy shapers. For the former, a long-term strategy for RES development, administrative capacity, and quality grid infrastructure seem to be essential conditions for sustained RES promotion and growth.

CONCLUSION

The implementation of the EU 2020 policy targets for renewable energy in Bulgaria illustrates the potential for unintended outcomes in carrying out ambitious top-down policy goals. This case underscores several key elements in designing and implementing RES policy goals. First, it is important to develop a long-term national strategy for RES development in line with socio-economic, political, and cultural realities. Second, there is a need for a long-term vision in the implementation of RES policy tools. Third, investments and upgrades in electricity transmission infrastructure are essential in the process of RES market development. Lastly, administrative capacity, a functioning free market for energy, and public support are cornerstones of effective policy implementation.

Many of the policy relevant lessons identified in the case of Bulgaria are echoed in the European Commission's progress report on the implementation of the EU Renewable Energy Directive (2009/28/EC). The report notes that while there has been "an overall strong initial start in EU renewables growth", there are still barriers, including: administrative capacity, economic barriers pertaining to the rising costs of capital, delayed investment, slow infrastructure development, market failures, delays in connection, and grid operational rules that disadvantage renewable energy producers (EC, 2013). The progress report further finds that member states' deviations from their National Renewable Energy Action Plans reflect changes that "reduce clarity and certainty for investors, [and] increase exposure to regulatory risk." (EC, 2013, p.13).

When properly designed, institutional rules can reduce risk and create predictability in governance regimes (Ostrom, 2005). Institutions that are based on shared understanding and backed by adequate monitoring and enforcement (i.e. rule of law) can ensure stability and predictability of human interactions. By extension, such institutional arrangements reduce the potential for crisis and boomerang effects. In the case of Bulgaria, it can be argued that shared understanding of the importance of increasing the share of RES consumption was hardly a bottom-up goal backed by public support and commitment. Given Bulgaria's political and economic history, the trajectory of policy implementation and subsequent policy outcomes may not seem too surprising. While the lessons from our study remain temporally and spatially bound, we believe that they contribute to a shared understanding of RES policy implementation in a multi-level governance system.

ACKNOWLEDGMENT

This research was supported by Appalachian State University Board of Trustees International Travel Grant, 2013.

REFERENCES

2009/28/EC. (2009). Renewable Energy (RE) Directive 2009/28/EC of the European Parliament and of the Council of 23 April 2009 on the promotion of the use of energy from renewable sources and amending and subsequently repealing Directives 2001/77/EC and 2003/30/EC (Text with EEA relevance). *Official Journal of the European Union, 140*, 16–62. Accessed May 6, 2018 from: http://data.europa.eu/eli/dir/2009/28/oj

Agrawal, A., Wollenberg, L., & Persha, L. (2014). Governing Mitigation in Agriculture-Forest Landscapes. *Global Environmental Change, 29*, 270–326. doi:10.1016/j.gloenvcha.2014.10.001

Brehm, S., & Brehm, J. W. (1981). *Psychological reactance: a theory of freedom and control*. New York: Academic Press.

Brid, L., Cochran, J., & Wang, X. (2014). *Wind and solar curtailment: Experience and practices in the United States*. National Renewable Energy Laboratory, Technical Report NREL/TP-6A20-60983. Accessed on March 10, 2018, from: http://www.nrel.gov/docs/fy14osti/60983.pdf

Coleman, E. (2015). Common property endowments in the trust game: Experimental evidence from Bulgaria. *Journal of Theoretical Politics, 28*(1), 27–43. doi:10.1177/0951629814568400

Couture, T., & Cory, K. (2013). *State Clean Energy Policies Analysis (SCEPA) Project: An Analysis of Renewable Energy Feed-in Tariffs in the United States*. National Renewable Energy Laboratory (NREL). Accessed October 27, 2018 from: https://www.nrel.gov/docs/fy09osti/45551.pdf

Davidescu, S. (2017). The Europeanization of renewable energy policy in Romania. In I. Solorio & H. Jorgens (Eds.), *A guide to EU renewable energy policy: Comparing Europeanization and domestic policy change in EU member states* (pp. 204–223). Cheltenham, UK: Edward Elgar Publishing. doi:10.4337/9781783471560.00021

EC - European Commission. (2013). *Report from the Commission to the European Parliament, the Council, the European Economic and Social Committee and the Committee of the Regions* (SWD-175 final). Accessed October 24, 2018, from: http://ec.europa.eu/transparency/regdoc/rep/1/2013/EN/1-2013-175-EN-F1-1.Pdf

EC - European Commission. (2016). *Proposal for a Directive of the European Parliament and of the Council on the promotion of the use of energy from renewable sources*. Accessed October 20, 2018 from: http://eur-lex.europa.eu/legal-content/EN/ALL/?uri=CELEX%3A52016PC0767R%2801%29

EC - European Commission. (2018a). *Europe Leads the Global Clean Energy Transition: Commission Welcomes Ambitious Agreement on Further Renewable Energy Development in the EU* (EC-Statement). Accessed on February 8, 2019 from: http://europa.eu/rapid/press-release_STATEMENT-18-4155_en.htm

EC - European Commission. (2018b). *EU Energy in Figures: Energy Statistical Pocketbook*. Accessed October 25, 2018 from: https://ec.europa.eu/energy/en/data/energy-statistical-pocketbook

EP - European Parliament. (2009). *Directive 2009/28/EC of the European Parliament and of the Council of 23 April 2009 on the promotion of the use of energy from renewable sources and amending and subsequently repealing Directives 2001/77/EC and 2003/30/EC. EUR-Lex.* Accessed on October 25, 2018 from: https://eur-lex.europa.eu/legal-content/EN/ALL/?uri=CELEX:32009L0028

Eurostat, European Commission. (2014). Smarter, Greener, More Inclusive? - Indicators to support the Europe 2020 strategy. In I. Savova (Ed.), *Climate change and energy* (pp. 73-91). Accessed on October 26, 2018 at: https://ec.europa.eu/eurostat/documents/3217494/5777461/KS-02-13-238-EN.PDF/1a6fa7e5-85b7-40aa-987e-6a6d049ad723

Eurostat, European Commission. (2018). *Share of renewable energy in gross final energy consumption by sector.* Accessed October 25, 2018 from: https://ec.europa.eu/eurostat/web/products-datasets/-/sdg_07_40

Frederickson, G. H., Smith, K. B., Larimer, C. W., & Licari, M. J. (2012). *The Public Administration Theory Primer* (2nd ed.). Boulder, CO: Westview Press.

Hiteva, R. (2009, August). *Untangling the puzzle of energy policy in Bulgaria.* Paper presented at Energy Vulnerability and Urban Transitions, University of Manchester, UK. Accessed on November 3, 2014 at: http://urban-energy.org/2013/08/29/guest-contribution-untangling-the-puzzle-of-energy-policy-in-bulgaria

Hiteva, R., & Maltby, T. (2017). Hitting the target but missing the point: Failing and succeeding in the Bulgarian energy sector. In I. Solorio & H. Jorgens (Eds.), *A guide to EU renewable energy policy: Comparing Europeanization and domestic policy change in EU member states* (pp. 224–246). Cheltenham, UK: Edward Elgar Publishing. doi:10.4337/9781783471560.00022

Iychettira, K., Hakvoort, R. A., & Linares, P. (2017). Towards a comprehensive policy for electricity from renewable energy: An approach for policy design. *Energy Policy, 106*, 169–182. doi:10.1016/j.enpol.2017.03.051

Jones, L. (2017). *Renewable energy integration: Practical management of variability, uncertainty, and flexibility in power grids* (2nd ed.). Academic Press, Elsevier Inc.

Kinzig, A., Ehrlich, P., Alston, L., Arrow, K., Barrett, S., Buchman, T., ... Saari, D. (2013). Social Norms and Global Environmental Challenges: The Complex Interaction of Behaviors, Values, and Policy. *Bioscience, 63*(3), 164–175. doi:10.1525/bio.2013.63.3.5 PMID:25143635

Kiser, L., & Ostrom, E. (1982). The Three Worlds of Action: A Metatheoretical Synthesis of Institutional Approaches. In E. Ostrom (Ed.), *Strategies of Political Inquiry* (pp. 179–222). Beverly Hills, CA: Sage.

Kondarev, G. (2013). Bulgaria: energy efficiency and renewable sources. Position of the Bulgarian Coalition for Sustainable use of the EU Funds and CEE Bankwatch network: Funding for clean energy through the Structural and Cohesion Funds for the programming period 2014-2020. *CEE: Bankwatch network*. Accessed September 9, 2014 from: http://bankwatch.org/sites/default/files/shadow-BG-EE-RES.pdf

Lynn, L. Jr, Heinrich, C. J., & Hill, C. J. (2000). Studying Governance and Public Management: Challenges and Prospects. *Journal of Public Administration: Research and Theory*, *10*(2), 233–262. doi:10.1093/oxfordjournals.jpart.a024269

McGinnis, M. (2011). An introduction to IAD and the language of the Ostrom Workshop: A simple guide to a complex framework. *Policy Studies Journal: the Journal of the Policy Studies Organization*, *39*(1), 169–183. doi:10.1111/j.1541-0072.2010.00401.x

MEE - Ministry of Economy and Energy. (2011, June). *Energy Strategy of the Republic of Bulgaria until 2020 for reliable, efficient, and clean energy*. Republic of Bulgaria. Accessed October 13, 2014: https://www.me.government.bg/files/useruploads/files/epsp/23_energy_strategy2020Eng_.pdf

MEE - Ministry of Economy and Energy. (2013, December). *Second National Report on Bulgaria's Progress in the Promotion and Use of Energy from Renewable Sources drawn up under Article 22(1) of Directive 2009/28/EC on the promotion of the use of energy from renewable sources and in accordance with the Template for Member States' progress reports under Directive 2009/28/EC*. Republic of Bulgaria. Accessed October 12, 2014 from: http://www.etipbioenergy.eu/images/Article_22_Bulgaria_report_EN.pdf

MEET - Ministry of Economy, Energy, and Tourism. (2011). *Re-submitted National Renewable Energy Action Plan (NREAP) Drawn up in Accordance with the Template for National Renewable Energy Action Plans as Set Out in Directive 2009/28/EC of the European Parliament and of the Council*. Accessed May 13, 2014 from: https://me.government.bg/library/index/download/lang/en/fileId/575

Nikolaev, A., & Konidari, P. (2017). Development and assessment of renewable energy policy scenarios by 2030 for Bulgaria. *Renewable Energy*, *111*, 7922–802. doi:10.1016/j.renene.2017.05.007

North, D. C. (1990). *Institutions, Institutional Change and Economic Performance*. Cambridge, UK: Cambridge University Press. doi:10.1017/CBO9780511808678

Novinite. (2014, October 31). *Bulgaria's National Electric Company Reports BGN 425M Loss in end-Sept.* Accessed October 20, 2018 from: https://www.novinite.com/articles/164465/Bulgaria's+National+Electric+Company+Reports+BGN+425M+Loss+in+end-Sept

NSI - National Statistical Institute. (2017). Overall Energy Balance Sheets for 2011, 2012, 2013, 2014, 2015, 2016. *Bulletin on the state and development of the energy sector in the Republic of Bulgaria.* Accessed October 22, 2018 at: http://www.nsi.bg/sites/default/files/files/data/timeseries/Energy4.1_en.xls

NSI - National Statistical Institute. (2018, August). *Bulgaria's 2016 final energy consumption by type of fuel (Total=9.7 Mtoe) (EU Country data sheets for Bulgaria).* Accessed August 18, 2017 from: https://ec.europa.eu/energy/en/data-analysis/energy-statistical-pocketbook*)*

O'Toole, L. J., & Meier, K. (1999). Modeling the Impact of Public Management: Implications of Structural Context. *Journal of Public Administration: Research and Theory*, *9*(4), 505–526. doi:10.1093/oxfordjournals.jpart.a024421

Ostrom, E. (1990). *Governing the Commons: The evolution of institutions for collective action.* New York: Cambridge University Press. doi:10.1017/CBO9780511807763

Ostrom, E. (1999). Institutional Rational Choice: An Assessment of the Institutional Analysis and Development Framework. In P. Sabatier (Ed.), *Theories of the Policy Process* (pp. 21–64). Boulder, CO: Westview Press.

Ostrom, E. (2005). *Understanding Institutional Diversity*. Princeton University Press.

Ostrom, V., & Ostrom, E. (1971). Public Choice: A Different Approach to the Study of Public Administration. *Public Administration Review*, *24*(2), 203–216. doi:10.2307/974676

Parks, R., & Oakerson, R. (2000). Regionalism, Localism, and Metropolitan Governance: Suggestions from the Research Program on Local Public Economies. *State & Local Government Review*, *32*(3), 169–179. doi:10.1177/0160323X0003200302

Roggero, M., Bisaro, A., & Villamayor-Tomas, S. (2018). Institutions in the climate adaptation literature: A systematic literature review through the lens of the Institutional Analysis and Development framework. *Journal of Institutional Economics*, *14*(3), 423–448. doi:10.1017/S1744137417000376

Ryan, S., Hebdon, C., & Dafoe, J. (2014). Energy research and the contributions of the social sciences: A contemporary examination. *Energy Research & Social Science*, *3*, 186–197. doi:10.1016/j.erss.2014.07.012

Solorio, I., & Jörgens, H. (2017). *A guide to EU renewable energy policy: Comparing Europeanization and domestic policy change in EU member states*. Cheltenham, UK: Edward Elgar Publishing. doi:10.4337/9781783471560

Theesfeld, I. (2004). Constraints on Collective Action in a Transitional Economy: The Case of Bulgaria's Irrigation Sector. *World Development*, *32*(2), 251–271. doi:10.1016/j.worlddev.2003.11.001

U.S. EIA. (2013, May). *Feedin tariff: A policy tool encouraging deployment of renewable electricity technologies*. NREL technical report, State Clean Energy. Policies Analysis Project. Accessed October 26, 2018 from: https://www.nrel.gov/docs/fy09osti/45551.pdf

Villamayor-Tomas, S. (2017). Disturbance features, coordination and cooperation: An institutional economics analysis of adaptations in the Spanish irrigation sector. *Journal of Institutional Economics*, *14*(3), 1–26.

Villamayor-Tomas, S., Grundmann, P., Epstein, G., Evans, T., & Kimmich, C. (2015). The Water-Energy-Food Security Nexus through the Lenses of the Value Chain and the Institutional Analysis and Development Frameworks. *Water Alternatives*, *8*(1), 735–755.

Vladimirov, M., Georgiev, A., & Kolarova, S. (2018). *Development of Small-Scale Renewable Energy Sources in Bulgaria: Legislative and Administrative Challenges*. Sofia, Bulgaria: Center for the Study of Democracy.

WB - World Bank. (2013). *Republic of Bulgaria: Power sector rapid assessment*. Washington, DC: World Bank. Retrieved from http://documents.worldbank.org/curated/en/391271468020330576/Republic-of-Bulgaria-power-sector-rapid-assessment

WB - World Bank. (2015). *Adapting to higher energy costs: Findings from qualitative studies in Europe and central Asia*. ACS12511. Washington DC: World Bank. Retrieved from http://documents.worldbank.org/curated/en/582021468245408944/pdf/ACS12511-WP-REPLACEMENT-Adapting-to-Energy-Costs-web.pdf

ADDITIONAL READING

Hiteva, R., & Maltby, T. (2017). Considering students' emotions in computer-mediated learning environments. In I. Solorio & H. Jorgens (Eds.), *A guide to EU renewable energy policy: Comparing Europeanization and domestic policy change in EU member states* (pp. 224–246). Cheltenham, UK: Edward Elgar Publishing. doi:10.4337/9781783471560.00022

Kinzig, A., Ehrlich, P., Alston, L., Arrow, K., Barrett, S., Buchman, T., ... Saari, D. (2013). Social Norms and Global Environmental Challenges: The Complex Interaction of Behaviors, Values, and Policy. *Bioscience*, *63*(3), 164–175. doi:10.1525/bio.2013.63.3.5 PMID:25143635

Ostrom, E. (2005). *Understanding Institutional Diversity*. Princeton University Press. Renewable Energy in the EU, European Commission: https://ec.europa.eu/energy/en/topics/renewable-energy

Ryan, S., Hebdon, C., & Dafoe, J. (2014). Energy research and the contributions of the social sciences: A contemporary examination. *Energy Research & Social Science*, *3*, 186–197. doi:10.1016/j.erss.2014.07.012

Yonk, R., Simmons, R., & Steed, B. (2012). *Green vs. Green: The Political, Legal, and Administrative Pitfalls Facing Green Energy Production*. Florence, KY: Routledge Research in Environmental Policy and Politics.

KEY TERMS AND DEFINITIONS

Boomerang Effects: Unintended consequences that reduce the prevalence of desired policy outcomes and the effectiveness of policy measures promoting those outcomes.

Curtailment: Curtailment is a reduction in the amount of output from wind and PV generators relative to what they can otherwise produce with available resources. Curtailment on an involuntary (unscheduled) basis occurs when the systems operator cuts down output due to the system's inability to accommodate the full dispatch of wind or PV plants. Voluntary curtailment, as when PV and wind producers reduce their output, is typically a result of differences in supply and demand or low power prices.

Feed-in-Tariff (FIT): A policy that provides a guarantee of payment to renewable energy producers, typically in the form of long-term (15-20 year) contracts or power purchase agreements. A FIT policy is a type of production-based incentive with payments based on actual electricity produced ($/kWh).

Governance: Laws, administrative rules, judicial rulings, and practices that constrain, prescribe, and enable government activity, where such activity is broadly defined as the production and delivery of publicly supported goods and services.

Institutional Rules: Prescriptive statements that require, prohibit, or permit some action or outcome.

Policy Implementation: A stage in the policy process that involves actions needed to put laws or policies into effect, or to solve a problem.

Policy Tools: Tools or instruments, such as incentives, regulations, or penalties intended to bring about change in individual or collective behavior in line with public goals and policies.

ENDNOTES

1 The EU defines energy from renewable sources (RES) as energy from "non-fossil sources, namely wind, solar, exothermal, geothermal, hydrothermal and ocean energy, hydropower, biomass, landfill gas, sewage treatment plant gas and biogases" (2009/28/EC, Art. 2(a)).

2 Note that there were 27 countries that were part of the EU in 2009. Iceland and Norway as members of the European Economic Area submitted national action plans, as well. Iceland's intended renewable energy target was 72%.

3 Directive 2009/28/EC of the European Parliament and of the Council on the promotion of the use of energy from renewable sources. Official Journal of the European Union, 5.6.2009. Preamble, Paragraph 13.

4 Anticipated net results from EU member states' national plans indicate that the EU's share of energy from renewable sources will reach 20.3% by 2020 (European Commission, Energy, Summary of the Member State Forecast Documents, available at: https://ec.europa.eu/energy/sites/ener/files/dir_2009_0028_article_4_3_forecast_by_ms_symmary.pdf)

5 *Final energy consumption* refers to energy that is supplied to the consumer for all final energy uses such as heating, cooling and lighting. This is different from *primary energy consumption, which* refers to the direct use of energy at the source, or crude energy that has not been subjected to any conversion or transformation process (Source: Eurostat, International Energy Agency, 2018, https://ec.europa.eu/eurostat/web/products-datasets/-/sdg_07_40).

6 A parastatal company or organization is one that is owned by a country's government and often has some political power. Definition from https://dictionary.cambridge.org/us/dictionary/english/own

7 In February 2018 the Czech Electricity producer CEZ announced the sale of its assets (including the EDC assets) to Intercom Bulgaria. (Reuters, Feb. 22, 2018: https://www.reuters.com/article/us-cez-divestiture/cez-set-to-sell-bulgarian-distribution-renewable-assets-idUSKCN1G627K)

8 Electricity System Operator - the owner and operator of Bulgaria's high and medium voltage electricity transmission grid. ESO administers the balancing market for electricity, which was launched on 1 June 2014, holds tenders for transmission capacity and provides the centralized dispatching of the national electric power generation system. https://www.bgenh.com/en/page/64/Electricity-System-Operator-EAD.html

APPENDIX

Table 3. Interviews

1.	Bulgarian Wind Energy Association, representative, May 9, 2014, Sofia
2.	Ministry of Energy and Economy, Energy Strategies and Policies for Sustainable Energy Development Director, May 12, 2014, Sofia
3.	Agency for Sustainable Energy Development, Executive Director, May 13, 2014, Sofia
4.	State Energy and Water Regulatory Commission, Representative, May 14, 2014, Sofia
5.	Bulgarian Photovoltaic Association, Administrative and program director, May 14, 2014, Sofia
6.	Association of Producers of Ecological Energy, Chairman, May 15, 2014, phone interview
7.	Alpiq/ Vetrocom, Asset Manager, May 13, 2014, Sofia; May 14, 2015, Sofia
8.	Ventus Bulgaria Ltd, Managing Director, May 16, 2014, phone interview
9.	SeeNews, Media representative, May 14, 2014, Sofia
10.	Penkov-Markov & Partners Attorneys at Law, Managing partner and an associate, May 15, 2014, Sofia
11.	AES-3C Maritza East I Ltd., Commercial officer, August 7, 2014, Sofia
12.	AES-3C Maritza East I Ltd., Commercial officer, May 14, 2015, Sofia
13.	Eolica Bulgaria EAD, Director, August 8, 2014, Sofia
14.	Eolica Suvorovo AD, Director, August 8, 2014, Sofia
15.	Lobom Energy Ltd., Manager, May 12, 2015, Sofia
16.	CEZ Bulgaria EAD, Member of the Management Board, May 13, 2015
17.	Center for the Study of Democracy, Senior Analyst, May 13, 2015, Sofia
18.	Bulgarian Electric Vehicles Association, President, May 15, 2015, Sofia
19.	Clean energy and energy efficiency consultancy and project management, Representative, May 15, 2015, Sofia
20.	EVN Bulgaria, Deputy-Chair of the Management Board, May 18, 2015, Sofia
21.	Bulgarian Solar Association, Chair of the Management Board, May 19, 2015, Sofia
22.	Bulgarian Photovoltaic Association, Chairperson of the Board, May 19, 2015, Sofia
23.	Energy and Development Company, Chairman Supervisory Board, May 19, 2015, Sofia
24.	Association of Producers of Ecological Energy, Chairman, Aug 18, 2015, Varna
25.	Electric vehicles, Koorteh, CEO, Aug 18, 2015, Varna
26.	AES, Commercial officer, August 23, 2015, Sofia
27.	TSO EAD, Management consultant and former electrical engineer at NEK, Sofia (2003-07), August 15, 2015, Sofia
28.	Ministry of Energy and the Environment, Energy Markets Senior Expert, August 13, 2015, Sofa
29.	Hydroelectric Power Plants, Consultant (previously at NEK), August 12, 2015, Sofia
30.	Capital Newspaper Journalist, Economics and Energy Division, August 12, 2015

Chapter 3
Promotional Policies and Legislative Support for Grid-Connected Renewable Energy Projects

Vasundhara Sen
Symbiosis International University (Deemed), India

Ashish Kulkarni
Gokhale Institute of Politics and Economics, India

EXECUTIVE SUMMARY

Renewable Energy Sources-Based Electricity (RES-E) plays a key role in sustainable development – of meeting current energy demands, without adding to global warming concerns. However, as of 2017, only 8.5% of the total electricity generation came from RES-E. To boost this contribution, countries rely on strong legislative and policy support/tools. This case focuses on studying the legislative or regulatory frameworks put in place by the top three developed countries, and compares it with three developing countries, each of which are forerunners in RES-E, as of 2017. The comparative study suggests that while no single policy can be credited with the success behind rising RES-E in these countries, two key incentives are most important – namely feed-in-tariffs and renewable purchase obligations. Feed-in-tariffs act as floor price guarantee to the generator and renewable purchase obligations assures the generator of quantum of sale of the RES-E generated. When combined, these two incentives remain the most trusted policy tools even today for countries starting their journey in increasing their RES-E footprints.

DOI: 10.4018/978-1-5225-8559-6.ch003

BACKGROUND INFORMATION

The future will either be green or not at all - Sir Jonathon Porritt, British environmentalist and writer.

The world has been and continues to be dependent on fossil fuels to meet its energy and electricity demands. Coal (30%), oil (33%) and natural gas (24%) together account for 85% (approximately) of total energy consumption in the world (World Energy Council [WEC], 2016). The resultant rise in carbon emissions, are now a matter a grave concern, the world over. Between 1990 and 2015, carbon emissions from fuel combustion have increased from 14 million tons to 32 million tons, recording an increase of 57.5%. The International Energy Agency (IEA) estimates that close to 42% of this increase has been brought about by electricity generation (based on fossil fuels) to meet lighting, heating and cooling demands, and another 24%, can be attributed to the use of fuels for transportation needs (International Energy Agency [IEA], 2017). Research dating back to the early 2000's found that higher uses of non-green energy forms which cause carbon emissions, have led to increasing temperatures, a problem now widely described as "climate change". Increased mortality and a fall in crop yields, thereby affecting food availability for humans, are some of the most common results of climate change (Usikalu, 2009). Additionally, increasing the footprints of coal fired power plants seriously damage not only adult health, but also impair the growth patterns in children (Perera, 2017). Rapidly increasing air pollution, direct and indirect effects of warmer temperatures, associated health hazards in adults and children, rising oil prices that result in higher expenditures for oil importing countries- all of these factors have led to serious considerations of adopting greener and cleaner forms of energy.

Penetration of Renewable Energy Sources (RES) has been the highest in grid connected electricity generation. The share of RES in total electricity generation has steadily increased from 1.5% in 2001 to 7.5% in 2016, globally (Statistical Review of World Energy, 2018). Of great success has been solar energy and wind energy, amongst other sources, with solar and wind capacity installations, measured in Megawatt (MW) growing at 44%, and 20% respectively, during the same period.

According to data from International Renewable Agency (IRENA), a total of 164 countries around the world had instituted renewable energy targets as of 2015. Within these, 150 countries have adopted specific goals for Renewable Energy Sources based Electricity (RES-E) generation. Due to at least in part to many such incentives, RES-E footprint has increased from 3500 Giga-watt-hours (GWh) in 2007 to 5885 GWh in 2017, globally, marking an increase of approximately 60%

(International Renewable Energy Agency [IRENA], n.d). Every country has a story to tell of their pursuit of cleaner power generation. The most successful story is presented by China. China's RES-E portfolio grew from a mere 11 Tera-watt-hours (TWh) to a massive 380.6 TWh in a matter of only 10 years between 2006 and 2016. Denmark is the "greenest" of all economies, with 68% of total power generation coming from renewable sources in 2017. Germany is applauded for being the first country to mark its place in policy maps that provided a fillip for large scale grid – connected renewable penetration. Other countries that are deserving of further study are USA, Brazil, India and Japan, amongst others. Put together, these set of countries contribute to 55% of the total RES-E generation base installations in the world, as of 2017 (Statistical Review of World Energy, 2018).

While the transition to renewable sources, in any country's developmental roadmap holds importance, research has proven that the knowledge about its benefits and how it is being used, are far less known than is deserved. According to a study conducted of over 107 students in Turkey, 33% of the students held the opinion that "battery" is a form of renewable energy, and 25% students even chose not to respond (Çoker, Cathoglu, & Birgin, 2010). Recommendations have been made to design an optimal curriculum that enhances knowledge of renewable sources (Kandpal & Broman, 2014). They suggest that the presence of renewable energy in education curricula is key to sensitize students (they being future decision makers in the industry) to bring awareness on issues of climate change, and how renewable sources should be harnessed to overcome related challenges. Therefore, the aim of this chapter will be to provide the reader, with knowledge of policy and regulatory tools that have played a pivotal role for some countries to rapidly advance their grid-connected RES-E footprints. This chapter will focus on case studies of countries that have recorded decadal growth rates in excess of 50%, in RES-E generation, between 2007 and 2017. Comparison of legislative frameworks adopted to boost RES-E in 3 developed countries (namely USA, Germany and Japan) will be made against that of 3 developing countries (India, China and Brazil). United Nations Conference on Trade and Development classification list has been followed to bucket countries into developed and developing respectively (United Nations Conference on Trade and Development, n.d). Apart from these 6 countries, the chapter also studies the Denmark case. Denmark pioneered the wind energy revolution, beginning as early as 1890s, and is currently the "greenest" country in the world – 68% of its total electricity is generated from RES-E.

What policy and legislative support was put in place by these countries could become best practices for countries that are just starting out on the "green electricity" journey.

Figure 1. Rising share of renewable energy generation between 2000 and 2017
(Statistical Review of World Energy, 2018)

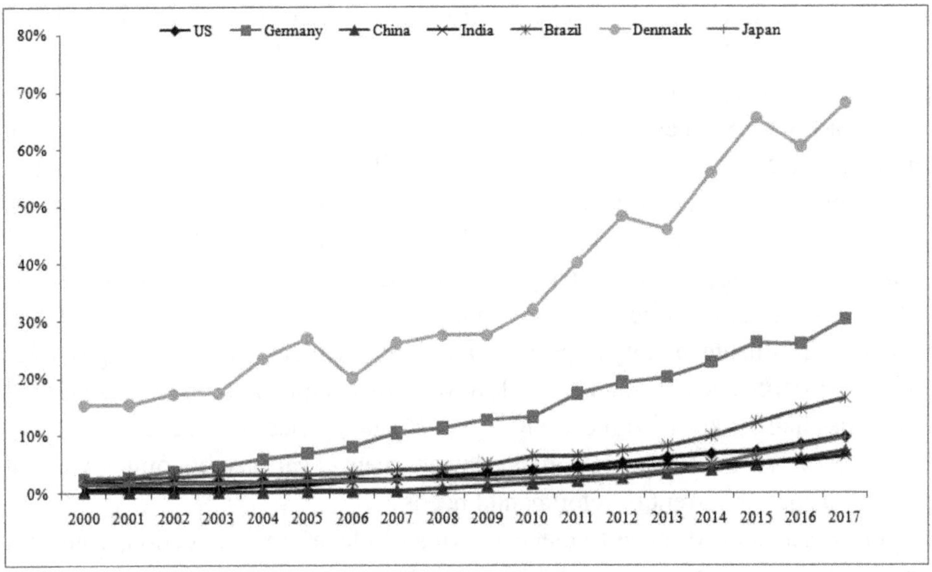

SETTING THE STAGE

The global oil crisis of 1974 brought with it several macro-economic worries. Rising oil prices led to inflated cost of oil imports for the importing nations and subsequently resulted in higher consumer prices and inflationary pressures increased for economies concerned. International Energy Agency (IEA) forecasts that oil prices will increase in the near future (Birol, 2010), and coal will continue to be the most polluting energy source. This has necessitated a change in energy profiles for many countries. A transition to renewable energy sources to meet energy needs is thus necessary. Move to RES-E will also help meet sustainable development goals. RES-E finds a vital connect with sustainable development (Dincer, 2000). Along with ensuring secured energy supply sources, RES-E also has several other benefits – such as creating jobs, help mitigate climate change impacts, improved productivity due to energy access, amongst others. Hence, the United Nations Sustainable Development Goals (UN SDGs) released in 2016 feature RES-E prominently. Wider RES-E adoption is envisaged to fulfill targets set under each goal, directly or indirectly (Mapping the Renewable Energy Sector to the Sustainable Development Goals: An Atlas, 2018). Biggest contributions of RES-E will be towards meeting SDG Goal numbers 7 (ending energy poverty and transition to cleaner energy forms), and SDG goal number 13 (urgent actions needed to combat climate change). Sub-goals set under the UN SDG goal 7 requires that by 2030, the share of renewable energy in the global energy mix should increase significantly (Sustainable Development Goal 7, n.d). Latest data suggests that the share of RES-E in final energy consumption has increased to 17.5% as of 2015, owing to large scale wind and solar energy based energy consumption (United Nations New York, 2018).

Increasing RES-E generation can also alter the pace of sustainable development for countries. Sinha, Shahbaz and Balsalobre (2017) study the impact of RES-E generation on environmental quality and economic growth, for N-11 countries (a set of countries that develop through use of better technology, enabled through cross – border technology transfer). They find that countries that started the transition to RES-E generation earlier have seen positive impacts on economic growth, and lead in environmental quality improvements. Emperical evidence also suggests that transition to RES-E to ensure sustainable development is practiced in many emerging economies like Turkey (Ozturk & Yuksel, 2016) and Africa (Bugaje, 2006). Hence use of RES-E will accelerate sustainable development of various economies, without compromising on current developmental needs.

For an accelerated move to cleaner forms of electricity production, countries have implemented either price based incentives such as feed-in-tariff, competitive bidding mechanism or quota/quantity based incentives such as renewable purchase obligations, green energy certificates, or both. Other fiscal incentives such as tax exemptions/ tax credits, carbon taxes that indirectly bolster the shift to RES-E investments also exist. Such robust legislative support frameworks have enhanced the contribution from RES-E in each country's total electricity generation over the years, as is evident from Figure 1. Such measures deserve in-depth understanding, and thus is the focus of this chapter.

This chapter is structured as follows. The next section details the most often used policy levers/regulatory support structures to promote RES-E generation, along with providing an overview of advantage and disadvantages of each incentive. Next, the path to achieve higher RES-E in each focus country will be detailed out. A comparison between the developed and developing countries will help identify efficient policies from the in-efficient ones. Further, problems associated with large scale grid integration of RES-E will be discussed, before presenting the conclusions at the end of the chapter.

POLICY ENABLERS FOR GRID CONNECTED RENEWABLE ENERGY

Renewable Energy Based Electricity (RES-E) generation comes with its own set of problems. Not only are the generation costs high, but there risks of uncertainty and variability. Uncertainty in RES-E refers to the lack of availability of the fuel source at all times of the day. Variability refers to the lack of guarantee, that even if the fuel source is available, it should be consistently available, both in terms of quantity and quality (Bessa, Moreira, Silva, & Matos, 2014). Hence, the success of RES-E generation rests on a combination of factors, and an enabling regulatory/ legislative environment is required to attract investments into the sector. Support frameworks have come in many forms- financial/fiscal, legislative and political intentions, amongst others. However, legislative support, which comes by way of policy instruments/ regulatory structures, have proven sufficient to attract investments, and have reduced the need for other forms of incentives (Abdmouleh, Alammari, & Gastli, 2015). In what follows, a few of the most commonly used tools/enablers have been detailed:

1. **Price Based Incentives:** Under this, investments in RES-E are invited by providing an attractive purchase price for the power generated. Price based incentives can be either in terms of Feed-in-Tariffs or by way of Competitive Bidding:
    a. **Feed-in-Tariffs:** "Feed-In-Tariffs" or FITs, refers to the price at which local power suppliers/utilities purchase the electricity generated from green/renewable sources. Such tariffs are usually higher than what is offered to thermal power generating plants, and for a fixed time period. The said tariffs are announced by a designated electricity regulator, state or central. The higher prices so offered also come with a suitable recovery mechanism such that the resulting financial burden on the purchasing authorities is minimized (covered in more detail under sub-section f). FITs offered have been Market Dependent and Market Independent as well (Stokes, 2013):
        i. **Market Dependent FITs:** These types of FITs offered by the local power utility/supplier are a function of the spot market price of electricity. FITs can be offered either in a premium price model (FIT is the premium paid over the spot market price); variable price model (utility pays the difference between the fixed FIT and the spot market price) or as a percentage of the spot price.
        ii. **Market Independent FITs:** In this case, the FITs are fixed and are not related to the market price. The regulator fixes the FIT, which may be partially or fully indexed to changes in inflation rates. India and Germany have followed this model in the past. Literature provides that while market independent FITs secure the revenue streams for the generators, market dependent prices provide the incentive to generate more when demand is higher, and supply is low. Most

countries have adopted the market independent models, in order to provide a safe and secure proposition for RES-E investors.

b. **Competitive Bidding:** Under competitive bidding, the RES-E purchase price (offered by the local power supplier) is discovered through a bid, placed by generators. The amount of RES-E that the power utility wants to purchase is pre-fixed and interested generators then bid to sell the power they generate. The generator that offers to accept the lowest price wins. United Kingdom (UK), France and India are examples of this framework. Competition among the bidders ensures that the purchase price for the power utility is significantly reduced. Beck & Martinot (2004) found that the rates of power purchase for renewable power was reduced significantly in the UK in the 1990s, after implementing the competitive bidding process.

2. **Quantity Based Incentives:** Such incentives create a demand for RES-E and hence increase the quantum of greener electricity in the total electricity portfolio mix. They can take the following form:

a. **Renewable Purchase Obligations (RPO)/ Renewable Portfolio Standards (RPS):** Renewable Portfolio Standards (RPS) (alternatively also called renewable purchase obligations) are mandates levied on the local power suppliers, wherein the obligated entity must procure a share of its total electricity supply needs from RES-E. This way, a secured market and revenue is promised to the generators, while also achieving the "green energy" targets set by countries/state. Historically and in many countries, feed-in-tariffs have been complimented with a RPS/RPO framework to attract investors in renewable generation industry. When combined, the 2 incentives ensure the demand for RES-E generated, as well as the price. Amongst the noted success stories of using RPS to boost RES-E is that of the United States of America (USA). As of 2016, 29 states in USA had RPS targets, which have undergone upward revisions as well (Barbose, 2017). While RPOs guarantee a demand market for RES-E, they come at a cost to the utilities/ obligated entities. US utilities have spent close to one billion dollars per year between 2010 and 2013, to comply with the targets (Barbose et al., 2016). However, the large social benefits in terms of reduced GHG emissions, air pollution and water use, effect on job creation, wholesale electricity prices and natural gas prices, outweigh the costs. The monetary value of reduction in air pollution and GHG emission for all RES-E to meet RPS set in 2013 stood at USD 7.4 billion dollars and close to 200,000 jobs were created as well (Barbose et al., 2016). RPS has also been the go-to method for many European countries such as Sweden, UK and Italy, also known alternatively as "quotas", "renewable obligations" and "green energy certificates" (Wiser, Namovicz, & Smith, 2007).

b. **Tradable Green Energy Certificates/ Renewable Energy Certificates:** Green Energy Certificates (GEC)/ Renewable Energy Certificates (REC) are a measure to increase the generators/ renewable energy investor's revenue stream. Compared to conventional generation methods, investors are found to be more risk-averse to renewable generation, and hence need more surety of investment returns (Fagiani, Barquin, & Hakvoort, 2013). Each unit generated from renewable energy sources is allotted a GEC/REC. The generator thus has a tradable instrument that can be sold at a price in a market (mostly power exchanges), and is treated like a financial instrument. The generator is also able to sell the physical electricity generated separately. Hence, the generator gets revenues from 2 sources – 1st from the sale of physical energy generated in a physical energy market and 2nd, by sale of the certificate from the sale of the certificate in designated financial markets. In order to ensure that there is demand for such certificates, national RPS/ RPO targets are announced. Consumers/ power suppliers (also referred to as obligated entities) are then allowed to purchase these certificates, and show compliance to their mandates. The GEC/RECs system transfers the financial burden to promote RES-E from the tax payers to the electricity consumer. Traditionally, the higher FITs offered by the power utilities would be recovered from the tax payer. However, under the GEC/REC system, the consumer (either end consumer or the local power utility) ends up paying the premium for RES-E (Jeppesen & Nielsen, 2003), by paying the certificate purchase price.

3. **Other Incentives:** Countries have often relied on "Policy packages" – a combination of direct measures, and indirect measures as well, to boost RES-E. Some of the indirect incentives put in place are:

   a. **Tax incentives:** Tax exemptions, tax credits and environmental taxes on "brown electricity" (fossil fuel based generation) are common complimentary fiscal incentives. Investments made in production of RES-E have been credited with the Production tax credit and renewable energy production incentives in USA, as reported by Bird et al., (2005) and in India (through the accelerated depreciation allowance to the extent of 80% on the investment made) (Shrimali, Trivedi, Srinivasan, Goel, & Nelson, 2016). Through a carbon tax, fossil fuel based generation is penalized and hence RES-E is encouraged. Based on the pigouvian polluter pays principle (costs of pollution should be borne by those who cause it), carbon taxes have been successful in so far as they promote technological innovation, and provide sustained income for the implementing government. However, carbon taxes have met with sustained political opposition (Zhang, Wang, Liang, & Chen, 2016), like in the case of France and USA.

b. **Renewable energy cess/ surcharges:** Higher FITs offered for RES-E call for a recovery system, in order to avoid burdening the utilities. One way is to charge a green energy tax on the final consumer of the local utility, like the "Global Adjustment Fee" in Canada (Bohringer, Rivers, Rutherford, & Wigle, 2012), "EEG surcharge" in Germany (Busgen & Durrschmidt, 2009) and surcharge under the "cost sharing mechanism" in China (Schuman & Lin, 2012). The pool of funds created, helps local utilities to purchase green power/RES-E generated. Other recovery forms include cross – subsidy between the final consumers of electricity. Cross subsidy refers to the practice, where one consumer category pays more than the rest, for their electricity consumption. However, since the practice of cross-subsidies has met with significant political friction, it is often avoided (Thakur, Deshmukh, Kaushik, & Kulshrestha, 2005).

Figure 2 summarizes the policy enablers, detailed above.

Figure 2. Summary of policy enablers used for promotion of RES-E

Type of Regulatory Policy Lever	Name of Policy Lever	Features	Advantages	Disadvantage
Price based incentives	Feed in Tariffs (FIT)	Provides an attractive price to the renewable energy generator. Offered by the local power supplier/ utility	Provides security of revenues for the generator. Historically, has proven successful to attract investments in RES-E	Proves expensive for the purchasing authority
	Market Dependent	FIT is a function of the spot market electricity price		
	Market Independent	FIT is not related to the spot market electricity price		
	Competitive Bidding	RES-E generators bid to sell the power generated. The lowest bidder wins. The power utilities thus procure RES-E at lower prices	Encourages competition amongst generators and hence more cost effective technology is adopted	Purchase prices can fall, and make the RES-E project financially infeasible for the investors.
Quantity based incentives	Renewable Purchase Standards (RPS)/ Renewable Purchase Obligations (RPO)	Mandates local power suppliers/ consumers to procure a share of their total electricity consumption from renewable sources	Provides a guaranteed demand for the RES-E generated and hence secures the investor's risk	Compliance costs are high for the obligated entities
	Tradable Green Energy Certificates (TGCs)/ Renewable Energy Certificates (RECs)	Each unit of RES-E generated is allotted a tradable certificate, which can be traded on a power market	Generators get additional revenue from sale of certificate, in addition to revenues accruing from sale of physical power	Uncertainty associated with the price of the certificate, as it is market oriented.
Other Incentives	Tax Incentives/ Tax Credits	Provide tax exemptions/credits for investing in RES-E	Indirectly makes fossil fuel based generation costly and promotes the shift to "greener" generation	Has met with large-scale political opposition in many countries
	Carbon Tax	Penalize production generating from fossil fuels		
	Renewable Energy Surcharges/Cess	Premium to be paid by the end consumer for RES-E. This creates a pool of funds used by the power utilities to support offering high FITs.	Provides financial support for power utilities	Proves to be a burden on the end consumer

As is evident, many tools to promote RES-E projects have been available for countries to choose from. However, efficient measures can be separated from the inefficient ones, only through a more detailed study, with specific reference to adopting nations. The next section presents case studies from select developed and developing nations.

CASE DESCRIPTION

Developed Economies

Denmark pioneers wind energy technology innovations; USA progresses at the back of state level RPOs; Germany relies on FITs; Japan vows to replace all nuclear power with RES-E post 2011 nuclear disaster.

Denmark

Not only does 68% of the country's total electricity come from renewable sources as of 2017, 48% of the total renewable footprint is from wind energy alone (Statistical

Figure 3. Winds of Change in Denmark: 2001 marks policy change from fixed FITs to market based prices
(Statistical Review of World Energy, 2018)

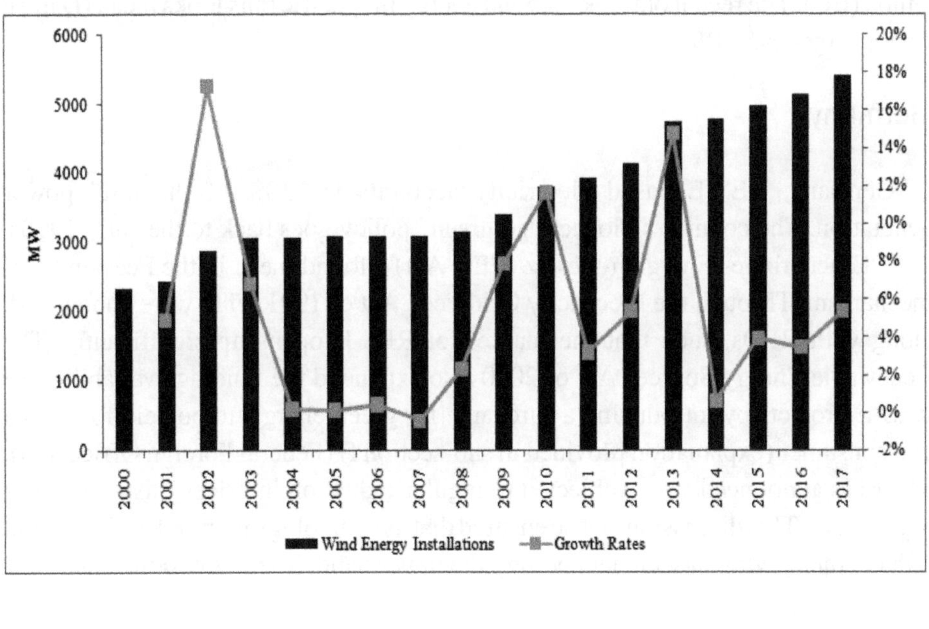

Review of World Energy, 2018). Innovations in wind power started as early as 1890s, and picked up pace in 1970s, post the global oil crisis, and a decision to reduce dependency on nuclear power (Lipp, 2007). Danish policy makers introduced renewable energy, not as competition to conventional energy, but as complementary to the same. Driven largely by technological research and innovations, Denmark soon became the pioneer in wind energy. Public awareness, created through community grounded NGOs, also played a pivotal role. Community participation, through discussions held by a small wind enthusiast groups, name RE-OVE, kept the pressure on policy makers, to accept emerging technologies. However the biggest push came in 1993, with the Feed-in-Tariff (FIT) mechanism introduced, mandating utilities to purchase wind power at 85% of the final price paid by consumers. The FIT regime, investment subsidy, tax exemptions and a dedicated wind power test station boosted growth in the sector. Focus was kept on technology up gradation as well, with equal participation from private and government entities. That helped in serial cost reductions. Denmark hence reaped the benefits of FIT, community involvement based growth and research in technology up gradation. However, the FIT regime was discontinued in 2001. Market based rates (spot prices as realized in power exchanges) along with a premium for environmental benefits for wind production, marked the new policy regime since 2001. Data suggests that between 1993 and 2001, Denmark rode high on the success of the wind sector, but the growth came down significantly post 2001 (Lipp, 2007), as can be seen from Figure 3. While the country had set an ambitious target of 100% replacement of coal by renewable sources by 2035 (Hvelplund, Østergaard, & Meyer, 2017), the plan has hit roadblocks due to discontinuation of subsidies and incentives from European Union (EU). The recent outlook for wind energy in specific thus looks bleak (Danish Energy Agency, 2017).

Germany

In Germany, RES-E based electricity accounts for 30% of the total power generation. The country's pioneer "go green" policy goes back to the early 1990s. The "Erneuerbare- Energien-Gesetz (EEG Act)", found merit in the Feed-in tariff mechanism. Through the Electricity Grid Feed Act of 1991, FITs were announced, along with RPOs, such that the market for RES-E opens up significantly. The Renewable Energy Sources Act of 2000 also expanded the funding availability for RES-E projects by introducing a surcharge for green energy, to be paid for by the end consumer (explanation provided in sub-section f) (Federal Foreign Office, n.d). The FITs announced were subjected to regular reduction in prices (also known as digression). This digression clause ensured that as technology matured; tariffs would reduce, thus giving nascent technology the upper hand. The digression clause also ensured that final tariff payable by end consumer is not too burdensome either. The EEG led to large scale job creation. Some wind project developers opine that the number of jobs created from wind projects may be close to 710,000 by 2030 (Aquila, Pamplona, Queiroz, Junior, & Fonseca, 2017). Germany's feed-in-tariff regime was instrumental in reaching the country's green goals, and soon became the new norm. The FIT remains one the most widely opted measure for promoting RES-E. Germany is often positioned as a success story to the European Union (EU) member countries (Busgen & Durrschmidt, 2009). The German EEG act was subsequently reformed to "Energiewende" in 2010, with steeper targets for emissions reductions and RES-E. Targets set suggest near replacement of all coal and nuclear based generation capacities, and set renewables to contribute to 80% of electricity consumption by 2050 (Nordensvärd & Urban, 2015). The EEG Act has undergone numerous revisions since then, and the latest amendment marks the shift away from FITs to competitive bidding for RES-E generation (Federal Ministry of Economic Affairs and Energy, n.d).

USA

Figure 4 suggests wind energy, to be a part of the green energy portfolio (early 2000s), much before solar energy (only after 2010).

Figure 4. Wind energy precedes solar energy, with state level FITs
(Statistical Review of World Energy, 2018)

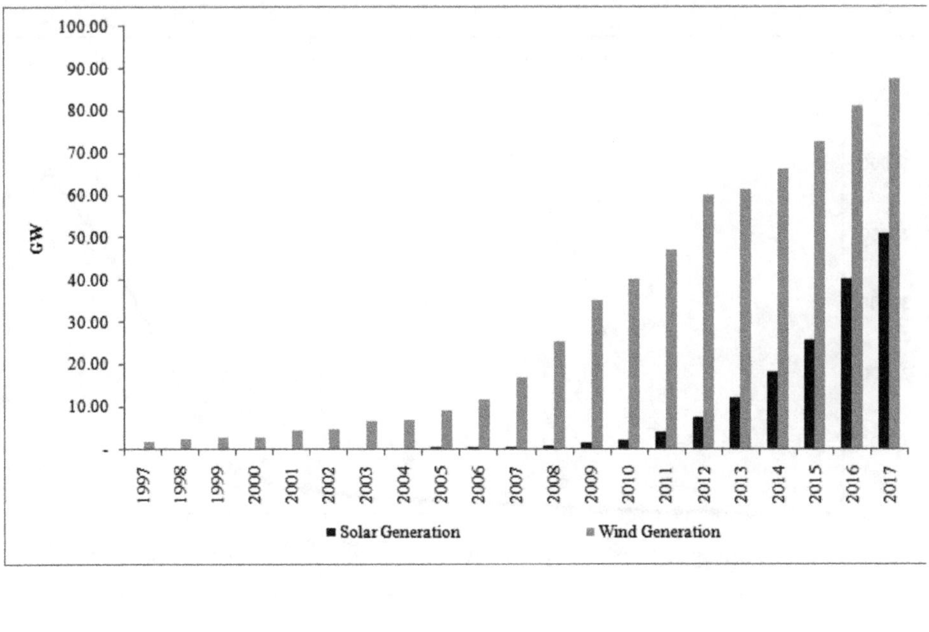

The Feed-in-tariff regime, while existing as an incentive, was discontinued in 2000. Renewable energy additions were thus supported through other incentives. USA adopted quantity-based incentives, moving away from the price-based incentive system that Germany had showcased. The economy widely opted for the "**Renewable Purchase Standards (RPS)"** scheme – a state offered incentives. "Renewable Purchase Standards" refers to a mandate, imposed on the local energy suppliers, which requires them to source a minimum percentage of the total electricity to be supplied by them, to come from renewable sources (Wiser et al., 2007). As of 2013, 29 states had implemented RPS mandates; with some states setting targets in the ranges of 25% - 40% (IRENA Abu Dhabi, 2015). Barbose and Wiser report that as of 2016, all states were close to reaching their targets (Barbose et al., 2016). The large scale accomplishment of RPS targets by state utilities has been the backbone for the success for the renewable industry in the USA. While the state government in the US provided the so called "quantity based incentives", the federal government of USA relied on the other forms. The **Federal Production Tax Credit** promised a relief from tax burden for investors. For renewable generators, a per unit tariff was announced – a product of total generation and announced per unit tariff rate would then reduce the generator's total tax liability. For all those units who did not come under the net of federal taxes, the **Renewable Energy Production Incentive** was also announced, that meant higher incomes/revenues for renewable generators (Bird et al., 2005).

Figure 5. Japan's RES-E gains pace post 2012 FIT regime
(Statistical Review of World Energy, 2018)

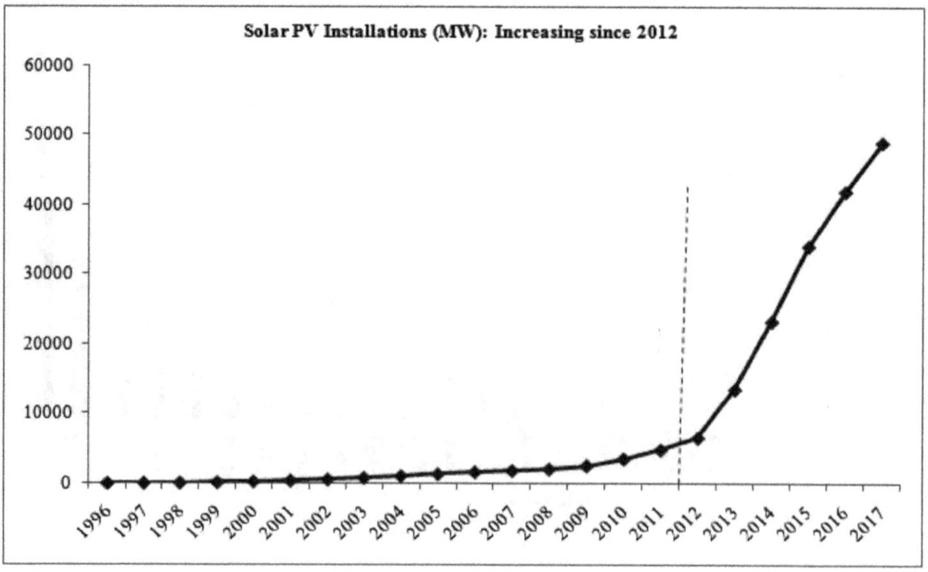

Japan

Before the Fukushima nuclear disaster of 2011, Japan relied upon nuclear power.

Japan set targets to increase share of nuclear power to 50% of total generation by 2030 (Huenteler, Schmidt, & Kanie, 2012). While subsidies, tax incentives and RPS existed in the 2000s, renewables contributed minimally. However, the growth rates in RES-E generation skyrocketed post 2011. RES-E installations grew by only 29% between 2000 and 2010, but shows aggressive progress in the next decade. Between only 7 years between 2011 and 2017, green electricity generation grew by a substantial 83% (Statistical Review of World Energy, 2018). Within that, priority was provided to solar energy. Literature suggests that Japan's focus on Solar Photo-Voltaic (PV) started at the decentralized level. As early as 1994, Japan announced the residential solar PV program, wherein excess solar energy generated by residences, was bought by the local power utilities at 23 Yen/kWh (0.23 USD/kWh)[1] – which was also the ongoing retail electricity price, as well as provided subsidy to cover installation costs (Muhammad-Sukki, et al., 2014). With that in place, solar energy installations gained pace, leaving behind the growth in wind energy footprints. While the country boasts of close to 49 GW of solar power as of 2017, wind energy's presence is limited to 3.3 GW (Statistical Review of World Energy, 2018).

RPS levels were announced in 2003. However the lack of interest in RES-E (as the local power utilities inherently preferred nuclear until 2011), resulted in non-compliance of RPS. Subsequently, and to achieve the set RE targets, FITs were announced in 2009 for solar, and for other forms only in 2012. The 2012 FIT announcements, renewed the FITs significantly, and thus marked an era of exponential growth in the sector, as can be seen from Figure 5. Higher purchase costs for the utilities, were subsequently transferred to the end consumer (in monthly electricity bills). The **"Renewable Energy Power Promotion Surcharge",** meant an additional spending of 2.25 yen per Kwh (0.019 USD/kWh[2], as of 2016), and is expected to rise even further by 2020 (Tanaka, Chapman, Sakurai, & Tezuka, 2017). This practice has come to be known as the "social burden" of the Japanese FIT regime. As for wind energy and other forms of RES-E, delays in environmental clearances, and an already burdened electrical system (from the solar energy injections) have kept the growth in those sectors damp (Matsubara, 2017).

Developing Economies

China races ahead after the Renewable Energy Law 2005 and constant technology innovations; India and Brazil move to competitive bidding from FITs.

Amongst the developing countries, India and China have been the most aggressive with their RES-E targets, but yet can't find a match for the sector's growth in these countries, with countries like Denmark and Germany. While Denmark and Germany boast of double digit contribution from RES-E in the total electricity generated (68% and 30% respectively), India and China see a contribution of only 6-7%. The two fastest growing economies in Asia have relied extensively on coal fired generation in the past, to meet their economic developmental plans. The same dependence on coal has resulted in making China the world's largest emitter of CO2 emissions, and India - the 4th largest, as of 2016 (Janssens-Maenhout, et al., 2017). Being relatively new entrants in the green energy space, the focus to greener sources came in only in the late 2000s. Although with a late start, both the countries today have set steep targets for RES-E and are on good pace to achieve the same. At the back of strong policy and legislative support, China today leads in RES-E generation, worldwide.

China

China declared the Renewable Energy Law in 2005 (REL, 2005), under its Renewable Energy Power Generation (REPG) program. Wind installations, which was the core focus of the program, have recorded a Compounded Annual Growth Rate (CAGR) of close to 150% in a span of 12 years between 2005 and 2017 as is evident from Figure 6. China now holds installations of 164 GW of wind energy and 131 GW of solar energy, the highest of all countries, as of 2017.

Figure 6. Renewable Energy Law (2005) brings RES-E to the forefront
(Statistical Review of World Energy, 2018)

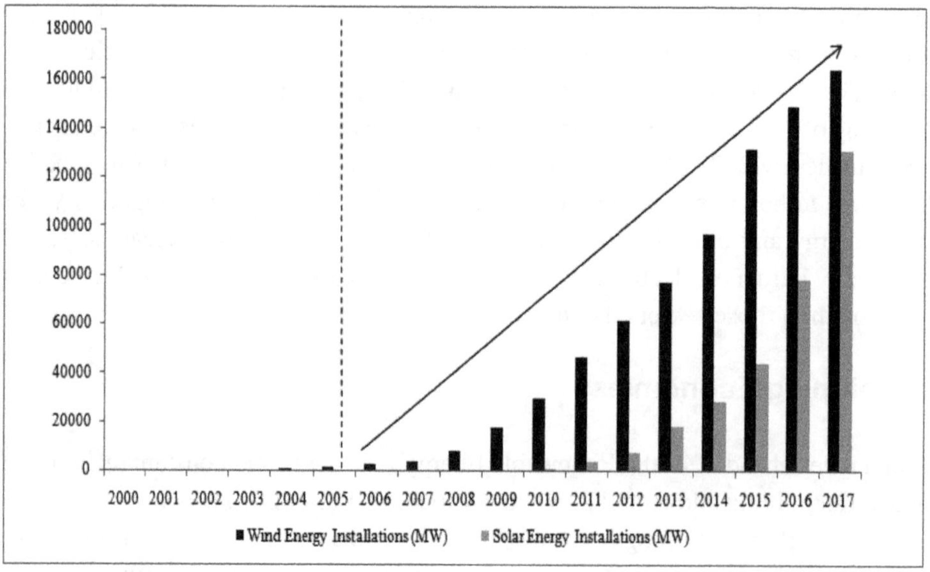

The Chinese government, apart from adopting the run of the mill, quantity and price based incentive such as the RPS and the Feed-in-tariff mechanism, also laid equal, if not higher, emphasis on Research and Development (R&D) initiatives. The Renewable Energy Development Special Fund was established, as a means to provide financial assistance for such research and developmental activities, done for wind and solar PV technologies. Zhao, Chen & Chang (2016) report that close to 13.9 billion Yuan (or 2 billion USD[3]) may have been spent towards such R&D initiatives. Fiscal incentives such as lower than normal tax rates and the restrictive tax policy (which imposed a tax on consumption of energy generated from fossil – fuels), also featured big in the REL 2005. Additionally, feed-in-tariffs were declared by region/location of project installation, such that the tariffs offered are aligned with technological progress. Costs recovery, for the higher FITs offered, were recovered by a renewable energy surcharge on sale of electricity to end consumers, coined under "cost-sharing". This surcharge was applied at varying rates on different categories of consumers, since 2006, and subsequently increased over the years 2006-2011. Initially charged at 0.001 RMB/unit in 2006 (0.0001 USD/kWh)[4], the surcharge was increased to 0.008 RMB/unit (0.0012 USD/kWh[5]), applicable for all consumer categories in 2011 (Schuman & Lin, 2012).

The RPS in China took the form of a full purchase guarantee (Schuman & Lin, 2012) –the local utilities were obligated to purchase the full amount of electricity that

Figure 7. Falling wind energy installations due to curtailment
(Statistical Review of World Energy, 2018)

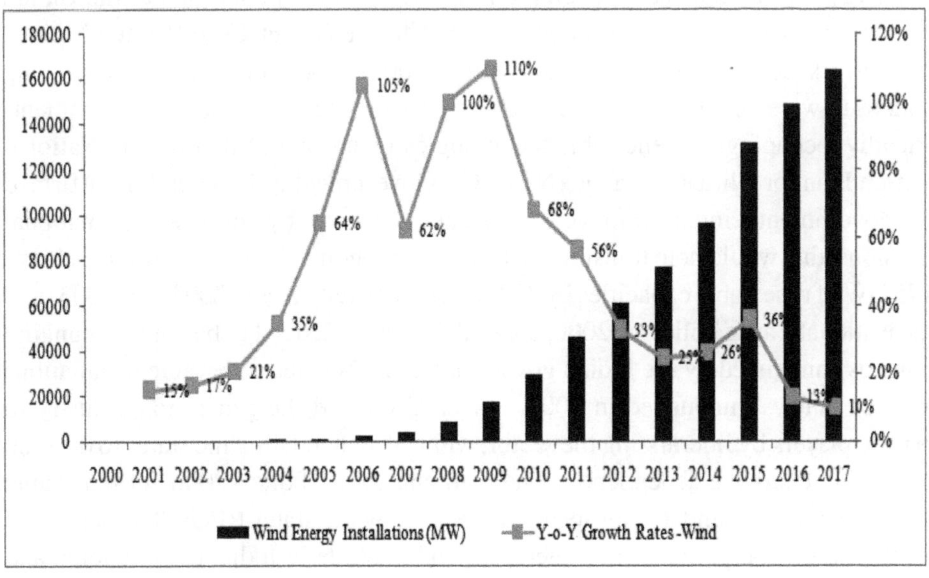

is generated, from renewable generators. However, this policy while having proven beneficial for the sector's growth is now faced with increasing resistance from grid operators. The large scale integration of RES-E poses threat to the health of the grid (due to variability and uncertainty issues), and is now a grave problem in China.

As a solution, China is now undertaking large-scale curtailment of wind and solar power projects. Curtailment refers to the inability of the transmission network to consume the entire electricity generated at a given point in time (Fink, Mudd, Porter, & Morgernstern, 2009). Generators are asked to reduce output for their generating plants by the utilities/grid operators, mostly on account of lack of transmission capacities or over-generation. Curtailment rates for wind power were as high as 40% since 2014, in some regions of North-West China (Luo, Dan, Zhang & Guo, 2018). Estimates put wind curtailment losses at 17.8 TWh in 2015 (He, Xu, Pang, Tian, & Wu, 2016).The practice of curtailment has deterred further investments from generators and has severely impacted the year –on –year growth rates in renewable installations, as is evident from Figure 7. However, even with such problems, China's consumption of RES-E in total electricity consumption is poised to grow to 40% by 2030 (He et al., 2016).

India

Carbon emissions in India, owing to its continued dependence on coal, increased by 5.4% in 2015 as compared to 2014, while all other major emitting countries showed reduction (Janssens-Maenhout, et al., 2017). While countries like Denmark and Germany started the transition to RES-E much earlier, India's focus on cleaner form of energy was announced through the Bali Action Plan in 2008 (under the United Nations Framework Convention on Climate Change (UNFCCC)) (United Nations Framework Convention on Climate Change, 2008). Under the plan, all developing countries were required to cut emissions growth rate by adopting environment-friendly technologies (renewable technologies being one of them). The National Action Plan for Climate Change (NAPCC) (Government of India, n.d) re-affirmed the government's intent to move to a cleaner economy, by announcing 8 national missions, that would help transition to a greener economy. India now aims to achieve 175 GW of renewable capacities by 2022. The Electricity Act of 2003 (EA 2003) and the National Tariff Policy of 2006, are widely acknowledged to be course changing policies announced by the Indian government, that boosted renewable installations in the country. Announced in 2003, EA 2003 opened the generation industry to private players by liberalizing the sector, while also mandating the states to develop renewable energy independently. The National Tariff Policy 2006 set renewable energy purchase mandates for local utilities, set by the states RPOs. The generated RES-E was to be procured at preferential FITs. While in India, both national and state level policies co-existed, state level push has proven to be more instrumental than the national level incentives, similar to USA (Schmid, 2012).The Indian story presents a good case of using market linked incentives (feed-in-tariffs) as well as command and control measures (renewable purchase obligations). However, the growth in this sector is now constrained on several grounds. Luthra, Kumar, Garg & Haleem (2015) studied literature to identify seven main barriers to larger renewable energy adoption in India. High initial capital costs, absence of large-scale subsidies and insufficient consumer awareness are important deterrents. Also, while governmental plans to expand renewables exist, lack of political commitment is a major growth dampener. Further, the high FITs offered by local power utilities, were never matched by a commensurate recovery mechanism, like it was done in Germany, China and Japan. The local utilities continued to pay high costs for power purchase without matching revenue increases. As a result, the local utilities are now financially burdened. The accumulated losses of all state power suppliers were USD 38.4 Billion in 2012, which increased to USD 64.1 Billion in 2014-2015 (Power Finance Corporation (PFC), 2016). High cost of power supply, non-revision of end consumer tariffs, non-payment of subsidies, and high distribution losses fueled the financial worries (Ministry of Power, 2012).The cost of purchasing electricity, as a percentage of total costs for the utilities, increased from 75.16% in 2012-2013 to 78.56% in 2014-2015 (PFC, 2016). To remedy rising costs, India moved to competitive bidding, from fixed FITs. Adopted in 2015 for wind and solar energy projects, purchase prices offered by utilities have already fallen below Rs. 3/Kwh (0.04 USD/kWh[6]) (Press Information Bureau, 2017). Lower tariffs, while providing financial relief to the power utilities, favor only big and financial stable bidders (since they have the financial backing to bid), and may render projects unviable (Gambhir, Sant & Deshmukh, 2011). Inspite of these stumbling roadblocks, India remains committed to achieving 175 GW of renewable energy installation by 2022.

Brazil

Brazil generated 98 TWh of RES-E (16% of total electrical generation), closely following Japan and just ahead of India. 63% of the total installed capacity of 138 GW came from hydro power in 2016, owing to its large water resources. However since the early 2000s, the focus has moved to developing alternate renewable sources, due to problems of power blackouts, caused by droughts and resulting low water levels at the reservoirs (Faria Jr., Trigoso, & Cavalcanti, 2017). The concentrated policy effort towards this came with the announcement of the alternate sources incentive program (PROINFA) in 2002 (Aquila et al., 2017). PROINFA aimed to promote 3 main RES – wind, biomass and small hydro, and solar kept out of the purview, and was divided into 2 phases. Phase 1 aimed at getting 3300 MW of RES (1100 MW from these 3 sources) by 2007, through the FIT route. The FIT system was market based, and assured price was offered by the local utility ELECTROBRAS. The tariffs applicable for small hydro, biomass and wind were set at 50%, 70% and 90% of the average retail price, only in the last 12 months of the project (Global CCS Institute, n.d). The higher prices paid by ELECTROBRAS came from the "Energy Development Account", constituted from higher bills paid by end consumers of electricity (International Energy Agency, n.d). The Brazilian National Development Bank (BNDES) was the primary lending authority, and the tariffs offered to generators ranged between USD 0.10/kWh and USD 0.90/kWh ($102.53/MWh and $90.40/MWh).

However, due to the higher tariffs paid to the generators, the retail electricity rates shot up. This changed the policy focus for the 2nd phase, and in 2003, technology specific auctions were introduced. With this, all RES-E competed at the same time to sell, at lower prices (Dutra & Szklo, 2008). Further, local content requirements stood at 60% – 70% of equipment value, to encourage job creation. However RE investors could not ensure the same. The first bid for wind energy projects was held in 2009, and tariff offered to wind plants fell to USD 0.077/kWh for 1800 MW (Faria Jr. et al., 2016). PROINFA was subsequently discontinued from 2011.

Solar net metering regulation was introduced much later, only in 2012. Excess generation, over and above self consumption by investors, could be fed back to the system. However, utilities showed reluctance to allow large amount of electricity being fed into the grid. Additionally, the responsibility to install the net meter/ bi-directional meter rested with the generators themselves, which owing to bureaucratic delays, discouraged large scale adoption (Faria Jr. et al., 2016). Even though much is being done to promote solar, policy advocates suggest more number of fiscal incentives for this sector. The Brazilian market still suffers from lack of domestic production capacity of solar equipments. Local utilities do not promote large scale distributed solar adoption in the fear of loss of income (since distributed solar will reduce sales of electricity to final consumers) and fear of grid ill-health, caused by excess solar power fed into the grid.

The RES-E growth stories for the select seven countries selected for this study can be found summarized in Figure 8 (developed economies) and Figure 9 (developing economies), as under:

From the above it appears evident that FITs and RPS/RPOs complimented with Tradable Green energy certificates have been the driving policy levers in most countries. However, the results of such policies have not been without challenges as well, which are briefly described in the next section.

Figure 8. Comparison of legislative measures taken to boost RES-E in developed economies

Country	Legislative Support
USA	**Feed-in-Tariff** • State-wise FITs. Local power supplier/utility level FITs also exists (Toby Couture, 2009) **Renewable Purchase Standards and Green Energy Certificates:** • Main driver for the sector. 37 states have RPS, strengthened in 2016 and 2017 (Barbose, 2017). Like China, RPS mandates can be met through purchase of RECs. **Other Incentives** • The Federal Investment Tax Credit (ITC): A % of the investment made in renewable projects is eligible for tax deductions, which can go up to 30% of investment value, like in Solar and wind projects (Department of Energy) • Renewable Energy Production Incentive (REPI): Effective 1992, generators were provided additional revenue @ 1.5 cents/kWh, aligned with inflation.
Germany	**Feed-in-Tariff and Competitive Bidding:** • Amended German Energiewende 2017 marks the shift away from FITs to Competitive bidding. RES-E procurement price ceilings are also in place (Federal Ministry of Economic Affairs and Energy, n.d). **Renewable Purchase Standards and Green Energy Certificates:** • The Energiewende 2017 announces quantum of RES-E to be purchased. However no REC market exists in the country (Devenyi, 2012). **Other Incentives** • "Green energy account", collects funds from the EEG surcharge (paid by the end consumer). The funds collected are directed towards financing aid for renewable projects. Surcharge stood in the range of 6 – 7 cents/ kWh in 2017 (Bundesverband Erneuerbare Energie). • Loans from developmental banks like KfW for RES-E project developments.
Japan	**Feed-in-Tariff and Competitive Bidding** • FITs for solar PV were announced in 2009. FIT for other RES-E implemented in 2012 (Muhammad-Sukki, et al., 2014). Projects more than 2 MW to undergo auction/bidding system. • Pre- approval from grid operators mandatory for availing the FIT benefit. **Renewable Purchase Standards and Green Energy Certificates:** • RPS announced in 2003, superseded by the FIT regime of 2012. "Green Power Certificates" could be purchased to show compliance (Ogasawara, 2008). **Other Incentives** • Net metering for residential solar installations, from 1992 onwards • Renewable energy surcharge on end consumers –100 yen per bill.
Denmark	**Feed-in-Tariff and Competitive Bidding:** • FITs announced in 1993, replaced by a "market price plus a premium" model in 2001. • Premium prices also fixed such that total FIT paid to generators did not exceed 6.1 cents/kWh (Mendonca, Lacey, & Hvelplund, 2009) • Currently, auctions decide the premium price to be paid, over and above the market price (Petrova, 2018). **Renewable Purchase Standards and Green Energy Certificates:** • Quotas and tradable RECs introduced in 2000 (Odgaard, 2000). **Other Incentives** • Electricity balancing costs are refunded (International Renewable Energy Agency, n.d) • Carbon taxes implemented since 1992 (Andersen, 2017) • Net metering regulations (International Renewable Energy Agency, n.d)

CURRENT CHALLENGES

Research suggests that the share of renewable energy in electricity generation is poised to grow from 23% in 2015 to 82% by 2050 (Ralon, Taylor, Ilas, Diaz-Bone & Kairies, 2017) with solar and wind based installations in the lead. However, increased integration of renewable energy based electricity has its own set of problems. Some of the common ones are:

1. Rising curtailment of RES-E due to intermittency issues, lack of infrastructure and supply demand mismatch.

Figure 9. Comparison of legislative measures taken to boost RES-E in developing economies

Country	Legislative Support
China	**Feed-in-Tariff and Competitive Bidding:** • FIT revisions since 2003, with FIT ranging from 0.30 RMB/kWh to 1 RMB/kWh for grid connected wind energy projects (Ming, Ximei, Na, & Song, 2013). • Likely replacement of wind FITs with auctions post 2018 (Kabeer, 2018). • For Solar PV, FITs announced for 3 separate regions, with no digression clause (Ye, Rodrigues, & Lin, 2017). **Renewable Purchase Standards and Green Energy Certificates:** Provincial level renewable quotas, planned for 2018 and 2020. Targets can be met by purchasing tradable RECs. Quotas range from 8.5% to 91% of total electricity consumption requirements (International Renewable Energy Agency and International Energy Agency). **Other Incentives** • Renewable energy surcharges implemented to recover high costs of FITs. Currently, for solar PV FITs, surcharge is to be paid by industrial and commercial users pay .019 CNY/kWh (Ye et al., 2017) • Favorable Value Added Tax (VAT), Income tax and import duty provisions. Subsidies, for grid connected RES-E development also extended (Zhao, Chen, & Chang, 2016).
India	**Feed-in-Tariff and Competitive Bidding:** • Competitive bidding replaced fixed FITs since 2015, for both wind and solar based projects. **Renewable Purchase Standards and Green Energy Certificates:** • Renewable Purchase Obligations exist at the state levels, and by technology type. Targets revised annually. • Renewable energy certificates announced in 2010, to help meet RPO targets. Price of REC to be realized by market demand and supply at designated trading platforms, and will range between the floor price (minimum) and the forbearance price (maximum), as has been set by the regulatory authority (Khaparde, 2015). **Other Incentives** • Tax incentives in forms of accelerated depreciation (that allowed tax deductions to the extent of 80% of investments) and tax holidays. • Capital subsidy for rooftop solar installations and Net Metering • Generation based incentive (similar to the REPI in USA).
Brazil	**Feed-in-Tariff and Competitive Bidding:** • PROINFA announced fixed FITs in 2002 • Competitive bidding/auctions in 2009 (for wind) replaced the FIT regime. **Renewable Purchase Standards and Green Energy Certificates:** • No framework announced. **Other Incentives** • Wind turbine component tax exemption for manufacturing units • Net metering for small scale renewable generator – to sell surplus power generated to local utilities (2012) • Invova Energia program (2013): Allows for ready finance for renewable projects.

Curtailment, refers to the involuntary rejection of RES-E generated, more specifically wind energy, by grid operators. Reasons could be the uncertain and variable nature of RES-E, or lack of infrastructure facilities, or both. In 2012, curtailment rates in china reached 17% due to lack of proper transmission corridors. 2015 and 2016 again saw surge in curtailment due to lack of demand owing to economic slump. As much as 11 provinces in China are suffering from this phenomenon (Ye, Jiaqi, & Mengye, 2017). Other countries, like United States witnessed more reasonable curtailment, in the range of 4% - 5% (Bird, Cochran, & Wang, 2014). Solutions are presented by enhanced transmission/ grid capacities. Operational mandates, such as forecasting and scheduling of power, where generators provide a forecast of the power that will be generated, are also necessary. Grid operators are then able to draw an accurate demand – supply estimate, and maintain a sound healthy grid.

2. High FITs have financially burdened purchasing authorities.

Most countries that once offered substantially high FITs are now moving to auctions/competitive bidding systems, since high FITs prove expensive for the utilities in the long run. India, Brazil, Germany and now China are some examples. However, the lower realized power purchase rates (revenues to the generators) have resulted in growing financial concerns for generators. For example, Indian projects have witnessed subsequent delay in projects, and projects have been rendered financially infeasible (Pathak & Bhaskar, 2018).

3. Renewable energy surcharges mean unhappy end consumers.

Renewable energy surcharges that end consumers pay in order to finance the high FITs offered to generators have been unpopular. In Germany, it came to be known as the 'sweet poison' (Buchan, 2012), and 'over-costly' (Quitzow et al., 2016).

4. Rising community issues in some countries makes RES-E expansion challenging.

Denmark, which pioneered wind energy, now faces increasing community concerns and the NIMBY (Not In My Backyard) syndrome. High density of wind energy installations, with limited land availability, has met with high local resistance. The Wind2050 investigations (Linowes, 2017), was specifically launched to understand the concerns of the local communities, towards rising wind energy footprints in the country. Trust issues with the local municipal bodies, and pre-sanctioning of projects without incorporating public concerns, were found as important concerns.

SOLUTIONS, RECOMMENDATIONS, AND CONCLUSION

Reliance on grid connected renewable energy based electricity generation features prominently in the energy transition for many nations. In the past, while many regulatory tools have been used, a few tools have been more successful than the others. Amongst all, RES-E generators have been assured of a risk free market by the use of Feed-in-Tariffs (guaranteeing a price) and Renewable Purchase Standards/Obligations (guaranteeing volumes). Other fiscal incentives, such as tax concessions and subsidy allowances, have also co-existed. However, over time, rising RES-E installations have met with varied problems as outlined above. Solutions are being worked upon alongside.

Grid curtailment issues are being solved by the requirement of forecasting and scheduling mandates. A switch from FITs to auctions will reduce power purchase rates and improve competition in the RES-E sector. Higher competition will ensure technology innovation and quality enhancements for better performance. Auctions will also reduce the power purchase costs, and hence renewable energy surcharges may also decline, if not eliminated, eventually. Finally, public participation based RES-E project development will cure local resistance issues (Langer, Decker, & Menrad, 2017).

Focus is now seen shifting to decentralized RES-E (RES-E in smaller units), which helps achieve dual purpose of improving electricity access, while increasing "green energy" penetration. Decentralized RES-E enjoy the benefits of being locally applicable, small scale and not dependent on the national network (Dincer, 2000). Socio-economic problems of poverty, health and education, can also be solved through energy access in rural areas, provided by decentralized RES-E (Bugaje, 2006). Since 2012, 34% of all new electricity connections, in energy deficient areas have come from decentralized RES-E, mostly via solar rooftop adoption (United Nations New York, 2018). This number is expected to increase to 60% by 2030 (International Energy Agency, 2017). Net metering regulations, where surplus power generated can be sold, has been a major driver. While for centralized RES-E projects, the trend has been to move from fixed FITs to competitive bidding/auctions, FITs offered in decentralized RES-E has been rising. Japanese utilities bought 1.4 billion kWh worth of surplus power, by the end of 2010, at twice the existing market rate for electricity, and surpassed Germany and Italy in Solar PV installations by 2011 (Chen, Kim, & Yamaguchi, 2014). Efforts are now taken to increase RES-E consumption, through mini grids and off-grids, in energy starved areas.

Inspite the RE sector being plagued with issues, RES-E targets are being revised, only upwards. The renewable energy legislative framework, in the future years will see suitable changes, to solve issues that may crop up. The reworked legislative framework will ensure that growth fosters in the sector, suitably.

REFERENCES

Abdmouleh, Z., Alammari, R., & Gastli, A. (2015). Review of policies encouraging renewable energy integration and best practices. *Renewable & Sustainable Energy Reviews*, *45*, 249–262. doi:10.1016/j.rser.2015.01.035

IRENA Abu Dhabi. (2015). *Renewable Energy Prospects: United States of America, REmap 2030 analysis*. Retrieved from www.irena.org/remap

Andersen, S. M. (2017). *Experiences with carbon taxation - Denmark, Sweden and Europe*. Retrieved from http://pure.au.dk/portal/files/111427005/CapeTown1_ANDERSEN.pdf

Aquilaa, G., Pamplona, E., Queiroz, A. R., Junior, P., & Fonseca, M. (2017). An overview of incentive policies for the expansion of renewable energy generation in electricity power systems and the Brazilian experience. *Renewable & Sustainable Energy Reviews*, *70*, 1090–1098. doi:10.1016/j.rser.2016.12.013

Barbose, G. (2017). *U.S. Renewables Portfolio Standards: 2017 Annual Status Report*. Retrieved from https://emp.lbl.gov/publications/us-renewables-portfolio-standards-0

Barbose, G., Wiser, R., Heeter, J., Mai, T., Bird, L., Bolinger, M., ... Millstein, D. (2016). A retrospective analysis of benefits and impacts of U.S. renewable portfolio standards. *Energy Policy*, *96*, 645–660. doi:10.1016/j.enpol.2016.06.035

Bayod-Rújula, A. A. (2009). Future development of the electricity systems with distributed generation. *Energy*, *34*(3), 377–383. doi:10.1016/j.energy.2008.12.008

Bayrakc, A. G., & Kocar, G. (2012). Utilization of renewable energies in Turkey's agriculture. *Renewable & Sustainable Energy Reviews*, *16*(1), 618–633. doi:10.1016/j.rser.2011.08.027

Beck, F., & Martinot, E. (2004). Renewable Energy Policies and Barriers. Encyclopedia of Energy, 5, 365-383.

Bessa, R., Moreira, C., Silva, B., & Matos, M. (2014). Handling renewable energy variablity and uncertainty in power system operations. *Wiley Interdisciplinary Reviews*, *3*(2), 156–178. doi:10.1002/wene.76

Bird, L., Bolinger, M., Gagliano, T., Wiser, R., Brown, M., & Parsons, B. (2005). Policies and market factors driving wind power development in the United States. *Energy Policy*, *33*(11), 1397–1407. doi:10.1016/j.enpol.2003.12.018

Bird, L., Cochran, J., & Wang, X. (2014). *Wind and Solar Energy Curtailment: Experience and Practices in the United States*. National Renewable Energy Laboratory. Retrieved from https://www.nrel.gov/docs/fy14osti/60983.pdf

Birol, F. (2010). *World Energy Outlook*. International Energy Agency. Retrieved from http://www.oecd.org/berlin/46389140.pdf

Bohringer, C., Rivers, N. J., Rutherford, T., & Wigle, R. (2012). Green Jobs and Renewable Electricity Policies: Employment Impacts of Ontario's Feed - in - Tariff. *The B.E. Journal of Economic Analysis & Policy*, *12*(1), 1–40. doi:10.1515/1935-1682.3217

British Petroleum. (2018). *Statistical Review of World Energy*. Retrieved June 10, 2018, from BP Global: https://www.bp.com/en/global/corporate/energy-economics/statistical-review-of-world-energy.html

Brudermann, T., Reinsberger, K., Orthofer, A., Kislinger, M., & Posch, A. (2013). Photovoltaics in agriculture: A case study on decision making of farmers. *Energy Policy*, *61*, 96–103. doi:10.1016/j.enpol.2013.06.081

Buchan, D. (2012). *The Energiewende - Germany's Gamble. The Oxford Institute of Energy Studies*. Retrieved from https://www.oxfordenergy.org/wpcms/wp-content/uploads/2012/06/SP-261.pdf

Bugaje, I. (2006). Renewable energy for sustainable development in Africa: A review. *Renewable & Sustainable Energy Reviews*, *10*(6), 603–612. doi:10.1016/j.rser.2004.11.002

Bundesverband Erneuerbare Energie. (n.d.). *BEE forecast: slightly lower EEG surcharge in 2018*. Retrieved October 26, 2018, from: https://www.bee-ev.de/home/presse/mitteilungen/detailansicht/bee-prognose-leicht-sinkende-eeg-umlage-2018/

Busgen, U., & Durrschmidt, W. (2009). The expansion of electricity generation from renewable energies in Germany - A review based on Renewable Energy Sources Act Progress Report 2007 and the new German feed - in legislation. *Energy Policy*, *37*(7), 2536–2545.

Chen, W.-M., Kim, H., & Yamaguchi, H. (2014). Renewable energy in eastern Asia: Renewable energy policy review and comparative SWOT analysis for promoting renewable energy in Japan, South Korea, and Taiwan. *Energy Policy*, *74*, 319–329. doi:10.1016/j.enpol.2014.08.019

Çoker, B., Cathoglu, H., & Birgin, O. (2010). Conceptions of students about renewable energy sources: A need to teach based on contextual approaches. *Procedia: Social and Behavioral Sciences*, *2*(2), 1488–1492. doi:10.1016/j.sbspro.2010.03.223

Cory, K., Couture, T., & Kreycik, C. (2009). *Feed-in Tariff Policy: Design, Implementation, and RPS Policy Interactions*. Retrieved from https://www.nrel.gov/docs/fy09osti/45549.pdf

Danish Energy Agency. (2017). *Denmark's Energy and Climate Outlook*. Retrieved from https://ens.dk/sites/ens.dk/files/Analyser/denmarks_energy_and_climate_outlook_2017.pdf

Department of Energy. (n.d.). *Business Energy Investment Tax Credit*. Retrieved October 26, 2018, from https://www.energy.gov/savings/business-energy-investment-tax-credit-itc

Devenyi, R., & Mladenova, I. (2012). *International Markets for Renewable Energy Certificates (RECs). Sustainable Real Estate Roundtable: Member Briefing*. Retrieved from http://sustainround.com/library/sites/default/files/SRER_Member%20Briefing_International%20Markets%20for%20Renewable%20Energy%20Certificates_2012-07-16.pdf

Dincer, I. (2000). Renewable energy and sustainable development: A crucial review. *Renewable & Sustainable Energy Reviews*, *4*(2), 157–175. doi:10.1016/S1364-0321(99)00011-8

Dutra, R. M., & Szklo, A. S. (2008). Incentive policies for promoting wind power production in Brazil: Scenarios for the Alternative Energy Sources Incentive Program (PROINFA) under the New Brazilian electric power sector regulation. *Renewable Energy*, *33*(1), 65–76. doi:10.1016/j.renene.2007.01.013

Evans, A., Strezov, V., & Evans, T. J. (2012). Assessment of utility energy storage options for increased renewable energy penetration. *Renewable & Sustainable Energy Reviews*, *16*(6), 4141–4147. doi:10.1016/j.rser.2012.03.048

Fagiani, R., Barquin, J., & Hakvoort, R. (2013). Risk-based assessment of the cost-efficiency and the effectivity of renewable energy support schemes: Certificate markets versus feed-in-tariffs. *Energy Policy*, *55*, 648–661. doi:10.1016/j.enpol.2012.12.066

Faria, H. Jr, Trigoso, F. B. M., & Cavalcanti, J. A. M. (2017). Review of distributed generation with photovoltaic grid connected systems in Brazil: Challenges and prospects. *Renewable & Sustainable Energy Reviews*, *75*, 469–475. doi:10.1016/j.rser.2016.10.076

Federal Foreign Office. (n.d.). *The German Energiewende*. Retrieved from https://www.auswaertiges-amt.de/blob/610620/5d9bfec0ab35695b9db548d10c94e57d/the-german-energiewende-data.pdf

Federal Ministry of Economic Affairs and Energy. (n.d.). *Information portal Renewable Energies*. Retrieved October 26, 2018, from https://www.erneuerbare-energien.de/EE/Navigation/DE/Recht-Politik/Das_EEG/das_eeg.html

Fink, S., Mudd, C., Porter, K., & Morgernstern, B. (2009). *Wind Energy Curtailment Case Studies, May 2008 – May 2009*. Retrieved from https://www.nrel.gov/docs/fy10osti/46716.pdf

Gambhir, A., Sant, G., & Deshmukh, R. (2011). India's Solar Mission: Procurement and Auctions. *Economic and Political Weekly*, *46*(28), 22–26.

Global C. C. S. Institute. (n.d.). *PROINFA*. Retrieved from https://hub.globalccsinstitute.com/publications/analysis-regulatory-framework-wind-power-generation-brazil-summary-report/1-proinfa

Government of India. (n.d.). *National Action Plan for Climate Change*. Retrieved from http://www.moef.nic.in/downloads/home/Pg01-52.pdf

He, Y., Xu, Y., Pang, Y., Tian, H., & Wu, R. (2016). A regulatory policy to promote renewable energy consumption in China: Review and future evolutionary path. *Renewable Energy*, *89*, 695–705. doi:10.1016/j.renene.2015.12.047

Huenteler, J., Schmidt, T. S., & Kanie, N. (2012). Japan's post Fukushima challenge – implications from the German experience on renewable energy policy. *Energy Policy*, *45*, 6–11. doi:10.1016/j.enpol.2012.02.041

Hvelplund, F., Østergaard, A., & Meyer, N. (2017). Incentives and barriers for wind power expansion and system integration in Denmark. *Energy Policy*, *107*, 573–584. doi:10.1016/j.enpol.2017.05.009

International Energy Agency. (2017a). *Energy Access Outlook 2017 – From Poverty to Prosperity. World Energy Outlook Special Report*. Retrieved from https://www.iea.org/publications/freepublications/publication/WEO2017SpecialReport_EnergyAccessOutlook.pdf

International Energy Agency. (2017b). *CO2 emissions from fuel combustion*. Retrieved from http://www.indiaenvironmentportal.org.in/files/file/CO2 Emissions from Fuel Combustion 2018 Highlights.pdf

International Energy Agency. (n.d.a). *Programme of Incentives for Alternative Electricity Sources – PROINFA*. Retrieved from https://www.iea.org/policiesandmeasures/pams/brazil/name-21963-en.php?s=&s=

International Energy Agency. (n.d.b) *Renewable Electricity Quota and Assessment Method*. Retrieved October 26, 2018, from https://www.iea.org/policiesandmeasures/pams/china/name-170554-en.php

International Renewable Energy Agency. (n.d.). Retrieved September 24, 2018, from International Renewable Energy Agency: http://resourceirena.irena.org/gateway/dashboard/index.html

Janssens-Maenhout, G., Crippa, M., Guizzardi, D., Muntean, M., Schaaf, E., … Schure, K.M. (2017). *Fossil CO2 and GHG emissions of all world countries*. Retrieved from http://edgar.jrc.ec.europa.eu/booklet2017/CO2_and_GHG_emissions_of_all_world_countries_booklet_online.pdf

Jeppesen, T., & Nielsen, L. (2003). Tradable Green Certificates in selected European countries - overview and assessment. *Energy Policy*, *31*(1), 3–14. doi:10.1016/S0301-4215(02)00112-X

Kabeer, N. (2018). *China Ends Wind Feed-In Tariffs and Opts for Auctions*. Retrieved October 26, 2018, from https://mercomindia.com/china-ends-wind-feed-in-tariffs/

Kandpal, T., & Broman, L. (2014). Renewable energy education: A global status review. *Renewable & Sustainable Energy Reviews*, *34*, 300–324. doi:10.1016/j.rser.2014.02.039

Langer, K., Decker, T., & Menrad, K. (2017). Public participation in wind energy projects located in Germany: Which form of participation is the key to acceptance? *Renewable Energy*, *112*, 63–73. doi:10.1016/j.renene.2017.05.021

Linowes, L. (2017). *Denmark's Anti-Wind Problem: Wind News Update*. Retrieved October 27, 2018, from https://www.masterresource.org/uncategorized/wind-news-update-denmarks-anti-wind-problem/

Lipp, J. (2007). Lessons for effective renewable electricity policy from Denmark, Germany and the United Kingdom. *Energy Policy*, *35*(11), 5481–5495. doi:10.1016/j.enpol.2007.05.015

Luo, G., Dan, E., Zhang, X., & Guo, Y. (2018). Why the Wind Curtailment of Northwest China Remains High. *Sustainibility*, *10*(3), 570. doi:10.3390u10020570

Luthra, S., Kumar, S., Garg, D., & Haleem, A. (2015). Barriers to renewable/sustainable energy technologies adoption: Indian perspective. *Renewable & Sustainable Energy Reviews*, *41*, 762–776. doi:10.1016/j.rser.2014.08.077

Mapping the Renewable Energy Sector to the Sustainable Development Goals: An Atlas. (2018). Retrieved from http://unsdsn.org/wp-content/uploads/2019/01/190108-mapping-renewables-report_WEB.pdf

Matsubara, H. (2017). *Renewable Energy Status in Japan and the World, Sustainable Zone 2016, April 2017*. Retrieved from https://www.japanfs.org/en/news/archives/news_id035824.html

Mendonca, M., Lacey, S., & Hvelplund, F. (2009). Stability, participation and transparency in renewable energy policy:Lessons from Denmark and the United States. *Policy and Society*, *27*(4), 379–398. doi:10.1016/j.polsoc.2009.01.007

Ming, Z., Ximei, L., Na, L., & Song, X. (2013). Overall review of renewable energy tariff policy in China: Evolution, implementation, problems and countermeasures. *Renewable & Sustainable Energy Reviews*, *25*, 260–271. doi:10.1016/j.rser.2013.04.026

Ministry of Power. (2012). *Scheme for Financial Restructuring of State Distribution Companies (DISCOMS)*. Retrieved from http://powermin.nic.in/sites/default/files/uploads/Financial_restructuring_of_State_Distribution_Companies_discoms_Oct2012.pdf

Muhammad-Sukki, F., Abu-Bakar, S. H., Munir, A. B., Yasin, S. H., Ramirez-Iniguez, R., McMeekin, S. G., ... Tahar, R. M. (2014). Feed-in tariff for solar photovoltaic: The rise of Japan. *Renewable Energy*, *68*, 636–643. doi:10.1016/j.renene.2014.03.012

Nordensvärd, J., & Urban, F. (2015). The stuttering energy transition in Germany: Wind energy policy and feed-in-tariff lock-in. *Energy Policy*, *82*, 156–165. doi:10.1016/j.enpol.2015.03.009

Odgaard, O. (2000). Workshop on Best Practices in Policies and Measures. *The Green Electricity Market in Denmark: Quotas, Certificates and International Trade*.

Ogasawara, J. (2008). *Overview of the Green Power Certification System*. Retrieved from https://eneken.ieej.or.jp/en/data/pdf/493.pdf

Ozturk, M., & Yuksel, Y. E. (2016). Energy structure of Turkey for sustainable development. *Renewable & Sustainable Energy Reviews*, *53*, 1259–1272. doi:10.1016/j.rser.2015.09.087

Pathak, M., & Bhaskar, U. (2018). *India defers largest global solar tender on bidders concerns*. Retrieved from https://www.livemint.com/Industry/uhzOpWbfficUpzEpLujveL/India-defers-largest-global-solar-tender-on-bidders-concerns.html

Perera, F. P. (2017). Multiple Threats to Child Health from Fossil Fuel Combustion: Impacts of Air Pollution and Climate Change. *Environmental Health Perspectives*, *125*(2), 141–148. doi:10.1289/EHP299 PMID:27323709

Petrova, V. (2018). *EC approves plan for mixed solar/wind tenders in Denmark*. Retrieved October 29, 2018, from https://renewablesnow.com/news/ec-approves-plan-for-mixed-solarwind-tenders-in-denmark-623981/

Power Finance Corporation (PFC). (2016). *The Performance of State Power Utilities for the years 2012-13 to 2014-15*. Retrieved from http://www.pfc.gov.in/Default/ViewFile/?id=1490186954263_Report%20on%20Performance%20of%20State%20Power%20Utilities%202012-13%20to%202014-15.pdf&path=Page

Press Information Bureau. (2017). *India gets Lowest Wind Tariff of Rs. 2.64 per kWh in second Wind Auction of 1000 MW*. Retrieved from http://pib.nic.in/newsite/PrintRelease.aspx?relid=171394

Quitzow, L., Canzler, W., Grundmann, P., Leibenath, M., Moss, T., & Rave, T. (2016). The German Energiewende - What's Happening? Introducing the Special Issue. *Utilities Policy*, *41*, 1–9. doi:10.1016/j.jup.2016.03.002

Ralon, P., Taylor, M., Ilas, A., Diaz-Bone, H., & Kairas, K. P. (2017). Electricity Storage and Renewables: Costs and Markets to 2030. International Renewable Energy Agency, Abu Dhabi. Retrieved from file:///C:/Users/Vasundhara/Downloads/IRENA_Electricity_Storage_Costs_2017.pdf

Schmid, G. (2012). The development of renewable energy power in India: Which policies have been effective? *Energy Policy*, *45*, 317–326. doi:10.1016/j.enpol.2012.02.039

Schuman, S., & Lin, A. (2012). Chinese Renewable Energy Law and its impact on renewable power in China: Progress, Challenges and recommendation for improving implementation. *Energy Policy*, *51*, 89–109. doi:10.1016/j.enpol.2012.06.066

Shereef, R. M., & Khaparde, S. A. (2015). A unified REC market and composite RPO scheme for promotion of renewable energy in India. *International Journal of Sustainable Energy*, *36*(6), 606–618. doi:10.1080/14786451.2015.1075988

Shrimali, G., Trivedi, S., Srinivasan, S., Goel, S., & Nelson, D. (2016). Cost-effective policies for reaching India's 2022 renewable targets. *Renewable Energy*, *93*, 255–268. doi:10.1016/j.renene.2016.02.062

Sinha, A., Shahbaz, M., & Balsalobre, D. (2017). Exploring the relationship between energy usage segregation and environmental degradation in N-11 countries. *Journal of Cleaner Production*, *168*, 1217–1229. doi:10.1016/j.jclepro.2017.09.071

Stokes, L. (2013). The politics of renewable energy policies: The case of feed-in tariffs in Ontario, Canada. *Energy Policy*, *56*, 490–500. doi:10.1016/j.enpol.2013.01.009

Sustainable Development Goal. (n.d.). Retrieved from https://sustainabledevelopment.un.org/sdg7

Tanaka, Y., Chapman, A., Sakurai, S., & Tezuka, T. (2017). Feed-in Tariff Pricing and Social Burden in Japan: Evaluating International Learning through a Policy Transfer Approach. *Social Sciences*, *6*(4), 27. doi:10.3390ocsci6040127

Thakur, T., Deshmukh, S. G., Kaushik, S., & Kulshrestha, M. (2005). Impact assessment of the Electricity Act 2003 on the Indian Power Sector. *Energy Policy*, *33*(9), 1187–1198. doi:10.1016/j.enpol.2003.11.016

United Nations Conference on Trade and Development. (n.d.). Retrieved October 24, 2018, from United Nations Conference on Trade and Development Classifications: http://unctadstat.unctad.org/EN/Classifications.html

United Nations Framework Convention on Climate Change. (2008). *Report of the Conference of Parties on its thirteenth session*. Retrieved from https://unfccc.int/resource/docs/2007/cop13/eng/06a01.pdf

United Nations New York. (2018). *The Sustainable Development Goals Report*. Retrieved from https://unstats.un.org/sdgs/files/report/2018/TheSustainableDevelopmentGoalsReport2018-EN.pdf

Usikalu, M. R. (2009). Health impact of climate change due to combustion of fossil fuel. *International Journal of Physical Sciences*, *4*(13), 880–884.

Wiser, R., Namovicz, C., & Smith, M. G. (2007). The Experience with Renewable Portfolio Standards in the United States. *The Electricity Journal*, *20*(4), 8–20. doi:10.1016/j.tej.2007.03.009

World Energy Council. (2016). *World Energy Resources 2016 Summary*. Retrieved from https://www.worldenergy.org/wp-content/uploads/2016/10/World-Energy-Resources-Full-report-2016.10.03.pdf

Ye, L.-C., Rodrigues, J. F., & Lin, H. X. (2017). Analysis of feed-in tariff policies for solar photovoltaic in China 2011 -2016. *Applied Energy*, *203*, 496–505. doi:10.1016/j.apenergy.2017.06.037

Ye, Q., Jiaqi, L., & Mengye, Z. (2017). *Wind Curtailment in China and Lessons from the United States*. Brookings: Tsinghua Centre for Public Policy. Retrieved from https://www.brookings.edu/wp-content/uploads/2018/03/wind-curtailment-in-china-and-lessons-from-the-united-states.pdf

Zhang, K., Wang, Q., Liang, Q., & Chen, H. (2016). A bibliometric analysis of research on carbon tax from 1989 to 2014. *Renewable & Sustainable Energy Reviews*, *58*, 297–310. doi:10.1016/j.rser.2015.12.089

Zhao, Z.-Y., Chen, Y.-L., & Chang, R.-D. (2016). How to stimulate renewable energy power generation effectively? - China's incentive approaches and lessons. *Renewable Energy*, *92*, 147–156. doi:10.1016/j.renene.2016.02.001

KEY TERMS AND DEFINITIONS

Giga-Watt: Measure of output of a power station, equals 1,000 Mega-watts.

Giga-Watt Hour (GWh): Measures generation output of a power station and equals 1,000,000,000 watt hours.

Grid-Connected RES-E: Refers to the cleaner electricity that is used to supply to a national and/or state or regional grid for consumption.

Mega-Watt: Measure of output of a power station, equals 1,000,000 watts.

RES-E: Refers to the electricity generated from clean energy sources such as photovoltaic, hydro, tidal/wave, wind, geothermal, and renewable biomass.

Solar Energy: Radiant light and heat energy from the sun is harnessed using technology such as solar photovoltaic, solar water heaters, and solar thermal, etc.

Tera-Watt Hours (TWh): Another measure of output of a power station and equals 1000 GWh.

Wind Energy: Electricity generated from harnessing the wind resources using wind turbines.

ENDNOTES

1 Conversion rate of 1 USD = 99 JPY as on 28th Dec 2016

2 Conversion rate of 1 USD = 117 JPY as on 28th Dec 2016

3 Conversion rate of 1 USD = 6.94 Chinese Yuan as of 30th Dec 2016

4 Conversion rate of 1 USD = 7. 80 RMB as of 29th Dec 2006

5 Conversion rate of 1 USD = 6. 29 RMB as of 30th Dec 2011

6 Conversion rate of 1 USD = 63. 87 INR as of 29th Dec 2017

APPENDIX: ANALYSIS AND DISCUSSION QUESTIONS

1. What have been some of the most successful legislative tools/policy measures supporting large scale deployment of grid-connected Renewable Energy Sources based Electricity (RES-E) generation?

   Legislative/ policy support to encourage RES-E have been plenty, and as detailed in the chapter, can either be price based, or quantity based. The study of leading countries (when looked at in terms of growth in RES-E installations) confirms that a combination of Feed-in-Tariffs (FITs) and Renewable Purchase Obligations (RPOs) have proven instrumental to attract investments in RES-E projects. While the FITs have proven as a price guarantee for RES-E generators, RPOs have created demand, and a market, for such clean electricity. However, owing to problems associated with historically high FITs (of proving expensive for the power utilities and final consumers), many countries are now switching to auction/ competitive bidding based FIT realization. RPOs however still remain widely adopted.

2. How will the move to competitive bidding/ auctions impact the growth in RES-E installations?

   Under auctions/ competitive bidding system, generators of RES-E (also most often the investors in the projects) bid to sell the power generated. The lowest tariff bid wins. Thereby, one can argue, that this mechanism forces the generators to sell the power generated at lower rates. While such a process proves beneficial for the power purchasing authorities since it reduces the cost of power purchase (who are most commonly, the local power suppliers), it dampens the returns to generators. In a few countries (like in India), such competitive bidding process have meant very low tariffs, and have resulted in financially infeasible projects. One way to counter this effect (whereby generators lose interest to invest in RES-E projects) is to compliment the process with some other form of guarantee, in interest of the generators. Renewable Purchase Obligations (RPO), in which obligated entities, are mandated to purchase RES-E, provides one such complimenting solution to falling tariffs. Given this, even while many countries are seen switching from fixed preferential FITs to an auction system, RPOs remain valid and applicable.

3. Why is distributed RES-E gaining attention? What purposes does it solve and how?

Distributed generation (DG) refers to electricity generation through small/modular power generating plants that are located close to the consumption point. This is in contrast to Centralized power plants that generate power at a point, and then transmitted over large distances to the consumption point. Centralized power generation, while being in surplus availability, is not able to reach remote areas, due to lack of Transmission and Distribution (T&D) networks. In such cases, DG systems can help provide access to electricity. In many rural areas, especially in the Asian and African sub-continent, RES-E through DG systems is the new answer to end energy poverty. Solar Photovoltaic (Solar PV) based rooftop applications (where solar PV modules are installed on rooftops of residences) are becoming popular, with evidences of success coming from rural Africa. Such applications, not only guarantee access to electricity, but are also clean/green in nature. Policies like Net Metering, that allow sale of excess power generated back to the grid, are helping in increasing solar rooftop adoption. Japan presents a case of declaring substantially high FITs fro excess power generated through the Solar rooftop modules, which has been instrumental to the growth of rooftop adoption in the country.

4. What are some of the problems of increasing RES-E footprint?

While being climate friendly, RES-E suffers from "intermittency"- thereby not providing any certainty around the amount of generation and of its quality. These intermittency issues have given birth to reluctance on part of power suppliers to induct more RES-E into the grid, owing to fears of grid fluctuations. Curtailment, where the RES-E generators are asked to not produce more power, is now becoming a popular practice, in countries with large RES-E presence. Local resistance to higher RES-E installations is another problem, and suffering countries are now devising methods to do "inclusive" RES-E projects – a practice that engages all affected stakeholder in the development of RES-E projects. This way, stakeholders get a voice in the process, and helps in reducing local resistance to a significant extent.

Finally, the lack of electricity storage options, are increasingly dampening investor emotions. Battery backed solutions to store RES-E, is forming the centre of Research and Development in many countries.

5. Given the problems associated with higher penetration of RES-E in the grid, what can be some of the expected regulatory/legislative and research based developments in the near future?

a. Intermittency: Due to irregular and uncertain quantum of injection of RES-E to the connected grid, grid planning (demand and supply of power) is greatly impacted. In view of this, countries are now mandating forecasting and scheduling of RES-E, with the responsibility resting with the RES-E generators. Under this, generators are required to forecast the quantum of RES-E that will be generated, with a respectable degree of accuracy (using past generation records and expected weather data, as prime input variables). This enables the grid managers to plan injection and withdrawal of power accordingly, and avoid surprises.
b. Storage: The inability to store RES-E is one of the major limitations, hampering a wider adoption of RES-E. Energy storage systems, enabling storage of excess generation of RES-E, and making it available for consumption during lean generating hours, provides a suitable answer to the problem. Research is now increasing focusing on developing such storage systems, which can be Mechanical, Electrical, Thermal or Chemical in nature (Evans, Strezov, & Evans, 2012). Of these, currently for the forms of RES-E in wide use (such as wind and solar PV), Chemical storage in form of batteries (Lithium-ion and Lead-acid) are most prevalent. However, higher costs of such batteries have restrained the adoption rates.
c. Decentralized RES-E: Decentralized RES-E/small generating units which are located close to the consumption point are on the rise. However, with this, operation and handling of the transmission network becomes more complicated, due to numerous points of power injection. Upgrades in the transmission networks are thus warranted, such that it does not hamper the efficiency of the grid. Such solutions are presented through Micro-Grids, Virtual Grids and Active distribution networks (Bayod-Ru´jula, 2009).
d. RES-E for agriculture: Sustainable agriculture includes reducing the carbon footprint generated through agricultural processes. Renewable sources such as Solar, Biomass, Geothermal and Wind energy can be used for many farm activities (Bayrakc & Kocar, 2012), and research shows that sustainability concerns are an important concern amongst farmers. However the role played by economic incentives should not be undermined (Brudermann, Reinsberger, Orthofer, Kislinger & Posch, 2013). In the coming years, most research will be focused on aligning the use of cleaner energy forms, to conduct farm activities, while also enhancing crop yields/ farm productivity.

Chapter 4
Scaling up Renewable Energy Investment for Sustainable Development

Abderrahim Assab
European Bank for Reconstruction and Development, UK

EXECUTIVE SUMMARY

Renewable energy generation is a fundamental component of the transition to a low-carbon economy. The world needs to invest up to USD 600 million annually to meet the electricity demand in a sustainable way whereas the current investment level stands at USD 280 billion. Scaling-up the current level of investment requires a larger implication of the private sectors and a different role for the public sector. The challenge lays in the fact that different investors are motivated by a different risk and return profiles. The current chapter presents the trends in renewable energy financial flows and investment vehicles. It looks at the risks associated with the investment in renewable energy and the relevant risk mitigation instruments. Finally, it applies these concepts to the case of the Lake Turkana wind farm in Kenya, a project that faced many challenges and involved more than 15 investors.

BACKGROUND INFORMATION

The global financing needed to meet the Sustainable Development Goals (SDGs) is estimated at USD 3.9 Trillion per year. Today, only USD 1.4 trillion are invested leaving a gap of USD 2.5 trillion globally.

DOI: 10.4018/978-1-5225-8559-6.ch004

Clean energy plays a central role in delivering the transition to a low-carbon and climate resilient economy and investment has to scale up in order to meet a growing global energy demand. In 2017, the global investment in renewable energy reached a record USD 279.8 billion thanks to the increasing competitiveness of solar and wind technologies. The current investment level is however far from the estimated USD 600 billion needed to meet the global growing energy demand.

SETTING THE STAGE

Renewable Energy Financing: Understanding the Risks

Increasing investment in renewable energy will bring substantial opportunities to the global economy and is an essential element to the decarbonisation of the electricity generation sector necessary to reach the targets set by the Paris Agreement. As mentioned previously, the current level should be scaled up to reach at least the double by 2030.

According to the International Renewable Energy Agency (2016), the private sector is supposed to provide the highest share of this financing as the share of public financing is not likely to increase above the current level of 15%.

The involvement of the private sector is necessary to scaling up the financing for renewable energy. When mentioning the private sector, the literature is generally referring to institutional investors: insurance companies, pension funds and sovereign wealth funds that can more than USD 90 trillion in the real economy (OECD, 2016).

The narrative behind contribution of institutional investors is based on the fact that the liability profile of these investors matches the characteristics of renewable energy as an asset. The fact that revenues from renewable energy projects can be stable over a long term period is particularly attractive. At the same time, renewable energy projects need long-term financing.

Corporations are also important players in the financing of renewable energy projects. More and more companies around the world are voluntarily and actively investing in self-generation of renewable energy. This trend is driven by both the decrease in renewable energy generation prices as well as the demand for corporate sustainability among investors and consumers (IRENA, 2018).

Developing markets have the fastest growing energy demand and will require the largest increase in investment (IRENA, 2016). Development Financial Institutions (DFIs), such as the World Bank, historically played a very important role in scaling up renewable energy investment in developing countries. Multilateral Development Banks are banking institutions owned by governments and whose aim is to help the financing of projects in the developing economies. In 2017, Multilateral Development

Figure 1. Risks faced by investors in renewable energy projects
Source: "Instruments to mitigate financial risk in renewable energy investments", Shrimali and Reicher, 2017, Stanford Precourt Institute for Energy.

PHASE	RISK	DESCRIPTION
Financing	Foreign exchange risk	Currency risk due to uncertain currency movements and high cost involved with market based currency hedging solutions.
	Offtaker credit risk	The risk that the buyer/off-taker will not fulfill its contractual obligations. It is a key contributor to the overall credit risk of a power project.
	Quality of renewable energy Projects	The credit rating of the operational renewable energy assets may be low overall, leading to operational assets not meeting investment criteria.
	Lack of instruments for investment	Lack of financial instruments (or pathways) – illiquid or liquid – to invest in renewable energy.
	Low returns compared to expectations	Renewable energy projects not being able to meet the risk-return expectations of investors.
	Limited availability of debt capital	Limited availability of debt capital due to capital market conditions, either domestically or internationally
Completion	Construction risk	Risks related to increase in overall financing cost due to construction related issues – esp. due to delays in construction due to permitting.
	Land acquisition issues	Issues faced in land acquisition, esp. if there is no single window clearance in place, or if the time taken to obtain clearances is high.
	Transmission evacuation	The lack of availability of transmission evacuation infrastructure, and time taken to get the clearances and permitting
Operational	Curtailment issues	Wind developers may face this issue during high wind seasons when higher than expected generation creates oversupply situations as well as congestion
	Contract enforceability risk	Drastic reduction in cost of solar power generation may result in poor contract enforceability in the long-term
Others	Lack of trusted intermediaries	Lack of trusted financial intermediaries may result in new and/or smaller investors staying away from the sector
	Limited understanding of sector	Many investors are not aware of renewable energy sector and, therefore, prefer to make investments in mainstream asset classes
	Regulatory/policy risk	The risks related to uncertainty in availability of incentive schemes, poor implementation of policies and non-uniform policies across states.
	Net metering policies	The net metering policies across states may lack coherency as well as poor implementation.

Banks (MDBs) alone financed USD 9.2 billion in renewable energy projects. Beyond direct financing, MDBs are increasingly positioning themselves as "private sector mobilisation engines", institutions that act as corner stone investors taking some of the most important risks in a project, using blended finance structures, and therefore facilitating the investment from private financing sources in difficult contexts.

Figure 1 presents in detail the risks associated with renewable energy financing. The better these risks are mitigated and protected against, the more attractive renewable energy projects become for investors.

Figure 2 describes the various levels at which risks can be mitigated in renewable energy projects (i) the enabling policies and tools, (ii) the financial risk mitigation instruments, (iii) and the structured finance mechanisms and tools. The next section is going to explore further all these elements with a focus on the structured finance mechanisms and tools.

The Lake Turkana case study discussed in the case study section will dive further into the risk mitigation instruments and the way the various stakeholders mentioned earlier can collaborate to finance a renewable energy project in a developing country.

Renewable Energy Financing Models and Trends

As of the end of 2017, the global cumulative investment in renewable energy since 2004 reached USD 2.9 trillion; in 2017 alone investment reached USD 279 billion (FS-UNEP & BNEF, 2018). Still in 2017, developing economies committed USD 177 billion to renewables against USD 130 billion for developed countries – China alone accounted for USD 126.6 billion.

When looking at types of sources of finance for energy assets, the International Energy Agency (IEA) categorises financing into *Balance sheet financing* and *Project financing*.

Balance sheet financing refers to the financing of assets on company's balance sheet using retained earnings and equity issuances in capital markets; the funds available from this type of financing depend on the business performance and creditworthiness of the entire corporate entity rather than on an individual energy project (IEA, 2018).

Figure 2. Policies, tools, and instruments that reduce barriers and mitigate risks in renewable energy projects
Source: "Unlocking Renewable Energy Investment: The Role of Risk Mitigation and Structured Finance", International Renewable Energy Agency, 2016.

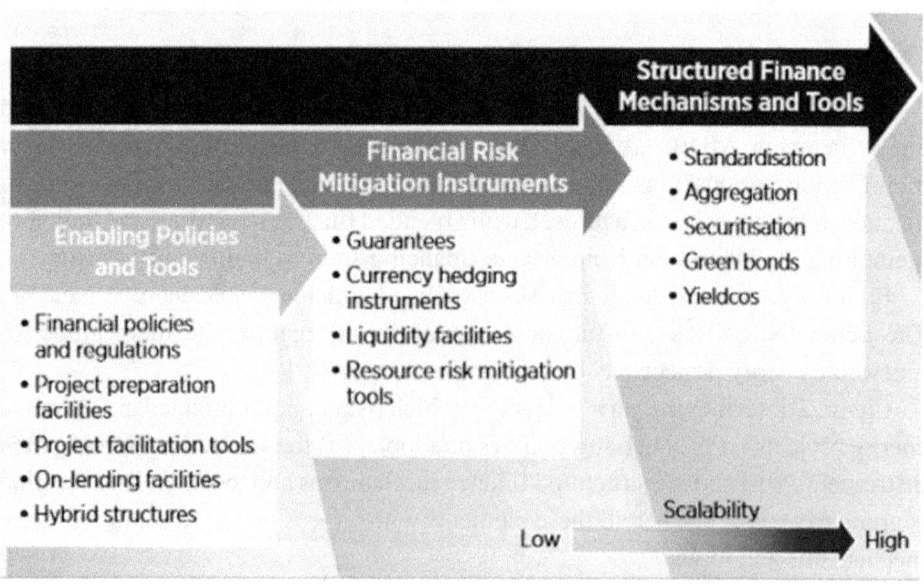

The IEA defines *Project financing* as the financing involving external lenders (commercial banks, development banks, or investment funds) sharing the risks linked to the project with the sponsor of the project. These structures are used in complex large projects when the government policy that underpins the business model is clear (IEA, 2018). One important aspect of the project finance model is that it often involves non-recourse of limited-recourse loans where lenders provide funding on a project's future cash flow and the project company has no limited liability (IEA, 2018).

In 2017, project financing represented 20% of the total financing for renewable energy generation at USD 60 billion. This share is steadily growing in comparison with fossil-fuel electricity generation where the share of project finance is decreasing, the share increased by 13% in 2017. It is also interesting to note the difference in the use of project finance for renewable electricity investment by region. In Europe project finance represented 28%, it represented 28% in the United States, 20% in Latin America, 7% in Developing Asia, and 4% in Africa.

Innovation is an important component in the development of renewable energy. Therefore it is important to look at the financing of the Research and Development activity in clean technologies.

Venture Capital (VC) financing refers to money invested by venture capital and Private Equity (PE) funds in specialist companies developing renewable energy technology (FS-UNEP & BNEF, 2018) with high growth potential in their earliest

Figure 3. Renewable energy financing models
Source: "Global Trends in Renewable Energy Investment", UN Environment and Bloomberg New Energy Finance, 2018.

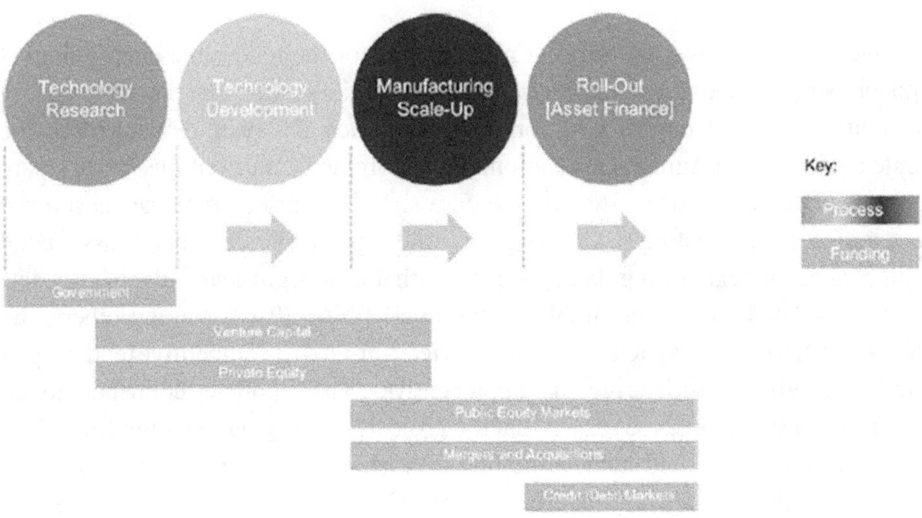

and riskiest stage (IEA, 2017). According to Bloomberg New Energy Finance, in 2017, VC and PE financing dropped by 33% to USD 1.8 billion, the lowest since 2005. VC capital financing has progressively dried as a result of lower returns and higher failure rates, especially for energy-related technologies that need early-stage financing such as biofuels and batteries (Gaddy, et al., 2016). VCs progressively shifted investments from early-stage financing for hardware and materials to later-stage investments and less capital intensive sectors such as software development potentially creating financing gaps within the clean technology value chain (OECD, 2017) – energy storage for instance, a critical area of investment, seems to be underserved by the VC model.

When looking at scaling up the financing to renewable energy, it is important to consider alternative financial sources such as pension funds and insurance companies. Institutional investors are increasingly interested in participating in the renewable energy financing driven both by Environmental Social and Governance (ESG) consideration, as well as a search for yield in the current low-interest rate environment.

In addition to large solar power projects and wind farms, small scale projects are also important. Small-scale renewable energy is an important element in the decarbonisation of the residential and commercial buildings sector, as well as industry. According to Bloomberg New Energy Finance (2018), small-scale solar attracted USD 40 billion in financing representing ~ 20% of the total renewable energy financing flows.

Renewable Energy Investment Barriers

Enabling Policy Environment

The lack of clarity over the support mechanisms for the renewable energy market and industry is one of the strongest impediments to the scaling up renewable energy financing. National targets and commitments, such the ones developed in the context of the Paris Agreements Nationally Determined Contributions (NDCs), are strong signals to investors. Beyond the pure signalling effect, renewables support mechanism are mitigating risks such policy and regulatory risk which is associated with changes in legal or regulatory measures that have significant adverse impacts on project development and implementation (IRENA, 2016). Among others, the IRENA (2016) includes such intervention as feed-in-tariffs, competitive tendering or auction schemes, net-metering, and tax incentives. These policies contribute to the creation of a stable revenue stream that the projects can rely on to secure financing.

Even with a sound and stable policy environment investors might find it difficult to source bankable renewable energy projects. One of the reasons could be the lack of information about opportunities in new renewable energy markets or the lack of the technical capacity to assess renewable energy projects. Project preparation facilities can provide grant financing and technical assistance to accelerate the investment process.

DFIs play a very important role in creating the enabling investment environment in developing countries. DFIs provide financing at competitive cost in environment where local banks are not willing to finance renewable energy projects. Local banks lack expertise to assess renewable energy projects bankability and risk profile. Through co-financing structures and intermediated finance, DFIs can help local lenders build the necessary capacity to autonomously support renewable energy projects.

Financing Mechanisms and Tools

Mazzucuto and Semieniuk (2018) analysed Bloomberg New Energy Finance (BNEF) renewable energy data from 2004 to 2014 using a heuristic risk measure to study the heterogeneity in financial actors in the renewable energy space challenging the simplistic split between public and private actors when looking at renewables financing. According to the study the choice of the financing instruments sets the direction of the renewable energy innovation. If policies favour a subset of financial actors, it indirectly supports a specific technology. It is therefore important to understand the historical trends of financiers and the way they invested in the various renewable technologies in order to orient policies in a strategic way.

More importantly, Mazzucuto and Semienuk (2018) point out the importance of coordinating between the different types of financing in the development phase. The Lake Turkana case study is an exceptional example of coordination between the various stakeholders as well as risk mitigation.

CASE DESCRIPTION

Lake Turkana Wind farm project (Lake Turkana Wind Power, 2018) is the biggest wind farm in Africa (Bloomberg, 2018). It is also the single largest private investment in Kenya's history. The wind farm is located in Marsabit County in Kenya. It is comprised of 365 wind turbines of 850 kW each and 310 MW in total installed capacity capable of supplying 330,000 households. It is estimated that the project helped increase the share of renewable energy in Kenya's electricity mix by up to 10%.

Situated in Kenya, the project initially faced a number of risks ranging from financial and policy risks to technical risks linked to electricity transmission.

Every aspect of the Lake Turkana project reveals the challenge associated with investing in renewable energy in a developing country. The project's development started in 2006 and lasted 8 years to reach financial close in 2014. The construction of the project started in 2015 and took another four years ending in August 2018.

Mazzucuto and Semienuk (2018) pointed out the importance of coordination between the various renewable energy financing actors. The financing of the Lake Turkana project is of particular interest as it involved more than sixteen stakeholders of different nature and used various structures. It is an example of how risk is shared between the different stakeholders in order to deliver the necessary financing for renewable energy in developing countries. It is also a great example the use of blended finance which is defined as "*the strategic use of development finance for the mobilisation of additional finance towards sustainable development in developing countries*" (OECD, 2018).

The project sponsor is a consortium of seven companies that provided the equity proportion of the financing. Aldwych International, who owns an equity stack of 30% is an energy company owning and operating power generation in Africa is backed by Shell Foundation and the Dutch development bank FMO. KP&P the second equity holder in the project is also a Dutch company created to invest in renewable energy in Africa.

Google also has a 12.5% equity stake in the project. Over the years Google became the biggest corporate buyer of renewable energy on the planet with 3 gigawatts of output from renewable energy projects as of April 2018.

Figure 4. Lake Turkana wind farm lenders and equity investors
Source: Infrastructure Journal data, 2018. Authors calculations.

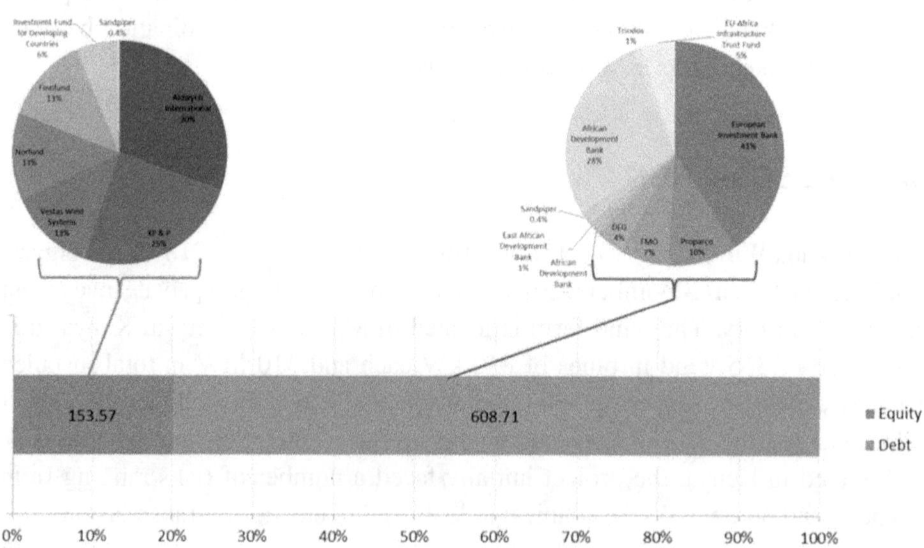

Figure 5. Example of the impact of risk categories on financing costs for wind energy: Belarus
Source: "Belarus: Derisking Renewable Energy Investment", Henrich and Waissbein, 2017.
Note: these estimations are based on interviews with investors.

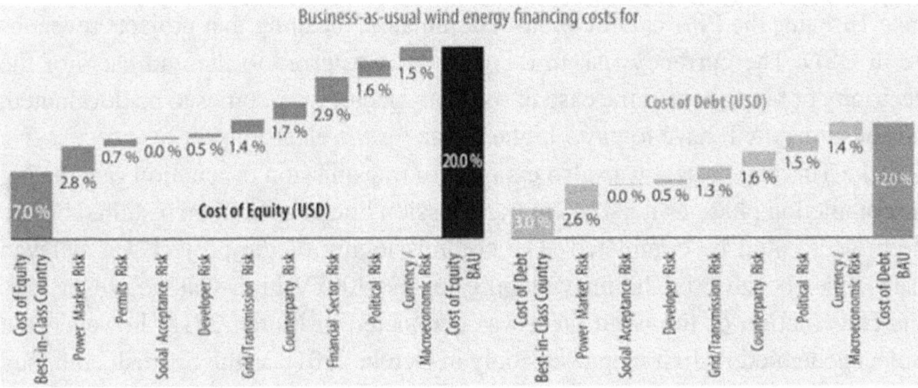

The USD 608 million debt components were very largely provided by various MDBs including the European Investment Bank (EIB), the African Development Bank (ADB), East African Development Bank (EAfDB), Eastern and Southern African Trade and Development Bank. In addition to MDBs, bilateral DFIs participated in the debt financing; this included the Dutch FMO and the German DEG (a subsidy of kfW). Export Agencies also participate in the debt financing such as Proparco as well multilateral infrastructure funds represented by the EU-Africa Infrastructure Trust.

CURRENT CHALLENGES FACING THE ORGANIZATION OR SOCIETY

The management and organizational concerns for late Turkana are linked to the necessity to mitigate the financial and technical risks that can hinder the project's access to financing.

The previous sections introduced policy and regulatory risk and the important of incentive schemes in order to ensure that projects offers attractive and stable cash flows. The policy and regulatory risk are linked to the uncertainty around the availability of incentive schemes. In the case of the Lake Turkana project, the Kenyan government provided a Power Purchase Agreement (PPA) purchase the electricity produced by the wind farm over a period of 20 years. Such a mechanism provides certainty to investors over the future cash flows generated by the project.

The other particularity of the Power Purchase Agreement is that it includes a mechanism to mitigate currency risk. Currency risk rises from a mismatch between (i) the payments made by the electricity consumers in local currency and (ii) the debt repayment by the project developers that are in hard currency. In the case of Lake Turkana, the PPA is US Dollar-denominated, meaning that projects revenues are in USD. The currency risk in this case is transferred to the end-users of the electricity or taxpayers; in the case of the Kenyan currency comes to be devaluated, the consumers will have to pay a higher price for the electricity.

Lake Turkana project was also exposed to transmission evacuation risk during the completion phase as it a 428 km transmission line needed to be installed before the project could be completed. The commissioning of the project was delayed many times because of the unavailability of the 400kV high-voltage power line. The construction of the wind farm was completed in March 2018; however, the project generated its first megawatts only in October 2018 as the Spanish company in charge of completing the transmission line, Grupo Isolux Corsan, encountered financial difficulties. Isolux Corsn delayed signing the contract for the transmission line and later went bankrupt; the national transmission system operator was then in charge of completing the high-voltage line.

Limited information is available on the curtailment risk associated with Lake Turkana; however, the action of the World Bank during the financing phase of the project suggests it is a possibility. Wind project can face curtailment issues during high wind seasons when higher than expected generation creates oversupply situations as well as congestion. The World Bank was initially committed to the project but pulled out in 2012 as it had concerns over the profitability of the project. The World Bank's Director at the time said that the country could end up with a surplus of up to USD 100 million worth of electricity (Business and Financial Times, 2018).

In terms of financing structure, Lake Turkana saw the participation of a very large number of investors through various structures of which two are particularly notable.

The African Development Bank facilitated the participation of the Netherlands based bank Triodos through an A/B loan structure providing a great example of blended finance and cooperation between financiers of different investment profile - Triodos is a global leader in sustainable banking. The AfDB provides an "A" portion of the loan from its own resources (USD 167 million) and a "B" portion where it partners with Triodos (USD 6 million).

It is also interesting to note that the EU-Africa Infrastructure Trust provided a USD 31 million of quasi-equity debt. This is an unsecured and flexible loan regarding the repayment schedule. The repayment relays on the future cash-flows of the project. De facto, when the cash flows are lower than expected the investors get a lower revenue and vice-versa when the cash-flows are higher than expected.

Lake Turkana gives the opportunity to contemplate the positive side effects of the transition to a low-carbon economy and renewable energy investment in particular. According to the Business and Financial Times (2018), the project will reduce Kenya's dependence on imported fossil fuel for electricity generation that currently cost taxpayers USD 150 million per year. The replacement of fossil fuel generation will generate revenue of USD 35 million per year and USD 673 million in tax revenue over the next 20 years. The project also employed 2,500 people during the construction phase and will continue to employ 200 people after the construction.

Figure 6. The global benefits of a decisive shift to a low-carbon economy when compared with business-as-usual

Note: The results cited for the US$26 trillion in direct economic benefits are cumulative for the 2018-2030 period, whereas the other data points reported are for the year 2030.
Source: Garrido, L., Fazekas, D., Pollitt, H., Smith, A., Berg von Linde, M., McGregor, M., and Westphal, M., 2018. Forthcoming. *Major Opportunities for Growth and Climate Action: A Technical Note*. A New Climate Economy contributing paper. To be available at: http://newclimateeconomy.net/content/technical-notes-and-fact-sheets.

SOLUTIONS AND RECOMMENDATIONS

The Lake Turkana project offers both a view of the challenges associated with the financing of renewable energy projects in developing countries as well as the best way to address them.

It is first important to prepare the proper legal and regulatory environment for the project to attract investors. Well-structured Power Purchase Agreements with payments denominated in hard currency can help mitigate the regulatory and the currency risks which are the main concerns of international investors.

With the appropriate regulatory environment and an economically viable project, it is possible to attract a large and diversified pool of investors. In the case of Lake Turkana wind farm the involvement of various Development Finance Institutions facilitated the involvement of other investors through blended finance structures such as the A/B loan structure. Figure-9 describes existing risk mitigation instruments addressing different types of risks at different stages.

More broadly, understanding renewable energy as an asset class is very important to scale the financing and attract large investors such institutional investors. Data on the risk and return profiles of renewable energy projects in different geographies is essential in order to allow these investors to assess the diversification benefits of

Figure 7. Examples of risk mitigation through contractual arrangements and risk mitigation instruments

Source: "Selected Good Practices for Risk Allocation and Mitigation in Infrastructure in APEC Economies", Della Croce, Paula, Assab, 2017.

Note: The blue boxes are guarantee and credit enhancement instruments and orange boxes contact-based risk mitigation measures.

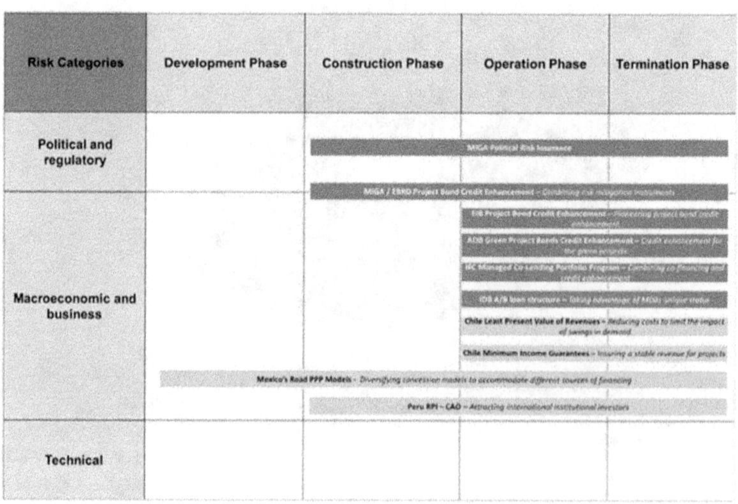

renewable energy investment in the context of a strategic asset allocation (OECD, 2017).

The development of wind technology, and a project such Lake Turkana, is tributary to the improvement in costs of renewable energy technologies. The rapid scaling up of solar power generation is the perfect example of this effect. Governments have an important role to play in supporting renewable energy technologies until these reach the development stage where they become attractive for developers and investors.

The development in the international banking regulation might also play an important role in scaling up renewable energy financing. Indeed, the Taskforce for Climate Financial risk Disclosure (TCFD) issued recommendations to financial institutions urging them to assess climate risk in their portfolios. Similarly, insurance companies and pension funds are exploring the possibility of adjusting the notion of fiduciary duty to address climate risks and environmental and social considerations in general (OECD, 2017).

Finally, the international community's ability to deliver on the commitments made under the Paris Agreement will be a key signal to investors. For instance, the extent to which countries define renewable energy installed capacity targets in their Nationally Determined Contribution (NDCs) will guide the investors looking for opportunities, as well the development financial institutions looking for opportunities to assist their countries of operations.

REFERENCES

Bloomberg. (2018, October 25). *Africa's biggest wind farm nears transmission as line is tested.* Retrieved from https://www.bloomberg.com/news/articles/2018-08-31/africa-s-biggest-wind-farm-nears-transmission-as-line-is-tested

Bloomberg New Energy Finance. (2018). *Clean energy investment trends, 2Q 2018.* BloombergNEF.

Business and Financial Times. (2018). *Lake Turkana wind farm produce first megawatts.* Business and Financial Times.

CEPA. (2014). Policy risk in renewable energy investments in developing countries. London: UK Department of Energy and Climate Change (DECC).

Climate Bonds Initiative. (2018, April 5). *From billions to trillions – Your 2-minute takeaway on our Annual Conference 2018.* Retrieved from Climatebonds.net: https://www.climatebonds.net/2018/04/billions-trillions-%E2%80%93-your-2-minute-takeaway-our-annual-conference-2018

EY. (2015). *The YieldCo structure - Unlocking the value in power generation assets.* EY.

Frangoul, A. (2018, April 05). *Google says it's biggest corporate buyer of renewable energy on the planet.* Retrieved from https://www.cnbc.com/2018/04/05/google-says-its-the-biggest-corporate-buyer-of-renewable-energy-on-the-planet.html

FS-UNEP & BNEF. (2018). *Global trends in renewable energy investment 2018.* Author.

Gaddy, B., Sivaram, V., & O'Sullivan, F. (2016). *Venture capital and cleantech: The wrong model for clean energy innovation.* Author.

Henrich, C., & Waissbein, O. (2017). *Belarus: Derisking renewable energy investment - Selecting public instruments to promote wind energy investment in Belarus*. Minsk: GEF.

IEA. (2017). *Early-stage venture capital for energy innovation: Financing models, trends and implications for policy.* IEA.

IEA. (2018). *World energy investment 2018*. IEA.

IRENA. (2016). *Unlocking renewable energy investment: The role of risk mitigation and structured finance*. International Renewable Energy Agency.

IRENA. (2018). *Corporate Sourcing of Renewables: Market and Industry Trends.* IRENA.

Joint MDBs Report. (2018). *Joint Report on Multilateral Development Banks Climate Finance.* Author.

Kaminker, C., & Stewart, F. (2012). *The role of institutional investors in financing clean energy*. Paris: OECD Publishing.

Konrad, T. (2016, May 13). *The YieldCo boom and bust: The consequences of greed and a return to normalcy.* Retrieved from greentechmedia.com: https://www.greentechmedia.com/articles/read/the-yieldco-boom-and-bust-the-consequences-of-greed#gs.0D19Jx0

Lake Turkana Wind Power. (2018, October 25). Retrieved from https://ltwp.co.ke/: https://ltwp.co.ke/

Mazzucato, M., & Semieniuk, G. (2018). Financing renewable energy: Who is financing what and why it matters. *Technological Forecasting and Social Change*, *127*, 8–22. doi:10.1016/j.techfore.2017.05.021

OECD. (2013). *The role of banks, equity markets and institutional investors in long-term financing for growth and development: Report for G20 leaders*. OECD Publishing.

OECD. (2015). *Mapping channels to mobilise institutional investment in sustainable energy*. Paris: OECD Publishing.

OECD. (2016). *Annual survey of large pension funds and public pension reserve funds: Report on pension funds' long-term investment*. Paris: OECD Publishing.

OECD. (2017a). *Breaking silos: Actions to develop infrastructure as an asset class and address the information gap - an Agenda for G20*. Paris: OCED Publishing.

OECD. (2017b). *Investing in climate, investing in growth*. Paris: OECD.

OECD. (2017c). *Investment governance and the integration of environmental, social and governance factors*. Paris: OECD Publishing.

OECD. (2017d). *Selected good practices for risk allocation and mitigation in infrastructure in APEC economies*. Paris: OECD Publishing.

OECD. (2018). *Making blended finance work for the Sustainable Development Goals*. Paris: OECD Publishing.

Shrimali, G., & Reicher, D. (2017). *Instruments to mitigate financial risk in indian renewable energy investments*. Stanford Precourt Institute for Energy.

The New Climate Economy. (2018). *Unlocking the inclusive growth story of the 21st Century*. Retrieved from: https://newclimateeconomy.report/2018/wp-content/uploads/sites/6/2018/09/NCE_2018_FULL-REPORT.pdf

U.S. Partnership for Renewable Energy Finance. (2017). *Tax credits, tax equity and alternatives to spur clean energy financing*. ACORE.ORG. Retreived from: https://acore.org/wp-content/uploads/2017/12/Tax-Credits-Tax-Equity-for-Clean-Energy-Financing.pdf

Zogo, B., Cedrick, E., & Long, W. (2017). International motivation in renewable energy: A PPP approach. *Energy Procedia*, 229–238.

Zurich, E. T. H. (2017). *The role of multilateral development banks as enabler for new power generation technologies*. Zurich: ETH Zurich.

KEY TERMS AND DEFINITIONS

Bond: A bond is fixed income instrument that represents a loan by an investor to a borrower (typically corporate or government).

Consortium: A group of companies that cooperate and share resources in order to achieve a common objective.

Debt: Debt is an amount of money borrowed by one party from another. Debt is used by many corporations and individuals as a method of making large purchases that they could not afford under normal circumstances. A debt arrangement gives the borrowing party permission to borrow money under the condition that it is to be paid back at a later date, usually with interest.

Equity: Equity is typically referred to as shareholder equity (also known as shareholders' equity) which represents the amount of money that would be returned to a company's shareholders if all of the assets were liquidated and all of the company's debt was paid off.

Institutional Investor: An institutional investor is a nonbank person or organization that trades securities in large enough share quantities or dollar amounts that it qualifies for preferential treatment and lower commissions.

Multilateral Development Bank: A multilateral development bank is an international financial institution chartered by two or more countries for the purpose of encouraging economic development in poorer nations.

Project Sponsor: The sponsor is responsible for securing the financing and overall resource budget approval and owns the opportunities and risks related to the financial outcome of the project.

APPENDIX 1: ACRONYMS

DFI: Development Financial Institution
FIT: Feed-in-Tariff
IEA: International Energy Agency
IRENA: International Renewable Energy Agency
MDB: Multilateral Development Bank
OECD: Organisation for Economic Cooperation and Development
PPA: Power Purchase Agreement
SDG: Sustainable Development Goals

APPENDIX 2: ANALYSIS AND DISCUSSION QUESTIONS

1. What is the main risk for investors when considering renewable energy investments in Sub-Saharan Africa?

   Construction risk is one of the main risks for investors considering renewable energy investments. Once a project has been launched there is a high uncertainty over the number of years needed to complete the project. In the case of the lake Turkana, it took eight years to secure the financing and another four years to complete the project once it started construction.
   Governments have an important role to play in protecting investors against the risk of the project no being complete. This is possible for instance through guarantee mechanisms that allow investors to get compensation from governments or a third party if the project is not completed.

2. What government support mechanism makes a project such as the lake Turkana interesting for investors?

   The lake Turkana wind far benefited from a Power Purchase Agreement (PPA). This is a contractual agreement through which the government of Kenya commits to purchasing the electricity produced by the wind farm over the next 20 years. Such a contract allows investors to be sure that they will receive a steady stream on revenues from the project once it is constructed.
   Beyond the lake Turkana example, PPAs are often considered as one of the most important ways a government can encourage private investment in renewable energy projects.

3. What is the role of Development Financial Institutions in supporting the scaling of renewable energy?

   Development Financial Institutions (DFI) are financial institutions, or banks, owned by states either by one country or by multiple countries. This shareholding structure gives DFIs a special legal status by which they have privileges that other investors don't have. For instance, if a borrower owes money to many lenders including a DFI and he cannot pay back his debt entirely, the DFIs most often has a priority over the other lenders. The preferential status of DFIs allows them to take more risks than usual investors and therefore provide lending during the risky construction period for instance that we mentioned previously. DFIs are increasingly asked by shareholding countries to take those risks that other investors are not able to take in order to efficiently channel financing towards renewable energy projects.

4. What sort of social issues can be caused by renewable energy support schemes?

   Renewable energy support schemes called "feed-in-tariffs" can create inequalities if not properly designed. Subsidies to renewable energy are one of the most popular ways to insure that renewable energy generation is competitive and cheap enough for users. If the subsidies are too high, they can create new oligarchies as the energy producers can get richer through state support.
   As the cost of the renewable energy generation is dropping, subsidies are replaced by auction systems where renewable energy generation competes directly with other types of generation to provide electricity. The question of the "just transition" is at the heart of the transition to the low-carbon economy as it is necessary to make sure that renewable energy policies do not have rebound effects on wealth distribution.

Chapter 5
Microfinance and Polycentric Governance as Strategies for Renewable Energy Deployment in Urban Sub-Saharan Africa

Dumisani Chirambo
https://orcid.org/0000-0003-2310-9896
Brandenburg University of Technology Cottbus – Senftenberg, Germany

EXECUTIVE SUMMARY

Sub-Saharan Africa (SSA) is one of the least electrified regions in the world and also a region that is characterized by poverty and inequality due to high levels of climate change vulnerability. In order to reduce greenhouse gas emissions and facilitate the attainment of the Sustainable Development Goals, SSA policymakers are compelled to devise new innovative strategies and policies to enhance investments in renewable energy technologies (RETs). Accordingly, this chapter provides an assessment of some strategies to accelerate RET deployment and the potential of polycentric governance systems to improve RET deployment. The assessment concluded that even though renewable energy investments through climate finance and microfinance modalities are not at a level sufficient to ensure that universal energy access can be attained in the region, SSA can still accelerate its progress on RET deployment by utilizing nationally determined contributions as instruments to direct South-South aid, trade, and investments into priority renewable energy sub-sectors.

DOI: 10.4018/978-1-5225-8559-6.ch005

BACKGROUND INFORMATION

A lack of access to modern energy and electricity is noted to be one of the factors that exacerbates poverty and inequality in Sub-Saharan Africa (SSA) and makes SSA to be the most physically and economically backward developing and poverty-stricken region in the world (Suberu et al., 2013a). Some studies have estimated that the global electrification rate stands at 80.5% of which the rural electrification rate is 68% and the urban electrification rate is 93.7%. However, SSA is estimated to have an electrification rate of 30.5% of which the rural electrification rate is 14.2% and the urban electrification rate is 59.9%, and North Africa is estimated to have an electrification rate of 99.0 of which the rural electrification rate is 98.4% and the urban electrification rate is 99.6 (Javadi et al., 2013). Some of the reasons that have been cited as the causes of the low rates of electrification in SSA include limited capital investment, lack of technological knowledge on renewable energy development, constricted power generation planning, high cost of electrical energy generation and high transmission losses (Suberu et al., 2013b). This therefore means that renewable energy deployment is constrained by various financial and management aspects which could consist of a combination of internal and external factors. Accordingly, internal factors are constraining factors that lie within a firm's environment and are largely controllable by the firm whilst on the other hand, external constraining factors consist of market and infrastructure aspects that lie outside the direct influence of the firm such as the policy and regulatory environment, infrastructure policy, consumer awareness and technology availability (Abdmouleh et al., 2015).

An estimated 68% of current total anthropogenic greenhouse gas emissions emanate from energy related-activities (Suberu et al., 2013a) hence there is a great threat that increases in energy access and demand in SSA can potentially lead to rises in anthropogenic emissions of greenhouse gases which result in climate change (Lau et al., 2013). Climate change has already been cited as a factor perpetuating poverty, food insecurity, migrations and social conflict in Africa, and could potentially put an additional 100 million people in extreme poverty by 2030 (Hallegatte et al., 2014; Hallegatte et al., 2016). Consequently, the Sustainable Development Goals (SDGs) are calling for state and non-state actors to put in place policies and mechanisms that can reduce the vulnerability of communities to climate related extreme events, and other economic, social and environmental shocks and disasters (UN, 2015). It is in the light of these developments that renewable energy technologies (RETs) are now considered to be vital for climate change mitigation as they can secure energy supply and access whilst contributing to social and economic development and reducing negative impacts of energy supply on the environment and health (Edenhofer, 2011).

RET deployment in SSA has historically been constrained due to a number of reasons. However, there is now potential for RET deployment to be undertaken at an unprecedented rate due the synergies between RET deployment and climate change mitigation. Climate change is already perceived to be, or becoming a priority issue whereby in a number of countries substantial legislation is already in place to facilitate climate change preparedness (mitigation and adaptation) (Tompkins & Amundsen, 2008), and Development Finance Institutions have made progress in designing approaches and business models to drive private investment in climate resilience (Abramskiehn et al., 2017; Trabacchi & Mazza, 2015). Arguably, there have been increases in awareness about climate change and increases in sources of funding and technical assistance for climate change mitigation and renewable energy deployment. For example, South-South Climate Finance (SSCF) modalities and Sino-African aid, investments and trade through the Forum on China–Africa Cooperation (FOCAC) will provide Africa with new sources of investments and technologies to enhance RET deployment and climate change mitigation (Chirambo, 2016a). In addition to this, the international climate change architecture through the Nationally Determined Contributions (NDCs) shows that both developed and developing countries need to revise upwards their climate change mitigation ambitions as the current ambition of NDCs falls far short of reaching any of the goals in the Paris Agreement and the global goal to limit temperature increase to 2°C (Röser et al., 2016; Day et al., 2016). This therefore suggests that more efforts and investments should go into increasing investments into RETs and infrastructure not only to achieve universal energy access but also to limit global temperatures to an optimum level.

On the other hand, a lack of bankable projects and effective financing mechanisms for incentivising investments into Africa have also been cited as aspects that constrain funding for (climate change and renewable energy) projects in Africa (APP, 2014; Trabacchi & Mazza, 2015). This is also compounded by social and economic issues such as poverty, lack of political will and wrong approaches in addressing the energy problem (Gamula, 2013). In other words, RET deployment is being constrained not by a lack of availability of appropriate RETs but by weaknesses in institutions, business models, and legal and regulatory frameworks to support and incentivise RET deployment (Bazilian et al., 2012). Therefore, despite the positive signs that there is more funding and financing available for RET development, many African states might still not be able to access and utilise these mechanisms to increase electrification rates particularly through RETs.

SETTING THE STAGE

Climate change is usually considered as a global development challenge that will make the attainment of the SDGs unattainable more particularly in the developing world.

However, since reducing climate change impacts are priority considerations in most research, business and public policy domains, there has been an impetus by many stakeholders to enhance climate change mitigation and adaptation efforts through technology transfers. In this regard, it can be argued that climate change threats have not only enhanced innovation in RETs to augment climate change mitigation, but they have also created new financing channels and programmes to ramp up investments in enhancing the deployment of RETs across the world. To this effect, some estimates have pointed out that climate change mitigation activities accounted for an average of 93% of climate finance between 2015 and 2016 and that 74% of these mitigation investments were for renewable energy generation (Buchner et al., 2017). To put it differently, the global climate flows for the period 2015 and 2016 amounted to US$410 billion of which an average of US$381.3 billion was channelled towards mitigation efforts and US$282.2 billion was specifically targeted towards renewable energy generation (Buchner et al., 2017). At a disaggregated level, the most popular RETs being deployed for climate change mitigation were solar technologies and wind technologies as estimates for 2016 show that solar power investments amounted to US$133 billion (52 gigawatts (GW)) and wind power investments followed on with US$103 billion (61 GW) (Buchner et al., 2017). However, whilst the presence of new climate change financing mechanisms and programmes are a welcome development in the renewable energy domain, challenges still remain in ensuring that SSA attracts more climate change mitigation projects in the renewable energy sector. For example, a significant proportion of climate finance renewable energy projects are in developed countries such as China, the United States of America and Japan, and out of approximately US$410 billion in climate finance that was mobilised in the 2015 and 2016 period, SSA received an average of US$12 billion in comparison to US$132 billion for East Asia and Pacific, US$107 billion for Western Europe and US$52 billion for America. Accordingly, with SSA receiving only 3% of total global climate finance flows, it can be extrapolated that climate change renewable energy investments are equally meagre. Therefore, since Africa spends about US$8 billion annually on energy investments and infrastructure but the region requires to spend US$41 billion to US$55 billion annually until 2030 to ensure that universal access can be attained (Johnson et al., 2017; Schwerhoff & Sy, 2017), it can be argued that whilst climate finance will have an important role in facilitating the deployment of RETs in many marginalised countries in SSA, the current levels of energy investments through climate finance modalities are not at a level significant enough to singly ensure that universal energy access can be attained in the region.

Various studies have been undertaken in-order to determine how researcher, businesses and policy makers can support efforts to promote the deployment of RETs in various geographic regions. Such studies include de Oliveira (2009) who looked at the implementation of climate change policies at local level. In his analysis, de Oliveira (2009) concluded that subnational governments (cities, states, counties) have taken the lead to tackle climate change even in countries where national governments have been reluctant to support international efforts for controlling the greenhouse gas emissions; and the abilities of subnational governments to deal with climate change may be strengthened by improving the capacity of government mobilisation with other non-governmental local actors to implement voluntary actions. Ramli & Twaha (2015) analysed the implementation of renewable energy Feed-in Tariffs (FITs) in various countries and concluded that Feed-in Tariffs increase renewable energy resource development in pertinent countries as they improve investment security in the renewable energy sector. Müller (2016) assessed the potential of 'share of proceeds' from national and sub-national emission trading schemes and crowdfunding as two unconventional sources of climate finance that would enhance both the predictability and magnitude of multilateral climate finance. Müller (2016) concluded that the international climate change governance architecture needs to revive efforts aimed at developing new sources of finance beyond existing bilateral and multilateral sources. Cobbinah et al. (2015) assessed the implications of rapid urbanisation on the sustainable development of Africa. Cobbinah et al. (2015) discovered that there was very limited meaningful guidance available to African governments, planning institutions, and policy makers regarding how best to address the socio-economic and environmental challenges associated with urbanisation. Park (2016) looked at clean energy entrepreneurship in SSA and concluded that institutional capacity that is necessary to support entrepreneurship, innovation and new venture development for renewable energy development was weak in many SSA countries. Gupta et al. (2015) analysed innovation for social enterprises in Africa. In their conclusion, Gupta et al. (2015) stated that in the absence of a favourable environment for innovation, the impact of social entrepreneurship and social enterprises in Africa is constrained by existing institutional voids and market inefficiencies which drive up transaction costs.

Olstrom (2010) suggested that climate change was a complex multi-level problem that would adequately be addressed by complex multi-level systems such as polycentric governance systems. Similarly, even though the Intergovernmental Panel on Climate Change (IPPC) considers that in the context of Africa, improved climate change governance can best be addressed by replacing hierarchical governance systems with integrated, multilevel, and flexible governance approaches (Niang et al., 2016), there are knowledge gaps on how climate change governance and renewable energy deployment can be enhanced through polycentric governance systems, more particularly in urban areas. Africa has a 37% urban population rate and is a region that is currently experiencing the highest rate of urbanisation at approximately 4% per annum (Gumbo, 2014) hence there is a need for more studies on how Africa's urban population can be provided with energy in an environmentally sustainable manner. Additionally, Africa's urban population was anticipated to increase from about 409 million in 2010 to approximately 1,364 million in 2050 (Scarlat, et al., 2015), and globally the number of people relying on traditional biomass was projected to increase by 10%, from 585 million in 2009 to 645 million in 2030 under a business-as-usual scenario, as the rate of electricity connections would not be able to keep pace with population growth (Glemarec, 2012). Arguably, these statistics show that there are knowledge gaps on effective RET deployment strategies and policies to facilitate universal energy access in Africa. This chapter therefore aims to investigate the strategies that can accelerate RET deployment in Africa and explore the potential of using polycentric governance systems to improve renewable energy deployment, increase climate change mitigation, and reduce climate change vulnerability in Africa's urban environments.

CASE DESCRIPTION

Technology Concerns

The types of RETs to be deployed in a particular region and/or the potential for a country to diversify its energy supply systems are influenced by various factors such as an individual country's demand profiles, energy resource endowments, socio-economic profiles, country development level, and institutional, technical and human resource capacities. With reference to the renewable energy endowments in Africa, in general terms, solar resources are abundant everywhere; biomass and hydropower potential are more plentiful in the wet, forested central and southern regions; wind resources are of the highest quality in the north, the east, and the southern regions; and geothermal energy is concentrated along the Great Rift Valley (IRENA, 2015). Accordingly, the efforts to harness such endowments have culminated in solar photovoltaic having a cumulative installed capacity of 1,334 megawatt (MW) at the end of 2014 (more than ten times larger than in 2009 (127 MW)); Concentrated Solar Power (CSP) at an installed capacity of just over 180 MW, wind power installed capacity at a total of 2,462 MW at the end of 2014; hydropower at 28,000 MW of installed capacity at the end of 2014; and as of 2014 there was 606 MW of geothermal capacity installed in Africa, of which 579 MW was in Kenya (IRENA, 2015).

With the advent of growing climate change threats and the need to increase the deployment of RETs to mitigate climate change, NDCs will arguably become significant policy tools to simultaneously promote the deployment of RETs and augment global efforts to mitigate climate change. NDCs are climate change policy documents that national governments develop in-order to highlight their domestic priorities, capabilities, and responsibilities with regard to climate change. As of December 2015, it was estimated that approximately 156 NDCs were submitted to the United Nations Framework Convention on Climate Change (UNFCCC). However, even though many countries took the initiative to produce NDCs and demonstrate their commitment to mitigate and adapt to climate change, the current ambition of NDCs falls far short of reaching any of the goals in the Paris Agreement and the global goal to limit temperature increase to 2°C (Röser et al.,2016; Day et al., 2016). Consequently, even if all countries fully implemented their NDC pledges, global temperatures would increase by an estimated 2.7°C and thereby fall short of the desired 2°C and 1.5°C goals (Röser et al., 2016). However, since the deployment of renewable energy is integral to facilitating climate change mitigation and reducing greenhouse gas emissions from the energy sector, it can be anticipated that more funding, technology transfer, and research and development will be directed towards the renewable energy sector so that the climate change mitigation commitments in NDCs are ramped-up and more RETs are deployed globally. Africa might therefore stand to benefit from this scenario as African countries can choose to increase their renewable energy commitments as presented in their NDCs and receive financial and technical support from the global community to achieve this.

Historically, financial and technical assistance for climate change mitigation, adaptation, capacity building, and research and development to Africa has usually been provided by western countries or the Global North countries through bilateral agreements or contributions to multilateral development banks such as the World Bank. For example, the World Bank has over the past decades initiated numerous programmes aimed at reducing various barriers to RET deployment in Africa. Such programmes include the Scaling Solar Programme which aims to help developing countries to develop utility-scale solar power plants through the provision of advice and auxiliary services to enable developing countries to reduce risk perceptions and transaction costs related to their proposed solar projects. The Scaling Solar Programme has been credited with providing Zambia with an additional 76 MW of solar power and Ethiopia will add another 200-700 MW of solar power to its grid through this Programme (World Bank, 2019a). Similarly, the World Bank's Energy Sector Management Assistance Program (ESMAP) aims to help low and middle-income countries to reduce poverty and boost growth through environmentally sustainable energy solutions. Through ESMAP, World Bank financing for mini grid and off-grid solutions has increased from an average of less than US$200 million per year between FY2014-17 to US$600 million in FY2018 and in 2018 alone ESMAP supported the mobilisation of US$301.3 million through the Green Climate Fund (GCF) (World Bank, 2019b).

However, there are signals to indicate that the global development paradigms for trade, aid and investments are transforming since emerging economies (e.g. Brazil, China, Russia, India, etc.) are now providing significantly larger amounts of trade, aid and investments to African countries to facilitate their economic growth ambitions. For example, investment commitments in Africa by these emerging financiers jumped from less than US$1 billion per year before 2004 to US$8 billion in 2006, and by 2012 this had exceeded US$20 billion (Ubi, 2014). China has also emerged as Africa's largest trading partner as between 2003 and 2011, Foreign Direct Investment (FDI) from China to Africa increased thirty-fold, from US$491 million to US$14.7 billion (Ubi, 2014) and as of end-2013, China had more Outward Direct Investment (ODI) in Africa (US$26 billion) than in the United States of America (US$22 billion) (Chen et al., 2015). Consequently, some contemporary development analysts like Ha et al. (2016) and Yu (2014) are of the opinion that emerging economies such as China have the potential to become major contributors of climate finance to other developing countries through SSCF modalities. According to Ha et al. (2016) SSCF takes four major forms: i) developing countries' contributions to established multilateral funds; ii) bilateral initiatives; iii) new Southern-led international organisations like the New Development Bank and the Asian Infrastructure Investment Bank; and iv) private sector investments. It is therefore conceivable that if the concept of SSCF is embraced and developed, it can greatly benefit Africa as it can enable Africa to leverage climate finance, aid and investments from the Global North with those from the Global South, hence this can ultimately increase funding and financing for renewable energy projects in Africa.

Additionally, China has reinforced its commitment to assist Africa's socio-economic development aspirations by establishing the FOCAC as an official forum and consulting mechanism between China and the 53 African nations that it has diplomatic ties with. Accordingly through the FOCAC Johannesburg Action Plan (2016-2018) China committed US$60 billion of assistance to Africa in the form of grants, loans, export credits, development funds, and scholarships and training for Africans (FOCAC, 2016; Onishi, 2015). Through this arrangement, Sino-African cooperation will focus on i) political cooperation; ii) economic cooperation (i.e. agriculture and food security, industry partnering and industrial capacity cooperation, infrastructure development, energy and natural resources, tourism, investment and economic cooperation, trade, and finance); iii) social development cooperation (i.e. medical care and public health, education and human resources development, science and technology cooperation and knowledge sharing, and environmental protection and tackling climate change); iv) cultural cooperation and people-to-people exchanges; and v) security cooperation (FOCAC, 2016). It can therefore be seen that within this framework, there are opportunities for African countries to receive technical and financial support for the deployment of renewable energy directly through energy sector infrastructure development initiatives and indirectly through projects in other sectors such as health and education that require electricity as a facilitator. In the energy sector, the strengthened relations between China and various African countries have potential to significantly accelerate the pace of electrification in Africa and cement the prominence to which Chinese actors have in Africa's energy sectors. As it stands, approximately 30% of new capacity additions in SSA's energy sector in 2010-15 were through Chinese companies operating as the main contractors and Chinese contractors have built or are contracted to build 17 GW of generation capacity in SSA from 2010 to 2020, equivalent to 10% of existing installed capacity in the region. All in all, over the period 2010 to 2020, a total of 120 million people will gain access to electricity through the power grid, enabled by grid development and increasing power generation capacity, of which Chinese contractors are responsible for 30% (IEA, 2016). Additionally, within this portfolio, it can also be argued that the development of renewable energy is being prioritised as RET based projects account for 56% of total capacity added by Chinese projects between 2010 and 2020, including 49% from hydropower. Some projects implemented through Chinese actors include the 400 MW Bui hydro dam (Ghana) which was built by Sinohydro at a cost of US$790 million. In this project, partial funding was provided by China Exim Bank (i.e. a concessional loan of US$263.5 million and a buyer's credit of US$298.5 million). In Ethiopia, some Chinese promoted projects include the 100 MW Fincha hydropower project by Gezhouba, the 300 MW Tekeze River hydropower project by Sinohydro, the 50 MW Rappie waste-to-energy project by China National Electric Engineering Company (CNEEC); and the 51 MW Adama wind farm constructed by HydroChina Corporation (IEA, 2016). Since these projects are also of different scales and encompass different technologies they can potentially reduce the risk perceptions that some investors have when considering using novel technologies in SSA. In addition to these investments in grid-connected energy supply systems, Chinese investors and state actors have also been involved in facilitating off-grid electrification where for example, in 2014 Chinese Aid to Rwanda included the delivery of solar kits to provide electricity access to 2,000 villagers in remote areas (IEA, 2016).

Arguably, SSA in the post 2015 development era now has the NDC framework and strengthened Sino-African cooperation mechanisms as two new modalities to promote technology transfer and climate change mitigation, thereby creating synergies between these two mechanisms could enable drastic transitions in SSA's renewable energy sector as never seen before. More importantly, since some SSA country NDCs such as that for Malawi explicitly provide conditional and unconditional targets related to increasing the deployment of solar water heaters, solar photovoltaic systems, hydro power, energy saving cookstoves, etc. (GoM, 2015), it can be argued that NDCs are providing guidance on how renewable energy investments can be undertaken in some countries to avoid an unbalanced focus on one particular RET type and NDCs can also provide guidance on renewable energy sector investment priorities for which Sino-African cooperation mechanisms and emerging economy renewable energy focused trade, aid and investments can focus on. These approaches can therefore ensure that urban and peri-urban poor households will have increased chances of accessing solar kits and other off-grid RETs to enhance their livelihoods.

Management and Organizational Concerns

Unlike other world regions, small-scale entrepreneurial ventures are one of the main sources of livelihood in African communities (Shahidullah & Haque, 2014); and Africa has a larger number of its population involved in the informal sectors (i.e. micro and small businesses account for 90% of all firms and make up about 25% of Gross Domestic Product in SSA) (Soubeiga & Strauss, 2013). Not surprisingly, the SDGs are calling for policymakers to develop policies that promote entrepreneurship and encourage the formalisation of micro, small, and medium sized enterprises (MSMEs), including through access to financial services (UN, 2015). However, some studies have pointed out that existing capital markets and regulatory systems are poorly adapted to the realities of the informal economy and the needs of those who depend on it (Brown & McGranahan, 2016; IIED, 2016; APP, 2014). Consequently, limited access to finance and market information are widely cited as main contributors to the limited impact that African entrepreneurs and the informal sector have on renewable energy deployment and job creation (Amorós & Bosma, 2014). However, it might be argued that creating stronger synergies between microfinance and entrepreneurship-led off-grid electrification models can potentially enable policymakers to improve electrification rates in communities whilst creating new employment and livelihood opportunities for people in the informal sector.

Policymakers have various options available to them to facilitate the enhanced deployment, affordability and adoption of RETs and the services they provide. In the case of SSA, even though off-grid systems are often cheaper and quicker to deploy than large centralised infrastructure as they require less investment costs and regulatory approvals, their use is minimal because traditional financing intermediaries, such as the multilateral development banks, are less able to finance small-scale projects directly, given the higher transaction costs (Soanes et al., 2017). Consequently, in countries like Tanzania, between 2009/10 and 2016/17 the government of Tanzania allocated nearly US$2 billion to energy access, of which US$40 million – or 2% – was targeted to off-grid energy projects (Kaijage et al., 2017). However, there are now new prospects that by using entrepreneurship-led business models such as the Avon model, the utilisation of off-grid electrification models can be increased since local job creation will be happening in tandem with increasing electrification rates and this can also empower marginalised sectors of the society such as women and the youth. With the Avon model, people within the communities are trained to become door-to-door salespersons or direct salespersons of renewable energy products and services (Amankwah-Amoah, 2015). For example, Solar Sisters uses the Avon model to recruit and train women on renewable energy products and services thereby providing opportunities for women without gainful employment to start their own social enterprises in their communities (Solar Sisters, 2019). The Avon model may be considered to be successful as in the case of Solar Sisters, the organisation is recorded to have started with two female entrepreneurs in 2010 and by 2015 the number of its female solar entrepreneurs from Uganda, Tanzania and Nigeria had reached 1,200 (IRENA, 2015). More importantly, an investment of US$1 in a Solar Sister entrepreneur is reported to generate more than US$46 of economic benefit in the first year, through income for the entrepreneurs, reinvestment and customers' avoided kerosene expenses (IRENA, 2015).

Since access to finance is a major limiting factor for business and entrepreneurship activities, the need to establish sustainable financing mechanisms to complement off-grid solutions such as the Avon model are required so that the financing requirements of the entrepreneurs can be met easily. In this regard, microfinance may then be considered as an essential complement to off-grid RET solutions. Microfinance is regarded as a dynamic development tool that can empower Base of Pyramid (BOP) entrepreneurs and reduce (rural and urban) poverty, as in contrast to conventional development aid, microfinance involves and often even focuses on the informal sector (including those in slum areas) and may be an alternative to macro-economic solutions that are often used in development aid programmes (Mutisya & Yarime, 2014). Various microfinance programmes have also been used to increase RET deployment and climate change mitigation through the provision of various RET products and credit facilities (Hogarth, 2012; Asif & Barua, 2011). Microfinance is also used to enhance climate change adaptation and reduce climate change vulnerability through the provision of savings and insurance products for marginalised people and communities (Hallegatte, 2014; Chirambo, 2016b; Marincioni et al., 2013). However, unlike other world regions, Africa has a low rate of financial inclusion and this arguably limits the impact of microfinance on socio-economic development in the region. According to Aga & Peria (2014), in SSA an average of only 24% of the population has an account with a formal financial institution (in contrast to 55% of adults in East Asia, 35% in Eastern Europe, 39% in Latin America, and 33% in South Asia), subsequently leading to reduced female empowerment and productive investment in the region. Other reports indicate that most Africans are disconnected from the formal financial system as over two-thirds of the adult population of Africa – 316 million people – have no bank account and fewer than 15% of the adults in Africa have accounts at a formal financial institution (APP, 2014). This therefore means that the operations of informal enterprises, entrepreneurs, and MSMEs are constrained as their use of formal savings and credit institutions is low. Since the provision of financial services and credit plays a vital role in unlocking productivity gains and expanding markets and enabling people to invest in their homes (APP, 2014). It can be argued that the low penetration of formal financial services, savings and credit facilities in Africa has reduced impact of microfinance on RET deployment and has made microfinance to have a marginal impact on RET deployment in comparison to other developing regions.

According to Scrieciu et al. (2015), measures for coping with climate change may be enhanced by taking a polycentric approach at multiple local, regional, and national levels involving different stakeholders rather than focusing on single top-down policies. Ostrom (2008; 2009; 2010), stated that polycentric governance is characterised by an organisational structure where multiple independent actors mutually order their relationships with one another under a general system of rules. As opposed to monocentric hierarchies, polycentric systems can function independently or form an interdependent system of relations in-order to address collective action problems, free-rider problems and social dilemmas such as climate change mitigation and adaptation (Ostrom, 2008; 2009; 2010). There are now great opportunities for polycentric governance approaches to be used in the microfinance sector due to the emergence of mobile money technologies which can facilitate branchless banking, loan disbursement and loan repayment through mobile phone based platforms (Musoni, 2016). Polycentric governance systems can also be seen to be applicable in the microfinance sector where a microfinance institution can increase its outreach at minimal cost by having agents or affiliates (i.e. third party organisations be it other microfinance institution or any institution not necessarily in the microfinance sector) who assist it with monitoring of loans, communicating with beneficiaries, and training beneficiaries on the use of mobile money services (SOO, 2016). Since microfinance institutions already have some established business models for improving the affordability of renewable energy products, it can be argued that the use of polycentric governance systems in the microfinance sector can improve financial inclusion and access to capital thereby enabling more African entrepreneurs to engage into the renewable energy sector. Figure 1 depicts a polycentric governance approach for a microfinance institution whereby a microfinance institution utilises agents and affiliates in various constituencies of a country in-order to increase its outreach and enhance financial inclusion.

Figure 1. Microfinance beneficiary led development framework (M-BLDF)
Source: Author

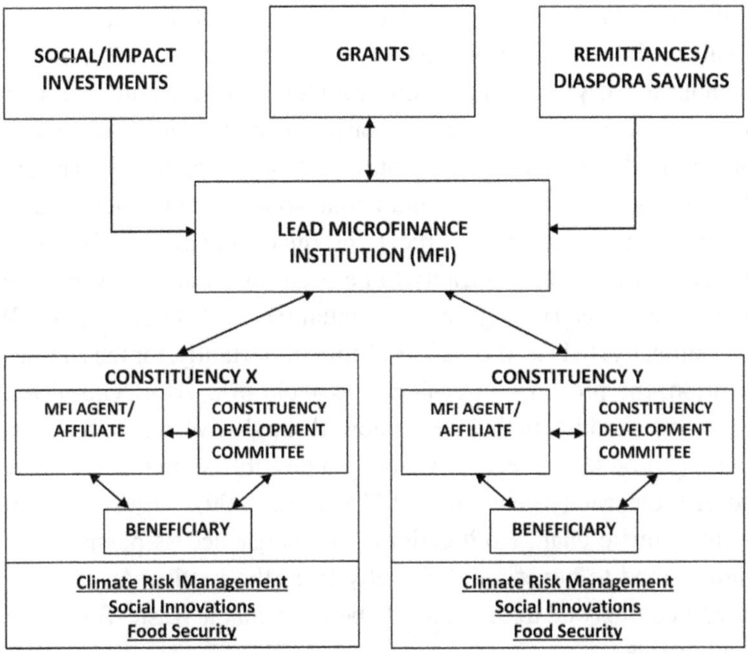

SOCIETAL CHALLENGES FOR URBAN ELECTRIFICATION WITH RETS

Many cities and urban centres in Africa are characterised by a general lack sound municipal solid waste management due to challenges in the collection, transportation, disposal, and treatment of waste (Gumbo, 2014). Consequently, such cities and urban centres are noted to have environmental problems as the waste is indiscriminately dumped at uncontrolled dumpsites and on river banks, street corners, passageways, and the backs of buildings (Gumbo, 2014). However, municipal solid waste has a significant potential to be used for producing biogas which can be converted into electrical energy. Furthermore, converting waste into biogas and electricity can minimise the levels of greenhouse gas emissions into the atmosphere thereby mitigating climate change. A study by Scarlat et al. (2015) on the energy potential of municipal solid waste from African urban areas indicated that the theoretical potential of electricity produced from waste and its contribution to electricity consumption would reach 62.5 TWh in 2012 and 122.2 TWh in 2025, in comparison with a total electricity consumption of 661.5 TWh at continental level in 2010. Scarlat et al. (2015) further estimated Africa's theoretical electricity production from biogas from all generated waste to reach 27.5 TWh in 2012 and 51.5 TWh in 2025. Waste-To-Energy technologies can therefore provide numerous economic and environmental win-win scenarios due to their potential to convert materials that can perpetuate environmental pollution into a vital energy resource.

For example, in Ivory Coast, the Abidjan Municipal Solid Waste-To-Energy Project is an illustration of how waste is being converted into a renewable energy source in-order to satisfy the energy requirements of the urban population. The project was designed to manage 200,000 tons of municipal solid waste per year and the biogas derived from the waste to be used as fuel to produce annually 25 GWh of renewable electricity (i.e. 4.5 GWh of electricity to be used for on-site consumption and the 20.5 GWh of excess electricity generated annually is sold at €20.00 per MWh to the state utility company) (SITRADE, 2009). More importantly, for the first seven years of its operation, the project was designed to avoid 502,318 tCO_2e of greenhouse gas emissions by avoiding methane emissions through diverting organic waste from landfills and by producing electricity from a renewable source that displaces fossil fuel-based grid electricity generation (SITRADE, 2009). Arguably, from a waste management, climate change mitigation, and energy access perspective, African urban centres stand to benefit significantly from the replication of such projects in other urban centres due to the multiple benefits that a Waste-To-Energy project like this can instigate.

Developing and scaling-up Waste-To-Energy projects in many African urban centres faces many challenges. Many municipal governments have institutional capacity limitations in that the municipal governments lack the human resources capacity (i.e. lack technical and planning expertise in waste management) and financial resources to enable the municipalities to invest in appropriate technologies and waste management processes (Lemaire & Kerr, 2016). Africa is noted to have a large portion of its population in the informal sector (i.e. the informal economy employs 66% of the population in SSA and 45% of the population in the Middle East and North Africa) and to have high rates of youth unemployment (i.e. some reports have indicated that developing countries will need to create 1 billion more jobs by 2030 to match their growing young populations) (IIED, 2016). However, there is potential for the Waste-To-Energy sector to provide avenues for employment and income generation for a majority of Africa's urban dwellers and subsequently reduce the logistical constraints that municipalities have in collecting and sorting out waste. The 2012 Informal Economy Monitoring Study (IEMS) on urban informal workers discovered that street vendors made up between 2% and 24% of all urban informal employment and approximately three quarters of the IEMS sample relied on waste picking as their main source of income (IIED, 2016). In Bogota, the city's waste pickers are officially recognised for their work (i.e. waste pickers are paid as public service providers) and included in the city's recycling and waste management processes thereby creating an income for the waste pickers and making the amalgamation of waste for recycling, re-use, and energy production easier (IIED, 2016). Similarly, many African cities can emulate the initiative taken by Bogota and utilise people in the informal economy to be used for improving waste management in cities. Arguably, African cities can be a step closer to developing efficient Waste-To-Energy projects and facilities by engaging with people in the informal economy and waste pickers, and supporting their organisation into recognised formal associations so that their concerns can easily be integrated in urban development plans and they can be utilised for waste management processes.

Similarly, in SSA the use of solar energy has not been adequately exploited compared to its naturally endowed potential (Suberu et al., 2013a). Even though across Africa, solar energy potential is unevenly distributed across the region, the intensity of solar radiation (varying between 4,000 and 7,000 Wh/m^2) in the SSA region is potentially capable of sustaining the needed domestic electricity to power home appliances (i.e. in the entire region of Africa, solar irradiation is higher than the typical daily domestic load requirement of 2,324 Wh/m^2 in urban and rural areas (Suberu et al., 2013a). The amount of solar energy in Africa is arguably higher than that of most developed countries in the Global North. However, in developed countries, solar energy is significantly being utilised due to the implementation of renewable energy policies that encompass incentives and flexible energy pricing mechanisms. For example, in the United States of America, Denmark and Australia, net metering is being used as a mechanism to encourage households to utilise RETs as the system allows households and institutions to reduce both their carbon foot print and electricity bills (Poullikkas, 2013; Tan & Chow, 2016). A FIT scheme provides a guaranteed premium price to the renewable electricity producer and puts an obligation on the grid operators to purchase the generated electricity output whilst net metering is an electricity policy which allows utility customers to offset some or all of their electricity use with self produced electricity from RETs (Poullikkas, 2013).

In the absence of a sound regulatory framework to facilitate net metering as the case is in many African cities, it can be argued that an alternative approach is for large urban developments to generate renewable energy for their consumption and also for neighbouring households. For example, in Kenya, the Two Rivers Development is a private development that has an integral 12 MW sub-station (Centrum, 2016). The Two Rivers Development comprises of a mixed-use commercial development that includes a mega shopping mall, a hotel, office blocks and apartments. The development was designed to have its power supply from a solar and diesel power supply system that will generate and supply 12 MW of electricity to the development. Since most African cities are growing due to urbanisation and population growth, it is conceivable that the model used for the Two Rivers Development can be modified so that shopping malls and large urban developments can be designed to have integrated RETs to supply renewable energy to the mall/development as well as surrounding dwellings. Such an approach can prove to be more effective and practical as most African countries already have legislation and regulations that governs self generation but do not yet have legislation and regulations governing net metering.

SOLUTIONS AND RECOMMENDATIONS

Even though Africa has lagged in the development of its energy infrastructure, there is now a great potential that Africa can improve its energy infrastructure and achieve sustainable energy transitions in its urban environments due to heightened concerns to increase climate change mitigation which are then motivating more actors to support and facilitate RET deployment in developing countries. There is no single identified factor that can singularly have a significant positive effect on the successful integration of RETs into Africa's energy infrastructure but rather the association of benefits from supportive measures that determine the extent to which a RET may or may not be successfully exploited (Abdmouleh et al., 2015). Therefore, for SSA to achieve its ambitions of achieving universal energy access by 2030 various nations have to design and implement different strategies and innovative policies in keeping with the different aid, trade and investment needs of particular actors.

Achieving universal access as desired by SDG 7 will require significant policy innovation. This follows that whilst SDG 7 has the ambition for the world to achieve universal energy access by 2030 (UN, 2015), some studies project that universal electrification can be achieved by 2050 by countries with at-least 60% current electrification and that countries below this level can achieve at-least 80% electrification (Sanoh et al., 2014). SSA with an estimated electrification rate of 30.5% will therefore have insurmountable challenges in-order to attain SDG 7 or be close to attaining this Goal. However, with more policy innovation in the climate change and renewable energy domains, this challenge can partly be overcome. Policy innovation can be defined as changes to existing policy practices which introduce non-status quo, if not necessarily entirely novel, policy components or combinations of components which often result in new outcomes (Howlett, 2014). As it stands, a lack of innovation in climate change and sustainable development policies is arguably limiting the rate of RET deployment in Africa. For example, integrated policies and strategies for climate change and sustainable development which are used in both developed and developing countries are only managing to shape perceptions, enable governments to meet international obligations and provide communication tools to outline a vision for society, but are failing to build momentum for political commitment which can shape governmental agendas or major political decisions in favour of more effective climate change action (Casado-Asensio & Steurer, 2014; Nagoda, 2015; Park & Brooks, 2015).

On the other hand, a lack of policy innovation also limits the amount of funds that are spent on Africa's energy infrastructure. For example, SSA countries on average spend less than 3% of their GDP on their energy sector with about 75% of these spending used as operating costs (Gujba et al., 2012). This translates into a mere 0.75% of GDP being used in expanding energy infrastructure (Gujba et al., 2012). Therefore, SSA spending in the energy sector is much less than required, even for conventional development and energy use patterns. However, Africa still has great prospects to attract investments into its energy sector and leapfrog towards expansion of the energy sector and a low carbon trajectory. According to APP (2014), while returns to investment in secure assets in Organisation of Economic Cooperation and Development (OECD) countries have been close to zero, the potential social and economic returns to investment in Africa's infrastructure are very high. The APP has therefore reported that the returns to foreign investors in energy projects in SSA are higher than in any other developing region and investments in Africa's cross-border power transmission projects have exceptionally high returns, typically paying for themselves in less than a year (APP, 2014). Since the returns on investment in energy projects are already attractive, what is therefore plausible is that improving the implementation of policies and embracing innovative approaches for renewable energy deployment could be vital in attracting investments and spending into the renewable energy sector.

Many national governments and local authorities in SSA are unable to improve the provision of environmental services and social services due to various resource constraints. National governments and local authorities in SSA therefore need to adopt new flexible approaches and innovations for renewable energy deployment and climate change management so that they can be able to leverage their capacities with those of other non-state actors such as microfinance institutions and impact investors. Africa needs to increase its electrification rates which makes the adoption of technological innovations in generating renewable energy from sources such as waste imperative in the continent, and this is a predicament that demands not only immediate and concerted efforts from many stakeholders, but also a complete shift in approaches and techniques of problem solving (Gumbo, 2014). Consequently, polycentric governance systems could provide the alternative approaches to enhance climate change management and renewable energy deployment in Africa. In this paper, the use of polycentric governance systems in the microfinance sector was illustrated. Even though Africa has a low penetration rate of microfinance services in comparison to other developing regions, renewable energy deployment can still benefit from microfinance services as mobile money technologies and (mobile phone based) renewable energy pay as you go solar home systems are increasing in popularity and becoming more affordable (Musoni, 2016; Azuri Technologies, 2015). Additionally, microfinance can also augment entrepreneurship-led off-grid electrification models such as the Avon model, which is increasing in popularity due to its focus on empowering women. This suggests that some of the factors that were constraining the growth of the microfinance sector and the integration of microfinance for renewable energy applications in the region are gradually being eradicated.

Since Africa is experiencing rapid population growth and urbanisation leading to the growth in shopping malls and housing developments, large urban developments and shopping centres can now be utilised to enhance RET deployment in urban environments. The Two Rivers Development in Nairobi was highlighted as a development that generates renewable energy for self consumption. It was then hypothesised that since the rates or urbanisation and urban growth in Africa are on the increase, many shopping centres and urban developments may also incorporate renewable energy systems to generate and supply electricity for their own consumption and surrounding areas. The potential of waste for generating renewable energy was also illustrated through an elaboration of the Abidjan Municipal Solid Waste-To-Energy Project which provides 20.5 GWh of electricity to the electricity grid.

Population growth, urbanisation, and economic development are expected to produce increasing quantities of waste that can either be considered as an overburden on existing waste management systems or a valuable resource for renewable energy generation and employment. Africa's urban population grew from 56 million in 1960 to 409 million in 2010 and it is projected to further increase to 672 million in 2025 and 1,364 million in 2050 (Scarlat et al., 2015). Africa's rates of urbanisation could reach 47% in 2025 and 62% in 2050 (Scarlat et al., 2015), thereby putting strain on existing waste management systems and energy infrastructure. However, the presence of landfill sites in the vicinity of urban areas therefore creates good opportunities for producing electricity from landfill sites and creating jobs for people in the informal sector as waste pickers and recyclers.

Even though there are various Waste-To-Energy projects in Africa which can be used as models and benchmarks for Waste-To-Energy project implementation and operation in SSA, the replication and wide-scale deployment of Waste-To-Energy projects is minimal. Some factors constraining the replication of such projects in other urban environments include the technical and financial limitations which many local governments and municipalities have. However, the successful implementation of Waste-To-Energy projects in Abidjan (Ivory Coast) and eThekwini Municipality (South Africa) (Gumbo, 2014; SITRADE, 2009) suggests that Waste-To-Energy projects become more investable and viable in countries that have FITs and mechanisms to provide additional revenue streams through carbon credits. This is similar to the notions of Amatayakul & Berndes (2012) who stated that power purchase guarantee rather than guarantee of access to sell carbon credits (carbon credit purchase guarantee) is more decisive to attract investments in renewable energy projects. With the prices of carbon credits decreasing in value and the long term operations of multilateral carbon crediting mechanisms like the Clean Development Mechanism (CDM) being uncertain, some would be quick to assume that the avenues for accessing additional revenue streams for such projects to be diminishing. However, other bilateral carbon crediting systems like Japan's Joint Crediting Mechanism (JCM) could provide a sustainable means for generating carbon credits and enhancing the diffusion of advanced low carbon technologies, products, systems, services and infrastructures in developing countries (PMR, 2013).The JCM is similar to the CDM in that it intends to contribute to the ultimate objective of the UNFCCC by facilitating global action for greenhouse gas emissions reductions or removals hence provides carbon credits for renewable energy projects and other projects with potential to mitigate the emissions of greenhouse gases. The JCM can therefore be utilised by countries in SSA to increase the deployment of Waste-To-Energy projects.

SSA is in a fortunate predicament as the post-2015 development era now has new global development paradigms and governance structures that can promote inclusive growth and accelerate progress towards achieving universal energy access. As elaborated earlier, the NDC framework, FOCAC framework and other emerging economy led trade, aid and investments to Africa all have the potential to mobilise new forms of technical and financial support to assist African countries with their climate change and renewable energy challenges. However, the implementation of all these frameworks simultaneously might also create new implementation challenges hence it could be argued that there is now a new onus on African governments to develop additional policies and institutional frameworks to monitor and evaluate the impacts to which these frameworks are having on different development parameters and institute changes where the impacts are negative. It therefore still remains to be seen as to which African countries, local governments and municipalities will take the lead in successfully utilising these frameworks to improve climate change management and renewable energy access for the benefit of the environment and urban dwellers. However, once the pacesetters and successful African countries, local governments and municipalities have been identified, the next step could then be to use South-South cooperation modalities to enable the pacesetters to offer good practice guidance and capacity building to other countries and non-state actors on how they can successfully simultaneously utilise the NDC framework, FOCAC framework and other emerging economy led trade, aid and investments for their benefit with due regard to their socio-economic and political contexts.

ACKNOWLEDGMENT

An abridged version of this paper was presented at the Strategies for Sustainable Energy Transitions in Urban Sub-Saharan Africa International Conference (Accra, 2017). The author is grateful to the organisers of the Urban Sub-Saharan Africa International Conference (i.e. the Supporting Sub-Saharan Africa's Municipalities with Sustainable Energy Transitions (SAMSET) Project) for providing a platform for this research to be disseminated.

REFERENCES

Abdmouleh, Z., Alammari, R. A. M., & Gastli, A. (2015). Review of policies encouraging renewable energy integration & best practices. *Renewable & Sustainable Energy Reviews*, *45*, 249–262.

Abramskiehn, D., Hallmeyer, K., Trabacchi, C., Escalante, D., Netto, M., Cabrera, M., & Vasa, A. (2017). *Supporting National Development Banks to Drive Investment in Nationally Determined Contributions of Brazil, Mexico and Chile*. Washington, DC: Inter-American Development Bank.

Aga, G. A., & Peria, M. S. M. (2014). *International remittances and financial inclusion in Sub-Saharan Africa*. Policy Research Working Paper 6991. Washington, DC: World Bank Group (Development Research Group).

Amankwah-Amoah, J. (2015). Solar energy in Sub-Saharan Africa: The challenges and opportunities of technological leapfrogging. *Thunderbird International Business Review*, *57*(1), 15–31.

Amatayakul, W., & Berndes, G. (2012). Determining factor for the development of CDM biomass power projects. *Energy for Sustainable Development*, *16*(2), 197–203.

Amorós, J. E., & Bosma, N. (2014). *Global entrepreneurship monitor 2013 global report: Fifteen years of assessing entrepreneurship across the globe*. London: GEM.

APP (Africa Progress Panel). (2014). *Finance and banking in Africa: Extracts from the Africa progress report 2014*. Geneva: Africa Progress Panel.

Asif, M., & Barua, D. (2011). Salient features of the Grameen Shakti renewable energy program. *Renewable & Sustainable Energy Reviews*, *15*(9), 5063–5067.

Azuri Technologies. (2015). *Pay as you go solar systems*. Retrieved from http://www.azuri-technologies.com/info-hub/azuri-quad

Bazilian, M., Nussbaumer, P., Rogner, H., Brew-Hammond, A., Foster, V., Pachauri, S., ... Kammen, D. (2012). Energy access scenarios to 2030 for the power sector in sub-Saharan Africa. *Utilities Policy*, *20*(1), 1–16.

Brown, D., & McGranahan, G. (2016). The urban informal economy, local inclusion and achieving a global green transformation. *Habitat International*, *53*, 97–105.

Buchner, B. K., Oliver, P., Wang, X., Carswell, C., Meattle, C., & Mazza, F. (2017). *Global Landscape of Climate Finance 2017*. Venice: Climate Policy Initiative.

Casado-Asensio, J., & Steurer, R. (2014). Integrated strategies on sustainable development, climate change mitigation and adaptation in Western Europe: Communication rather than coordination. *Journal of Public Policy*, *34*(03), 437–473.

Centrum. (2016). *Centum to build 12MW plants for Two Rivers Mall*. Retrieved from http://www.businessdailyafrica.com/Corporate-News/Centum-to-build-12MW-plants-for-Two-Rivers-Mall/539550-3047216-5a3rylz/index.html

Chen, W., Dollar, D., & Tang, H. (2015). *Why is China investing in Africa? Evidence from the firm level*. Washington, DC: The Brookings Institution.

Chirambo, D. (2016a). Moving past the rhetoric: Policy considerations that can make Sino-African relations to improve Africa's climate change resilience and the attainment of the sustainable development goals. *Advances in Climate Change Research*, *7*(4), 253–263.

Chirambo, D. (2016b). Integrating microfinance, climate finance and climate change adaptation: A Sub-Saharan Africa perspective. In Climate Change Adaptation, Resilience and Hazards (pp. 195-207). Heidelberg, Germany: Springer.

Cobbinah, P. B., Erdiaw-Kwasie, M. O., & Amoateng, P. (2015). Africa's urbanisation: Implications for sustainable development. *Cities (London, England)*, *47*, 62–72.

Day, T., Röser, F., & Kurdziel, M. (2016). *Conditionality of Intended Nationally Determined Contributions (INDCs)*. Berlin: International Partnership on Mitigation and MRV and New Climate Institute.

de Oliveira, J. A. P. (2009). The implementation of climate change related policies at the subnational level: An analysis of three countries. *Habitat International*, *33*(3), 253–259.

Edenhofer, O., Pichs-Madruga, R., Sokona, Y., Seyboth, K., Matschoss, P., Kadner, S., … von Stechow, C. (Eds.). (2011). IPCC special report on renewable energy sources and climate change mitigation. Cambridge, UK: Cambridge University Press.

FOCAC (The Forum on China-Africa Co-operation). (2016). *The Forum on China-Africa Cooperation Johannesburg Action Plan (2016-2018)*. Retrieved from http://www.focac.org/eng/ltda/dwjbzjjhys_1/t1327961.htm

Gamula, G., Hui, L., & Peng, W. (2013). Development of renewable energy technologies in Malawi. *International Journal of Renewable Energy Technology*, *2*(2), 44–52.

Glemarec, Y. (2012). Financing off-grid sustainable energy access for the poor. *Energy Policy*, *47*(S1), 87–93.

GoM (Government of Malawi). (2015). *Republic of Malawi Intended Nationally Determined Contribution*. Environmental Affairs Department, Lilongwe. Retrieved from http://www4.unfccc.int/submissions/INDC/Published%20Documents/Malawi/1/MALAWI%20INDC%20SUBMITTED%20TO%20UNFCCC%20REV%20pdf.pdf

Gujba, H., Thorne, S., Mulugetta, Y., Rai, K., & Sokona, Y. (2012). Financing low carbon energy access in Africa. *Energy Policy*, *47*(S1), 71–78.

Gumbo, T. (2014). Scaling up sustainable renewable energy generation from municipal solid waste in the African continent: Lessons from eThekwini, South Africa. *Journal of Sustainable Development*, *12*(1), 46–62.

Gupta, S., Beninger, S., & Ganesh, J. (2015). A hybrid approach to innovation by social enterprises: Lessons from Africa. *Social Enterprise Journal*, *11*(1), 89–112.

Ha, S., Hale, T., & Ogden, P. (2016). Climate finance in and between developing countries: An emerging opportunity to build on. *Global Policy*, *7*(1), 102–108.

Hallegatte, S., Bangalore, M., Bonzanigo, L., Fay, M., Kane, T., Narloch, U., ... Vogt-Schilb, A. (2016). *Shock waves: Managing the impacts of climate change on poverty. Climate Change and Development Series*. Washington, DC: World Bank.

Hallegatte, S., Bangalore, M., Bonzanigo, L., Fay, M., Narloch, U., & Rozenberg, J. (2014). *Climate change and poverty: An analytical framework*. Policy Research Working Paper 7126. Washington, DC: World Bank Group.

Hogarth, J. R. (2012). Promoting diffusion of solar lanterns through microfinance and carbon finance: A case study of FINCA-Uganda's solar loan programme. *Energy for Sustainable Development*, *16*(4), 430–438.

Howlett, M. (2014). Why are policy innovations rare and so often negative? Blame avoidance and problem denial in climate change policy-making. *Global Environmental Change*, *29*, 395–403.

IEA (International Energy Agency). (2016). *Boosting the Power Sector in Sub-Saharan Africa: China's involvement*. Paris: IEA.

IIED. (2016). *Informality and inclusive green growth: Evidence from 'The biggest private sector' event 2016*. London: IIED.

IRENA (International Renewable Energy Agency). (2015). *Roadmap for a renewable energy future*. Abu Dhabi: IRENA.

Javadi, F. S., Rismanchi, B., Sarraf, M., Afshar, O., Saidur, R., Ping, H., & Rahim, A. (2013). Global policy of rural electrification. *Renewable & Sustainable Energy Reviews*, *19*(C), 402–416.

Johnson, O., Muhoza, C., Osano, P., Senyagwa, J., & Kartha, S. (2017). *Catalysing investment in sustainable energy infrastructure in Africa: overcoming financial and non-financial constraints*. Stockholm Environment Institute Working Paper No. 2017-03. Nairobi: Stockholm Environment Institute.

Kaijage, E., Nyagawa, S., Best, S., Cosmas, R., Temba, S., Mtwanga, B., & Mahanga, N. (2017). *Money is Power: Tracking finance flows for decentralised energy access in Tanzania*. IIED Working Paper. London: The International Institute for Environment and Development (IIED).

Lau, L. C., Lee, K. T., & Mohamed, A. R. (2012). Global warming mitigation and renewable energy policy development from the Kyoto Protocol to the Copenhagen Accord – a comment. *Renewable & Sustainable Energy Reviews*, *16*(7), 5280–5284.

Lemaire, X., & Kerr, D. (2016). *Waste Management: Innovative solutions for municipalities. SAMSET Policy Brief.* London: UCL Energy Institute.

Marincioni, F., Appiotti, F., Pusceddu, A., & Byrne, K. (2013). Enhancing resistance and resilience to disasters with microfinance: Parallels with ecological trophic systems. *International Journal of Disaster Risk Reduction*, *4*, 52–62.

Müller, B. (2016). *Two unconventional options to enhance multilateral climate finance: Shares of proceeds and crowdfunding*. Oxford, UK: European Capacity Building Initiative.

Musoni. (2016). *Musoni microfinance system*. Retrieved from http://musonisystem.com/

Mutisya, E., & Yarime, M. (2014). *Microcredit for the development of the bottom of the pyramid segment: Impact of access to financial services on microcredit clients, institutions and urban sustainability*. Working Paper Series N° 199. Tunis: African Development Bank.

Nagoda, S. (2015). New discourses but same old development approaches? Climate change adaptation policies, chronic food insecurity and development interventions in northwestern Nepal. *Global Environmental Change, 35*, 570–579.

Niang, I., Ruppel, O. C., Abdrabo, M. A., Essel, A., Lennard, C., Padgham, J., & Urquhart, P. (2014). Africa. In *Climate Change 2014: Impacts, adaptation, and vulnerability. Part B: Regional Aspects. Contribution of Working Group II to the Fifth Assessment Report of the Intergovernmental Panel on Climate Change*. Cambridge, UK: Cambridge University Press.

Onishi, N. (2015). *China pledges $60 billion to aid Africa's development*. Retrieved from https://www.nytimes.com/2015/12/05/world/africa/china-pledges-60-billion-to-aid-africas-development.html

Ostrom, E. (2008). *Polycentric systems as one approach for solving collective-action problems*. doi:10.2139srn.1304697

Ostrom, E. (2009). *A polycentric approach for coping with climate change*. World Bank Policy Research Working Paper Series 5095. Washington, DC: World Bank.

Ostrom, E. (2010). Beyond markets and states: Polycentric governance of complex economic systems. *The American Economic Review, 100*(3), 1–33.

Park, J. (2016). Clean energy entrepreneurship in Sub-Saharan Africa. *Global Entrepreneurship: Past, Present & Future, 29*, 257-277.

Park, J., & Brooks, C. (2015). Local flood resiliency in an era of global climate change: Understanding the multi-sectoral policy dimensions. *Vermont Journal of Environmental Law, 17*(2), 160–177.

PMR (Partnership for Market Readiness). (2013). *Recent development of the Joint Crediting Mechanism (JCM)/Bilateral Offset Credit Mechanism (BOCM)*. Retrieved from https://www.thepmr.org/system/files/documents/JCM.pdf

Poullikkas, A. (2013). A comparative assessment of net metering and feed in tariff schemes for residential PV systems. *Sustainable Energy Technologies and Assessments, 3*, 1–8.

Ramli, M. A. M., & Twaha, S. (2015). Analysis of renewable energy feed-in tariffs in selected regions of the globe: Lessons for Saudi Arabia. *Renewable & Sustainable Energy Reviews, 45*(C), 649–661.

Röser, F., Day, T., & Kurdziel, M. (2016). *After Paris: What is next for Intended Nationally Determined Contributions (INDCs)?* Berlin: International Partnership on Mitigation and MRV and New Climate Institute.

Sanoh, A., Kocaman, A., Kocal, S., Sherpa, S., & Modi, V. (2014). The economics of clean energy resource development and grid interconnection in Africa. *Renewable Energy*, *62*(C), 598–609.

Scarlat, N., Motola, V., Dallemand, J. F., Monforti-Ferrario, F., & Mofor, L. (2015). Evaluation of energy potential of Municipal Solid Waste from African urban areas. *Renewable & Sustainable Energy Reviews*, *50*, 1269–1286.

Schwerhoff, G., & Sy, M. (2017). Financing renewable energy in Africa – Key challenge of the sustainable development goals. *Renewable & Sustainable Energy Reviews*, *75*, 393–401.

Scrieciu, S. S., Barker, T., & Ackerman, F. (2015). Pushing the boundaries of climate economics: Critical issues to consider in climate policy analysis. *Ecological Economics*, *85*, 155–165.

Shahidullah, A. K. M., & Haque, C. E. (2014). Environmental orientation of small enterprises: Can microcredit-assisted microenterprises be "green"? *Sustainability*, *6*(6), 3232–3251.

SITRADE (Société Ivoirienne de Traitement des Déchets). (2009). *Abidjan Municipal Solid Waste-To-Energy Project. CDM Project Design Document for the Abidjan Municipal Solid Waste-To-Energy Project (Version 5)*. SITRADE.

Soanes, M., Rai, N., Steele, P., Shakya, C., & Macgregor, J. (2017). *Delivering real change: Getting international climate finance to the local level*. IIED Working Paper. London: The International Institute for Environment and Development (IIED).

Solar Sisters. (2019). *About solar sisters*. Retrieved from https://solarsister.org/about-us/

SOO (Seeds of Opportunity). (2016). *Beneficiary led climate change resilience building programme*. Retrieved from http://seedsofopportunity.org/activities-programmes/beneficiary-led-climate-change-resilience-building-programme-blccrbp/

Soubeiga, S., & Strauss, J. (2013). *Financial sector policy note: Financing small and medium-sized businesses in Burkina Faso*. Washington, DC: World Bank.

Suberu, M., Mustafa, M., & Bashir, N. (2013a). Status of renewable energy consumption and developmental challenges in Sub-Sahara Africa. *Renewable & Sustainable Energy Reviews*, *27*, 453–463.

Suberu, M. Y., Mustafa, M. W., Bashir, N., Muhamad, N. A., & Mokhtar, A. S. (2013b). Power sector renewable energy integration for expanding access to electricity in sub-Saharan Africa. *Renewable & Sustainable Energy Reviews*, *25*, 630–642.

Tan, R. H. G., & Chow, T. L. (2016). A comparative study of feed in tariff and net metering for UCSI University North Wing Campus with 100kW solar photovoltaic system. *Energy Procedia*, *100*, 86–91.

Tompkins, E. L., & Amundsen, H. (2008). Perceptions of the effectiveness of the United Nations Framework Convention on Climate Change in advancing national action on climate change. *Environmental Science & Policy*, *11*(1), 1–13.

Trabacchi, C., & Mazza, F. (2015). *Emerging solutions to drive private investment in climate resilience*. Venice: Climate Policy Initiative.

Ubi, E. N. (2014). Foreign aid and development in Sino-African relations. *Journal of Developing Societies*, *30*(3), 243–272.

UN (United Nations). (2015). *Transforming our world: the 2030 Agenda for Sustainable Development*. New York: United Nations.

World Bank. (2019a). *Scaling solar active engagements*. Retrieved from https://www.scalingsolar.org/

World Bank. (2019b). *Energy sector management assistance program annual report 2018*. Washington, DC: World Bank.

Yu, Y. (2014). Climate finance, Africa and China's role. *Africa East-Asian Affairs*, *1*, 36–57.

ADDITIONAL READING

Barnard, S. (2015). *Climate finance for cities: How can international climate funds best support low-carbon and climate resilient urban development? ODI Paper 419*. London: Overseas Development Institute.

Chirambo, D. (2018). Towards the achievement of SDG 7 in sub-Saharan Africa: Creating synergies between Power Africa, Sustainable Energy for All and climate finance in-order to achieve universal energy access before 2030. *Journal of Renewable & Sustainable Energy Reviews*, *94*(C), 600–608.

Griffiths, S. (2017). A review and assessment of energy policy in the Middle East and North Africa region. *Energy Policy*, *102*, 249–269.

Sweerts, B., Longa, F. D., & van der Zwaan, B. (2019). Financial de-risking to unlock Africa's renewable energy potential. *Renewable & Sustainable Energy Reviews*, *102*, 75–82.

Taylor, A., & Peter, C. (2014). *Strengthening climate resilience in African cities: A framework for working with informality*. Rondebosch: African Centre for Cities.

APPENDIX: ANALYSIS AND DISCUSSION QUESTIONS

1. What are some of the issues constraining the deployment of RETs in African cities?

   A combination of various factors constrain the deployment of RETs to different extents in different countries/contexts. However, in general terms the lack of affordable finance for project developers, utilities and consumers, and the lack of technological knowledge on renewable energy development have led to the underinvestment and underutilisation of RETs in African cities.

   The socioeconomic context of African cities also means that there are a significant number of small businesses operating and creating jobs, and a majority of people living in informal settlements. However, the needs of informal businesses and people in informal settlements are not always addressed by local and central government structures and as such this limits the potential of informal businesses and people in informal settlements to be successful stakeholders in energy transition paradigms and policies.

2. What institutional and technological innovations can improve the deployment of RETs in African cities?

   The use of hierarchical governance systems for local and central government limits the actions to which state and non-state actors can implement climate change mitigation projects, and arguably this also affects the deployment of renewable energy climate change mitigation projects. Consequently, by promoting the use of polycentric governance systems to enhance climate change governance, there could arguably be improvements in the implementation of renewable energy climate change mitigation projects in some cases.

   Various types of RETs such as photovoltaic technologies and Waste-To-Energy technologies exist in Africa and globally. The presence of applicable technologies however does not always lead to their use or adoption in various contexts. The issue therefore becomes how can the private sector and non-state actors help state actors to acquire or utilise such technologies. This can be done by the private sector and non-state actors facilitating the provision of finance and technical assistance to state actors and/or the private sector and non-state actors being the owners or service providers using such technologies so that consumers can still make use of the technologies or the services that they provide.

3. Which non-state actors can support climate change mitigation and the deployment of RETs in African cities?

   Developing countries have remarkable differences in their innovation capacities, natural resource endowments; GDP per capita and potential to manufacture RETs. Consequently, there is now potential to harness South-South Climate Cooperation modalities and South-South Climate Finance (SSCF) modalities for enhanced RET deployment where well-off developing countries such as China, Brazil and South Korea can provide support to other not well to do developing countries, or the Least Developed Countries to support them in their ambitions to mitigate climate change through RETs. Since SSCF modalities occur through new Southern-led international organisations like the BRICS bank/New Development Bank and the Asian Infrastructure Investment Bank; and private sector investments, the new entrants into the RET deployment sector are therefore developing country renewable energy project developers and developing country financial services providers that are seeking partnerships and markets in other developing countries.

4. What policies would you recommend to international policymaker and state actors to be implemented in-order to further enhance the deployment of RETs in African cities?

   RETs are in constant competition with various forms of fossil fuel energy supply systems. Therefore, new fossil fuel energy supply systems will continue to be developed as the demand for energy increases and economic concerns override environmental concerns.

   Some policies that may be tried or suggested include for local governments to have quotas on how much renewable energy should be used in their jurisdiction. For now, the quotas are usually applied to energy utilities rather than the cities themselves. Arguably, if the cities have renewable energy quotas, they could have more power to direct new urban developments to utilise or create on site renewable energy systems.

   African cities need to attract local and international RET project developers to increase energy access. In some cases this is not possible due to the mandates to which central and local governments have usually had. However, with more calls for decentralised governance and polycentric governance systems, local governments should be striving to have more autonomy on attracting local and international RET project developers. International policymakers can therefore be supporting central governments to device mechanisms that can improve the collaboration between RET project developers and local government structures.

Chapter 6
The Boulder Breakup

Kate Clark
Western Colorado University, USA

Keriann F. Conroy
Western Colorado University, USA

EXECUTIVE SUMMARY

The City of Boulder, Colorado has for 10 years attempted to break up with its electric utility, Xcel Energy, in favor of forming its own municipal utility. Environmental proponents of the separation argue that a democratically accountable, local utility would be better suited to achieve Boulder's ambitious environmental and climate action goals. However, other environmentalists disagree and instead argue that Xcel Energy is a willing and capable environmental partner. This case examines this conflict in order to illustrate a divide in Boulder's environmental community, which mirrors a divide in the larger environmental movement, between structural environmentalists on the one hand and neoliberal environmentalists on the other. The case offers a review of the theoretical work that informs these conflicting perspectives. Finally, it analyzes structural and neoliberal sentiments expressed in the opinion pages of the city's newspaper in order to demonstrate how they intervene and shape Colorado electricity politics.

BACKGROUND INFORMATION

On July 17, 2014, the City of Boulder, Colorado moved to acquire the local assets of electric utility giant, Xcel Energy, a taking unprecedented in the history of electricity in the U.S. It would be the first time a city expropriated a private company's property for the purpose of addressing the environmental costs and climate changing effects of coal-powered electricity. It was the most confrontational moment in a decade-long conflict between the city and its investor-owned electricity provider.

DOI: 10.4018/978-1-5225-8559-6.ch006

Over the prior ten years, environmentalists from Boulder had engaged the company repeatedly, appealing to Xcel Energy to move beyond coal in favor of renewable sources of electricity, using tactics ranging from collaborative policymaking to rowdy protest. For its part, Xcel Energy developed a reputation as one of the greenest utilities in the nation, complementing its coal-dominant fuel portfolio with natural gas, and to a lesser extent, renewable sources of energy.

Despite the utility's efforts to adopt a greener fuel portfolio, some environmentalists in Boulder were left unsatisfied. In 2013, 56% of Boulder's electricity was still coming from coal-burning power plants (Xcel Energy, 2014) and 21.6% came from natural gas. In total, over three-quarters of Xcel Energy's portfolio remained tied to fossil fuels. Xcel Energy's commitment to environmental objectives and renewable energy was questioned further when it built Colorado's largest coal-fired power plant in 2009. In seeking permission to charge ratepayers for the cost of the new plant, the utility argued that it would run for 60 years, which meant Colorado would burn coal until at least 2069.

When the City of Boulder announced its intention to separate from its utility to form its own, Boulder's environmental community showed signs of a deep divide. While Boulder environmentalists generally agreed that greenhouse gas emissions associated with electricity production had to be curtailed, the community was split on whether Xcel Energy would be the best partner in achieving the goal. Some activists approached Xcel Energy antagonistically, using messaging and tactics that highlighted the utility's environmental shortcomings and arguing that Xcel Energy was structurally unable to adopt renewable energy as fast as the climate crisis required. They pointed, for example, to the utility's mandate to maximize returns to shareholders and its continuing investments in fossil fuels. In their view, the market was insufficiently competitive. As a result, they argued, the power of consumer demand was severely constrained. These environmentalists argued that the utility lacked motivation to pursue large amounts of renewable energy, and called for substantial change to transform the rules within which electricity utilities like Xcel Energy operate. Beyond renewable energy, these environmentalists were mistrustful of the utility and wanted more decision-making power themselves.

Yet, the sentiments of these *structural* environmentalists did not capture all of Boulder's environmental community. Others celebrated the strides the utility had taken to increase renewable energy in its fuel portfolio. These environmentalists argued that working with the experienced utility was more likely to garner larger, quicker action on climate change. This form of environmentalism, identified as *neoliberal* environmentalism, is rooted in an affinity for market-based solutions to environmental problems. In this view, individual consumers already have substantial power to demand renewable energy and competitive firms are motivated to respond to that demand.

SETTING THE STAGE

This case examines the dichotomy between the neoliberal and structural approaches to environmentalism to better understand diverging strategies within the environmental movement broadly and how they appear within the realm of renewable energy advocacy specifically. The case proceeds as follows. The neoliberal-structural divide in the broader environmental movement is described in order to demonstrate the distinct characteristics of each approach, and their connections to two theories in environmental sociology. Next, the authors describe how environmentalists helped shape renewable electricity policy in Colorado before returning to the conflict in Boulder. Finally, environmental discourse is analyzed to demonstrate how structural and neoliberal arguments were mobilized in favor of and against a breakup between the City of Boulder and Xcel Energy.

First, however, the authors explain why the environmental movement has focused on electricity production as a major target for climate action. The section also introduces the governance of electric utilities and the role of regulatory policy.

Electricity and the Utility Structure

The environmental movement has targeted electricity production as among the most environmentally destructive. In 2016, electricity production in the U.S. accounted for 28% of the nation's climate changing greenhouse gas emissions (it was 34% in 2010). Transportation accounted for 28% of emissions and Industry, Commercial/Residential uses and Agriculture followed at 22%, 11%, and 9% respectively (Environmental Protection Agency, 2018). Extraction of the country's primary source of electricity, coal, involves massive land disruption, particularly when surface mining techniques such as mountaintop removal are utilized. Mercury pollution from coal production costs the U.S. $5.5 billion a year due to negative health impacts, such as retardation and cardiovascular disease, and lost productivity (Epstein et al., 2011, p. 6). Furthermore, the extraction of natural gas – often considered a cleaner substitute for coal – pollutes both the air and ground water, with hydraulic fracturing of natural gas releasing, in some cases, greater concentrations of greenhouse gases than coal (Howarth, Santoro, & Ingraffea, 2011). While renewable sources of electricity, such as wind, solar hydro power, are also associated with environmental problems, the environmental costs of these fuels are far less than nuclear power or fossil-fuel based electricity. Thus, decisions about electricity fuel sourcing have major implications for the environment (EPA, 2012).

As the environmental movement has learned more about the environmental consequences of the electricity sector, movement actors also had to learn about the governance structure of the sector. In the U.S., the vast share of electricity is delivered through one of three types of utilities. Municipal electric utilities (also called munis) are organized at the municipal – or city – scale. Elected or appointed boards, overseen by the community's City Council, manage these not-for-profit utilities, and they serve about 49 million people (Public Power, 2018). They have more independence from state-level public utility commissions and the Federal Energy Regulatory Commission (FERC). The next type of electric utility is the Rural Electric Coop (REC). RECs, which originally grew out of farmer cooperatives aided by the Rural Electrification Administration, are also not-for-profit entities and managed by a board of directors elected by the utility's members. RECs provide electricity to over 42 million people (America's Electric Cooperatives, 2018). Finally, investor-owned utilities, which serve about 220 million people, are for-profit firms owned and managed by shareholders (Edison Electric Institute, 2018).

All three utility models are monopolies. Electricity has historically been considered a natural monopoly, meaning that due to extremely high infrastructure costs, it would be too costly to have multiple and overlapping electricity providers. As a result, a single monopoly utility generally serves a given service territory with no competitors.. Due to concerns of improper, monopolistic behavior – such as overcharging or failing to provide reliable service in the absence of competition – these utilities are designed for, or regulated with, public accountability in mind. For munis and RECs, the ability to vote for the given utility's board of directors (either directly, or in the case of munis, through city council) is intended to provide accountability and serve as a check on poor performance. For investor owned utilities, which are ultimately accountable not to their consumers but to their shareholders, oversight is provided by a state-level public utility commission, whose board is either elected or appointed (in most cases by the state's governor). These dynamics make the electricity sector somewhat distinct from other services exchanged on the market, and compel environmental advocates to look for strategic levers for changing the sector.

In Boulder, Colorado, the conflict introduced above involves replacing an investor-owned utility (IOU), Xcel Energy, with a municipal utility (a muni). This process is called *municipalization.* Like *nationalization* on the federal level, *municipalization* entails transferring privately held assets into public hands, at the municipal level. Xcel Energy is the larger parent company of Colorado-based Public Service Company of Colorado, which is the state's largest electric utility, providing electricity to 1.5 million ratepayers in the state (Bloomberg, 2018). PSCo was founded in 1869 and bought by Xcel Energy in 2001 (Xcel Energy, 2017). Xcel Energy, an electric utility and natural gas company, is based in Minneapolis and operates in Minnesota, Michigan, Wisconsin, North Dakota, South Dakota, New Mexico, Texas and Colorado. It is a vertically integrated electric utility, meaning that its operations include electricity generation, transmission and distribution, as well as the purchase and sale of electricity. Though PSCo is the legal name of the utility's operations in Colorado, it uses the name Xcel Energy as its public face in the state. As such, the utility is referred to as Xcel Energy.

The Colorado Public Utilities Commission (PUC) is the primary state regulator for electric utilities like Xcel Energy. In addition to electric utilities, the PUC regulates a broad array of other utilities, most of them granted monopolies, ranging from intrastate natural gas pipelines to telecommunications utilities and taxis. The PUC falls under the executive branch's Colorado Department of Regulatory Agencies, and its three commissioners are appointed by the Governor's Office. The PUC's mission is to serve "the public interest by effectively regulating utilities and facilities so that the people of Colorado receive safe, reliable, and reasonably-priced services consistent with the economic, environmental and social values of our state" (Department of Regulatory Agencies, 2014).

As the reader will see, while the conflict centered on how to best transition the city's fuel portfolio to renewable energy sources, the case uncovers the complicated social and political landscape of electricity decisions. In the process of fighting over the most strategic way to usher in a renewable energy transition, Boulder environmentalists found themselves fighting over the role of democracy, the functioning of political bodies like the PUC, and the responsiveness of market forces to consumer preference. As a result, Boulder's potential separation with Xcel Energy offers an exceptionally illuminating space to investigate a broader movement divide between *structural environmentalists* on one side and *neoliberal environmentalists* on the other.

The next section describes the structural-neoliberal divide in the broader environmental movement and highlights the distinct worldview that shape these environmentalists' varied approaches to achieving their goals.

Structural vs. Neoliberal: The Environmental Movement Divide

Many environmentalists represent the structural emphasis introduced above. They work toward environmental policy intended to change the social-structural context in which businesses and the public operate (Gottlieb, 2005; van Huijstee et al., 2011, p. 44). More and more, however, environmentalists are attracted to a newer form of environmentalism, which aims to harness the power of individuals to effect change through voluntary consumer choices in a self-regulating market populated by independent, competitive, and responsive firms (Barry, 2005; Rondinelli & London, 2003; Stafford et al., 2000). While *structurally*-oriented environmentalism calls for public policy to address major environmental problems, the *neoliberal* approach calls for market solutions. Figure 1 identifies these two perspectives as falling on a spectrum. That is, an individual or organization may be identified as more or less neoliberal or more or less structural.

Figure 1. The characteristics of neoliberal and structural environmentalists

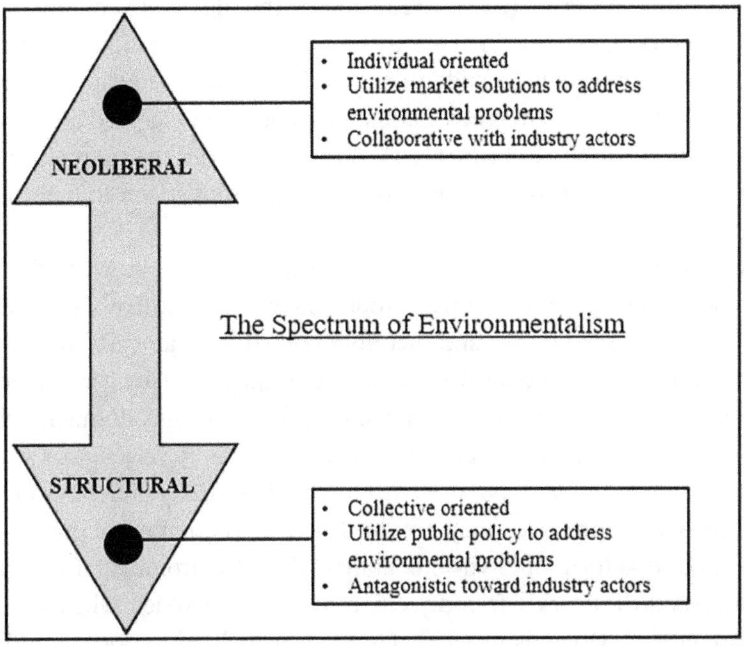

As opposed to the individually focused and relatively atomized view of society held by neoliberals, structuralists conceive of environmental problems as shared and collective. They focus less on the aggregation of individuals' choices and behaviors and instead draw attention to the social fabric (e.g. relationships and structures) that, in their view, shape the range of options available to individuals. For example, 350.org's cofounder Bill McKibben (2012) downplays the importance of individual demand for fossil fuels, writing that, "[Fossil fuel] companies don't simply exist in a world whose hungers they fulfill – they help create the boundaries of that world." He and other structuralists point to concentrated power in industry, or connections between economic and political power, as they seek to understand and explain the causes of environmental problems. Moreover, it is not just the aggregated consequences of *ordinary* individuals that concern the structuralists, but the power of certain individuals (e.g. pro-fossil fuel Senators or CEOs of coal companies), which derives from their unique positions within the social structure. Characteristics of the social structure itself, such as the motivation for profit-making entities to reduce their costs by externalizing those costs onto the environment, are among the targets of these environmentalists.

Yet other structuralists have gone even deeper in their critique of the market and market-friendly approaches, further illustrating the gulf between structural and neoliberal environmentalists. The structurally-focused environmental group Rising Tide North America (2011) argues that, "the idea that economic growth is both desirable and inevitable" is one of the many "false solutions" getting in the way of environmental health. In addition to their skepticism toward the strength of consumer demand and the degree of competitiveness among producers, structural environmentalists adopt more adversarial positions toward their targets in industry and have less faith in market-oriented solutions, in comparison to their neoliberal counterparts.

Conversely, neoliberal environmentalists emphasize the power of individual action in the market. They assert that producers are competitive with one another and accountable to consumers, and therefore consumers can shift markets toward environmentally friendly products and processes if they so desire (Anderson & Leal, 2001, p. 4). The neoliberal approach is a rising and increasingly dominant viewpoint in U.S. environmentalism. In place of structuralists' more adversarial orientation toward industry actors, neoliberal environmentalists seek collaboration with the private firms whose practices they aim to modify. Voluntary and market-friendly programs mark this form of environmentalism. The Environmental Defense Fund (EDF) pioneered the market-friendly environmental approach when it initiated a partnership with McDonalds in 1990. The first of its kind, according to EDF, the partnership substantially reduced waste by, "eliminating more than 300 million pounds of packaging," and provided a model which others in the industry soon replicated (Environmental Defense Fund, 2012). In 2008, one of the oldest U.S. environmental organizations, the Sierra Club, celebrated Earth Day by announcing its partnership with Clorox, a top producer of bleach cleaning products. In exchange for the right to post the Sierra Club's logo on Clorox's *Green Works* brand of bleach cleaning products, a portion of *Green Works* sales would be received by the Sierra Club. The Sierra Club received $1.3 million over the course of the four-year deal with Clorox (Sahagun, 2011). Then executive director of the Club, Carl Pope, explained "Instead of just saying, let's boycott somebody who's making a toxic product, let's find a good product and help people who are trying to help consumers."

Similarly, since 2004, the World Wildlife Fund for Nature (WWF) and the Nature Conservancy have joined Monsanto, Syngenta, Cargill, Shell and BP in the Roundtable on Sustainable Soy, which promotes a 'responsible soy' certification developed by the group. The environmentalists' presence lends the roundtable considerable legitimacy and the relationship has earned WWF millions of dollars from Monsanto, since acquired by Bayer. A former Director of Marketing and Communication at WWF explained that the organization's credo was, "The more money for nature, the better" (van Huijstee, 2011, p. 62). The billion-dollar carbon trading market (World Bank, 2009, p. 13) further reflects the prominence of market approaches to solving environmental problems.

Because of its emphasis on the potentially environmentally-friendly power of the market, and the market's role in encouraging technological advances, neoliberal environmentalism is complemented by the sociological theory of ecological modernization, which has had a significant influence on the contemporary environmental movement. Beginning in the 1980s, ecological modernization developed as an explanation for the incorporation of some environmental concerns into mainstream society, primarily in Western European countries and later in the United States (Mol, 2000, p. 46). The theory states that continued modernization, with technological innovation and *increasing economic growth*, provide a path to environmentally sound outcomes (Barry, 2005, p. 306).

[The ecological modernization] discourse quintessentially reconciles the (perceived) contradictions between economy and ecology. It assumes a crucial role for technology in combating the negative environmental effects of economic activity, and emphasizes the role of good-willed and well-informed producers and consumers in giving ecological rationality a role. (van Huijstee, 2011, p. 46)

Ecological modernization's earlier formulations preferred market-oriented solutions over ones with heavy-state involvement, and this market orientation has largely survived throughout the theory's evolution. Later, an emphasis on institutions and cultural changes also arose (Mol, 2000; Mol, 1997; Spaargaren & Mol, 1992). The role of civil society, for example, gained prominence within the theory, recognizing the role of environmental groups and government agencies in spreading environmental values. As individuals adopted an ecological perspective, this change in culture would be incorporated into the economy, according to the theory. Consumer power, in other words, would increasingly be ecologically-oriented. Importantly, ecological modernization, does not expect advanced economies to incorporate environmental concerns automatically (Cole, 2000, p. 89, as cited by Barry, 2005). Instead, ecologically-focused institutions are required, and the state is expected to take a market-friendly role in encouraging the incorporation of environmental concerns into the economy (Barry, 2005, p. 307). Whereas the free market environmentalism (Anderson & Leal, 2001) approach identifies almost anything outside of the bounds of voluntary market exchange as political, and therefore problematic, many environmentalists that embrace neoliberal solutions reflect ecological modernization's call for strong institutions that guide the market over time. Indeed, environmental organizations themselves constitute the very institutions anticipated by ecological modernization theorists. Like neoliberal environmentalism, moreover, ecological modernization, turns "collective ecological problems for society as a whole into selective economic opportunities for market actors (aided by the state)" (Barry, 2005, p. 310). In that way, it follows free market and neoliberal environmentalism's implicit requirement that environmental goals be consistent with economic growth and amenable to the interests of private firms. By focusing on consumer power, furthermore, neoliberal environmentalism promotes individual choice for consumers and voluntary action for firms.

While neoliberal environmentalism aligns with ecological modernization theory, the treadmill of production theory informs structural environmentalism. The treadmill of production theory, first developed by Allan Schnaiberg in 1980, was originally formulated to explain the sharp increase in environmental problems following World War II. It has since been further elaborated, linking modern-day environmental degradation to trends in social and economic change. With roots in Marxism, the treadmill model is a class-based theory of economic, social and environmental change. It utilizes a political-economy approach to explain why environmental degradation is likely to increase over time in capitalist societies *and* why the majority of the population, though they are hurt by such environmental degradation, has no choice but to accept the logic and costs of the treadmill (Schnaiberg et al., 2000). According to the theory, the treadmill is a metaphor for society running in place. The owners of capital (labeled 'treadmill elites') invest their capital in new labor-saving technologies, which are chemical, material or energy-intensive, in order to increase profits. Replaced by such technologies, laborers are thrown off the ever-moving treadmill. The working class is marginalized by diminished employment, the majority of the population bears the environmental, and public health costs associated with heavy chemical use and fossil fuel-intensive machinery. Yet, the treadmill continues as elites require increased productivity to cover the high costs of technological investments. Workers cannot challenge the treadmill but must accept and embrace it because its forward momentum (propelled by continued economic growth) is the only opportunity for continued demand for labor (Gould et al., 2004, p. 296). The state, meanwhile, embraces the treadmill because of the increasing tax revenues brought by economic growth. The state also comes to see economic growth as the only way to solve social problems and maintain its legitimacy through the collection of revenue and the creation of new jobs.

For treadmill theorists, environmental degradation is not simply ancillary to the growth-driven trajectory of modern societies, but the inevitable result. Consequently, new technology and new institutions, even green ones, cannot effectively confront the treadmill's requirement for ever-more natural resources. Instead, environmental degradation is built into the system's structural imperative for economic growth and the use of chemical, material or energy-intensive technologies. Thus, green technologies, anticipated by proponents of ecological modernization, are expected to speed up the treadmill. "Once in place, the expanded production of the new technologies substantially increased both the volume of production waste and the toxicity of wastes (due to increased use of chemicals)" (Gould et al. 2004, p. 300).

The limits to technology, as described by structuralists, relate to efficiency as well. The contemporary fixation with efficiency gains, generally obtained through technological innovation, finds a counterargument in Jevons Paradox. In *The Coal Question*, Jevons (1875) found that efficiency gains in the generation of coal fired power led to an *increasing* use of the natural resource. The finding and its implications for efficiency more generally have led many to question efficiency as a panacea. As Missemer (2012, p. 97-98) explains,

A rebound effect happens when an energy efficient technology is implemented instead of an old one, and when it creates an opposite effect to what could be expected: the new technology does not reduce the total quantity of energy consumed, but increases it. Being more efficient, it cheapens the energy costs and calls for larger uses. On the macroeconomic stage, the new technology not only does not solve the energy scarcity issue, but it also may even make it worse.

In other words, gains in efficiency will be outstripped by increased demand, structuralists argue, because of the growth imperative. Meadows et al. (1972) argue "the application of technology to apparent problems of resource depletion…has no impact on the essential problem, which is exponential growth in a finite and complex system."

Where ecological modernization advocates and neoliberal environmentalists focus on individuals' power to green the economy through voluntary consumer choices in the market, treadmill theorists and structural environmentalists focus on the decision-making power of producers to explain environmental degradation. According to the theory, consumers have little power to shape business practices specifically or the economy more generally. "Decisions about the types of technologies, the use of labor, and volumes of production are made outside the realm of consumer decision-making," for several reasons (Gould et al., 2004, p. 300). First, consumers only interact with the "outputs of a given production technology…Although consumers can accept or reject these products, they have no influence over the allocation of capital to productive technologies" (ibid). The extraction phase, though it has significant environmental consequences, remains invisible to the consumer. Second, where neoliberal advocates would argue that supply is responsive to demand, "…it is in the decision to *provide* supply, and the means by which that supply is provided, where social systems and ecosystems first collide" (ibid). This point is particularly relevant when considering green alternative products. Whether a consumer is able to opt for an ecologically-friendly product, and display effective demand for it, has much to do with whether producers have made alternatives available. Furthermore, "greenwashing" could misdirect ecologically-minded consumers, even if genuine alternatives are available. Outside of the treadmill theory, Downey (2015) has also challenged the idea of consumer power, arguing that it requires perfect information, consumers' ability to rationally process a large amount of information, and a wide range of affordable consumer options, factors that are not consistently present in the real world.

These two perspectives, elaborated and refined by social theorists, also manifest in the real world of environmental advocacy, as does the conflict between them. Facing a lack of recent successes (Hansen, 2012; Speth, 2008; Nordhaus & Shellenberger, 2004), the environmental movement is in need of direction. The movement is caught between efforts to embrace the neoliberal emphasis on voluntary, individual action in the market or to refocus on structural contributors to environmental problems. A large part of the problem for environmentalists is that these approaches are based on competing expectations of how markets and society operate. If consumers are powerful and producers are highly competitive, we should expect that widespread support for environmental values will be reflected in market products and practices. If individuals and firms are constrained by structural factors, such as a lack of competitiveness among producers, widespread support for environmental values will not necessarily be reflected in market outcomes. Given the importance of solving the environmental crisis, this case attempts to shed light on these two distinct approaches to environmental advocacy so that students of the environmental movement can better understand its internal dynamics, and locate their own perspectives along this spectrum. The next section offers a review of Boulder and Colorado's history with renewable energy policy and highlights the role of structural and neoliberal interventions along the way.

Though the conflict between Xcel Energy and the City of Boulder came to a head in July of 2014, when the city signaled its intention to start the process of separation and move toward municipalization, Boulder already had a long history of advocating for environmental and climate change action. For instance, the city became the first to establish a voter-approved carbon tax in 2006 (Kelley, 2006), and has since developed a thorough climate action plan (City of Boulder, 2018). However, due to the city's relationship with its utility, environmentalists came to see the Colorado PUC, and state-wide legislation that regulates all electric utilities, as strategic forums for pushing for a greener electric fuel portfolio. As a result, some Boulder environmentalists joined with individuals and organizations across Colorado to advocate for a transition to renewable energy at the state level.

For example, Colorado-based Western Resource Advocates and its predecessor, Land and Water Fund of the Rockies, was highly involved in the 1990s in developing Xcel Energy's voluntary renewable energy purchasing program, Windsource, in partnership with the utility. Started in 1998 and one of the first programs like it in the country, Windsource allows ratepayers to pay a premium so that their electricity consumption is offset by new renewable energy added to the grid. From the start, residential ratepayers enrolled in the program were able to offset between a fraction and one hundred percent of their energy consumption through wind. The program was immediately well received, serving 9,019 customers in Colorado in its first year (Proctor, 2005). In Fall 2005, enrollment spiked to over 34,000 customers, when Windsource participants were paying less than conventional ratepayers due to the then high price of natural gas. However, after a series of changes to Windsource rates, Windsource participants currently pay about $2.16 more per 100 kilowatt-hours of energy use (Xcel Energy, 2015), and according to Xcel Energy, there are about 50,000 subscribers to the program as of 2018 (personal correspondence, Oct 22, 2018). Importantly, neoliberal environmentalists point to the collaboration between environmental organizations and Xcel Energy as the source of this policy success, while structuralists point to the program's limitations.

Colorado advocates also helped the state pass the nation's first voter-approved renewable portfolio standard (RPS) in 2004 with Amendment 37. The amendment required Colorado utilities to source a certain percentage of their electricity from renewable sources. Its first iteration required the state's larger utilities (those serving over 40,000 ratepayers) to source three percent of their electricity from renewable sources by 2007 and ten percent by 2015. It has been amended several times and today requires Xcel Energy to source 30% of its electricity from renewable sources by 2020.

Importantly, the first RPS iteration, Amendment 37, was actively opposed by Xcel Energy, the Intermountain Rural Electric Association (the state's third largest utility) and Tri-State Generation and Transmission Company (Sands, 2004). The opposition argued that the mandate was unnecessary and would lead to higher electricity prices. However, Xcel Energy met the original requirement eight years ahead of schedule (in 2007) and then supported the expansion of the RPS that same year (Denver Post, 2008). Ever since, the utility has lauded the mandate. Thus, while Windsource represents a relatively market and utility-friendly approach to encouraging renewable energy by relying on consumer power and choice, the RPS is more of a structural intervention that changes the rules of the market, requiring Xcel Energy to adopt renewables. Despite the notable gains in renewable energy acquisition that accompanied these policies, Xcel Energy actually increased its reliance on fossil fuels between 2004 and 2011.

Just as the neoliberal-structural spectrum is well represented in Colorado electricity policy, so too are the divisions between neoliberal and structural environmentalists working in the state. Two other electricity decisions further highlight this underlying neoliberal-structural divide: the building of the state's largest coal plant (referred to as Comanche III) in 2009 and the Clean Air, Clean Jobs Act of 2010. In 2004, after Xcel Energy proposed building a new coal plant (Comanche III), several of Colorado's leading environmental groups (Environment Colorado, Environmental

Defense Fund, Western Resource Advocates, Colorado Conservation Voters, Colorado Environmental Coalition and the Sierra Club) engaged in a settlement agreement with Xcel in which those groups committed to refrain from challenging the coal plant in exchange for certain environmental concessions (Colorado Public Utilities Commission 2004). Greenpeace, Wild Earth Guardians, and Clean Energy Action are organizations that did not join the settlement. While those who settled have argued that it was a realistic and pragmatic decision, others suggest that it was overly appeasing and a sign of a weakened, compromised environmental community. Similarly, the same organizations that settled on Comanche III also supported the Clean Air, Clean Jobs Act. In addition to those environmental organizations, the bill was also supported by Xcel Energy, Anadarko Petroleum Corp, EnCana Oil & Gas (USA) Inc., and the Colorado Oil & Gas Association. The bipartisan bill was designed to make Colorado comply with federal air quality standards, and involved the decommissioning of 900 MW of coal-fired electricity generation. While its environmental supporters celebrate the Act's focus on transitioning the state from coal to natural gas, other environmental organizations oppose continued investments in fossil fuels and point to provisions which will extend the lives of two coal plants.

These choices were the final straw that sorted the environmental community into two groups: those who sought compromise and collaboration with Xcel Energy and those who did not. With this contentious context in mind, the next section returns to how these dynamics manifested within Boulder's environmental community, as the city proceeded down a path of separation with its partner.

CASE DESCRIPTION

Around 2009, amidst the record of mixed success introduced above, that some Boulder activists began thinking of a future without Xcel Energy, and the environmental community's split became increasingly stark. The city has since experienced four rounds of voting on the question of separating from Xcel Energy and proceeding with the process of municipalization. Those votes occurred in 2011, 2013, 2014, and 2017. In each year, proponents of municipalization have secured the voters' support to proceed, though only barely in some cases. For each of the votes, Xcel Energy and its supporters outspent pro-municipalization advocates between 2 to 1 and 10 to 1, depending on the vote. Further, there is evidence that voters' support is waning as the arduous legal process takes a toll on the patience and pocketbooks of city residents. Millions of dollars have already been spent and the city anticipates that a municipal utility will ultimately cost $275 to $375 million (Burness, 2017) with Xcel Energy arguing the final cost will be much higher.

What is especially interesting, though, is that while the specific concerns and challenges have evolved with the PUC's rulings – sometimes in favor of the City and other times in favor of Xcel energy – on given components of the possible separation, the nature of arguments for and against municipalization have remained remarkably consistent. In other words, opponents of municipalization have closely tracked the neoliberal environmental approach described above, while proponents' arguments have stayed in line with the structuralists perspective. More than anything else, the divide comes down to whether one believes that a private market actor like Xcel Energy has the sophistication and motivation to satisfy Boulder customers' desire for renewable energy or if such an investor-owned utility's need to maximize profits for shareholders will hamstring a rapid transition to renewables.

In the next section, this characteristic focus of the divide is demonstrated using selections from Boulder's local newspaper, The Daily Camera. Using LexisNexis Academic, the authors reviewed hundreds of articles, opinion pieces and letters to the editor published in The Daily Camera between January 1, 2010 and September 30, 2018, to identify common themes articulated by proponents and opponents of municipalization. This analysis considered only environmental arguments. For instance, Xcel Energy and some residents of Boulder opposed municipalization for non-environmental reasons. Yet the arguments of these actors are included when their arguments are framed with reference to the environment. Neoliberal opposition to separation with Xcel Energy is covered first, followed by structuralists' support for municipalization.

Over the course of the eight years considered in the analysis, opponents of municipalization cast government management as inefficient, wasteful, and, in some cases, naively idealistic. Xcel Energy, on the other hand, was characterized as sophisticated, professional, and competent. The private utility's existing environmental efforts were celebrated, and the firm was expected to have far more expertise than a government-run utility. Furthermore, opponents expressed confidence in the expertise of the firm's personnel. While opponents did not convey concern about the monopoly utility's ability to force its will on ratepayers, they frequently argued against a government-run monopoly's right to do so.

Xcel Energy itself conveyed neoliberal environmental arguments. For instance, in an opinion piece the President and CEO of Xcel Energy in Colorado, David Eves, lauded partnership and cooperation between the utility and the city,

Together, we've already been able to move in a positive direction...Plus, we've had the chance to bring some leading initiatives and best practices to Boulder, including SmartGridCity, on-site and utility scale renewable generation, and conservation and energy efficiency programs. A continued partnership with Boulder to address issues on the local, state and national levels is the best way to address electricity production and its impact...All the residents of our state stand to gain by our collaborative efforts and we stand a better chance of addressing these issues together, rather than separately... We will also continue to lead the nation on our environmental commitments for the benefit of our customers in Boulder and throughout Colorado. (2010)

Here, the neoliberal reference to the willing market actor is clear. Further, it was not just the utility itself that worked to characterize Xcel Energy as an environmental ally. Its experience, and its position as an independent, competitive market actor, opponents to municipalization argued, could be mobilized in the interest of rapid renewable energy adoption. Bob Greenlee, a Daily Camera columnist and Boulder city councilor made such references as well. While it is important to note that Greenlee does not identify as an environmentalist and in many letters refers to environmentalists disparagingly, he too demonstrates neoliberal environmentalism is his arguments against separation.

What's most amazing about the whole municipalization gambit is how Xcel Energy, one of those dreadful investor-owned corporate entities, has actually matured in its thinking about the current and future status of energy generation and distribution. To its credit, Xcel is looking beyond generating electricity using dwindling and polluting fossil fuels. They've become one of the nation's leaders in advocating and supplying wind power...In addition, local activist consumers are demanding more and more responsible actions concerning how electric energy is produced, distributed and used. (Greenlee, 2008)

As seen above, Greenlee only sarcastically refers to the utility's requirements to serve its shareholders. Further, his arguments convey the neoliberal appeal to consumer power and highlight consumers' ability to direct the actions of the utility. Two years later, Greenlee further conveys the neoliberal perspective, referring the city leaders pushing for municipalization as "overzealous local politicians and bureaucrats" who are much less capable of managing Boulder's energy future than the, "professionals" at Xcel Energy (Greenlee, 2010). He thus presents a faith in private market actors and a disdain for inefficient and foolish public actors.

The same references to Xcel Energy's history and anticipated future as an environmental partner was repeated dozens of times. One of the main environmental voices opposing municipalization wrote, for instance, "It would be one thing if Xcel were not proving themselves willing partners for [large] sized change. But the thing is: They truly are" (Christ-Janer, 2013). Further, in an editorial titled, "For real climate change impact, work with Xcel," another voice against municipalization argued that, "Working with Xcel to create a renewable energy choice model that can be replicated in many communities across the country would be faster, cheaper and more effective" than the alternative of municipalization. She goes on to say,

Xcel is a large, investor-owned monopoly utility. This does not mean that the company is by nature an evil organization. What it does mean is that the utility has the resources, experience and regulatory requirement to provide reliable, affordable and even renewable energy to the citizens of Boulder and millions of other customers... the utility is also clearly motivated to pursue an increasing amount of renewable energy regardless of what happens in Boulder. (Hartman, 2013)

In this case, the size of the utility is characterized as one of its advantages. Further, Hartman's argument includes the explicit claim that Xcel Energy has the right motivation to satisfy demand for renewables. Building on the preference for voluntary market action over a regulatory or governmental approach, another resident concerned about municipalization wrote,

I'm all for everyone making personal decisions about and contributions to the protection of our environment, but I'm not so much in favor of the city government making those decisions for its utility customers. (Burris, 2013)

Interestingly, Burris sees a lack of choice from a potential government-owned monopoly but does not highlight the lack of choice he experiences with his privately-owned monopoly. Moreover, this writer's letter to the editor represents a broader lack of concern among opponents of municipalization about Xcel Energy's ability to force decisions on their captured customers.

In 2017, former Boulder mayor and Boulder County Commissioner Will Toor, the CEO of the Boulder Chamber of Commerce John Tayer, and a board member of the Association of Energy Service Professionals Beth Hartman wrote a joint opinion piece, which in many ways summarized the neoliberal case against separation. Their piece highlighted the uncertainty and costs associated with pursing municipalization, and the advantages of working with Xcel Energy. They introduce their piece expressing their deep concern about climate change and the need to, "act as quickly and boldly as possible to reduce local emissions that contribute to this global crisis." The writers then go on to laud Xcel Energy's Colorado Energy Plan, its announcement of the early retirement of two Colorado coal plants, and renewable energy acquisitions. "In contrast," they write, "the deliberations from our state utility commission lay out a very expensive and slow path for the city of Boulder to create a municipal utility, a path which also is fraught with great uncertainties." Referring to the inefficiencies of the separation process, the three add,

We should work with Xcel to leverage their expertise on what might be possible for Boulder to achieve, especially considering the ideas they already are exploring in partnership with Denver. And by remaining a city served by Xcel Energy, Boulder can ... thereby [achieve] greater benefit for the planet than Boulder simply forming its own utility. (Hartman, Tayer & Toor, 2017)

Here again, we see the consistent and nearly unwavering faith in Xcel Energy to serve as a collaborative partner for positive change. This is in stark opposition to the structural environmentalists' characterization of the utility.

Whereas opponents to municipalization conveyed the neoliberal preference for voluntary action among private market actors, proponents characterized the private market actor – Xcel Energy – as incapable of satisfying Boulder's demand for renewable energy. These structural environmentalists presented the utility as beholden to its shareholders and motivated to continue investing in fossil fuels. Where the neoliberals express faith in the professionalism and good will of the utility, the structuralists frame Xcel Energy as outdated and dishonest. Furthermore, the proponents of municipalization package their goals in the language of democracy; in addition to arguing that Xcel Energy could not deliver sufficient renewable energy, these environmentalists also argued that the utility could not offer Boulder decision-making power. Like their counterparts, the nature of the pro-municipalization arguments remained very consistent throughout the eight years this analysis considers.

Early on, for instance, Boulder City Councilors pushing for consideration of municipalization characterized Xcel Energy as a poor partner. One such city councilor, Lisa Morzel, argued in a council meeting that Boulder was not "[getting] rid of coal... with Xcel," and described the utility as an uncooperative and unwilling partner" (Meltzer, 2010). For some proponents of municipalization, it was not so much that Xcel Energy was especially obstructionist in terms of achieving renewable energy goals; instead, many environmentalists called attention to the challenges presented by monopolies more generally. For example, Daily Camera columnist Anne Butterfield wrote in the newspaper's series on 'Boulder's Energy Future,' the following.

Boulder should pursue municipalization because delivery of electricity is by nature a monopoly, and to this American's thinking the only ethical way to allow a monopoly is to make it as low cost and publicly accountable as possible. In short, it must belong to the people so they can exert influence on the rates and energy mix by speaking with and electing local decision makers, and by efficiently calling up all data they wish on the utility's operations. None of that can be done with an investor owned utility which by definition is obliged first to its shareholders, which with the company executives are handsomely paid through a tradition of making over ten percent profit on bulky, central, base load plants that interfere with the usefulness of renewable energy modes. (2011)

The above captures several common themes expressed by municipalization advocates. First, Butterfield challenges Xcel Energy as a monopoly, and argues that legitimate monopolies must be accountable to the public and subject to public transparency. Here, she inserts the pro-municipalization appeal for a democratic utility by suggesting that profit maximization for private shareholders is incompatible with environmentalists' goals. Finally, she references the utility's incentive to continue investing in fossil fuel capital investments to serve shareholders at the cost of Boulder's desire for renewables. This, too, was a common refrain from this side of the conflict. Judy Amabile wrote in the same series that, "monopolistic, for profit companies like Xcel have little incentive to innovate or change. Their customers have no control over the services they pay for" (2011). Yet another member of the newspaper's editorial advisory board wrote in the series his preference for municipal utilities' ability to, "meet public accountability standards, hold open meetings and budget hearings, and have competitive bidding" (Raizman). Two years later, another supporter of municipalization wrote, "Yes, a muni carries risks, but consider the risk of staying with Xcel: rate increases with virtually no representation (as demonstrated by the Public Utilities Commission), being forced into a model with a vested interest in energy consumption, an archaic legislative environment resistant to change and far fewer renewable options" (Wilson, 2013). Here, the utility is seen as unaccountable to ratepayers and driven by perverse incentives out of line with Boulder's environmental goals.

These negative characterizations of Xcel Energy increased further as the election of 2013 approached. In that year, the utility backed an effort to get an initiative on the ballot that would end the municipalization process. It would ultimately be the most expensive ballot fight in Boulder's history with Xcel Energy, with supporters outspending pro-municipalization efforts 10-1. While the first election in 2011 passed with a very slim margin, municipalization advocates saw a huge increase in voter support in 2013. Some used Xcel Energy's efforts in the election to frame the utility as an outsider willing to flood Boulder politics with a misleading and expensive campaign. Debra Biasca, for example, wrote about the build up to the election as follows. "The whole experience was extremely distasteful and upsetting but proved this to me: Xcel Energy is ruthless, has infinite resources and will stop at nothing to hang onto its lucrative Boulder customers - even to the point of paying petition circulators to lie for signatures [to get their measure on the ballot]" (2013). Likewise, in referring to the utility's efforts to win over voters in its campaign against municipalization, another writer linked Xcel Energy's to larger concerns about corporate power. "On the heels of the disastrous "Citizens United" case, should I trust corporations at all?" Edward Arnold wrote, "Should I trust a corporation that has a hired gun…to speak for them? (Arnold, 2013).

When Xcel Energy moved to form a working group with the city to consider alternatives to separation – an example of neoliberal environmental action – another Daily Camera columnist and former Boulder city councilor, Steve Pomerance, characterized the gesture as a cynical attempt to derail forward momentum of municipalization.

Xcel Energy, in what appears to be an attempt to gain a PR victory to try to stop Boulder from creating a municipal electric utility to escape from Xcel's coal-intensive monopoly, asked the city to engage in yet another last-moment process. I fully expect Xcel to use this to try to kill the muni process so it can keep its Boulder customer base. (Pomerance 2013)

Whereas neoliberal environmentalists would understand such as gesture as evidence of the utility's good faith efforts to partner with the city, structural environmentalists responded with distrust and skepticism.

More recently, the themes continue among proponents of municipalization. Five years later, another opinion piece includes concerns about Xcel Energy's ability to externalize the costs of its decisions on a public with little ability to participate in those decisions. Danielle Gronhovd wrote for example, of her,

impression that Xcel's decisions are not being made in the public's best interest. It is absurd that Xcel is proposing to reallocate [ratepayer money] to compensate for shutting down coal plants...Another major concern is that Xcel would have a high percentage of ownership. This structure would drive competitors away, thus increasing cost to customers. Lastly, Xcel is looking to acquire more natural gas... But this change does not fit with Boulder's climate change goals, since it is a fossil fuel and contributes to climate change. (2018)

As demonstrated above, the structuralists in favor of municipalization engaged with Xcel Energy antagonistically. However, their animosity was also directed at the system and context in which the utility operated. Just as they characterized Xcel Energy as unwilling to transition to renewable energy, they also argued that the Colorado PUC was effectively captured by the utility, thus rendering it incapable of providing meaningful oversight. Highlighting the need for structural change and the disdain for the current political system overseeing the utility, for example, Anne Butterfield sarcastically referred to the Colorado PUC as "a haven of democracy and choice," mocking suggestions that Boulder ought to push for renewables through the regulatory body (2010).

In summary, the structuralists in favor of municipalization confronted the utility with antagonism and distrust. They argued that consumer power was limited, as the monopoly utility had no competition to push it to better serve the interests of its customers. They also described the political system as incapable of effectively regulating the utility and argued that more democratic decision-making would serve a more effective check on a monopoly provider. Thus, they argued for broad structural change that would change the roles within which electricity decisions are made.

CURRENT CHALLENGES

In Boulder, the saga continues. After four elections, the conflict has moved to the courts with a larger role for the Colorado PUC. The PUC has made several rulings on the technicalities of the separation, with wins and losses for both parties. For now, the city remains on its path toward municipalization. Whether Boulder will ultimately separate from Xcel Energy remains to be seen. Much of the uncertainty hinges on what the city will owe Xcel Energy in the event of separation. As of the end of 2018, the city anticipated that Boulder voters would have the final chance to approve or deny a new municipal utility – once the full costs are known – in 2020. If Boulder successfully navigates the legal components of the acquisition and formation of the utility, it will be the first municipal utility created for the primary purpose of decreasing climate changing greenhouse gas emissions.

SOLUTION AND RECOMMENDATIONS

As demonstrated above, the positions of opponents of municipalization closely aligned with neoliberal environmental perspectives described earlier in the case, while proponents of municipalization tracked with structural environmentalism. The conflict in Boulder, and the messages mobilized by either side, help illuminate the neoliberal-structural divide in the larger environmental movement. By looking closely at these conflicting perspectives, students of the movement can better understand internal movement conflicts. Environmental movement leaders would be well served by investigating and testing the claims embedded in both perspectives to direct movement participants toward the most effective strategies.

While this case has framed the divide as a strict dichotomy – between neoliberal environmentalism and structural environmentalism – both scholars and environmental advocates ought to consider what one perspective could learn from the other. For example, just as structuralists criticized free market discourse arguing that consumer power and choice did not exist in the context of a monopoly, these environmentalists also expressed the desire to see more market activity. Future research should continue to put structural critiques in conversation with market discourse. For example, how can social scientists honestly and assertively identify the problems associated with existing market dynamics and look for ways to encourage more competitive market functioning? Similarly, are there ways to rescue the promise of markets from the admittedly dismal (at least in the context of an electricity industry populated by monopolies) experience of existing market dynamics? Stated otherwise, some environmentalists characterize markets as a cure-all, while others cast markets as inevitably manipulated, skewed and destructive. Future research should go beyond the state-market dichotomy and help identify what sort of political or social contexts facilitate the type of market functioning neoliberal and structural environmentalists really desire. If both sides take seriously the opportunities and challenges recognized by their counterparts, the environmental movement may be rewarded with a more unified, strategic, and efficacious approach.

AUTHOR DISCLAIMER

One of the authors of this case was involved with the fight over separation from 2010 to 2014. Specifically, they campaigned for the ballot initiatives 2B and 2C in 2011, and against the ballot initiative 310 in 2013. Further, they served on the board of Clean Energy Action, an organization that advocated for municipalization, from Fall of 2013 through 2014. The author believes the embedded nature of their role as a researcher comes with advantages and disadvantages. On one hand, their personal experiences and deep involvement offered substantive, on-the-ground exposure to the discourse and strategies mobilized by either side of the conflict. In other words, they were immersed in the conflict in ways that led to connecting more directly with the various actors involved. Like a participant-observer, this immersion opened up doors unavailable to a more distant scholar. On the other hand, it comes with an even greater responsibility on the part of the researcher to check their own partiality. As a scholar, and an advocate committed to identifying the most strategic paths toward environmental sustainability, it required a readiness to question one's own assumptions. Acknowledging these dynamics, the author offers this disclosure for the sake of transparency.

REFERENCES

Amabile, J. (2011). Boulder's Energy Future [Editorial Advisory Board]. *The Daily Camera.* Retrieved from LexisNexis Academic.

Americas Electric Cooperatives. (2018). *We Are America's Electric Cooperatives.* Retrieved from https://www.electric.coop/our-mission/we-are-americas-electric-cooperatives/

Anderson, T. L., & Leal, D. R. (2001). *Free Market Environmentalism.* New York, NY: Palgrave. doi:10.1057/9780312299736

Arnold, E. R. (2013). Xcel Energy Can't Answer the Hard Questions [Opinion]. *The Daily Camera.* Retrieved from LexisNexis Academic.

Barry, J. (2005). Ecological Modernization. In J. Dryzek (Ed.), *Debating the Earth, 3030-321.* York, UK: Oxford University Press.

Biasca, D. (2013). Don't Sign Xcel's Misleading Petition [Opinion]. *The Daily Camera.* Retrieved from LexisNexis Academic.

Bloomberg. (2018). *Electric Utilities: Public Service Company of Colorado.* Retrieved from https://www.bloomberg.com/research/stocks/private/snapshot.asp?privcapId=298400

Burness, A. (2017). Boulder says municipalization ruling adds $23M in city costs. *The Daily Camera.* Retrieved from http://www.dailycamera.com/news/boulder/ci_31325514/boulder-says-municipalization-ruling-adds-23m-city-costs

Burris, P. (2013). We Should Re-vote on the Muni [Opinion]. *The Daily Camera.* Retrieved from LexisNexis Academic.

Butterfield, A. (2010). Franchise Agreement, Pay the Same Get More [Opinion]. *The Daily Camera.* Retrieved from LexisNexis Academic.

Butterfield, A. (2011). Boulder's Energy Future [Editorial Advisory Board]. *The Daily Camera.* Retrieved from LexisNexis Academic.

Christ-Janer, K. (2013). Municipal Energy is Not the Way to Create Real Change [Letter to the editor]. *The Daily Camera.* Retrieved from http://www.dailycamera.com/letters/ci_24418805/karey-christ-janer-municipal-energy-is- not-way

City of Boulder. (2018). *Boulder's Climate Commitment.* Retrieved from https://bouldercolorado.gov/climate

Colorado Public Utilities Commission. (2004). *Decision No. C05-0049 – Docket No.04A-214E: In the Matter of the Application of PSCo for Approval of its 2003 Least-Cost Resource Plan; Docket No. 04A-215E: In the Matter of the Application of PSCo for an Order Approving a Regulatory Plan to Support the Company's 2003 Least-Cost Resource Plan; Docket No. 04A-216E.* Author.

Department of Regulatory Agencies. (2014). *Mission and History. About Us.* Retrieved from http://cdn.colorado.gov/cs/Satellite/DORA- PUC/CBON/DORA/1251627010991

Downey, L. (2015). *Inequality, Democracy and the Environment.* New York, NY: NYU Press. doi:10.18574/nyu/9781479850723.001.0001

Edison Electric Utilities. (2018). *Our Members.* Retrieved from http://www.eei.org/about/members/Pages/default.aspx

Environmental Defense Fund. (2012). Strategic Plan: Leading Transformational Change: Strategic Plan 2010-2014. *Environmental Defense Fund.* Retrieved from http://www.edf.org/sites/default/files/10585_strategic_plan_2010-2014_0.pdf

Environmental Protection Agency. (2012). How Does Electricity Affect the Environment? *Environmental Protection Agency.* Retrieved from http://www.epa.gov/cleanenergy/energy-and-you/affect/index.html

Environmental Protection Agency. (2018). Sources of Green House Gas emissions. *Environmental Protection Agency.* Retrieved from https://www.epa.gov/ghgemissions/sources-greenhouse-gas-emissions

Epstein, P. R., Buonocore, J. J., Eckerle, K., Hendryx, M., Stout, B. M. III, Heiberg, R., ... Glustron, L. (2011). Mining Coal, Mounting Costs: The Life Cycle Consequences of Coal. *Annals of the New York Academy of Sciences, 1219*(1), 73–98. doi:10.1111/j.1749-6632.2010.05890.x PMID:21332493

Eves, D. (2010). Let the People Vote on Xcel Franchise [Opinion]. *The Daily Camera.* Retrieved from LexisNexis Academic.

Gottlieb, R. (2005). *Forcing the Spring: The Transformation of the American Environmental Movement. NW.* Washington, DC: Island Press.

Gould, K. A., Pellow, D. N., & Schnaiberg, A. (2004). Interrogating the Treadmill of Production: Everything You Wanted to Know About the Treadmill but were Afraid to Ask. *Organization & Environment, 17*(3), 296–316. doi:10.1177/1086026604268747

Greenlee, B. (2008). People Power [Opinion]. *The Daily Camera.* Retrieved from LexisNexis Academic.

Greenlee, B. (2010). Power to the People [Opinion]. *The Daily Camera.* Retrieved from LexisNexis Academic.

Gronhovd, D. (2018). Xcel's Electric Resource Plan Conflicts with Boulder's Climate Goals [Opinion]. *The Daily Camera.* Retrieved from LexisNexis Academic.

Hansen, S. (2012). *Cultivating the Grassroots: A Winning Approach for Environment and Climate Funders*. National Committee for Responsive Philanthropy.

Hartman, B. (2013). For Real Climate Change Impact, work with Xcel [Guest column]. *The Daily Camera*. Retrieved from http://www.dailycamera.com/guest-opinions/ci_24340299/guest-column-real-climate-change-impact-work-xcel

Hartman, B, Tayer, J. & Toor, W. (2017). Time to Change Course on Muni [Opinion]. *The Daily Camera*. Retrieved from LexisNexis Academic.

Howarth, R. W., Santoro, R., & Ingraffea, A. (2011). Methane and the greenhouse-gas footprint of natural gas from shale formations. *Climatic Change*, *106*(4), 679–690. doi:10.100710584-011-0061-5

Jevons, W. S. (1875). *The Coal Question: An Inquiry Concerning the Progress of the Nation, and the Probable Exhaustion of Our Coal Mines*. London, UK: Macmillan & Co.

Kelley, K. (2016). City Approves 'Carbon Tax' in Effort to Reduce Gas Emissions. *The New York Times*. Retrieved from https://www.nytimes.com/2006/11/18/us/18carbon.html

McKibben, B. (2012). Global Warming's Terrifying New Math. *Rolling Stone*. Retrieved from http://www.rollingstone.com/politics/news/global-warmings-terrifying-new-math-20120719

Meadows, D. H., Meadows, D. L., Randers, J., & Behrens, W. W. III. (1972). *The Limits to Growth*. Hanover, NH: Signet.

Meltzer, E. (2010). *Boulder Leaders Divided on Xcel Agreement. The Daily Camera*. Retrieved from LexisNexis Academic.

Missemer, A. (2012). William Stanley Jevons' *The Coal Question* (1865), beyond the rebound effect. *Ecological Economics*, *82*, 97–103. doi:10.1016/j.ecolecon.2012.07.010

Mol, A. P. J. (1997). Ecological Modernization: Industrial Transformations and Environmental Reform. In R. Michael & G. Woodgate (Eds.), The International Handbook of Environmental Sociology (pp. 138-149). Northampton, MA: Edward Elgar.

Mol, A. P. J. (2000). The Environmental Movement in an Era of Ecological Modernization. *Geoforum*, *31*(1), 45–56. doi:10.1016/S0016-7185(99)00043-3

Nordhaus, T., & Shellenberger, M. (2004). *The Death of Environmentalism: Global Warming Politics in a Post-Environment World*. The Breakthrough Institute.

Pomerance, S. (2013). Working Group Process is More Show than Tell [Opinion]. *The Daily Camera.* Retrieved from LexisNexis Academic.

Proctor, C. (2014). Xcel Proposes New Solar Power Option for Colorado Customers. *Denver Business Journal.* Retrieved from http://www.bizjournals.com/denver/blog/earth_to_power/2014/04/xcel-proposes- newsolar-power-option-for-colorado.html?page=all

Public Power. (2018). *Stats and Facts: Three Types of Electric Utilities*. Retrieved from https://www.publicpower.org/public-power/stats-and-facts

Raizman, M. (2011). Boulder's Energy Future [Editorial Advisory Board]. *The Daily Camera.* Retrieved from LexisNexis Academic.

Rising Tide North America. (2011). Confronting the Root Causes of Climate Change. *A Primer from Rising Tide North America.*

Rondinelli, D. A., & London, T. (2003). How Corporations and Environmental Groups Cooperate: Assessing Cross-sector Alliances and Collaborations. *The Academy of Management Executive*, *17*(1), 61–76.

Sahagun, L. (2011). Sierra Club Leader Departs Amid Discontent Over Group's Direction. *Los Angeles Times.* Retrieved from http://articles.latimes.com/2011/nov/19/local/la-me-sierra-club-20111119

Sands, W. (2004, September 29). Green Power Push Stirs up Sparks: Amendment 37 Mixes it Up Statewide. *The Durango Telegraph.* Retrieved from http://www.durangotelegraph.com/04-09-23/cover_story.html

Schnaiberg, A., Pellow, D. N., & Weinberg, A. (2000). The Treadmill of Production and the Environmental State. The Environmental State Under Pressure, 10, 15-32.

Spaargaren, G., & Mol, A. P. J. (1992). Sociology, environment, and modernity: Ecological modernization as a theory of social change. *Society & Natural Resources*, *5*(4), 323–344. doi:10.1080/08941929209380797

Speth, J. (2008). Environmental Failure: A Case for a New Green Politics. *Yale Environment360.* Retrieved from http://e360.yale.edu/feature/environmental_failure_a_case_for_a_new_green_politics/20 75/

Stafford, E. R., Polonsky, M. J., & Hartman, C. L. (2000). Environmental NGO-Business Collaboration and Strategic Bridging: A Case Analysis of the Greenpeace-Foron Alliance. *Business Strategy and the Environment*, *9*(2), 122–135. doi:10.1002/(SICI)1099-0836(200003/04)9:2<122::AID-BSE232>3.0.CO;2-C

The Denver Post. (2008). Xcel's new 'green' plan has cost savings in mind [Editorial]. *The Denver Post*. Retrieved from http://www.denverpost.com/editorials/ci_10180088

van Huijstee, M., Pollock, L., Glasbergen, P., & Leroy, P. (2011). Challenge for NGOs Partnering with Corporations: WWF Netherlands and the Environmental Defense Fund. *Environmental Values*, *20*(1), 43–74. doi:10.3197/09632711 1X12922350166030

Wilson, A. (2013). Xcel's Energy Model Outdated [Opinion]. *The Daily Camera*. Retrieved from LexisNexis Academic.

Xcel Energy. (2014). *2013 Owned and Purchased Energy: Public Service Company of Colorado Power Supply Mix*. Retrieved from http://www.xcelenergy.com/About_Us/Our_Company/Power_Generation/Power_Generation_Fuel_Mix_-_PSCo

Xcel Energy. (2015). Xcel Energy is the nation's No. 1 wind power provider. *Wind Power on our System*. Retrieved from http://www.xcelenergy.com/Environment/Renewable_Energy/Wind/Wind_Power_on_Our_System

Xcel Energy. (2017). Our Company. *About Us*. Retrieved from http://www.xcelenergy.com/About_Us/Our_Company/Service_Areas

ADDITIONAL READING

Barry, J. (2005). Ecological Modernization. In J. Dryzek (Ed.), *Debating the Earth, 3030-321*. York, NY: Oxford University Press.

Dyer, J. (2017). Municipalization: More Important Now Than Ever. *Boulder Weekly*. Retrieved from https://www.boulderweekly.com/opinion/municipalization/

Gould, K. A., Pellow, D. N., & Schnaiberg, A. (2004). Interrogating the Treadmill of Production: Everything You Wanted to Know About the Treadmill but were Afraid to Ask. *Organization & Environment*, *17*(3), 296–316. doi:10.1177/1086026604268747

Johnson, N. (2018). Lessons from Boulder's Bad Breakup. Grist. Retribed home

Sahagun, L. (2011). Sierra Club Leader Departs Amid Discontent Over Group's Direction. *Los Angeles Times*. Retrieved from http://articles.latimes.com/2011/nov/19/local/la-me-sierra-club-20111119

KEY TERMS AND DEFINITIONS

Municipalization: The transfer of private assets to public ownership at the city level.

Neoliberalism: A perspective that calls for market-based solutions as opposed to aggressive government intervention.

Renewable Portfolio Standard: Sometimes called a renewable electricity standard, a regulation that requires a certain percentage of electricity to be sourced from renewable energy.

Structuralism: An understanding of the causes of environmental problems as built into the social structure

Taking: The expropriation of property, meaning the forced transfer of private property into government ownership.

Utility: An entity that provides a service such as electrification.

APPENDIX: ANALYSIS AND DISCUSSION QUESTIONS

1. Do both environmental approaches offer long term and short term advantages? Why or why not?

   Yes.
   The neoliberal environmental approach offers significant short-term advantages by encouraging partnerships with the largest, and therefore most impactful, power suppliers. This allows for more rapid sustainable transitions which also lends itself to a long-term advantage, by reducing emissions from large emitters, benefiting climate change mitigation for the future.
   The structural environmental approach offers significant short-term advantages by prioritizing the role of grassroots public participation in utility decision making and ownership. Thus, structural environmentalism increases energy literacy in the short-term, offering long-term solutions for democratic governance within the energy sector, fostering durable solutions to climate change mitigation.
   Challenges: Neoliberal environmentalism embraces environmental goals which benefit large, share-holder run firms, but may avoid other goals. Structural environmentalism uses radical tactics which may polarize the environmental movement or the broader public.

2. What is the importance of developing a "Spectrum of Environmentalism"?

   Understanding environmentalist approaches on a spectrum does two things: 1) It prevents the creation of a concrete dichotomy between neoliberal and structural environmentalists, allowing the opportunity for both approaches to learn from each other's tactics and deliberate respectfully, even with a refusal to adopt opposing strategies. 2) The spectrum opens up space for various kinds of environmental approaches, furthering environmentalism's success in places distinctly different than Boulder, CO.

Chapter 7
Sustainable Land Development Using Permaculture

Jody M. Luna
Conscious Designs, USA

EXECUTIVE SUMMARY

This multi-faceted case study investigates sustainable land development using permaculture as the design tool. Permaculture, coined by Bill Mollison and David Holmgren, is a sustainable design theory that builds off three ethical principles used to produce a set of guidelines to follow in order to create an ecologically focused project. Permaculture, a contraction of perma-nent and initially agri-culture, has evolved to perma-nent and culture, understanding that without agriculture, culture is impossible. This chapter begins with an overview of the environmental issues followed by a description and brief history of sustainable development, with emphasis placed on the United Nations (UN) Sustainable Development Goals (SDGs). The focus will be a three-part case study examining different scales (urban, suburban, and rural) of permaculture land development in the midwestern United States (U.S.). These permaculture designs will illustrate how SDGs can be achieved to forge a sustainable future.

INTRODUCTION

The United Nation's (U.N.) Intergovernmental Panel on Climate Change (IPCC), released a report "on the impacts of global warming of 1.5°C above pre-industrial levels and related global greenhouse gas emission" in response to the invitation from "the 21st Conference of Parties of the United Nations Framework Convention on Climate Change to adopt the Paris Agreement" (IPCC, 2018, p. 4). The report

DOI: 10.4018/978-1-5225-8559-6.ch007

seeks to understand what will happen to the earth should global warming continue. Worland (2018) states, "to keep temperatures from rising more than 1.5°C humans need to shift the trajectory of carbon dioxide emissions so that we either stop emitting by around 2050 or pull more carbon out of the atmosphere than we release" (para. 5). The U.N. made this call to world leaders by developing "the 2030 Agenda for Sustainable Development, adopted by all United Nations Member States in 2015, provides a shared blueprint for peace and prosperity for people and the planet, now and into the future" (Sustainable Development Goals, n.d., para. 1). According to Walsh (2014),

Focusing on the impacts of climate change – ranging from the effects on endangered species to changes in agriculture – the new report demonstrates just how wide-ranging the effects of a warming world will be. "We have assessed impacts as they are happening in natural and human systems on all continents and oceans," said Rajendra Pachauri, the chair of the IPCC, which was jointly established by the U.N. and the World Meteorological Organization. "No one on this planet will be untouched by climate change." (para. 2)

The chapter begins with a discussion on climate change and global warming in an effort to set the stage and understand the urgency behind incorporating sustainability efforts into land development design. The objective of this chapter is to call attention to permaculture as a design philosophy in order to provide a template for sustainable development. The SDGs are discussed as they relate to each design to illustrate how permaculture can be incorporated into future developments.

BACKGROUND

To date there is a long list of environmental problems that must be rectified in order to maintain life on earth. One of the major issues at hand is climate change and global warming. According to the National Aeronautics and Space Administration's (NASA) Jet Propulsion Lab (2018), climate change is, as the name suggests, a change in climate that can include but is not limited to global warming.

Climate change refers to a broad range of global phenomena created predominantly by burning fossil fuels, which add heat-trapping gases to Earth's atmosphere. These phenomena include the increased temperature trends...also encompass changes such as seal level rise; ice mass loss in Greenland, Antarctica, the Artic and mountain glaciers worldwide; shifts in flower/plant blooming; and extreme weather events. (NASA Jet Propulsion Lab, 2018, para. 5)

Global warming results in increased global temperatures and is dangerous because as the oceans get warmer, the ice sheets and glaciers melt, causing raised sea levels which can lead to extreme weather events and ocean acidification (NASA Jet Propulsion Lab, 2018). These global changes affect how humans live and interact with the world around them. Consideration must be given as to how climate change can affect future land development and responsible decisions must be made to prevent loss of life. The United Nations Climate Change News (UNCC) has recorded

global temperatures for the first five months of 2018 to have been the highest on record...higher temperatures lead to more frequent and long-lasting heat waves causing adverse environmental impacts. Extreme temperatures, for instance, affect human health in the form of heat strokes, heat cramps, and even death; the agricultural sector faces retarded crop growth; and the energy sector experiences difficulties coping up with the increasing use of cooling during high temperatures. Moreover, extreme weather affect(s) water supplies, plants, animals and ecosystems, and forests. (World View, 2018, para. 1)

Regardless of these catastrophes, humanity continues to exist and those cities, homes, etc. that have been fortified to withstand these extreme external forces will persist.

HISTORY OF SUSTAINABLE DEVELOPMENT

The definition of sustainable development is vast and can include many aspects, however when one defines each word separately it comes together to mean the development of anything, a plot of land, educational curriculum, prison-reform program, a city, etc., that should be designed to withstand the test of time and serve the needs of the current population without sacrificing the needs of future generations (Sachs, 2015; Brundtland, 1987). "Sustainable development calls for a world in which economic progress is widespread; extreme poverty is eliminated; social trust is encouraged through policies that strengthen the community; and the environment is protected from human-induced degradation" (Sachs, 2015, p. 13).

In 1969, the first international commission was created to focus development efforts using research and knowledge, followed by establishment of the U.S. Environmental Policy Act (EPA), "one of the first countries to establish a national legislative framework to protect the environment" (International Institute for Sustainable Development, 2012). There were many other smaller initiatives taken by individual countries like Canada, as well as the United Nations (UN) however there were few significant global environmental efforts until Brundtland's 1987 *Report of the World Commission on Environment and Development: Our Common Future*. This World Commission focused on a global agenda for change and intended "to propose long-term environmental strategies for achieving sustainable development" (Brundtland, 1987, p. 5). Brundtland (1987) asserts that "the environment does not exist as a sphere separate from human actions, ambitions, and needs" (p. 7). In fact, the environment and humanity must co-exist in harmony with one another and human action/development needs to consider environmental factors.

Many of the development paths of the industrialized nations are clearly unsustainable... the development decisions of these countries, because of their great economic and political power, will have a profound effect upon the ability of all peoples to sustain human progress for generations to come. (Brundtland, 1987, p. 7)

The 1987 report was intended to inspire responsible and sustainable development globally and called for action:

The next few decades are crucial. The time has come to break out of past patterns. Attempts to maintain social and ecological stability through old approaches to development and environmental protection will increase instability. Security must be sought through change. The Commission has noted a number of actions that must be taken to reduce risks to survival and to put future development on paths that are sustainable. (Brundtland, 1987, p. 23)

Following the Brundtland Report, the UN has continued this work: the 2000 Millennium Development Goals, 2002 World Summit on Sustainable Development, 2005 Kyoto Protocol, 2009 G20 Summit, and 2015 Sustainable Development Goals that have continued to make a call to the world's leaders to make substantial changes in their countries' policies on development. The Sustainable Development Goals (SDGs)

are an urgent call for action by all countries – developed and developing – in a global partnership...The 17 Sustainable Development Goals and 169 targets... demonstrate the scale and ambition of this new universal Agenda. They seek to build on the Millennium Development Goals and complete what these did not achieve. They seek to realize the human rights of all and to achieve gender equality and the empowerment of all women and girls. They are integrated and indivisible and balance the three dimensions of sustainable development: the economic, social and environmental. (Sustainable Development Goals, n.d., para. 1)

These goals are: GOAL 1. End poverty in all its forms everywhere, GOAL 2. End hunger, achieve food security, improved nutrition and promote sustainable agriculture, GOAL 3. Ensure healthy lives and promote well-being for all - at all ages, GOAL 4. Ensure inclusive and equitable quality education and promote lifelong learning opportunities for all, GOAL 5. Achieve gender equality and empower all women and girls, GOAL 6. Ensure availability and sustainable management of water and sanitation for all, GOAL 7. Ensure access to affordable, reliable, sustainable and modern energy for all, GOAL 8. Promote sustained, inclusive and sustainable economic growth, full and productive employment and decent work for all, GOAL 9. Build resilient infrastructure, promote inclusive and sustainable industrialization and foster innovation, GOAL 10. Reduce inequality within and among countries, GOAL 11. Make cities and human settlements inclusive, safe, resilient and sustainable, GOAL 12. Ensure sustainable consumption and production patterns, GOAL 13. Take urgent action to combat climate change and its impacts, GOAL 14. Conserve and sustainably use the oceans, seas and marine resources for sustainable development, GOAL 15. Protect, restore and promote sustainable use of terrestrial ecosystems, sustainably manage forests, combat desertification, halt and reverse land degradation and halt diversity loss, GOAL 16. Promote peaceful and inclusive societies for sustainable development, provide access to justice for all and build effective, accountable and inclusive institutions at all levels, and GOAL 17. Strengthen the means of implementation and revitalize the global partnership for sustainable development. These goals include many different targets and the progress can be explored on the UN's website for the Division for Sustainable Development Goals at https://sustainabledevelopment.un.org/. Ongoing annual progress is reported, and targets and indicators help to outline action. This website is valuable with many resources for those interested in incorporating the goals into their development plans and provides a guide through which development can be assessed to ensure sustainability.

On October 8, 2018, an even more important call was made, the UN's Intergovernmental Panel on Climate Change (IPCC) published a special report in accordance with the 2015 Paris Agreement called *Global Warming of 1.5° C* (Reilly, 2018). This report detailed "the impacts of global warming of 1.5° C above pre-industrial levels and related global greenhouse gas emission pathways in the context of strengthening the global response to the threat of climate change, sustainable development, and efforts to eradicate poverty" (Intergovernmental Panel on Climate Change, 2018). The warnings were made very clear:

The world's carbon budget is nearly overspent, and we're facing a future fraught with climate change-driven death and disease. To reverse course and keep global warming below 1.5 degrees, we must take swift action to keep fossil fuels in the ground. Failure to stem both fossil fuel production and consumption will cause untold human suffering and condemn polar bears, coral and other species to extinction. (Wolf, 2018, para. 3)

The report outlines that without action, the following as potential environmental issues could occur: loss of biodiversity, rise of sea levels, extreme temperatures and weather events, increased heat-related health issues, and decreased crop yields. To mitigate these potential threats to humanity it is important for everyone to work together as part of "a global, coordinated effort to reduce emissions across all major sectors" (Wolf, 2018, para. 7). Using the SDGs is the key to future development outlined by the UN. The push to accomplish these goals is strong, and there are a number of efforts that can be made at both the macro- and micro-levels. At the macro-level, there needs to be governmental intervention which requires industry to reduce their carbon-emissions. It will create a central effort to decrease reliance upon non-renewable energy like natural gas, petroleum, and coal. Finally, pushing a focus on developing renewable energy sources like wind and kinetic energy. Additionally, encouraging energy-efficient construction and sustainable development by making governmental funding and tax breaks available to organizations and individuals. At the micro-level, citizens can focus on eating local, plant-based diets that are considered to be more sustainable.

It is clear that the earth is making a call to its inhabitants, it is time to consider sustainable development not as a specialty but as the standard. Ultimately, there needs to be a paradigm shift and there should be no difference between development and sustainable development with no specified difference that a project is considered to be "environmentally-friendly" or "green."

PERMACULTURE

Permaculture, coined by Bill Mollison and David Holmgren, is a sustainable design theory. It builds off three ethical principles used to produce a set of guidelines to follow in order to create an ecologically focused project. Permaculture, a contraction of *perma*-nent and initially agri-*culture*, has evolved to *perma*-nent and *culture*; understanding that without agriculture, culture is impossible.

In a permaculture design, the intention is to restore natural patterns instead of providing a manicured appearance. It is a functional design process where the property is designed in zones of usefulness. Suzuki (in Luna, 2014), stated:

What permaculturists are doing is (some of) the most important activity that any group is doing on the planet. We don't know what the details of a truly sustainable future are going to be like, but we need options. We need people experimenting in all kinds of ways and permaculturists are one of the critical groups that are doing that. (n.p.)

Glanzberg (n.d.) like many current permaculturists believe that the work being done now is vital to human existence. Permaculturists feel the urgency of the situation and realize that without significant and meaningful change humanity will go extinct (R. Nowicki, personal communication, August 10, 2018). They realize that systematic change is needed, not only to the way things are currently being done, but also to a shift in thought process and decision making. Permaculture needs to become part of the mainstream conversation as a possible answer to sustainable development. As humanity is faced with climate change, there needs to be a guide developed according to permaculture principles and ethics which details how to develop a more sustainable world.

Ethics are culturally evolved mechanisms that regulate self-interest, giving us a better understanding of good and bad outcomes…Permaculture ethics are distilled from research into community ethics, learning from cultures that have existed in relative balance with their environment for much longer than more recent civilisations. This does not mean that we should ignore the great teachings of modern times, but in the transition to a sustainable future, we need to consider values and concepts outside the current social norm. (Permaculture Ethics, n.d., para. 2)

The permaculture ethics are Care of People, Care of Earth, and Fair Share or Share of Surplus; these are the basis through which all decisions are made (Mollison, 1988; Permaculture Ethics, n.d.). From these ethics come the principles of permaculture in Mollison's *A Designers' Manual*, however the principles aren't as clearly numbered as the 12 principles that Holmgren details explicitly. Mollison (1988), discusses broader theories and addresses how to deal with each specific climatic factor and how they can affect the overall outcomes. He also addresses different sustainable concepts like aquaculture, soil health, and natural patterns. After completing his thesis, which was to become *Permaculture One: A Perennial Agriculture for Human Settlements* (1978), Holmgren, Mollison's graduate student, continued to develop his work on his own, documenting his work and sharing his resources on the *Permaculture Principles* website.

Holmgren's 12 principles of permaculture are: 1. Observe and interact, 2. Catch and store energy, 3. Obtain a yield, 4. Apply self-regulation and accept feedback, 5. Use and value renewable resources and services, 6. Produce no waste, 7. Design from patterns to details, 8. Integrate rather than segregate, 9. Use small and slow solutions, 10. Use and value diversity, 11. Use edges and value the marginal, and

12. Creatively use and respond to change (Permaculture Design Principles, n.d.). "These principles are seen as universal, although the methods used to express them will vary greatly according to the place and situation. They are applicable to our personal, economic, social and political reorganisation" (Permaculture Design Principles, n.d., para. 2).

In 2019 in the U.S., the academic research related to permaculture is limited but growing. According to Ferguson and Lovell (2013), "despite an international presence and high public profile, permaculture has remained relatively isolated from scientific research. Permaculture holds potential contributions to the transition to sustainability, but its value is constrained by its isolation from science…and a paucity of documentation" (n.p.). The majority of the publications exist for the public and not the scholarly audience (Ferguson & Lovell, 2013). It is important to spread this conversation on permaculture.

PERMACULTURE DESIGN

While a permaculture design can take many shapes, there are some common characteristics, these are: zones, guilds, food forests, land formations, and animal systems that will be discussed in this section. The idea of zones is to separate an area into zones of usefulness. Attention and energy are focused on the zones of major importance; one is closest to the center or the house and extends out to zone five which is the untouched area of the property left for future development. Zones two-four extend out as far as needed on the site based on the needs of the design. Zone one would hold the things that are used the most frequently, for example on a residential property, the house would be placed in zone one with a vegetable or kitchen garden placed close by for easy access. In this same residential property, zone two may house a chicken coop and potentially some fruit or nut tree centered guilds, zone three would be a fruit tree orchard or a food forest and zone four would be the outskirts of the property where a pond may be or possibly the septic runoff system. For a larger rural or farm-scale property the zones could be different as each property needs its own unique design.

Another feature that will typically be seen is a guild, which is an intentionally designed group of entities that work together to be productive, like plant species, insects, etc. A common example is the Three Sisters or Tres Hermanas guild from Native American Culture,

these sisters are corn, squash, and pole beans. Each plant has its own purpose, the corn grows into a tall stalk providing a "pole" for the pole beans to grow upon, the beans are nitrogen-fixers, which is highly coveted in soil fertility. The squash provide safety, because they grow low on the ground protecting the roots of the plants from large animals, like raccoons or rabbits, as well as provide shade to the soil, preventing scorching and over-drying. In a permaculture designed system, this is the goal, that all plants work together to create a "perfect" system. (Luna, Dávila, and Reynoso-Morris, 2018, p. 68)

Permaculturists have experimented with different plant guilds, Ruddock (2013) identifies different guilds, for example, the bee guild, designed to attract bees and other pollinators, is centered on a linden tree with hip bearing shrub roses like rose apple placed about 10 feet [3m] or more away, lovage is placed intermingled with mint, garlic, chives, dill, parsley, caraway, and/or echinacea. Each of these plants have their purpose like the three sisters, however they are less prescriptive, and many plants can be used for the same function. There are many different plant groupings designed by contemporary permaculturists. Additionally, this concept of a polyculture is not new and has been used by many indigenous and modern cultures. One popular guild-type function is planting marigolds next to tomatoes, which is seen in almost any garden visited globally.

Food forests, another typical permaculture designed system, is a system of plants that mimic the natural forest layers including edibles at each layer. The key here is having layers of vegetation: canopy, sub-canopy, shrub, herbaceous, groundcover/creeper, underground, vertical/climber, aquatic/wetland, and mycelial/fungal layers (Hahn, 2014). Food forests use guilds to create a high-density food producing space. One example of a flourishing food forest is the 0.68-acre [2751.8m^2] Winslow Food Forest in Portland, OR designed in the spirit of permaculture and dedicated to Toby Hemenway, the prolific permaculture author of *Gaia's Garden* (2009). The forest includes "a canopy of fruit and nut trees, a sub-canopy of berry bushes, and herbaceous layer of culinary and medicinal herbs, annual and perennial vegetables, and edible flowers" (Winslow Food Forest, n.d., para. 1). Established in November 2016, this urban food forest supplies food to a few local restaurants as well as has Community Supported Agriculture (CSA) memberships (Winslow Food Forest, n.d.). A CSA is a way for urban and sometimes suburban residents to support their local farmers by purchasing into the farm operations in exchange for a weekly share of the fruits and vegetables that are grown on that farm. This cooperative human exchange is similar to the beneficial exchanges seen in guilds, the two parties work together in a relationship that is beneficial to both. The members of the CSA help to fund the farm and ensure that the yields are consumed; the farm provides locally grown fruits and vegetables to the members who may not have the space or time to grow their own food.

The next permaculture system is land formations that can take the form of terraces, berms, or raised beds. First, terrace farming has been used by indigenous cultures for many years, by forming flat planes cut into the side of a mountain creating a space for plants, animals, and people to live. The water that runs down the side of the mountain is collected into the earth of the terrace; this water is gathered deep in the soil for the plants to use and is harvested for human consumption. Next, raised beds are used in many small kitchen gardens, however in a permaculture system they look a little different than the typical constructed wooden box. Permaculturists do not use the wood frame to build a box, they take a more natural approach and create raised beds in mounds placed on top of the ground. The mound could be built using Hügelkultur, "logs, branches, leaves, grass clippings, straw, cardboard, petroleum-free newspaper, manure, compost, or whatever other biomass" cover this mound with soil and plant directly into the soil, "the gradual decay of wood is a consistent source of long-term nutrients which should extend the growing season" (Keiren, 2013). Another method that could be used is sheet-mulching, commonly called lasagna gardening, which are layers of material that will break down over time creating a mound of organic matter. The layers can be composed of compost, seed-free hay or straw, stable bedding, newspapers, cardboard, manure, blood/bone meal, grass or other mown vegetation, or other organic matter that is on hand placed on top of moistened soil and topped with straw, leaves, or other seedless mulch (Keiren, 2013). Finally, almost all permaculture designs include an herb spiral, "productive, energy efficient, vertical gardens. The spiral creates a range of micro-climates for different herbs. Dry, sun loving species are plants at the top and moisture loving species towards the bottom near a small pond" (Hahn, 2014, p. 9).

Aquaculture and animal systems are also used throughout a permaculture designed system. There is a focus on attraction of beneficial insects and a humane approach to deterrence of pests. Beneficial insects and animals are considered when selecting the group of plants that are placed in a guild; close attention is paid to the different species that will be deterred and attracted. While some animals are integrated into a system, plans are made for deterrence as well; for example, deer can wreak havoc on a garden, some permaculturists have used a double-fence system or series of fences to keep them out of the garden. Other animals like goats and sheep are used to maintain overgrowth of plants and to increase the garden fertility or poultry to remove insects and pests while increasing garden fertility (Hahn, 2014).

While these are some of the common features that will be seen in a permaculture design, there are many other solutions. A permaculture design is only limited by the creativity of the designer and by using the principles and ethics. In the end, a permaculture design is ever changing and evolving. Year after year, the designer must take a step back and reevaluate using the principles and ethics to make sure it's a productive environment.

CASE STUDY

Methodology

The methodology chosen for this research is qualitative with the analysis of a single case study (Eisenhardt, 1989; Eisenhardt & Graebner, 2007; Patton, 1990; Yin, 2009). According to Yin (1984) "a case study is an empirical inquiry that investigates a contemporary phenomenon within its real-life context, especially when the boundaries between phenomenon and context are not clearly evident" (p.13). The case study method is used here because of the deliberate coverage of contextual conditions because of the pertinence to the phenomenon (sustainable development) explored in this chapter (Yin, 1984). "Case studies typically combine data collection methods such as archives, interviews, questionnaires, and observations. The evidence may be qualitative (e.g., words), quantitative (e.g., numbers), or both" (Eisenhardt, 1989, p. 534).

The three-part case study covered in this chapter uses a combination of archives, in-person interviews, site visits, and observations. This is a single case study despite having three distinct conditions that are investigated; ultimately, the properties or their locations are not the focus of the study but rather permaculture is the focus as it applies to the different situations, therefore the single case study focuses on permaculture as the central topic that is explored further.

The following case study will investigate different types of permaculture developments: urban, suburban, and rural. Each part will discuss a brief history of the location, the design elements that make it a permaculture design, and how these designs can help to achieve the SDGs.

Case Study Part 1: Urban Land

The first part of the case study focuses on urban-scale permaculture development in Chicago, IL. Urban permaculture can be a seemingly difficult task, especially in a city as dense as Chicago. The vacant land is minimal, small, and outwardly undesirable. However, Chicago is known for its gardens, parks, and greenery; some would even call it a "Garden City" which is an urban design concept that promotes the inclusion of gardens into the urban landscape to combat the hardness of urban life. The Chicago Sustainable Development Policy was implemented in 2004, "the goal of this policy is to enhance the sustainable performance of projects receiving City assistance…requires development projects that are receiving final assistance or special approvals from the City to include sustainable elements" (City of Chicago, 2017). As a result of this policy, a number of sustainable development grants have been offered to encourage projects such as green roofs, green buildings using the

Leadership in Energy and Environmental Design (LEED) rating system, rainwater collection, home renovations, etc. This governmental support has sparked other organizations to follow suit and focus their efforts on sustainable development.

In 2002, Christy Webber Landscapes installed urban-scale permaculture in Millennium Park, in an effort to create an urban food forest garden example in a highly visible tourist area. There are traditional looking plant beds filled with edible and perennial plants. In addition to this public installation, there have been many community gardening efforts throughout the city, some are seeming acts of rebellion where a guerilla-type garden is placed on the side of the railroad tracks while others are more calculated and involve non-profit organizations, educational aspects, and outreach efforts. Organizations like Growing Power, Growing Home, and Chicago Botanic Garden have partnered together to develop larger urban farms like The Edible Gardens at Lincoln Park Zoo, Wood Street Urban Farm in the Englewood neighborhood, and the Rooftop Farm at McCormick Place West.

Growing Home is Chicago's leading expert in farm-based training for people with employment barriers. By providing 25 hours per week of paid on-the-job experience and job-readiness training at our farms, plus the support to conquer issues like criminal records, medical needs, child-care, and housing, we have changed the lives of hundreds of workers, and thousands of their family members. (Growing Home, n.d., para. 1)

This organization not only trains individuals with barriers to employment but they have done so through urban agriculture. Building on the skills learned at the farm, the individuals have found successful careers focused in the food service industry, unfortunately though, Chicago lacks agricultural jobs so most of the graduates end up working in restaurants or grocery stores.

Another example of urban permaculture is a commercial building development called The Plant, "a collaborative community of small food businesses, all focused on growing, producing, and/or sourcing a variety of food products" (The Plant, n.d., para. 2). Bubbly Dynamics, the organization that initiated this development, created this concept for food production that epitomizes a permaculture design, "founded on a model of closing waste, resource, and energy loops" (The Plant, n.d. para. 4). The tenants work cooperatively, using the waste of one as a resource for another. In order to be considered as a tenant, companies must figure out how they can be included in this closed loop system. Additionally, the building is intentionally left up to 20% vacant as an effort to accommodate current tenants' future growth.

These are some of the primary efforts of sustainable development taking place in the city of Chicago, there are many more efforts being made throughout the city though. There are community gardens throughout the city, as well as other larger scale urban farms that provide job training opportunities. Additionally, there are efforts made through the Chicago Public Schools (CPS) with school gardens and a high school focused on agricultural sciences. These efforts fulfill many of the SDGs, specifically Goal 1. No poverty is met by providing educational and training opportunities, these organizations and programs help to mitigate the larger cycle of poverty by focusing on individuals that have barriers to employment. When investigating the targets and indicators in each goal, it will be found that these urban organizations meet the following SDGs: Goal 2. Zero Hunger, Goal 3. Good Health and Well-Being, Goal 4. Quality Education, Goal 5. Gender Equality, Goal 8. Decent Work and Economic Growth, Goal 9. Industry, Innovation and Infrastructure, Goal 11. Sustainable Cities and Communities, Goal 12. Responsible Consumption and Production, Goal 16. Peace, Justice and Strong Institutions, and Goal 17. Partnerships for the Goals.

Case Study Part 2: Suburban Land

The second part of the case study focuses on the suburban property of Ron and Vicki Nowicki, in Downers Grove, IL. This seemingly standard suburban property includes food production as well as propagation of plants for their clients. The Nowicki's are landscape architects, educators, and permaculturists. Ron Nowicki attended landscape architecture school at the University of Illinois, in the 1970's at the same time that Mollison and Holmgren were working in Australia. Little did he know that his thesis to design a suburban property with ecological consciousness was a permaculture design. The senior project ended up incorporating "a passive solar building design, a large lot for growing food, good school district, reasonable price, and access to public transportation, which was developed in Downers Grove, IL over the past 40 years" (Nowicki, n.d.). This passive solar building design included a south facing wall of windows to allow for solar heating in the winter and central fireplace and chimney that distributed heating throughout as well as incorporating overhangs and a canopy of trees to protect the south window wall from the intense summer sun. This intentional design was quite successful allowing for minimal air-conditioning to be used in the summer, which is usual for most residences in the U.S. This typical suburban property has evolved into the lush fruitful homestead it is today, during the height of the summer when the plants are flourishing it is difficult to see everything while walking through the garden. The images in Figure 1 show a comparison of the property during the winter and the summer from the top. This top view is ideal to be able to see the free-form beds for planting the guilds, in the upper right corner is a small gazebo made of found objects like an old wooden ladder that creates a respite in the summer. Growing on top of this gazebo is a variety of hardy kiwis, while it is surrounded on all sides by squash, tomatoes, peppers, herbs, and medicinal flowers. On their relatively small property (.35 acres [1416.4m^2]), they have created about four zones; with the farthest zone acting as their compost area. This is a corner of the property where they have piles of compost that will be broken down to use in the yard the following year.

The Nowicki's grow the majority of the fruits and vegetables that they eat, they also grow plants to be used in their business. They keep the majority of their waste on their property and recycle what cannot be reused or composted. With the exception of municipal power/sewage, they have managed to become a self-sustaining property in a suburban location. They can continue to invest into their property and purchase

Figure 1. The Nowicki's backyard, early summer view (top) and winter view (bottom) © *[2018]* ***[Ron Nowicki]*** *Used with permission.*

solar panels or a small wind turbine if they wanted to build upon their sustainability and go "off of the grid" as much as possible. The Nowicki's are contributing greatly to the SDGs and incorporate the following goals into their property: GOAL 2. Zero Hunger, GOAL 9. Industry, Innovation and Infrastructure, GOAL 11. Sustainable Cities and Communities, GOAL 12. Responsible Consumption and Production, GOAL 13. Climate Action, and GOAL 15. Life on Land.

Case Study Part 3: Rural Land

The third part of the case study will focus on rural opportunities for sustainable development, focusing on the efforts of Midwest Permaculture founded by Bill and Becky Wilson in 2007. "The purpose of our work is to support the transition of our society from a culture of consumption, into a culture of creation" (Wilson, 2017, para. 1). The Wilson's live in Stelle, IL, a small community started in the early 1970's. To create a sustainable community, the founders of Stelle, purchased farmland and designed a small suburban neighborhood. In essence, they turned the rural into suburban. This is one approach to creating a sustainable development. However, along with creating a community comes the problems that people bring with them: garbage and waste disposal, food production, education, electricity, etc.

Midwest Permaculture (MP) has over the years tackled these problems on their own property, throughout the Stelle community, and their designs. Through their permaculture connections and courses, they have built a team of individuals, each responsible for their unique specialty. Essentially, they have designed their lives using the permaculture principles and ethics and apply this when it comes to their projects. MP's current projects are all located in the midwestern U.S. states of Illinois, Missouri, and Ohio. The largest project is 320 acres [1.3km^2], the smallest, their own residence, .25 acres [1011.7m^2]. During their first year offering Permaculture Design Courses (PDCs) their students focused on designing their residential property. One of the courses designed a berm along the side of the house. Also, a front-yard rain garden, a water harvesting system that keeps the rain water from causing stress on the sewage system by storing it. The next course was tasked with designing the sewage system for the entire city of Stelle, and received some hands-on experience digging out the areas for the rain to gather and thus creating the berm. Based on the Wilson's adjustments, this property has evolved tremendously over the years, Figure 2, shows the contrast of the house before any development in 1988 and again in 2017 after the permaculture plan had been implemented.

MP worked with the Center for Sustainable Community (CSC) to develop the 8.7 acres [25307.7m^2] of land that Stelle, IL resides upon (Wilson, n.d.). The plan includes an orchard food forest, a seasonal hoop house, Hügelkultur swales with linear food forest, chickens and ducks, an experimental construction earthcamp village, and a natural pond with chinampas-style water gardens. This is a long-term project and at the time of this chapter, the development is in its initial stages, with the first house of the village almost completed, the hole dug for the pond, the Hügelkultur swales created and planted, the seasonal food forest has been started and the food forest has been initiated with edible tree varieties.

Many of these elements can be seen in the other project's MP has designed. Heal the Planet Farm is the largest property set in Koshkonong, MO. Jordan Rubin, owner of the farm, is a New York Times best-selling author of *The Maker's Diet*; his illness led him to permaculture while searching for a solution to "real healthy food" (Wilson, n.d.). The design for this 320-acre [1.3km^2] farm includes a food forest orchard, bees, chickens, cows, lambs, multiple south-facing greenhouses, south-facing gardens and raised garden beds, aquaponics, and cricket and worm farming. All of these systems in the end will work together to feed each other. They will use the calves or lambs to graze the area surrounding the food forest, this will assist in building fertility and reduce invasive plant species while providing food to the animals. Chickens will not only provide meat and eggs but will also fertilize and be fed from the farm's waste stream. Bees provide honey and assist in fruit and vegetable production through pollination, the greenhouses are south-facing to harvest the natural sunlight and include a rocket mass heater that uses the excess wood from the farm to burn the fire in the oven area, with the exhaust pipe running under the greenhouse, keeping the aquaponics systems area warm (Wilson, n.d.). This year-round growing system includes the production of the food for the fish as well as the closed-loop aquaponics system that grows fish and vegetables. The design includes a lot of land shaping efforts due to its 10'-0" [3m] elevation change, for example, swales and raised garden beds on the contours (Wilson, n.d.). This allows for the water to be collected in regular intervals for the plants to use at each level. This keeps all of the water on the land instead of running off the property, which could lead to flooding. By keeping all of the water on the property and putting it to use, the chance of flooding is significantly lessened. Interestingly, one of the key features of this vast property is the conversion of an old swimming pool into a natural pool that is cleaned by nature, not chlorine (Wilson, n.d.).

These properties provide numerous job opportunities as well as restore the natural environment while providing opportunities for vast amounts of food production. They also address a number of the SDGs, including: GOAL 1. No Poverty and GOAL 2. Zero Hunger which are the key goals that are met here with these rural-scale permaculture developments. Ultimately, the larger the property the more opportunity there are for jobs in food production. Heal the Planet Farm is a prime example of the amount of food that can be produced on a property. The variety of plant species, crop production, and animals will allow for an important diversity that will create a resilient farm. For example, if a farmer were to plant their entire 320 acres [1.3km^2]

Figure 2. Midwest Permaculture headquarters front yard, 1988 (top) 2017 (bottom)
*© **[2018] [Midwest Permaculture]** Used with permission.*

with a monoculture, the idea of most current farmers where only one crop is grown in an area, their farm would be vulnerable to tragedy. It's like the saying goes "don't put all your eggs in one basket" so this idea of a permaculture farm would provide that diverse income stream. The other goals that are achieved in these rural-scale properties are GOAL 3. Good Health and Well-being, GOAL 4. Quality Education, GOAL 8. Decent Work and Economic Growth, GOAL 9. Industry, Innovation and Infrastructure, GOAL 12. Responsible Consumption and Production, GOAL 13. Climate Action, GOAL 14. Life Below Water, and GOAL 15. Life on Land.

CURRENT CHALLENGES

The biggest challenge being faced in the U.S. in 2019 is the lack of governmental support and participation especially in the creation of environmental standards and regulations (Ferguson & Lovell, 2015). President Trump withdrew the U.S. from the Paris Agreement in 2017 (Shear, 2017). The U.S. is one of the largest polluters and without strict governmental regulations and efforts to reduce carbon emissions, the tragic events predicted by the IPCC may likely ring true. Those individuals who are working toward sustainable development are more likely to survive this tragic end while the remainder will perish. This is the same administration that has failed to provide adequate support to the victims of the tragic hurricanes, fires, and other natural disasters. This administration has dismantled the Environmental Protection Agency (EPA) whose purpose was to aid in the regulation of offshore drilling, pollution, and non-renewable resource use. There are only pockets of resistance where there are local government benefits that push for sustainable development.

Another challenge is that people often do not see the value in making changes, as the reality of climate change or global warming having a direct effect on them personally has not been felt. Kluger (2018) states:

You'd think the end of the world would be enough to get us scared. Humans have always been an exceedingly risk-averse species—which is how we came to survive as a species at all. If there are lions on one part of the savannah, we go to another. If crocodiles keep coming out of the river, we fish somewhere else. So, when it comes to the loss of the entire planet, well, we ought to take action. And yet we don't; we never do. (para. 1)

When looking at history, this idea of climate action rears its ugly head every 20-30 years, and it is clear from the UN reports that the time is now to act whether catastrophe is imminent or not.

SOLUTIONS AND RECOMMENDATIONS

This chapter has reviewed ways that this resistance can be made by using the projects as models, following the permaculture ethics and principles. Grassroots efforts need to be made to illustrate the citizen's desire to make a change in the world (Appadurai, 2000; Castells, 1983; Ferguson & Lovell, 2015; Seyfang & Smith, 2007; Veteto & Lockyer, 2008). The first solution is to start making the changes needed to create sustainable development by using permaculture as a guide. Another solution is to contribute to the academic and scientific research on permaculture to further the conversation on the theory. If there is a universal global effort towards a more sustainable end, it is possible to reverse the tragic prediction for the future. Ron Nowicki ended his interview with the following to consider:

The PLEDGE OF ALLIANCE is a pledge to the Earth from which we come. The Earth is alive, enables our lives and is where we are (quite literally) grounded. She gives us everything we need to survive and thrive IF we use those gifts wisely. She is very powerful, and we need her. She does NOT need us. It is truly in our best interests to become her ally and not remain the enemy we have become. I have a written a pledge of peace with the planet to help others commit to the urgent need for change:

I pledge alliance to the earth and to the living systems and cycles that give us life. One common direction, unchanging, with health and prosperity to all. (Ron Nowicki, personal communication, 2018)

REFERENCES

Akhtar, F., Lodhi, S. A., & Khan, S. S. (2015). Permaculture approach: Linking ecological sustainability to business strategies. *Management of Environmental Quality*, *26*(6), 795–809. doi:10.1108/MEQ-01-2015-0001

Akhtar, F., Lodhi, S. A., Khan, S. S., & Sarwar, F. (2016). Incorporating permaculture and strategic management for sustainable ecological resource management. *Journal of Environmental Management*, *179*, 31–37. doi:10.1016/j.jenvman.2016.04.051 PMID:27155728

Appadurai, A. (2000). Grassroots globalization and the research imagination. *Public Culture*, *12*(1), 1–19. doi:10.1215/08992363-12-1-1

Brundtland, G. H. (1987). *Our common future: Report of the world commission on environment and development (UN Publication No. A/42/427)*. United Nations.

Castells, M. (1983). *The City and the Grassroots: A cross-cultural theory of urban social movements*. Los Angeles, CA: University of California Press.

Christy Webber Landscapes. (n.d.). *Millennium park.* Retrieved from http://www.christywebber.com/projects/millennium-park/

City of Chicago. (2017, January). *Chicago Sustainable Development Policy*. Retrieved from https://www.cityofchicago.org/city/en/depts/dcd/supp_info/sustainable_development/chicago-sustainable-development-policy-update.html

Egelston, A. E. (2013). *Sustainable development: A history*. New York, NY: Springer. doi:10.1007/978-94-007-4878-1

Eisenhardt, K. M. (1989). Building theories from case study research. *Academy of Management Review*, *14*(4), 532–550. doi:10.5465/amr.1989.4308385

Eisenhardt, K. M., & Graebner, M. E. (2007). Theory building from cases: Opportunities and challenges. *Academy of Management Journal*, *50*(1), 25–32. doi:10.5465/amj.2007.24160888

FEMA. (n.d.). Retrieved from https://www.fema.gov/

Ferguson, R. S., & Lovell, S. T. (2013, April). Permaculture for agroecology: Design, movement, practice, and worldview. A review. *Agronomy for Sustainable Development*, *34*(2), 251–274. doi:10.100713593-013-0181-6

Ferguson, R. S., & Lovell, S. T. (2013, August). *Engaging ecological literacy from the ground up: The influence and impact of the permaculture movement.* Paper presented at 98th Ecological Society of America Annual Meeting: Sustainable pathways: Learning from the past and shaping the future, Minneapolis, MN.

Ferguson, R. S., & Lovell, S. T. (2015, December). Grassroots engagement with transition to sustainability diversity and modes of participation in the international permaculture movement. *Ecology and Society*, *20*(4), 39. doi:10.5751/ES-08048-200439

Glanzberg, J. (n.d.). *An open letter and plea to the permaculture community.*. Retrieved from http://patternmind.org/an-open-letter-and-plea-to-the-permaculture-community/

Growing Home. (n.d.). Retrieved from http://growinghomeinc.org/about-us/

Hahn, E. (2014). *A permaculture primer.* Retrieved from https://midwestpermaculture.com/wordpress/wp-content/uploads/2015/12/Midwest-Permaculture-Permaculture-Primer.pdf

Hathaway, M. D. (2015). Agroecology and permaculture: Addressing key ecological problems by rethinking and redesigning agricultural systems. *Journal of Environmental Studies and Sciences*, *6*(2), 239–250. doi:10.100713412-015-0254-8

Hemenway, T. (2009). *Gaia's garden*. White River Junction, VT: Chelsea Green Publishing.

Intergovernmental Panel on Climate Change (IPCC). (2018, October 6). *Global warming of 1.5° C* (UN Publication No. IPCC SR1.5). Retrieved from http://report.ipcc.ch/sr15/pdf/sr15_spm_final.pdf

Keiren. (2013, October 17). The many benefits of hügelkultur. *Inspiration Green and Permaculture magazine*. Retrieved from https://www.permaculture.co.uk/articles/many-benefits-hugelkultur

Kluger, J. (2018, October 8). Why we keep ignoring even the most dire climate change warnings. *Time*. Retrieved from http://time.com/5418690/why-ignore-climate-change-warnings-un-report/?utm_medium=social&xid=time_socialflow_facebook&utm_source=facebook.com&utm_campaign=time

Luna, J. M. (2014). *A mixed-methods study of sustainability education* (Doctoral dissertation).

Luna, J. M., Dávila, E. R., & Reynoso-Morris, A. (2018). Pedagogy of permaculture and food justice. *The Journal of Educational Foundations*, *31*(1 & 2), 57–85.

Mercader-Moyano, P. (Ed.). (2017). *Sustainable development and renovation in architecture, urbanism, and engineering*. New York, NY: Springer. doi:10.1007/978-3-319-51442-0

Mollison, B. (1988). *Permaculture: A designers' manual*. Tagari Publications.

Mollison, B., & Holmgren, D. (1978). *Permaculture one: A perennial agriculture for human settlements*. Melbourne, Australia: Transworld.

NASA's Jet Propulsion Lab. (2018). *Global climate change: Vital signs of the planet*. Retrieved from https://climate.nasa.gov/

Nowicki, R. (2016, April 12). *How to transition to a sustainable future at home*. Retrieved from https://www.youtube.com/watch?v=_ZIjgcdCjBc&feature=youtu.be

Nowicki, R. (n.d.). *The land office*. Retrieved from http://www.holdtonature.com/Hold_To_Nature/Home.html

Patton, M. (1990). *Qualitative evaluation and research methods* (2nd ed.). Newbury Park, CA: Sage.

Permaculture Design Principles. (n.d.). Retrieved from https://permacultureprinciples.com/principles/

Permaculture Ethics. (n.d.). Retrieved from https://permacultureprinciples.com/ethics/

Reilly, K. (2018, October 8). Here's what humanity must do immediately to prevent catastrophic climate change, according to the new U.N. report. *Time*. Retrieved from http://time.com/5418577/what-humanity-do-limit-climate-change/

RuddockB. (2013). *Plant guilds*. Retrieved from https://midwestpermaculture.com/eBook/Plant%20Guilds%20eBooklet%20-%20Midwest%20Permaculture.pdf

Sachs, J. D. (2015). *The age of sustainable development*. New York, NY: Columbia University Press. doi:10.7312ach17314

Seyfang, G., & Smith, A. (2007). Grassroots innovations for sustainable development: Towards a new research and policy agenda. *Environmental Politics*, *16*(4), 584–603. doi:10.1080/09644010701419121

Shear, M. D. (2017, June 1). Trump will withdraw U.S. from Paris Climate Agreement. *The New York Times*. Retrieved from https://www.nytimes.com/2017/06/01/climate/trump-paris-climate-agreement.html

Sustainable Development Goals. (n.d.) Retrieved from https://sustainabledevelopment.un.org/sdgs

The International Institute for Sustainable Development. (2012). *Sustainable development timeline*. Retrieved from https://www.iisd.org/pdf/2012/sd_timeline_2012.pdf

The Plant. (n.d.). Retrieved from https://www.bubblydynamics.com/the-plant/

United Nations Climate Change. (2018, June 20). *Extreme weather continues in 2018: A continuing call to climate action*. Retrieved from https://unfccc.int/news/extreme-weather-continues-in-2018-a-continuing-call-to-climate-action

U.S. Geological Survey (USGS). (n.d.). Retrieved from https://www.usgs.gov/

USDA Forest Service. (2018, August 5). *Active fire mapping program*. Retrieved from https://fsapps.nwcg.gov/#

Veteto, J. R., & Lockyer, J. (2008). Environmental anthropology engaging permaculture: Moving theory and practice toward sustainability. *Culture & Agriculture*, *30*(1&2), 47–58. doi:10.1111/j.1556-486X.2008.00007.x

Walsh, B. (2014, March 31). Warming world threatens us all, warns U.N. report. *Time*. Retrieved from http://time.com/43118/climate-change-global-warming-united-nations/

Wilson, B. (2017, August 15). *Residential design for Midwest permaculture home*. Retrieved from https://midwestpermaculture.com/2017/08/residential-design-for-midwest-permaculture-home/

Winslow Food Forest. (n.d.). Retrieved from https://www.winslowfoodforest.com/about/

Wolf, S. (2018, October 8). *Report details brutal cost of climate inaction.* Retrieved from https://medium.com/center-for-biological-diversity/report-details-brutal-cost-of-climate-inaction-8a5a038bd6e0

Worland, J. (2018, October 8). Scientists just laid out paths to solve climate change: We aren't on track to do any of them. *Time.* Retrieved from http://time.com/5418134/ipcc-climate-change-report-2030-crisis/

World Vision. (2018, August 1). *From the field: Facts, FAQs, and how to help.* Retrieved from https://www.worldvision.org/

Yin, R. K. (1984). *Case study research: Design and methods* (2nd ed.). Thousand Oaks, CA: Sage Publications.

KEY TERMS AND DEFINITIONS

Annual Plants: Plants that need to be replants with seeds year after year.

Closed-Loop System: A system that produces no waste because it is used within the system. For example, an aquaponics system is a closed-loop system, the waste from the fish is used to fertilize the plants while the plants produce the bacteria needed to break down the ammonia in the water.

Food Desert: An area that has no or minimal access to fresh food including fruits, vegetables, meat, etc. The stores that are typically located in the areas are limited to convenience or gas station stores.

Food Justice: The belief that everyone, no exceptions, has the right to fresh, wholesome food.

Guerilla Gardening: An underground gardening method where people garden as acts of resistance.

Perennial Plants: Plants that will regrow year after year without the need to replant from seed.

Sustainability: The ability of a society to be able to fulfill the needs of this generation without jeopardizing the ability of future generations to fulfill their needs.

Sustainable Development: A development that considers sustainability in aspects of its design.

Chapter 8
Green Operational Strategy for Airlines:
Content and Regional Analysis

Yazan Khalid Abed-Allah Migdadi
https://orcid.org/0000-0001-5687-5410
Qatar University, Qatar

EXECUTIVE SUMMARY

This chapter identifies the content of airlines' operations strategy and reports the strategy patterns adopted by the airlines of each region. A detailed configuration of the airlines' green operational strategy is developed, using the content analysis of the sustainability reports from 23 airlines in five regions (North America, Europe, Asia, the Middle East, and South America). The green operational strategy adopted by each region is identified; each region adopted a green pattern that was unlike those of any other region. The indicative models for each region and across regions are developed by using a simple and special tailored quantitative analytical technique. The results of this chapter raise a set of questions about the impact of contextual factors on whichever green strategy pattern is adopted, indicating the need to conduct more in-depth analysis of green actions. This is one of a few studies to have developed a comprehensive definition of airlines' green operations strategy and explore the green strategy patterns adopted by airlines from different regions.

BACKGROUND INFORMATION

Previous studies of airlines' green practices have investigated them from the perspectives of marketing, operations, information technology and technical

DOI: 10.4018/978-1-5225-8559-6.ch008

specifications, among others. The marketing perspective studies have investigated such aspects as the use of eco-labels (e.g. Baumeister & Onkila, 2017); customers' willingness to pay for airlines' green practices (e.g. Hagmann, Semeijn & Vellenga 2015; Mayer, Ryley & Gillingwater, 2015; Horio, Kumar, Levin & Sung, 2016; Han, Yu & Kim, 2019); and the willingness to purchase green aviation products (e.g. Hinnen, Hille, & Wittmer 2017). Most operational studies have investigated the impact of particular operational practices on green indicators (e.g. Smith, 2016; Teoh & Khoo, 2016; Will et al., 2016; Yan, Cui, & Gil, 2016). The green operations strategy aspect is rarely investigated (e.g. Lin, 2016; Teoh & Khoo, 2016; Lee, Tsai, Yang & Lin, 2017; Migdadi, 2018). Most studies have reported the practices in a single country or a limited number of countries (e.g. Lynes & Dredge, 2006; Harvey, Williams & Probert, 2013; Chapman, 2016; Horio, Kumar, Levin & Sung, 2016; Liu, Zhou, Zhou, & Wang, 2017), or in the context of a single region (e.g. Yan, Cui, & Gil, 2016).

The green operations strategy is a combination of different green operational actions taken by institutions to acknowledge green indicators (Migdadi, 2016). Some previous studies of airlines' green operational actions have classified the green operational actions as either technology based or process based. The technology based ones are related to aircraft design, whether of vehicles or engines, but the process oriented actions are related to such areas as route management, weight management, flight operating ... etc. Previous studies found that the process oriented actions positively affected the airlines' profits and efficiency (Yan, Cui, & Gil, 2016). The technology oriented actions have been a research issue for many scholars, who have focused on investigating the environmental impact of green fleet design (e.g. Teoh & Khoo, 2016). These studies find that the most important action taken by airlines is renewing the aircraft fleet (Will et al., 2016). The new generations of aircraft such as A350 XWB, A380 and Boeing 747-8 and 787 achieve lower CO_2 emissions and reduce noise (Szodruch, Grimme, Blumrich & Schmid, 2011). The adoption of other innovative technologies such as open rotor technology has the potential to reduce CO_2 emissions (Smith, 2016).

The process oriented actions have been put under the spotlight by scholars, who have investigated the use of bio-fuels (Will et al., 2016), the weight management, reduced use of APUs, air traffic management (Lee, Tsai, Yang, & Lin, 2017; Lee, Tsai, Yang & Lin, 2018), route distribution (the distance of flights), rate of fuel consumption, movements (the number of takeoffs and landings), the rate of use of each aircraft (Liu, Zhou, Zhou, & Wang, 2017), and the altitude restrictions (Williams, Noland & Toumi, 2002). These studies find that the use of bio-fuel is affected by economic rather than environmental factors: customers are not willing to pay extra for bio-fuel and it is not expected to make a significant contribution until 2025 (Will et al., 2016). In addition, these studies find that the most important factors in reducing CO_2 emissions are the routes distribution (Liu, Zhou, Zhou, & Wang, 2017) and the altitude restriction (Williams, Noland & Toumi, 2002). As discussed above, most of the previous studies have focused on investigating flight design and operating actions, or GHG1 and fuel saving actions. The scope of airlines' green actions is broader and covers many aspects. Therefore, this chapter defines in detail the content of airlines' green operations strategy.

THE CURRENT CHALLENGES FACING AIRLINES IN THE GREENING ERA

In September 2015 at an historic UN summit, the 17 Sustainability Development Goals (SDGs) were adopted by the world's leaders as part of the 2030 Agenda for Sustainability Development. Attaining these goals depends on tackling environmental climate change and fostering sustainable development as two sides of the same coin, according to the 2030 Agenda; many SDGs address the central causes of climate change (UN, 2018). SD forces governments to adopt the SDGs and make firms, whatever their size and whatever industry they represent, more accountable to ecology and the environment (Singh & Trivedi, 2016) and more willing to develop competitive green strategies (Bhardwaj, 2016; Kirchoff, Tate & Mollenkopf, 2016). The aviation industry, like other industries, started to adopt green practices and ecologically friendly strategies. Aircraft engines used to emit greenhouse gases such as carbon dioxide, water vapor, and hydrocarbons, and to causes negative environmental impacts such as heat and noise (Lee et al., 2010 and Brasseur et al. 2016). Around 2% of the world's emissions of carbon dioxide are produced by aviation (IPCC, 2015).

The increasing demand for air journeys over the past 25 years has increased the environmental impact of aviation (Lee et al., 2009; Macintosh & Wallace, 2009). For example, CO_2 emissions increased by 80% between 1990 and 2014 and the emissions of NO_X doubled. Both are expected to grow between 2014 and 2035 by a further 43% (EASA, 2016). While the literature of green air transport strategy reveals that no single strategy pattern can mitigate the prospect of ecological damage, a range of diverse green strategies for aviation, which include technological, operational and infrastructural improvement, have been devised to lead to better environmental impact (Teoh & Khoo, 2016). The future target of airlines by 2050 is to halve their 2005 total of CO_2 emissions (Brooks et al., 2016) and they are doing their best to manage the environmental impact of their operations. This raises the question of what are the effective green operational strategies adopted by airlines to reduce their negative environmental impact?

The green air transport literature reveals that there is no single strategy or approach that can be implemented by all to alleviate ecological damage (Teoh & Khoo, 2016). The aviation industry needs to improve strategies for managing aviation fleets in light of corporations' social responsibility (CSR) policies and environmental considerations. Green aviation fleet development involves complex relationships between technology, operations, infrastructure and economic performance (Teoh & Khoo, 2016). Previous studies have demonstrated the need to understand the key success factors of green airline performance (Abdullah Chew & Hamid, 2016), but if high performing green airlines could provide benchmarks, a green airline framework might supply a solution for the future of the airline industry;

THE CONTENT OF AIRLINES' GREEN OPERATIONS STRATEGY

This section was developed by using the content analysis of sustainability reports made by 23 airlines from different regions (Asia, Europe, North America, South America and the Middle East). Most of the reports were retrieved from the GRI database. The building blocks of green operations strategy are in two main parts: the green performance indicators, and green operations actions. Accordingly, the facts from each region related to these two parts are summarized in each case. Then the summary is analyzed as follows:

Identifying and Defining the Green Indicators

Table 1 identifies the green indicators reported by airlines over the period 2013-2016, and defines them. It can be seen that eight green indicators were reported: three related to GHG emissions (GHG1, 2, 3): three related to resources consumption (energy, fuel and water) and two indicators related to waste management (wants generated and recycling). Table 2 shows this summary. The most widely green indicator reported by airlines is GHG1 emissions, which was reported by 22 cases, or 95.7% of the study sample. The least reported indicator was noise, which was reported by 4 cases, or 17.4%.

Identifying the General Theme and Sub-Themes of the Summarized Actions

The general theme was determined by the green indicators listed above; however, the sub-themes were identified according to the titles used by airlines in the reports, and the most widely adopted titles were used in the classification scheme of this construct. The general themes and sub-themes are presented in Table 3. It can be

Table 1. The list and definition of airlines green indicators

Green indicator	Operational definition
GHG 1 emissions	Metric tons of direct CO2 emissions from jet fuel and ground support.
GHG 2 emissions	Metric tons of indirect CO2 emissions of electricity, power and heat from direct billing of owned or leased facilities
GHG 3 emissions	Metric tons of indirect CO2 emissions in the parts of the value chain out of GHG 2 emissions
Energy consumption	Kilowatts per hour of electricity
Fuel consumption	Gallons of fuel
Water consumption	m^3 of consumed water
Waste generated	Tons of waste generated
Recycling	Tons of waste recycled
Noise	Decibel level of aircraft sounds

Table 2. The frequency of green indicators reported by airlines across regions

Green indicator	Frequency of green indicators reported by airlines of each regions					number of cases reported the indicator	% of sample (N= 23)
	North America	Europe	Asia	South America	Middle East		
GHG 1 emissions	6	6	6	1	3	22	95.7%
Energy consumption	3	5	7	1	2	18	78.3%
GHG 2 emissions	6	3	3	1	2	15	65.2%
Fuel consumption	3	3	5	1	3	15	65.2%
Waste generated	4	4	5	0	1	14	60.9%
Water consumption	2	2	5	0	1	10	43.5%
Recycling	3	3	3	0	0	9	39.1%
GHG 3 emissions	5	1	0	1	1	8	34.8%
Noise	2	1	1	0	0	4	17.4%

seen that the general themes are five in number; GHG1 and fuel saving actions, GHG2 and energy saving actions, GHG3 actions, waste management and recycling actions, water management actions.

The sub-themes of GHG1 and fuel saving actions number six; these themes are related to airlines' core operations, the design of aircraft, the planning of routes, the operating of flights, and the management of their weight, fuel and maintenance. There are seven sub-themes, related to building, facilities, engines and vehicles, for

GHG2 and energy saving actions. The GHG3 actions have eight sub-themes, related to the upstream and downstream parts of the value chain. Hence, these sub-themes are related to non-core operations, which relate to preparing for operations such as maintenance, vehicles and engines design and operations.

Table 3. Airlines green operations actions categories

GHG1 and Fuel saving actions	GHG2 and Energy saving actions	GHG3 actions	Waste management and recycling actions	Water management actions	Noise
GF1: Aircraft design (4 actions)	**GE1:** Facilities and building ways to save energy (3 actions)	**GHG3-1:** Fights maintenance management (4 actions)	**WR1:** Onboard and ground waste recycling (10 actions)	**W1:** Recycling and the recovery of maintenance water (2 actions)	**N1:** Meeting international standards (1 action)
GF2: Route management for flights (7 actions)	**GE2:** Facilities and building accreditation (1 action)	**GHG3-2:**Vehicle and engine design (3 actions)	**WR2:** Onboard and ground waste upcycling (5 actions)	**W2:** Saving maintenance water (1 action)	**N2:** Flight procedures (3 actions)
GF3: Flights operating (15 actions)	**GE3:** Facilities and building sustainable energy use (3 actions)	**GHG3-3:** Vehicle and engine operating (4 actions)	**WR3:** Reusing onboard and ground waste (1 action)	**W3:** Recycling and the recovery of water for facilities and buildings (3 actions)	
GF4: Flight weight management (13 actions)	**GE4:** Upgrading and replacing facilities (3 actions)	**GHG3-4:** Vehicle and engine maintenance management (2 actions)	**WR4:** Reducing the use of paper (3 actions)	**W4:** Saving water in facilities and buildings (13 actions)	
GF5: Flight fuel management (4 actions)	**GE5:** Vehicle and engine design (3 actions)	**GHG3-5:** Transportation management (5 actions)	**WR5:** Recycling industrial waste (4 actions)		
GF6: Flight maintenance management (4 actions)	**GE6:** Vehicle and engine operating (4 actions)		**WR6:** Upcycling industrial waste (1 actions)		
	GE7: Vehicle and engine maintenance management (2 actions)		**WR7:** Reusing industrial waste (8 actions)		
	GE8: Transportation management (5 actions)		**WR8:** Recycling hazardous waste (2 actions)		
			WR9: Upcycling hazardous waste (1 action)		
			WR10: Using hazardous waste (1 action)		

The sub-themes concerning waste management and recycling actions fall into nine classes, to do with recycling, upcycling and the use or reuse of three kinds of waste. The types of waste are onboard and ground waste, hazardous waste and industrial waste. The sub-themes related to water management number four, touching the recovery, recycling and saving of two uses of water, either for maintenance or facilities and buildings. Finally, the actions (items) are grouped under the sub-themes identified previously. The detailed actions under each theme and sub-theme are listed in Appendix 1 (Tables 11 through 15).

RESEARCH METHODOLOGY

The sample in the present study was a convenient stratified sample determined by the availability of secondary published data; the sample contained reports of airlines' green actions and the green indicators in their sustainability reports, as well as social responsibility reports over the period 2013-2016. The airlines represented five regions; North America, Europe, Asia, the Middle East, and South America. The study sample consists of 23 cases: six airlines from North America, six airlines from Europe, seven airlines from Asia, three airlines from the Middle East and one airline from South America. This study is a time series, to allow the changes in green indicators and actions to be tracked. Sustainability reports of the selected cases (i.e., the initial report and the most recent) were retrieved. Most of them started to report their level of sustainability (social responsibility) in 2013 and the most recent reports were published in 2016. Thus, the study period is 2013-2016.

The sources of data were the Global Reporting Initiatives database (GRI) and the reports published on the corporations' websites. Most of these reports were retrieved from the GRI. The GRI is an independent international organization which was established in 1997 in the USA. It seeks to help organizations to report their effect on sustainability. It develops standards in reporting sustainability, and publishes reports via its website (GRI, 2018). Data on 17 of the airlines were retrieved from GRI and on the remaining six from the corporations' websites. To allow any changes to be traced, each airline should have published two reports, the first in 2013 and the second in 2016. Once retrieved for the present study, these reports were classified according to region and then the facts reported about each airline were summarized to cover the period 2013-2016.

The summary included the green indicators and the actions taken by each airline. To check the reliability of the data, the facts were in some cases collected for a second time. All such cases were randomly chosen from the various regions, and then initial and later collections of facts were compared. Furthermore, when there were two sources of data – websites and the GRI – data from both were compared,

Table 4. Example of how to compute action index

Action Category	Actions list	NA1
Aircraft design	fleet modernization	1
	split scimitar winglets	
	Sharklet wingtips	
	Engine modification	
	Total # of actions adopted by NA1	1
	Index = 1 / 4	0.25

to check their reliability. After the required data had been collected, the analysis began, in which different tools and techniques were used, as follows:

Phase 1: The data for each region were reported, the date of the first period being 2013 and of the last period 2016. Two separate lists, one for green indicators and another for green actions, were developed

Phase 2: the relative changes index in the green indices over the study period for each region were computed. The following formula was used for this purpose:

$(P_{2016} - P_{2013}) / P_{2013}$

P_{2013}: Realized performance 2013

P_{2016}: Realized performance 2016

Phase 3: Computing the action index for each case. The index reflects the degree to which each category of action had been adopted. Below is the detailed procedure for computing the index.

Step 1: Count the number of categories of action adopted by the case

Step 2: Divide the above number by the total number of adopted actions.

The result is the adoption index. See the example in Table 4.

Phase 4: the models indicating the relationship between green indicators and actions were developed for each region and across regions.

THE STAGES OF DEVELOPING THE INDICATIVE MODELS

The sample size of this study, 23 cases, is small. Moreover, the level of analysis is the regional level, so the sample size of each region is between 1 and 7 cases in each region. So it would be impossible to adopt statistical analysis techniques such as Linear Regression Analysis. Accordingly, it is more appropriate to develop indicative models by using simple data analysis techniques. The indicative models for the relationship between green indicators and actions were developed for each region as follows:

Step 1: The ranks of each region according to the improvement shown by each green indicator were identified. The ranks were assigned to green actions; see the example in Table 5.

Step 2: The actions which were better realized by one region than by all the others were identified.

Step3: The difference between the green action index of the best-ranking region and the average of the rest was identified.

Step 4: If more than one significant action was taken, the relative importance of these actions was identified by summing the indices to find a total actions index. Then, each action index was divided by the total. The result was the relative importance for each action.

Step 5: The difference in green indicators between the higher and the lower ranked regions was computed. The differences in green indices are further discussed in Appendix 2 (Table 16).

Step 6: The effect of the actions on each indicator was found by multiplying the relative impact of each action by the percentage of change in the green indicator. This calculation was made for each indicator of each region. The result is thus the indicative model for each region.

Table 5. Example of the way to develop the indicative models at regional level (water saving actions)

Step 1: Rank of regions		Water saving actions Step 2: the significant action taken by the best ranked region			
	region	Maintenance water recycling	Maintenance water saving	Facility water recycling	Facility water saving
NI	Asia	0.17	0.00	0.11	0.09
2	Europe	0.08	**0.17****	0.00	0.03
1	Middle East	0.00	0.00	0.00	**0.115***
NI	North America	0.00	0.00	0.00	0.01
NI	South America	0.00	0.00	0.00	0.00

NI: No Improvement

*: Middle East better than all other regions in this action

**: Europe better than Asia, North America and South America in this action

Table 6. Example of the way to compute the differences in the green action indices across regions

Rank	Regions	water consumption
NI	North America	1.83%
2	Europe	-3.75%
NI	Asia	0.56%
NI	South America	NR
1	Middle East	-14.16%
	Average of lower ranked ranges than the Middle East	-0.45%
	Difference between the percentage for the Middle East and for the average of other regions	-13.71%

Step 3: The change in the action index of the higher ranked region has the average of the lower ranked region deducted from it.

Here, the Middle East action index (0.115) – the average index of the lower ranked regions (0.032) = 0.083

Step 4: The relative impact of facility water saving action taken by the Middle East is 1.00 since it is the only action taken.

Step 5: The difference between the higher ranked region's green indicator index and the lower ranked regions is calculated. The example presented in Table 6 shows the difference in water consumption. The difference is –13.71%, which means that the airlines of the Middle East region reduced their water consumption by 13.71% in comparison with the other regions. The detailed calculations of the improvement in actions indices across regions are shown in Appendix 3 (Tables 17 through 23).

Step 6: the significant action taken the airlines of the Middle Eastern region is the saving in facility water consumption. The relative impact of this action is 1.00, so the impact of this action on water saving is:

1.0 x -13.71%= -13.71%. Further details of the calculations of the relative impact of actions are shown in Appendix 4 (Tables 24 through 30).

SOLUTIONS AND RECOMMENDATIONS FROM REGIONAL ANALYSIS

Table 7 shows the percentage of change in the green indicators of airlines in different regions. It can be seen that the improvement in the green indicators could be measured by a negative or positive percentage of change. The improvement in the indicators of GHG1 emissions, GHG2 emissions, GHG 3 emissions, fuel consumption, energy consumption, water consumption and noise was measured by negative percentages. However, the improvement of recycling, and waste generated was measured by

Table 7. The reported green performance of airlines across regions

Green indicator	Regions				
	North America	**Europe**	**Asia**	**South America**	**Middle East**
GHG 1 emissions	1.81%	**-4.26% (1) IP**	2.80%	**-2.55% (2) IP**	1.40%
Energy consumption	7.70%	2.93%	**-4.17% (1) IP**	**-7.26% (2) IP**	**-5.27% (3) IP**
GHG 2 emissions	**-3.46% (3) IP**	3.98%	**-12.26% (2) IP**	11.90%	**-25.41 (1) IP**
Fuel consumption	0.03%	**-5.73% (1) IP**	4.95%	**-2.53% (2) IP**	11.35%
Waste generated	1.83%	**-0.38% (1) IP**	13.15%	NRI	11.82%
Water consumption	1.83%	**-3.75% (2) IP**	0.56%	NRI	**-14.16% (1) IP**
Recycling	-3.47	**22.54% (1) IP**	**5.46% (2) IP**	NRI	NRI
GHG 3 emissions	1.66%	16.02%	NR	**-1.61% (1) IP**	12.90%
Noise	0.00%	**-5.50% (1) IP**	0.00%	NRI	NRI

IP: Improved Performance
NRI: Indicator Not Reported
(1),(2),(3): Regions' ranks according to the improved performance indicators

positive percentages. In addition, it can be seen that the regions did not improve in all indicators, and some regions improved in more indicators than others. For example, North America improved in one indicator, which was GHG2 emissions, whereas the European region improved in six indicators, namely, GHG1 emissions, fuel saving, waste generated, water consumption, recycling and noise. The regions differed in the ranking of the improved indicators; for example, European airlines were first in terms of GHG1 emissions. However, the Asian airlines came first in terms of energy saving. The Middle Eastern airlines were first in terms of GHG2 emissions and water consumption. European airlines were first in terms of fuel consumption, waste generated, recycling, and noise.

Table 8 shows the impact of significant green actions on green indicators across regions. It can be seen that the effect of actions taken by South America region's airlines to reduce GHG1 emission and fuel consumption was seen in flight route management and flight operating. The impact of route management was greater than that of flight operating on reducing GHG1 emissions and fuel consumption. The impact of route management was -3.17% on reducing GHG1 emissions and -5.54% on reducing fuel consumption. No shared actions with other regions and no significant actions taken by other regions related to reducing GHG1 emissions and fuel consumption.

The significant actions taken by Asia's airlines to reduce GHG2 emissions and energy saving were shown in energy saving actions in facilities and buildings; the impact of these actions on reducing GHG2 emissions was -16.4% and on energy saving was -9.49%. The same action was taken by the Middle East's airlines, but the impact of this action on reducing GHG2 emissions by the Middle Eastern airlines was greater. The reduction was -25.54%, but the impact on reducing energy saving was -7.42%. Generally speaking, the impact of energy saving in facilities and buildings on reducing GHG2 emissions was greater than on energy saving in both regions.

The airlines of North America took more diverse actions than those of Asia and the Middle East to reduce GHG2 emissions. These actions were facilities and buildings accreditation, vehicle and engine design, and vehicle and engine operating. The impact of these actions was -6.1%, -3.02% and -2.29 respectively. The total impact of these actions in reducing GHG2 emissions was -11.31%, which was less than the impact of the actions in facilities and buildings taken by the airlines of Asia and the Middle East.

Asia's airlines took two significant actions related to improved recycling. These actions were onboard and ground waste upcycling, and the upcycling of hazardous waste. The impact of these actions in improving recycling was 2.77% and 6.16% respectively. Onboard and ground waste upcycling action was taken by the Europe's airlines, but its impact on improving recycling was better than that of Asia's airlines, which was 21.55%.

The water saving actions taken by Europe's airlines were shown in saving maintenance water. The impact of this action was -4.95%, and this action was not shared with any other region. The water saving action taken by the Middle East's airlines was the saving of water in facilities and buildings. This action made a saving of -13.71%, and was also an action not shared with other regions. The only significant action related to noise management was taken by Europe's airlines. The action taken was meeting international standards, and its impact was -5.5%.

Table 8. The indicative model for the relationship between actions and green indicators across regions

Actions Categories	Actions	Regions				
		Asia	Europe	Middle East	North America	South America
Reducing GHG1 emissions actions	**GF2:** Flights route management	------	------	------	------	-3.17%
	GF3: Flights operating	------	------	------	------	-1.39%
Reducing fuel consumption actions	**GF2:** Flights route management	------	------	------	------	-5.54%
	GF3: Flights operating	------	------	------	------	-2.34%
Reducing GHG2 emissions actions	**GE1:** Facilities and building energy saving	-16.4%	------	-25.54%	------	------
	GE2: Facilities and building accreditation	------	------	------	-6.1%	------
	GE5: Vehicle and engine design	------	------	------	-3.02%	------
	GE6: Vehicle and engine operating	------	------	------	-2.29%	------
Energy saving actions	**GE1:** Facilities and building energy saving	-9.49%	------	-7.42%	------	------
Recycling actions	**WR2:** Onboard and ground waste upcycling	2.77%	21.55%	------	------	------
	WR9: Upcycling hazardous waste	6.16%	------	------	------	------
Water saving actions	**W2:** Saving maintenance water	------	-4.95%	------	------	------
	W4: Saving water in facilities and buildings	------	------	-13.71%	------	------
Noise	**N1:** Meeting international standards	------	-5.5%	------	------	------

------: actions with no impact

Note: only the regions that have better positions in terms of actions and green indicators were reported. The regions which are ranked higher in terms of indicators but are not ranked higher in actions are not reported.

THE SUMMARY OF GREEN OPERATIONAL STRATEGY PATTERNS

Table 9 shows the summary of green operations strategy patterns adopted by each region. It can be seen that each region has its own pattern, as follows. The pattern of Asia was GHG2, energy and recycling. The pattern of the Middle Eastern region is close to Asia's, being GHG2, energy and water. The pattern of North America was GHG2. The pattern of Europe was somehow unique since it includes different combinations of the green indicators recycling, water and noise. The pattern of South America was GHG1 and fuel saving. It can also be seen that, the patterns of Asia, Middle East and South America were consistent in terms of improved indicators; for example, there is a correlation between GHG2 and energy saving; the saving of energy almost always leads to reduced GHG2 emissions. In addition, the saving of fuel leads to a reduction in GHG1 emissions.

Table 10 shows the summary of the impact of green actions on the green indicators across regions. Different combinations of actions taken by airlines relate to different indicators. GHG1 and fuel saving actions are the same actions as GF2 and GF3, but the impact of these action on fuel saving is more than a reduction of GHG1 emissions. Four actions were adopted to reduce GHG2 emissions: the impact of the first action is within the range -16.4 to 25.54%, whereas the impact of the remaining actions differs: GE2 -6.1%, GE5 -3.02%, and GE6 -2.29%. The total impact of these actions together is within a range of -27.81% to - 36.95%. Only one action was taken to save energy, which was GE1. The impact of this action was -9.94%. Two actions were taken to improve recycling: the impact of the first one, WR2, is in a range of 2.77% to 21.55%, and the impact of the second action, WR9,

Table 9. Summary of the green operations strategy patterns adopted by each region

Asia		Middle East		North America		Europe		South America	
Indicator Action	**Impact**	**Indicator action**	**Impact**	**Indicator action**	**Impact**	**Indicator action**	**Impact**	**Indicator action**	**Impact**
GHG2 GE1	**-16.4%** -16.4%	**GHG2** GE1	**-25.54%** -25.54%	**GHG2** GE2 GE5 GE6	**-11.41%** -6.1% -3.02% -2.29%	**Recycling** WR2	**21.55%** 21.55%	**GHG1** GF2 GF3	**-4.56%** -3.17% -1.39%
Energy GE1	**-9.49%** -9.49%	**Energy** GE1	**-7.42%** -7.42%	-------	-------	**Water** W2	**-4.95%** -4.955	**Fuel** GF2 GF3	**-7.88%** -5.54% -2.34%
Recycling WR2 WR9	**8.93%** 2.77% 6.16%	**Water** W4	**-13.71%** -13.71%	-------	-------	**Noise** N1	**-5.5%** -5.5%	-------	-------

Table 10. Summary of the impact of green actions across regions

Indicator	Action	Impact	Indicator	Action	Impact
GHG1	GF2 GF3 **Total**	-3.17% -1.39% **-4.56%**	Recycling	WR2 WR9 **Total**	2.77% to 21.55% 6.16% **8.93%-21.55%**
Fuel	GF2 GF3 **Total**	-5.54% -2.34% **-7.88%**	Water	W2 W4 **Total**	-4.95% -13.71% **-18.66%**
GHG2	GE1 GE2 GE5 GE6 **Total**	-16.4% to -25.54% -6.1% -3.02% -2.29% **-27.81% to 36.95%**	Noise	N1 **Total**	-5.5% **-5.5%**
Energy	GE1 **Total**	-9.49% **-9.49%**	-----------	-----------	-----------

is 6.16%. Two actions were taken to save water, W2 and W4. The impact of these action is -4.95% and -13.71% respectively. However, one action was taken to reduce noise, namely, N1. Its impact is -5.5%.

PICTORIAL REPRESENTATION OF AIRLINES' GREEN STRATEGY

Figure 1 shows the pictorial representation of green strategy of airlines in Asia region, it can be seen that the green index of this region's airlines was about 43%. The green actions taken by the airlines of this regions were related to GHG 2 emissions, energy saving and recycling indicators. One action was taken to reduce GHG 2 emissions and energy saving, this action was facilities and building energy saving, the impact of this action in reducing GHG 2 emission was -16.40% and was -9.49% for saving energy. Two actions were taken to increase the percentage of recycling, these actions were onboard and ground wastes upcycling and upcycling of hazard wastes. The impact of these actions was 2.77% and 6.16% respectively, and the total impact of these two actions together was 8.93%

Figure 2 shows the pictorial representation of green strategy of airlines in Middle East region, it can be seen that the green index of this region's airlines was about 91%. The green actions taken by the airlines of this regions were related to GHG 2 emissions, energy saving and water saving indicators. One action was taken to

Figure 1. The pictorial representation of Asia region airlines green strategy

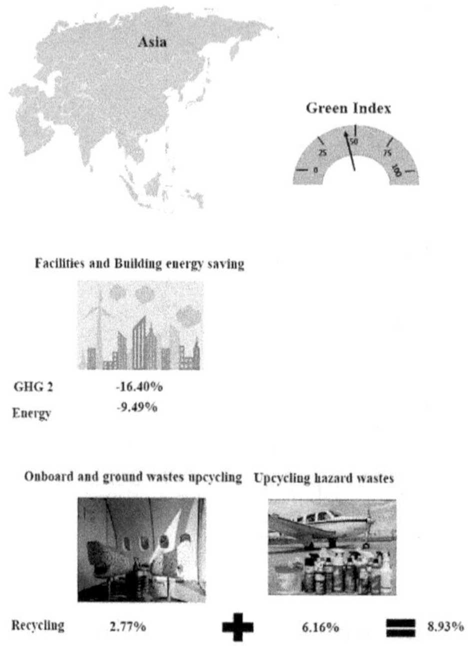

Figure 2. The pictorial representation of Middle East airlines green strategy

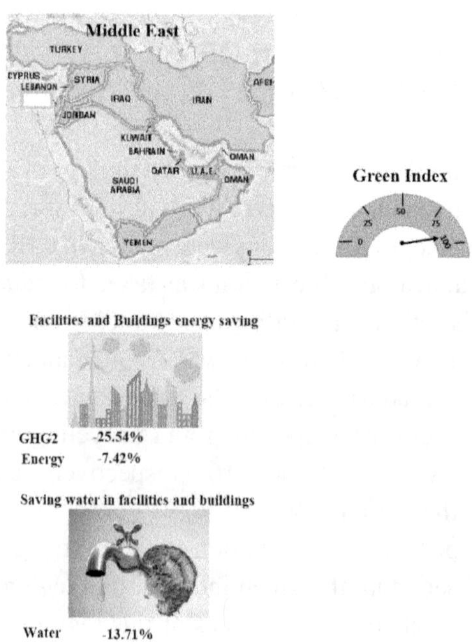

reduce GHG 2 emissions and energy saving, this action was facilities and building energy saving, the impact of this action in reducing GHG 2 emission was -25.54% and was -7.42% for saving energy. One actions was taken to reduce the water consumption, this action was saving water in facilities and buildings. The impact of this action was -13.71%.

Figure 3 shows the pictorial representation of green strategy of airlines in North America region, it can be seen that the green index of this region's airlines was about 14%. The green actions taken by the airlines of this regions were related to GHG 2 emissions. Three actions were taken to reduce GHG 2 emissions, these actions were building accreditation, vehicles design and vehicles operating, the impact of these actions was -6.10%, -3.02% and -2.29% respectively. The total impact of these actions together was -11.41%

Figure 4 shows the pictorial representation of green strategy of airlines in Europe region, it can be seen that the green index of this region's airlines was about 75%. The green actions taken by the airlines of this regions were related to recycling, water saving and reducing noise indicators. One action was taken to achieve each indicator. Onboard and ground wastes upcycling actions improved recycling by

Figure 3. The pictorial representation of North America airlines green strategy

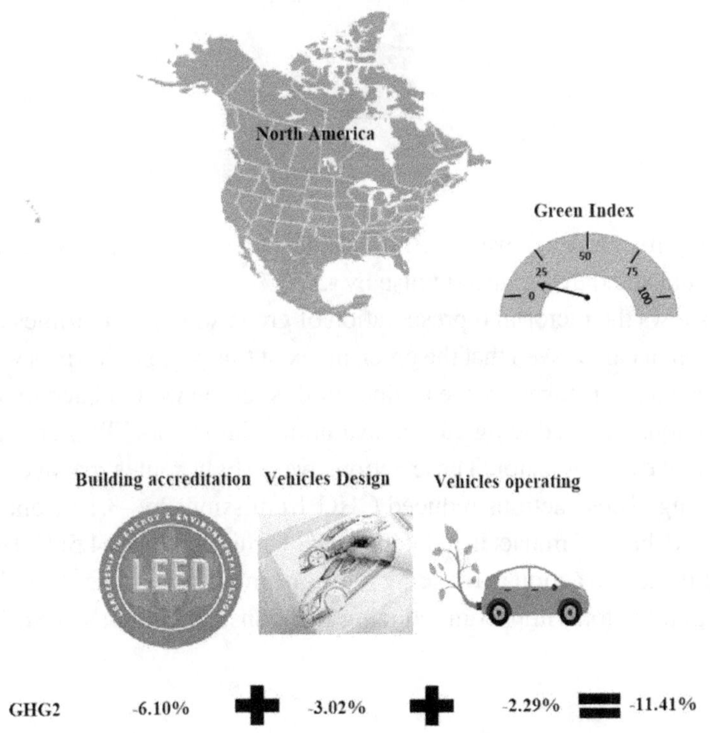

Figure 4. The pictorial representation of Europe airlines green strategy

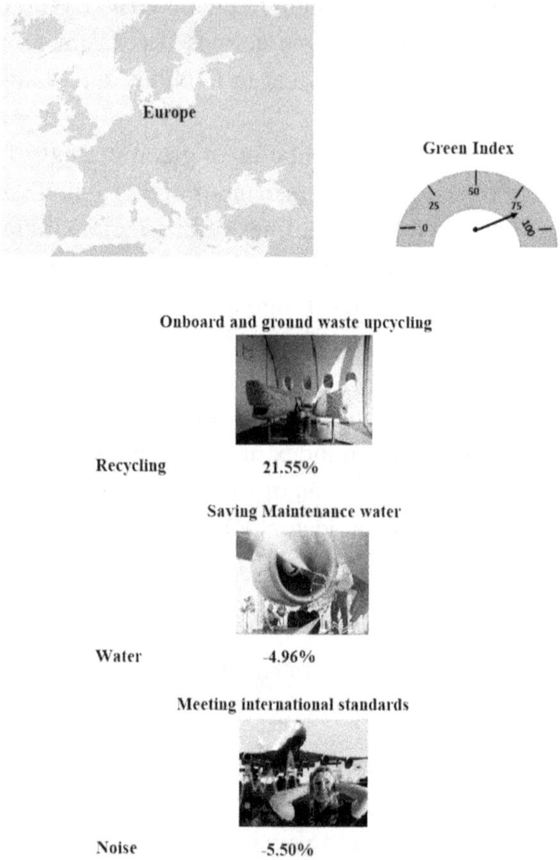

21.55%. Saving maintenance water action saved water by -4.96%, however meeting the international standards reduced noise by -5.50%.

Figure 5 shows the pictorial representation of green strategy of airlines in South America region, it can be seen that the green index of this region's airlines was about 52%. The green actions taken by the airlines of this region were related to reducing GHG 1 emissions and reducing fuel consumption indicators. Two actions were taken to achieve each indicator. These actions were flight routes management and flights operating. These actions reduced GHG 1 emissions by -3.17% and -1.39% respectively and the total impact in reducing GHG 1 emissions was -4.56%. However, the impact of these two action in reducing fuel consumption was -5.54% and -2.34% respectively and the total impact in reducing fuel consumption was -7.88%.

Figure 5. The pictorial representation of South America airlines green strategy

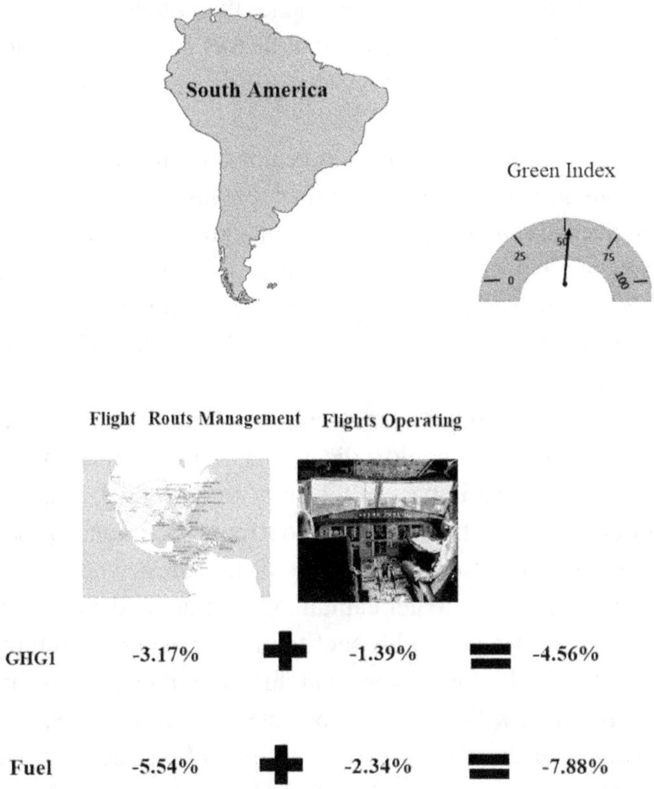

CONCLUSION

Each region's airlines demonstrate a green strategy pattern not duplicated by any other region. This may reflect the varied impact of contextual factors affecting the green practices of airlines in different regions. One of these factors is the impact of environmental regulations; for example, in Europe the regulations are greatly improved and very restrictive because the European Commission has imposed many green restrictions and regulations. The level of demand may also play a dynamic role in the effectiveness of airlines' green practices, which may explain the increase in GHG1 emissions, fuel and energy consumption, and waste generated by American airlines. The customers' awareness of green initiatives may also impact on the airlines' adoption of green practices. Furthermore, the role of NGOs such as Greenpeace or local NGOs in some regions could be more influential than that in others. For example in Europe and America NGOs as Greenpeace have more influential impact than elsewhere. The location of the region also plays a dynamic role in its environmental impact, whereby temperate regions such as the Middle East or Europe may cause less negative impact than others do. This result raises questions regarding the impact of contextual factors such as regulations, demand, customers' awareness and the role of NGOs on the green operations strategy of the airlines in each region.

Some regions improved in their green indicators, but did not adopt any significant operational actions in comparison with other regions. This raises the question of how these regions made such improvement in the green indicators. Answering this question requires more in-depth analysis of the actions themselves together with the categories of green actions, suggesting that the improvement may be a byproduct of contextual factors such as change in demand. This issue too may be a research issue in the future.

Most of the effective actions taken were operational (process oriented) rather than technology-oriented. This could indicate that the operational (process oriented) can lead to a more effective impact in the short to mid-term. It also requires less capital expenditure and hence most airlines adopt more operational (process oriented) actions; however, technology-oriented factors such as the newness of the aircraft, or changing the aircraft require greater capital expenditure and more time before the impact can be traced. This issue would benefit from the attention of future researchers. Some actions taken in different regions had differing impacts; this may be related to detailed actions taken within each action category; in response, future research might focus on tracing in detail the impact of each significant action in each category.

The results of this chapter could be beneficial for academics and industry decision makers. The documented content of airlines' green operations strategy in this chapter could benefit academic in teaching this topic and in developing more extended conceptual models of green operations strategy for future studies. Furthermore, the decision makers in each region have results about the effective green actions taken by airlines; also, they can adopt the most effective actions taken by airlines of each region. The decision makers could adopt the cross-regions indicative models as a decision support system.

REFERENCES

Abdullah, M. A., Chew, B. C., & Hamid, S. R. (2016). Benchmarking key success factors for the future green airline industry. *Procedia: Social and Behavioral Sciences*, *224*, 246–253. doi:10.1016/j.sbspro.2016.05.456

Baumeister, S., & Onkila, T. (2017). An eco-label for the airline industry? *Journal of Cleaner Production*, *142*, 1368–1376. doi:10.1016/j.jclepro.2016.11.170

Bhardwaj, B. R. (2016). Role of green policy on sustainable supply chain management: A model for implementing corporate social responsibility (CSR). *Benchmarking: An International Journal*, *23*(2), 456–468. doi:10.1108/BIJ-08-2013-0077

Brasseur, G. P., Gupta, M., Anderson, B. E., Balasubramanian, S., Barrett, S., Duda, D., ... Halthore, R. N. (2016). Impact of aviation on climate: FAA's aviation climate change research initiative (ACCRI), phase ii. *Bulletin of the American Meteorological Society*, *97*(4), 561–583. doi:10.1175/BAMS-D-13-00089.1

Brooks, K. P., Snowden-Swan, L. J., Jones, S. B., Butcher, M. G., Lee, G. S., Anderson, D. M., . . . Burton, F. (2016). Low-Carbon Aviation Fuel through the Alcohol to Jet Pathway. In Biofuels for Aviation (pp. 109-150). Academic Press. doi:10.1016/B978-0-12-804568-8.00006-8

Chapman, M. (2016). Sustaining Reductions in Aircraft Emissions for Canada's Major Airlines. In *Managing in a VUCA World* (pp. 175–193). Cham: Springer. doi:10.1007/978-3-319-16889-0_12

EASA (European Airline Safety Association). (2016). European Aviation. *Environment Reporter, 2016*. Available online https://ec.europa.eu/transport/sites/transport/files/european-aviation-environmental-report-2016-72dpi.pdf

GRI. (2018). *About GRI*. Available online: https://www.globalreporting.org/Information/about-gri/Pages/default.aspx

Hagmann, C., Semeijn, J., & Vellenga, D. B. (2015). Exploring the green image of airlines: Passenger perceptions and airline choice. *Journal of Air Transport Management*, *43*, 37–45. doi:10.1016/j.jairtraman.2015.01.003

Han, H., Yu, J., & Kim, W. (2019). Environmental corporate social responsibility and the strategy to boost the airline's image and customer loyalty intentions. *Journal of Travel & Tourism Marketing*, 1–13.

Harvey, G., Williams, K., & Probert, J. (2013). Greening the airline pilot: HRM and the green performance of airlines in the UK. *International Journal of Human Resource Management*, *24*(1), 152–166. doi:10.1080/09585192.2012.669783

Hinnen, G., Hille, S. L., & Wittmer, A. (2017). Willingness to Pay for Green Products in Air Travel: Ready for Take-Off? *Business Strategy and the Environment*, *26*(2), 197–208. doi:10.1002/bse.1909

Horio, B. M., Kumar, V., Levin, D. J., & Sung, P. E. (2016). Modeling Carbon Tax Policy Impacts on US Commercial Airlines using Agent-Based Modeling and Crowdsourced Data. In *AIAA Modeling and Simulation Technologies Conference* (p. 4302). AIAA.

IPCC. (2015). *Climate change 2014 synthesis report: fifth assessment report*. Geneva: IPCC.

Kirchoff, J. F., Tate, W. L., & Mollenkopf, D. A. (2016). The impact of strategic organizational orientations on green supply chain management and firm performance. *International Journal of Physical Distribution & Logistics Management*, *46*(3), 269–292. doi:10.1108/IJPDLM-03-2015-0055

Lee, D. S., Fahey, D. W., Forster, P. M., Newton, P. J., Wit, R. C., Lim, L. L., ... Sausen, R. (2009). Aviation and global climate change in the 21st century. *Atmospheric Environment*, *43*(22-23), 3520–3537. doi:10.1016/j.atmosenv.2009.04.024

Lee, D. S., Pitari, G., Grewe, V., Gierens, K., Penner, J. E., Petzold, A., ... Iachetti, D. (2010). Transport impacts on atmosphere and climate: Aviation. *Atmospheric Environment*, *44*(37), 4678–4734. doi:10.1016/j.atmosenv.2009.06.005

Lee, K. C., Tsai, W. H., Yang, C. H., & Lin, Y. Z. (2017). An MCDM approach for selecting green aviation fleet program management strategies under multi-resource limitations. *Journal of Air Transport Management*, *68*, 76–85. doi:10.1016/j.jairtraman.2017.06.011

Lee, K. C., Tsai, W. H., Yang, C. H., & Lin, Y. Z. (2018). An MCDM approach for selecting green aviation fleet program management strategies under multi-resource limitations. *Journal of Air Transport Management*, *68*, 76–85. doi:10.1016/j.jairtraman.2017.06.011

Liu, X., Zhou, D., Zhou, P., & Wang, Q. (2017). Dynamic carbon emission performance of Chinese airlines: A global Malmquist index analysis. *Journal of Air Transport Management*, *65*, 99–109. doi:10.1016/j.jairtraman.2017.09.009

Lynes, J. K., & Dredge, D. (2006). Going green: Motivations for environmental commitment in the airline industry. A case study of Scandinavian Airlines. *Journal of Sustainable Tourism*, *14*(2), 116–138. doi:10.1080/09669580608669048

Macintosh, A., & Wallace, L. (2009). International aviation emissions to 2025: Can emissions be stabilised without restricting demand? *Energy Policy*, *37*(1), 264–273. doi:10.1016/j.enpol.2008.08.029

Mayer, R., Ryley, T., & Gillingwater, D. (2015). Eco-positioning of airlines: Perception versus actual performance. *Journal of Air Transport Management*, *44*, 82–89. doi:10.1016/j.jairtraman.2015.03.003

Migdadi, Y. K. A. A. (2016). Identifying the best practices in the green operations strategy of leading mobile phone producers. *International Journal of Business Excellence*, *9*(1), 92–112. doi:10.1504/IJBEX.2016.073377

Migdadi, Y. K. A. A. (2018). Identifying the best practices of airlines' green operations strategy: A cross-regional worldwide survey. *Environmental Quality Management*, *28*(1), 21–32. doi:10.1002/tqem.21575

Singh, A., & Trivedi, A. (2016). Sustainable green supply chain management: Trends and current practices. *Competitiveness Review*, *26*(3), 265–288. doi:10.1108/CR-05-2015-0034

Smith, D. J. (2016). The sustainable and green engine (SAGE) – Aircraft engine of the future? *The International Journal of Entrepreneurship and Innovation*, *17*(4), 256–262. doi:10.1177/1465750316672601

Szodruch, J., Grimme, W., Blumrich, F., & Schmid, R. (2011). Next generation single-aisle aircraft –requirements and technological solutions. *Journal of Air Transport Management*, *17*(1), 33–39. doi:10.1016/j.jairtraman.2010.10.007

Teoh, L. E., & Khoo, H. L. (2016). Green air transport system: An overview of issues, strategies and challenges. *KSCE Journal of Civil Engineering*, *20*(3), 1040–1052. doi:10.100712205-016-1670-3

UN. (2018). *Suitable development agenda*. Available online: https://www.un.org/sustainabledevelopment/development-agenda/

Will, F., Tay, G., Becker, A., Carnelly, D., Eychenne, F., & Hornung, M. (2016). Green Airlines 2025: Environment and Sustainability in Commercial Aviation – A Scenario Study. In *16th AIAA Aviation Technology, Integration, and Operations Conference* (p. 3756). AIAA.

Williams, V., Noland, R. B., & Toumi, R. (2002). Reducing the climate change impacts of aviation by restricting cruise altitudes. *Transportation Research Part D, Transport and Environment*, *7*(6), 451–464. doi:10.1016/S1361-9209(02)00013-5

Yan, W., Cui, Z., & Gil, M. J. Á. (2016). Assessing the impact of environmental innovation in the airline industry: An empirical study of emerging market economies. *Environmental Innovation and Societal Transitions*, *21*, 80–94. doi:10.1016/j.eist.2016.04.001

ADDITIONAL READING

Aital, P., & Vijai, J. P. (2015). Operational practices and performances of green supply chain management in Indian firms. *International Journal of Process Management and Benchmarking*, *5*(3), 352–374. doi:10.1504/IJPMB.2015.070819

Florida, R. (1996). The environment and the new industrial revolution. *California Management Review*, *38*(Autumn), 80–115. doi:10.2307/41165877

Florida, R., & Davison, D. (2001). Gaining from green management: Environmental management systems inside and outside the factory. *California Management Review*, *43*(3), 64–84. doi:10.2307/41166089

Geffen, C., & Rothenberg, S. (2000). Sustainable development across firm boundaries: The critical role of suppliers in environmental innovation. *International Journal of Operations & Production Management*, *20*(2), 166–186. doi:10.1108/01443570010304242

Handfield, R., Walton, S. V., Sroufe, R., & Melnyk, S. A. (2002). Applying environmental criteria to supplier assessment: A study in the application of the Analytical Hierarchy Process. *European Journal of Operational Research*, *141*(1), 70–87. doi:10.1016/S0377-2217(01)00261-2

Huang, X., Tan, B. L., & Ding, X. (2015). An exploratory survey of green supply chain management in Chinese manufacturing small and medium-sized enterprises: Pressures and drivers. *Journal of Manufacturing Technology Management*, *26*(1), 80–103. doi:10.1108/JMTM-05-2012-0053

Khatua, D., Roymahapatra, G., & Maity, K. (2017). A green supply chain production inventory model with uncertain holding cost. *International Journal of Process Management and Benchmarking*, *7*(3), 277–304. doi:10.1504/IJPMB.2017.084905

Lin, W. (2017). Aviation and Climate Change: Practising Green Governmentality across the North-South Divide. *Geopolitics*, *22*(1), 129–150. doi:10.1080/14650045.2016.1210130

Migdadi, Y. K. A. A., & Elzzqaibeh, D. A. S. I. (2018). The evaluation of green manufacturing strategies adopted by ISO 14001 certificate holders in Jordan. *International Journal of Productivity and Quality Management*, *23*(1), 90–109. doi:10.1504/IJPQM.2018.088610

Sahu, A. K., Datta, S., & Mahapatra, S. S. (2013). Green supply chain performance benchmarking using integrated IVFN-TOPSIS methodology. *International Journal of Process Management and Benchmarking*, *3*(4), 511–551. doi:10.1504/IJPMB.2013.058272

Sarkis, J. (1995). Manufacturing strategy and environmental consciousness. *Technovation*, *15*(2), 79–97. doi:10.1016/0166-4972(95)96612-W

Shabani, A., & Farzipoor Saen, R. (2015). Developing a novel data envelopment analysis model to determine prospective benchmarks of green supply chain in the presence of dual-role factors. *Benchmarking: An International Journal*, *22*(4), 711–730. doi:10.1108/BIJ-12-2012-0087

Sharma, S., & Gandhi, M. A. (2016). Exploring correlations in components of green supply chain practices and green supply chain performance. *Competitiveness Review*, *26*(3), 332–368. doi:10.1108/CR-04-2015-0027

Singh, P., & Kumar, V. (2017). Quantitative analysis of drivers affecting green supply chain management in Rajasthan SMEs. *International Journal of Process Management and Benchmarking*, *7*(3), 332–353. doi:10.1504/IJPMB.2017.084906

Vinodh, S., Devadasan, S. R., & Rajanayagam, D. (2007). Roadmap for the lucrative greening of supply chains: Theoretical and practical perspectives. *International Journal of Process Management and Benchmarking*, *2*(1), 29–44. doi:10.1504/IJPMB.2007.013316

Zhang, H., & Yang, F. (2016). On the drivers and performance outcomes of green practices adoption: An empirical study in China. *Industrial Management & Data Systems*, *116*(9), 2011–2034. doi:10.1108/IMDS-06-2015-0263

KEY TERMS AND DEFINITIONS

Energy Consumption: Kilowatts per hour of electricity by facilities, buildings, engines, and vehicles.

Fuel Consumption: Gallons of fuel used to operate aircrafts.

GHG1 Emission: Metric tons of direct CO2 emissions from jet fuel and ground support.

GHG2 Emissions: Metric tons of indirect CO2 emissions of electricity, power and heat from direct billing of owned or leased facilities.

GHG3 Emissions: Metric tons of indirect CO2 emissions in the parts of the value chain out of GHG 2 emissions.

Green Actions: The adopted green decisions to realized green indicators as flights routes management, operating of flights, flights weight management, etc.

Green Indicators: Ecological performance indicators of airlines as GHG1 emissions, GHG2 emissions, GHG3 emissions, fuel consumption, energy consumption, water consumption, wastes generated, and recycling.

Green Strategy: Stream of actions adopted by airlines to realize green indicators.

Noise: Disabling level of aircraft sounds at landing and takeoff.

Recycling: Tons of waste recycled upcycled and reused.

Wastes Generated: Tons of waste generated, whether onboard, ground, industrial, or hazard.

Water Consumption: m^3 of consumed water whether used onboard, or ground.

APPENDIX 1: LISTS OF ACTIONS UNDER EACH THEME AND SUB-THEME

Table 11. GHG1 emissions and fuel saving green actions taken by airlines

Action category	Actions
GF1: Aircraft design	fleet modernization
	split scimitar winglets
	sharklet wing tips
	engine modification
GF2: Flight routes management	route optimization
	PC calculated flight path
	broadcasting technology for communication everywhere
	monitoring actual flight path and relevant data
	real time data text communication
	using GPS
	center of gravity
GF3: Flights operating	reduced use of APU
	single taxiing engine
	acceleration altitude
	accurate planning of flight
	altitude optimization
	reducing aircraft take-off time
	reducing flight launching time
	reducing use of extended landing lights
	reducing landing time
	optimal use of flap
	reduced use of idle reverse thrust when landing
	reduced aircraft waiting time
	landing continuous descent operations
	reduced taxiing time
	revised and improved fuel loading regulations
GF4: Flights weight management	optimal water potable carried
	using EFB
	using on board tablets
	Light meal trolleys
	eliminating magazines

continued on following page

Table 11. Continued

Action category	Actions
	digital route charts
	lighter seats
	lightweight cargo containers
	installing zonal driers in aircraft such as dehumidifiers which reduces moisture weight
	restricting fuel overload
	reducing number of flight operations manuals onboard
	Using lighter galley
	reduce pay load variance
GF6: Flights fuel management	using bio-fuels
	training in fuel saving
	installing fuel management and reporting system
	reducing contingency fuel planning for long haul routes
GF7: Flights maintenance management **GHG3-1**	engine washing
	maintaining interior aircraft plumbing
	washing aircraft body
	re-sorting the exterior of fan blades in the engines

Table 12. GHG2, GHG3 emissions and Energy saving actions taken by airlines

Actions category	List of actions
GE1: Facilities and building energy saving	air-conditioning energy saving
	installing electricity meter points to allow real reporting and auditing
	light energy saving
GE2: Facilities and building accreditation	certification of buildings according to LEED standards
GE3: Facilities and building sustainable energy	using sustainable energy at ground operations
	using heat pumps with aquifers to generate sustainable energy
	using solar panels
GE4: Upgrading and replacing facilities	upgrading the copiers, fax and printer to energy-friendly ones
	upgrading and adjusting building control and automation system to conserve electricity
	replacing power equipment at several radar sites
	upgrading lifts to more efficient models

continued on following page

Table 12. Continued

Actions category	List of actions
GE5: Vehicle and engine design **GHG3-2**	using electrical carts or tractors (trucks)
	using battery vehicles for ground operations
	using renewable energy-fueled equipment
GE6: Vehicle and engine operating **GHG3-3**	reducing vehicles' standby time
	management of office vehicle trips
	new forklifts
	loading more cargo at the same time
GE7: Vehicle and engine maintenance management **GHG3-4**	washing engines with less water
	cleaning vehicles and equipment
GE8: Transportation management **GHG3-5**	car sharing
	low CO_2 emissions cars
	subsidizing a shuttle bus
	pooling bicycles
	installing charging points for electrical vehicles

Table 13. Waste management and recycling green actions taken by airlines

Action category	Actions
WR1: Onboard and ground waste recycling	adopting a recycling tracking system
	auditing internal recycling
	increasing the number of recycling stations
	recycling electronic boarding passes
	recycling paper
	recycling plastic
	recycling cans
	using a video system to encourage passengers to sort their waste
	having a print smart campaign
	using the cabin intercom to encourage passengers to sort their waste
WR2: Onboard and ground waste upcycling	using outdated vests to manufacture travel amenity bags and kits
	Transforming tarpaulins into pencil cases
	modifying scrapped service carts
	shredding uniforms and using them for furniture blankets
	converting seats into leather products

continued on following page

Table 13. Continued

Action category	Actions
WR3: Reusing onboard and ground waste	donating blazers and trench coats to charities
WR4: Reducing the use of paper	using electronic airway bills for freight documents
	using electronic boarding passes
	providing crews with iPads instead of newspapers
WR5: Recycling industrial waste	recycling maintenance waste
	recycling aircraft parts
	recycling iron
	recycling of cargo shrink-wrap
WR6: Upcycling industrial waste	burning for energy recovery
WR7: Reusing industrial waste	keeping aircraft parts in warehouses as spare parts
	reusing wooden cargo pallets in land scrapping
	upcycling some parts of seats and converting them into leather seats
	collecting aviation fuel used during maintenance for later use
	collecting scraps of material and selling them for qualified companies to reuse
	reusing carpets
	keeping aircraft parts in warehouses as spare parts
	reusing the wooden cargo pallets in land scrapping
WR8: Recycling hazardous waste	recycling of hazardous waste
	inventory and tracking systems to track hazardous material early
WR9: Upcycling hazardous waste	burning it to recover energy
WR10: Using hazardous waste	Reducing the use of chemicals

Table 14. Water management actions taken by airlines

Action category	Actions list
W1: Recycling and recovery of, maintenance water	recycling and reusing of waste maintenance water
	refining of engine washing water
W2: Saving maintenance water	adopting the "Eco Shine" method to clean the exterior of aircraft using pads
W3: Recycling and recovery of water for facilities and buildings	using a rain water recovery system
	recovering waste drinking water
	recycling water used in the cooling towers of the building

continued on following page

Table 14. Continued

Action category	Actions list
W4: Saving water in facilities and buildings	optimization of swimming water
	reducing the frequency of cleaning buildings
	raising awareness of water saving among new employees
	training of all cleaning staff in minimal water use
	regular water tank maintenance
	modification of water equipment
	using minimal water for cooking and processing
	adoption of water saving equipment
	flow restriction and dual flash valves installed in all toilets
	installing meters
	eliminating least efficient dishwashing machines
	optimal use of water
	launching water saving campaigns

Table 15. Noise reduction actions taken by airlines

Action category	Actions list
N1: Meeting international standards	Meeting standards of ICAO ch.4 technology standards for noise
N2: flight procedures	Identifying the altitude of flight routes to avoid a sonic boom
	Using lower than maximum landing flap settings
	Adopting noise abatement takeoff procedures

APPENDIX 2: THE IMPROVEMENT IN GREEN INDICATORS ACROSS REGIONS

Table 16. The improvement in green indicators across regions

Green indicator	Regions				
	North America	Europe	Asia	South America	Middle East
GHG 1 emissions		-5.13%		-4.55%	
Energy consumption			-9.49%	-7.56%	-7.42%
GHG 2 emissions	-11.40%		-16.40%		-25.45%
Fuel consumption		-9.18%		-7.97%	
Waste generated		-9.31%			
Water consumption		-4.95%			-13.71%
Recycling		21.55%	8.93%		
GHG 3 emissions				-11.80%	
Noise		-5.50%			

APPENDIX 3: THE IMPROVEMENT IN GREEN ACTION INDEX ACROSS REGIONS

Table 17. Improvement in GHG1 and fuel actions indices

Ranks of regions according to the green indicators	Region	GHG1 and Fuel actions					
		GF1	GF2	GF3	GF4	GF5	GF6
	Asia	0.25	0.22	0.21	0.15	0.08	0.25
1	Europe	0.17	0.07	0.21	0.15	0.13	0.08
	Middle East	0.25	0.15	0.13	0.15	0.25	0.00
	North America	0.33	0.22	0.07	0.13	0.21	0.04
2	South America	0.25	**0.57***	**0.27***	0.15	0.00	0.25
	Europe improvement in performance						
	South America improvement in performance		0.39	0.17			

*signifies actions taken by the higher ranked region in comparison with lower ranking regions

Table 18. Improvement in GHG2 actions indices

Ranks of regions according to the green indicators	Region	GHG2 actions							
		GE1	GE2	GE3	GE4	GE5	GE6	GE7	GE8
2	Asia	**0.17***	0.17	0.06	0.00	0.06	0.06	0.00	0.00
	Europe	0.11	0.00	0.11	0.21	0.00	0.00	0.00	0.17
1	Middle East	**0.34***	0.00	0.00	0.13	0.00	0.00	0.00	0.00
3	North America	0.06	**0.33***	0.06	0.08	**0.17***	**0.13***	0.00	0.00
	South America	0.00	0.00	0.00	0.00	0.00	0.00	0.00	0.00
	Asia improvement in performance	0.11							
	Middle East improvement in performance	0.25							
	North America improvement in performance		0.33			0.17	0.13		

*signifies. actions taken by the higher ranked region in comparison with lower ranked regions

Table 19. Improvement in energy actions indices

Ranks of regions according to the green indicators	Region	energy management actions							
		GE1	GE2	GE3	GE4	GE5	GE6	GE7	GE8
3	Asia	0.17	0.17	0.06	0.00	0.06	0.06	0.00	0.00
	Europe	0.11	0.00	0.11	0.21	0.00	0.00	0.00	0.17
2	Middle East	**0.34***	0.00	0.00	0.13	0.00	0.00	0.00	0.00
	North America	0.06	0.33	0.06	0.08	0.17	0.13	0.00	0.00
1	South America	0.00	0.00	0.00	0.00	0.00	0.00	0.00	0.00
	Asia	0.08							
	Middle East improvement in performance	0.22							

*signifies. actions taken by the higher ranked region in comparison with lower ranked regions

Table 20. Improvement in GHG3 actions indices

Ranks of regions according to the green indicators	Region	GHG 3								
		GHG3-1	GHG3-2	GHG3-3	GHG3-4	GHG3-5	GHG3-6	GHG3-7	GHG3-8	GHG3-9
	Asia	0.17	0.17	0.06	0.00	0.06	0.06	0.00	0.00	0.25
	Europe	0.11	0.00	0.11	0.21	0.00	0.00	0.00	0.17	0.08
	Middle East	0.34	0.00	0.00	0.13	0.00	0.00	0.00	0.00	0.00
	North America	0.06	0.33	0.06	0.08	0.17	0.13	0.00	0.00	0.04
1	South America	0.00	0.00	0.00	0.00	0.00	0.00	0.00	0.00	0.25

Table 21. Improvement in waste management actions indices

Ranks of regions according to the green indicators	Region	waste management actions								
		WR1	WR2	WR3	WR4	WR5	WR6	WR7	WR8	WR9
	Asia	0.15	0.10	0.00	0.00	0.00	0.06	0.00	0.17	0.00
1	Europe	0.13	0.07	0.00	0.08	0.00	0.00	0.00	0.00	0.00
	Middle East	0.35	0.00	0.00	0.00	0.00	0.00	0.00	0.00	0.00
	North America	0.27	0.03	0.17	0.17	0.33	0.06	0.17	0.00	0.17
	South America	0.00	0.00	0.00	0.00	0.00	0.00	0.00	0.00	0.00

Table 22. Improvement in recycling actions indices

Ranks of regions according to the green indicators	Region	Recycling actions								
		WR1	WR2	WR3	WR4	WR5	WR6	WR7	WR8	WR9
1	Asia	0.15	**0.10***	0.00	0.00	0.00	0.06	0.00	**0.17***	0.00
2	Europe	0.13	**0.07***	0.00	0.08	0.00	0.00	0.00	0.00	0.00
	Middle East	0.35	0.00	0.00	0.00	0.00	0.00	0.00	0.00	0.00
	North America	0.27	0.03	0.17	0.17	0.33	0.06	0.17	0.00	0.17
	South America	0.00	0.00	0.00	0.00	0.00	0.00	0.00	0.00	0.00
	Asia improvement in performance		0.08						0.17	
	Europe improvement in performance		0.06							

*signifies actions taken by the higher ranked region in comparison with lower ranked regions

Table 23. Improvement in water actions indices

Ranks of regions according to the green indicators	Region	water saving actions			
		W1	W2	W3	W4
	Asia	0.17	0.00	0.11	0.09
2	Europe	0.08	**0.17***	0.00	0.03
1	Middle East	0.00	0.00	0.00	**0.115***
	North America	0.00	0.00	0.00	0.01
	South America	0.00	0.00	0.00	0.00
	Europe improvement in performance		0.17		
	Middle East improvement in performance				0.0829

*signifies actions taken by the higher ranked region in comparison with lower ranked regions

APPENDIX 4: DETAILED CALCULATIONS OF RELATIVE IMPACT OF ACTIONS

Table 24. Relative impact of GHG1 actions

Region	GHG 1 actions	Difference in index	relative impact	Reduction in GHG1
South America	**GF2:** Flights route management	0.39	0.695393759	-3.17%
	GF3: Flights operating	0.17	0.304606241	-1.39%
	Total	0.56		-4.55%

Table 25. Relative impact of fuel saving actions

Region	Fuel saving actions	Difference in index	relative impact	fuel saving
South America	**GF2:** Flights route management	0.39	0.695393759	-5.54%
	GF3: Flights operating	0.17	0.304606241	-2.43%
	Total	0.56		-7.97%

Table 26. Relative impact of GHG2 actions

Region	GHG 2 actions	Difference in index	relative impact	Reduction in GHG2
Asia	**GE1:** Facilities and building energy saving	0.11	1	-16.40%
Middle East	**GE1:** Facilities and building energy saving	0.25	1	-25.45%
North America	**GE2:** Facilities and building accreditation	0.33	0.534759358	-6.10%
	GE5: Vehicle and engine design	0.17	0.264705882	-3.02%
	GE6: Vehicles and engine operating	0.13	0.200534759	-2.29%
	total	0.62		-11.40%

Table 27. Relative impact of energy saving actions

Region	Energy saving actions	Difference in index	relative impact	Energy saving
Asia	**GE1:** Facilities and building energy saving	0.08	1	-9.49%
Middle East	**GE1:** Facilities and building energy saving	0.22	1	-7.42%

Table 28. Relative impact of recycling actions

Region	recycling action	Difference in index	relative impact	Recycling
Asia	**WR2:** Onboard and ground waste upcycling	0.08	0.310344828	2.77%
	WR9: Upcycling hazardous waste	0.17	0.689655172	6.16%
	Total	0.24		8.93%
Europe	**WR2:** Onboard and ground waste upcycling	0.06	1	21.55%

Table 29. Relative impact of water saving actions

Region	Water saving actions	Difference in index	relative impact	water saving
Europe	**W2:** Saving maintenance water	0.17	1	-4.95%
Middle East	**W4:** Saving water in facilities and buildings	0.082916667	1	-13.71%

Table 30. Relative impact of noise reduction actions

Region	Noise reduction action	Difference in index	relative impact	water saving
Europe	**N1:** Meeting standards of ICAO Ch.4 technology standards for noise	1	1	5.50%

Chapter 9
Sustainable Development in Family Firms

Angela Dettori
University of Cagliari, Italy

Michela Floris
University of Cagliari, Italy

Cinzia Dessì
University of Cagliari, Italy

EXECUTIVE SUMMARY

This chapter outlines the relevance of sustainable development as a key for family firm success and its ability to guarantee long-term survival and spread positive effects in social, economic, and natural environments. By particularly analyzing a single case study of a Sardinian family business, this work explores the intertwined relationships among sustainability, owner innovativeness, and firm success. Moreover, the importance of family businesses and the scarcity of the study conducted to date have suggested a focus on how these companies tackle sustainability challenges.

INTRODUCTION

Sustainability-related pressures for and benefits of engagement with environmental issues increased over time and this resulted in the intensified integration of non-financial goals related to environmental issues into the decision-making behaviour of firms. In this sense, firms need to ensure a better quality of life in the present without compromising that of future generations (Bansal, 2005). This requirement presents a significant challenge (Rimanoczy & Pearson, 2010) that, in order to be

DOI: 10.4018/978-1-5225-8559-6.ch009

won, firms must increase their efforts to meet their goal of improving economic, social and environmental performance (Sava, Rizea, & Flood, 2010) within new and innovative governance models. Many firms are actively involved in the sustainability debate with the goal of identifying methods to enhance performance and develop sustainable development strategies (Starik & Rands, 1995). In other words, firms are called to change their culture and structure to perceive and promote well-defined corporate sustainability as an invaluable tool that will focus on cost-cutting, risk management and new product realisation (Bansal, 2005). Therefore, the inclusion and integration of sustainability within a firm's structure and culture require a clear and shared vision, a high level of commitment and adequate leadership. Moreover, corporate sustainability entails a specific approach with an appropriate management framework that enables the design, management and communication of sustainable policies and strategies. Considerable effort has to be devoted to the planning, measurement and evaluation, steering and control as well as optimisation and communication processes of the holistically defined corporate value creation. Furthermore, a solution for enterprise sustainability management and its evaluation is necessary for ultimately balancing economic, ecological and social performance factors, to ensure optimized decision-making (Oertwig et al., 2017).

Several proposal have been made to promote sustainable development. At the European Union level, for example, the European Commission (EC) is actively involved in drawing up policies and practices to encourage corporate sustainability. These efforts range from 'polluter pays' legislation to 'producer responsibility' policies, from 'core labour standards' to 'social governance' and encompass all industrial sectors, from primary extraction to consumer products. In addition, by following its green paper, *Promoting a European Framework for Corporate Social Responsibility* (EC, 2001), the Commission has more recently issued a communication (EC, 2002) to further encourage the adoption of the Corporate Social Responsibility concept.

However, adopting a sustainable approach by embracing sustainability principles and considering them an integral part of business practice depends on the responsibility that firms have in making businesses more sustainable. To sustain firms in this renewed concept of business, the World Business Council for Sustainable Development (WBCSD) and the International Institute for Sustainable Development have recently identified a number of principles to address sustainable development concerns (WBCSD, 2000). Identified principles include, for example, cost savings from cleaner production methods; innovation and technology to improve material, energy and product efficiencies or to lower labour costs; and innovative solutions that provide good working conditions to improve motivation and productivity (Shrivastava, 1995).

Notwithstanding the mentioned efforts, achieving corporate sustainability is not an easy task and is accompanied by several challenges. First, a sustainability management system able to reflect both the specific characteristics of each business and the context in which it operates appears to be particularly relevant and difficult. Second, the proliferation of different approaches to corporate sustainability, which are difficult to compare, can create confusion for businesses, consumers, investors and the public and could lead to market distortion (EC, 2002). Third, the convergence and standardisation of corporate sustainability management approaches require an in-depth reflection to disentangle the complex and fragmented framework.

This chapter, which reinforces the relevance of developing and spreading sustainability in a firm context, focuses specifically on family firms as units of analysis. This choice draws on the awareness that this type of firm constitutes the major portion of the global economy (Heck & Stafford, 2001; Rowe, Haynes, & Stafford, 1999; Shanker & Astrachan, 1996), and its activities generate a significant impact on economic, social and natural environments. For this reason, sustainability concepts must be part of the organisational culture for guaranteeing successful long-term survival. In addition, family businesses are particularly interesting because their lifecycles tend to be passed from one generation to another. This transgenerational process, representing a physiological but contemporarily a destabilising step, can represent a suitable conjuncture able to stimulate new insights and new paths in the firm's physiognomy. In this sense, succession could be considered particularly apt for spreading new strategies with respect to sustainable development perspectives and to launch change processes according to a logic oriented to the future through the combination of continuity and renewal, tradition and innovation. This consideration is apt because, if the real objective of succession is to guarantee long-term business continuity, this objective can only be achieved by integrating the generational shift with the more general business development strategy (Cabrera-Suarez, Saa-Perez, & Almeida, 2001) of observing the past through the perspective of the future and finding the right balance between the 'tradition and innovation' (Sharma, Chrisman, & Chua, 2003) of the entrepreneurial formula hitherto adopted. Finally, family businesses represent a suitable unit of analysis because–despite the recognised importance of their role in social, economic and environmental development– very few studies have been developed in this direction. In fact, as underlined by Hall, Daneke and Lenox (2010), the literature on sustainable development in the entrepreneurship field of research and family businesses shows evident lacunae that have to be filled through further theoretical and empirical studies.

With the goal of contributing to filling the gap of literature on sustainable development in the entrepreneurship field of research and family businesses, and understanding and framing how family firms behave and meet sustainability concepts, this chapter analyses a single case study on Argiolas Formaggi. This firm, located in Sardinia (Italy), is a relevant case that shows how and why being sustainable provides benefits to firms, stakeholders and the natural environment.

THEORETICAL BACKGROUND

Sustainable Development Concepts: An Overview

Sustainable development was defined by the World Commission on Environment and Development (WCED, 1987) as the 'development which meets the needs of the present without compromising the ability of future generations to meet their own needs' (Brundtland, 1987, p. 43). In its broadest sense, this definition has been widely accepted and endorsed by thousands of governmental, corporate and other organisations worldwide (Gladwin et al., 1995). The satisfaction of human needs and aspirations is the main objective of development that can be considered 'sustainable'. More specifically, sustainable development guarantees the satisfaction of the basic needs of all people now and in the future or extends the opportunity to achieve a better quality of life for everyone (Gladwin et al., 1995).

In recent years, many scholarly definitions (Viederman, 1993; Springett, 2003) have been suggested, resulting in different interpretations. Kelly, Sirr and Ratcliffe (2004) defined sustainability as a multidimensional concept that incorporates different aspects of society and seeks environmental protection and the maintenance of natural capital to achieve economic prosperity and equity for present and future generations. Chichilnisky (1996) and Hove (2004) defined sustainability as a process of change in which the exploitation of resources, the direction of investments and the orientation of technological and institutional change are conducted according to future needs and considering present needs. Sustainable development has come to be considered a 'visionary development paradigm' with governments, businesses and civil society around the world acknowledging it as the guiding principle and normative goal (Drexhage & Murphy, 2010, p. 2) of a 'three pillar' approach to collective global futures i.e. economic development undergirded by greater social equity and environmental protection.

Several contributions have been offered also by environmental sociologists that in particular have development theories of Ecological Modernization: a macro-level framework that seeks to conceptualize the 'new rules of the game for the social organization of production and consumption' (Spaargaren & Van Vliet, 2000, p. 56). Ecological Modernization is argued to be the prevailing rationale for dealing with socio- environmental challenges, which emphasizes achieving the goals of continued development and environmental improvement through restructuring production and consumption processes. A particular emphasis is placed on technological innovations to improve efficiency and encourage new forms of technological intervention, as well as the role of markets and economic agents 'to introduce incentives for environmentally benign outcomes' (Sutton, 2007, p. 159). Some scholars linking sustainable development to ecological modernization (Spaargaren & Mol, 1992, p. 334) state that 'the concept of sustainable development, ecological modernization indicates the possibility of overcoming the environmental crisis without leaving the path of modernization'. The modernization equates to a form of 'sustainable capitalism' (Fisher & Freudenburg, 2001; Hamilton, Gemenne, & Bonneuil, 2015) which re-enshrines the existing neo-liberal order: an order that prioritizes economic growth and reductions in state activity, including the systematic dismantling of programs related to environmental, ecological and indeed social sustainability (Parr, 2014).

This brief overview presented highlights scholars' continuous enlargement of the meaning of sustainable development by considering it as an umbrella concept under which a set of inter-related issues can be gathered. In agreement with this ample perspective, sustainable development in this work is considered a variable and uninterrupted process of change that seeks the ultimate goal of sustainability itself. Similarly, sustainability could be defined as the ability of human beings to resist or adapt to change (Dovers & Handmer, 1992), represented as a goal or an end point (Hove, 2004; Malesios et al., 2018).

SUSTAINABILITY IN FAMILY FIRMS

Although a unique definition of the family business does not exist, it is quite obvious that a close connection exists between the lasting functionality of this type of firm and the evolution of the family engaged in managing the business. In general, the expression 'family business' refers to firms which are wholly owned and managed by members of the founding family with the intention to transfer their entities across generations, and they perceive that their business is based on family beliefs and values (Aronoff & Ward, 2011). Evidently, a family-owned firm is motivated by a long-term orientation (Lumpkin & Brigham, 2011), which implies increasing attention to strategies that can guarantee successful longevity.

Family businesses represent the majority of all businesses in most economies throughout the world (Brice & Richardson, 2009; Villalonga & Amit, 2010). Nevertheless, the academic study of family businesses is recent and still emerging (Heck & Stafford, 2001; Heck & Trent, 1999). Today, family businesses contribute significantly to economic and social sustainability and welfare, and their impact on local and national economies is substantial.

Miller and Le Breton-Miller (2005) maintained that successful family firms adopt practices intended to extend the enduring viability of their businesses. The authors emphasise the merits of continuity in investing in family firms and their offerings: sustaining a vibrant community culture, building long-term win–win relationships with stakeholders and courageously commanding leadership that renews the firm even in the face of challenges.

Sustainability practices may be particularly important to family-owned and managed businesses for a number of reasons. First, they can help the owning family develop a positive reputation in the community and establish relationships with stakeholders, especially those in the external environment. They can also facilitate continuity of a family firm across generations (Andersson, Carlsen, & Getz, 2002; Carlsen, Getz, & Ali-Knight, 2001; Getz & Carlsen, 2000, 2005; Getz & Petersen, 2005; Gomez-Mejıa et al., 2007; Le Breton-Miller & Miller, 2006).

In general, sustainable development is claimed to be an opportunity to enhance competitiveness and growth as a source of inspiration for efforts at innovation (Hall & Vredenburg, 2003; Hart & Milstein, 1999). Many studies clearly indicated that investing in accordance with sustainability principles possesses the capacity to create long-term value (Bebbington, 2001; Sage, 1999). A business strategy that improves sustainability and, thereby, increases firm value incorporates long-term social and ecological aims (Zadek, 2006). Such incorporation reflects a means of improving the firm's sustainability and increasing its long-term value (Maruffi, Petri, & Malindretos, 2013).

The longer time horizon derived from an intention to continue family control of a firm can help its leaders avoid managerial myopia, forgetting short-term earnings (James, 1999; Le Breton-Miller & Miller, 2008; Upton, Teal, & Felan, 2001) and directing their efforts toward developing patient capital and long-term investments such as sustainability practices. Furthermore, close monitoring and control by family owners with a long-term perspective can lead to confer higher priority to sustainability practices.

Craig and Dibrell (2006) investigated the effects that firm policies related to the natural environment produce on innovation and financial performance in family and non-family businesses and the results of their study demonstrated that family firms are in a favoured position with respect to their counterparts. More clearly, family businesses appear to be particularly able to enhance environmentally friendly policies with a positive impact on innovation and financial performance. In this view, more recent works (Berrone, Cruz, Gómez-Mejía, & Larraza-Kintana, 2010; Block & Wagner, 2014) emphasised the catching-up process of family firms relative to non-family firms with respect to their environmental behaviour.

Finally, an important theory to explain how the adoption of sustainability practices can be affected by family governance is the socioemotional wealth perspective, since family firm behavior is shaped by the family's desire to preserve socioemotional values (Chua et al., 1999; Gomez-Mejıa et al., 2007). Family firms tend to prefer strategies that can sustain these values across generations (Chrisman et al., 2012) and avoid strategies that can put them at risk (Chrisman & Patel, 2012; Gomez-Mejıa et al., 2007).

This perspective assumes that multiple non-economic goals exist in a family business (Kotlar & De Massis, 2013). Berrone et al. (2012) identify this specific kind of goal of owning families as the willingness to exercise authority and to influence the emotional value of owning a firm, family members' identification with the firm, the reputation of the family in the local community, and the renewal of family bonds with the firm through dynastic succession. Given the multiplicity of non-economic goals in the family business, long-term orientation is extremely crucial and it can help simplify judging criteria among strategic alternatives when facing complex situations. Thus, in a family business with marked long-term orientation (Lumpkin & Brigham, 2011), the decision-makers generally prefer strategies able to reconcile continuity and sustainability concerns, recognize and respond to internal and external stakeholders, and maintain a positive image and reputation.

METHODOLOGY

The methodology chosen for this research is qualitative. Through the analysis of a single case study (Patton, 1990; Yin, 2009, 2012), this chapter explores the sustainable orientation of Argiolas Formaggi, a Sardinian family business that is particularly meaningful to the aim of this study. The single case study is a methodology of empirical investigation that places the real phenomena at the centre of the analysis and observes them in their uniqueness as part of a particular scenario and its interactions in relation to the difficulty of distinguishing the boundaries of the phenomenon from the context in which it develops and acts (Patton, 1990; Yin, 2009, 2012). The single case study represents a form of qualitative investigation aimed at seeking the 'meaning' of reality in the experiential experience of people and organisations (Eisenhardt, 1989; Eisenhardt & Graebner, 2007; Patton, 1990; Yin, 2009, 2012). For the purposes of this work, the analysis of Argiolas Formaggi allows to contextualise the phenomenon of sustainability strategies within the analysed family firm. To collect data, the narrative approach has been adopted because, unlike quantitative methods, allows the generation of in-depth and contextualised empirical material (Suddaby, 2006). Studying processes in a family business using a narrative approach means participating in the social dynamics of relational constructs, such as roles, resources, projects, organisations and objectives. Given the importance of the narrative analysis, accurate data collection was completed that involved third-generation owners. These owners were contacted by e-mail and, after having agreed to narrate their history, they were called on the phone two times. The conversations were registered (approximately 40 minutes duration), then transcribed verbatim and finally analysed as a preliminary source of information. With this initial information and by continuously comparing the findings with the literature (Suddaby, 2006), the next step was framed particularly to explore gaps. This refinement represented the bases for the subsequent four face-to-face interviews (length approximately 85 minutes for each) that were halted when a reasonable level of theoretical saturation was attained (Walsh & Bartunek, 2011). Additionally, in this case, conversations were registered and transcribed. Moreover, to triangulate the data (Jick, 1979), the official website, social media pages (Facebook and Instagram), online videos, press conference reports, articles in the national and local press and others secondary sources were analysed.

Then, by systematising the narrative obtained with the two phone calls and the four face-to-face interviews and analysing secondary sources of information, the order of events was traced, the main actors were identified, the link between the events and the firm's sustainable orientation were investigated (Czarniawska, 1999).

Finally, following the suggestions of Czarniawska (1999), the data analysis was carried out through the hermeneutic trio (Hernadi, 1987) consisting of three phases: (1) *explication*–contextualisation, reconstruction and synthesis of the history; (2) *explanation*– identification, description and understanding of the meaning of the narrative and (3) *exploration*– discussion and identification of theoretical and practical implications. Moreover, to highlight congruencies and inconsistencies, the stories were analysed using the dramaturgical approach (Burke, 1985), which suggests highlighting the context of reference, the stories and sequence of the events, the main actors, the modalities of execution of the actions and the motivations underlying the actions.

The Explication

Contextualisation

Argiolas Formaggi is a family firm in Dolianova, a small village in Sardinia. Sardinia, an island and region of Italy with an area of approximately 24,000 square kilometres (km) and with coasts extending nearly 2,400 km, is an autonomous region with four provinces (Oristano, Nuoro, Sassari and South Sardinia) and the metropolitan city of Cagliari. The island is rich in history and agriculture, pastoralism, crafts, services and tourism are the cornerstones of the Sardinian economy (RAS, 2017). Dolianova is located in the south of the island and is characterised by a flourishing agricultural and pastoral economy. The choice of this geographical position, which is wholly immersed in the green, represents a strategic location for the sustainable development of this family business.

Reconstruction and Synthesis of the History

The Argiolas Formaggi is a dairy family firm at the forefront of the production of typical Sardinian cheeses and has, for more than 60 years, produced goat and sheep cheeses with skill and passion.

The firm was founded in 1954 through the initiative of two young brothers, Ennio and Eligio Argiolas.

During the first years of activity, the brothers bought from the shepherds the still fresh 'Fiore Sardo' cheese and let it ripen in cellars. They then sold the cheese to wholesalers throughout Italy.

'Fiore Sardo' is a sheep's milk cheese made with raw milk that is unique for its flavour and fragrance. Today, the cheese is produced by the company using the most modern technologies in full respect of the ancient original recipe.

A few years after founding the company, the two brothers decided to invest the proceeds of their activities to produce their own cheese and rented a small dairy. With the new plant, Ennio and Eligio began to experiment with new types of cheese and constantly improved its quality.

In 1966, the company built a warehouse in which to mature the cheese to meet increasing orders. The year 1970 proved crucial for the Argiolas brothers because they decided to build a dairy in the Dolianova industrial area, just below the hills where shepherds used to take the animals to pasture. This dairy formed the original core of the establishment at which Argiolas Formaggi operates today.

The first expansion took place in the 1980s, and the construction of new maturation warehouses that started in 1999 was completed in 2001. The new warehouses are managed and controlled by a technologically advanced central computer system which regulates functions such as ageing and moisture to ensure maximum product quality throughout the year. Tradition and innovation are the ingredients for the success of this family business. '*We start from all that our founders have taught us and then produce in an absolutely innovative way, with high hygiene standards and optimizing our resources*', stated Alessandra Argiolas (third generation), marketing manager of the company and daughter of the managing director, Antonello Argiolas. The firm has always been attentive to environmental protection and promoting sustainable development, which can balance the need for economic growth and protecting the environment with a view to full social responsibility. '*For us, sustainability and the defense of the environment are fundamental values, which have always represented a cornerstone of our company, right from the beginning of our history*', stated Alessandra.

In addition to a photovoltaic system that sustains production, the company has invested in the construction of a liquefied natural gas plant and the subsequent supply of fuel, which sustains the production processes at the Dolianova plant.

The plant also offers the possibility of producing electricity through cogeneration, increasing its efficiency. '*This innovation is sustainable both from an economic and environmental point of view guarantees a greater protection of our territory, enhancing even more the natural raw material that is the basis of the quality of our products*', emphasised Alessandra.

Then, through its traditional production activity, Argiolas Formaggi is committed to environmental protection and spreading farm welfare using different actions: (1) a photovoltaic system which reduces energy consumption; (2) process optimisation that reduces waste production; (3) a biological sludge treatment plant; (4) production of conjugated linolenic acid (CLA) and BIO sheep's milk cheeses based on the principles of organic farming; (5) packaging, glues and colours used that have low environmental impact and (6) a research project for the valorisation of traditional raw milk production that guarantees biopreservation.

Argiolas Formaggi's sheep's milk cheese CLA is a product of omega-3 (alpha-linolenic acid, ALA) and controls cholesterol without sacrificing taste and quality at the table. CLA sheep's milk cheese products are produced naturally with the addition of ALA, essential fatty acids which belong to the omega-3 group. ALA is a completely natural element that is particularly present in some Sardinian pastures, by selected, is odourless and tasteless and does not alter the organoleptic characteristics of the cheese, thus retaining its traditional and unique taste. At Argiolas Formaggi, twelve years of research on a healthy diet and controlled sheep have resulted in an innovative and high quality product that contributes– as part of a varied and balanced diet and a healthy lifestyle– to maintaining normal levels of blood cholesterol.

The involvement of the farmers who work on the project is significant. They play a powerful role in controlling the sheep and are committed to ensuring animal welfare.

'Thanks to the successful cooperation with scientific research, our firm is attentive to the wishes of its consumers, to environment and is able to offer highly innovative products and quality in respect of the Sardinian cheese-making tradition', asserted Alessandra.

'We are now evaluating the costs', explained Alessandra, 'to buy at least a thousand head of sheep to directly control the quality of the milk'.

Quality, innovation and environmental sustainability are the factors of business success. Through more than 40 innovative and traditional Sardinian products made with the highest technological and hygienic standards, and thanks to the sublime quality of its products, Argiolas Formaggi has received prestigious national and international awards and was the first dairy company in Sardinia to obtain ISO 9001 certification.

'The company is also equipped with a HACCP system, that is a systematic preventive approach to food safety from biological, chemical, and physical hazards in production processes that can cause the finished product to be unsafe and designs measures to reduce these risks to a safe level, and it is also currently certified according to the UNI EN ISO 22000: 2005, BRC and IFS, and our organic products are certified by ICEA. Argiolas Formaggi is a company at the service of tradition and quality, but above all the consumer. Always', stated Alessandra.

Finally, another fundamental aspect of the company is internationalisation, as Alessandra pointed out, '*For us, foreign is important both in the present and in the future. We currently export about 20% of our turnover but we aim, in the next ten years, to reach 50%. A very ambitious goal, our biggest market today is Japan, followed by Germany, where we export lots of organic products*'.

Explanation

Sustainability represents the fundamental driver to changing the mindset at Argiolas Formaggi. The history of this family firm is scattered by a marked propensity toward flexibility, ability to change, openness towards innovation and propensity for creative ideas since the first generation. These positive aspects have concurred in combining tradition with innovation by addressing new trajectories in the development of the entire production process of cheese. Their cheeses are born to enhance the quality and characteristics of Sardinian milk, and traditional and innovative recipes are combined in products with a unique and special taste. For this firm, the quality, first of all, is made up of values, such as respect for tradition, technological innovation and attention to research and constant commitment to environmental sustainability.

Several aspects have led to a growing interest in sustainable development, such as the sensitivity of the firm, the need to adapt to the regulatory evolution (the certifications of products), the need to increase the quality of the products and the simultaneous reduction in their production costs (the investments made in the construction of natural gas plants), customers' perception of improvements in image and reputation, increasing attention to environmental trends (the production of CLA and BIO sheep's milk cheese) and, finally, new market opportunities (the opportunity offered by internationalisation).

Therefore, promoting sustainable development means that this firm pursues a balance among the three dimensions of environmental, economic and social. Of course, maintaining this balance is a dynamic process that is continually challenged by the pressures from incessant changes and the improvements which are promoted by different subjects (public, social and private).

Exploration

Research on the literature has revealed a large number of studies on the relationship between sustainability and business activities in a broad sense (Rimanoczy & Pearson, 2010). Some studies identified how consumers' demand for quality products implicitly and explicitly includes environmental requirements (Hove, 2004; Rimanoczy & Pearson, 2010). The requests of civil society have led firms to recognise that environmental management can lead to profitable results. Specifically, being sustainable could represent a goal that pushes firms to produce high quality goods, rationalise resources, act in a socially responsible manner, respect the natural environment and, as a result, gain positive economic and financial performance (Reganold, Glover, Andrews, & Hinman, 2001).

This list reflects the mission of Argiolas Formaggi, which focuses on quality, innovation and sustainability while preserving tradition and values of firms (Chrisman et al., 2012). The case opens new stream of research within the socio-emotional wealth perspectives (Chua et al., 1999; Gomez-Mejıa et al., 2007). The owners, in fact, show a family proclivity toward the achieving of non-financial goals, but these goals, based on sustainability practices, stimulate family firm's long-term orientation (Lumpkin & Brigham, 2011), which leads to a less volatile behaviour, that is, more consistent pursuance of environment-related activities, innovations and performance.

In addition, the analysed firm has a production process that aims to promote the use of raw materials produced within the local context, reduce the environmental impact (including that of packaging), limit the use of energy (through recovery, recycling and modifications of production processes), maximise the use of renewable energy resources, reduce process and distribution costs to reach 'fair' consumer prices and minimise waste in the food chain. This family firm, as underlined by several authors (Howorth, Rose, Hamilton, & Westhead, 2010), was particularly able to innovate, adapt strategies and strengthen its competitive advantage. The analysis of its history, carried out through narratives, has revealed interesting and useful insights for both theory and practice, in particular because the firm operates in a traditional sector (Welter, Baker, Audretsch, & Gartner, 2017) that is often considered to be resistant to change. Argiolas Formaggi belongs to those family businesses that are anxious and excited about being entrepreneurially oriented and interested in preserving the past paths without compromising the possibility of tracing new and better development routes. In this sense, the analysis of this firm extends previous studies (König, Kammerlander, & Enders, 2013) that underlie the positive side of the attachment to tradition in terms of safeguarding an embedded identity that could be interpreted in innovative keys. In fact, this bond with tradition has stimulated the creativity of the owners who continue to produce and sell traditional products and have studied new and alternative ways to realise these products through the lens of innovation and modernity. Moreover, this firm demonstrates what has been emphasised by Fernández and Nieto (2005): that new generations have a propensity to internationalise because of their will and impatience with predecessors.

Finally, Argiolas Formaggi represents a case in which the third generation of entrepreneurs is not more conservative than the first, as has emerged in other studies (Garaud, Jain, & Tuertscher, 2008). In contrast, the three generation shows a clear propensity for innovation, but the latter has the merit to have concretely implemented new and important sustainable development strategies.

CONCLUSION

This chapter aimed to analyse sustainability policies and strategies in a specific type of firm: the family business. This analysis was done because family firms constitute a significant part of the economy in terms of creating new technologies, jobs and wealth (Berlemann & Jahn, 2015). This contribution depends on firms' ability to survive, in particular by adopting and implementing innovations and, thus, by taking risks to achieve and maintain competitive advantages. The relevance of this category of firms and the scarcity of the study carried out until now have suggested a focus on how these firms face sustainability challenges.

The Argiolas Formaggi case has revealed a reality that, born in a small village of Sardinia, has succeeded because of the passion and determination of the owners in creating a successful firm at the national and international levels by implementing the principle of sustainability within its business. This implementation has allowed the firm to create a process oriented towards efficiently exploiting natural resources, satisfying people's needs and consequently obtaining profitable results for the entire system.

This study shows important theoretical and practical implications. For what concerns theoretical implications, the research contributes to previous studies in three ways. First, it contributes to the debate around the topic of heterogeneity within family firms, by showing how governance, resources and goals play a relevant role in defining and implementing sustainable strategies. Second, findings demonstrate that sustaining and saving family socio-emotional heritage allow family firms to fulfil a privileged position in promoting sustainability due to the main attention done on non-financial rather than economic goals. Third, sustainable and green policies utilized by the analysed family firm guarantee the international success and the enhancing of a very recognized positive reputation in the market. In other words, sustainable strategies, in this case, become the main driver to create the basis for the long-term successful of the family firm.

On the other side, practical implications are essentially related to the important stimuli that the family firm can represent for other firms that operate in a complex and poor context. In fact, Argiolas Formaggi integrates sustainability in the core business and this choice guarantees the possibility to operate in oversee markets, overtaking local difficulties.

Obviously, as other exploratory researches, also this study shows limitations. The main is the reference to only one exemplary case. Future studies can be addressed to enlarge the analysis, by involving a large number of firm, also of different industries. Moreover, further researches could propose a set of propositions to develop a model to investigate sustainability in family firms.

Notwithstanding the mentioned limitations, the case Argiolas Formaggi represents an exemplary family firm able to show its ability in implementing strategies with positive effects both for the firm and for the environment. Finally, the analysed case can spur and motivate other firms to consider sustainability as a proper key to successful change.

REFERENCES

Andersson, T., Carlsen, J., & Getz, D. (2002). Family business goals in the tourism and hospitality sector: Case studies and cross-case analysis from Australia, Canada, and Sweden. *Family Business Review*, *15*(2), 89–106. doi:10.1111/j.1741-6248.2002.00089.x

Aronoff, C., & Ward, J. (2011). *Family business values: How to assure a legacy of continuity and success*. New York, NY: Palgrave MacMillan. doi:10.1007/978-1-137-51208-6

Bansal, P. (2005). Evolving sustainably: A longitudinal study of corporate sustainable development. *Strategic Management Journal*, *26*(3), 197–218. doi:10.1002mj.441

Bebbington, J. (2001). Sustainable Development: A Review of the International Development Business and Accounting Literature. *Accounting Forum*, *25*(2), 128–157. doi:10.1111/1467-6303.00059

Berlemann, M., & Jahn, V. (2015). Regional importance of Mittelstand firms and innovation performance. *Regional Studies*, *50*(11), 1819–1833. doi:10.1080/00343404.2015.1058923

Berrone, P., Cruz, C., & Gomez-Mejia, L. R. (2012). Socioemotional wealth in family firms theoretical dimensions, assessment approaches, and agenda for future research. *Family Business Review*, *25*(3), 258–279. doi:10.1177/0894486511435355

Berrone, P., Cruz, C., Gómez-Mejía, L. R., & Larraza-Kintana, M. (2010). Socioemotional wealth and corporate responses to institutional pressures: Do family-controlled firms pollute less? *Administrative Science Quarterly*, *55*(1), 82–113. doi:10.2189/asqu.2010.55.1.82

Block, J. H., & Wagner, M. (2014). The effect of family ownership on different dimensions of corporate social responsibility: Evidence from large US firms. *Business Strategy and the Environment*, *23*(7), 475–492. doi:10.1002/bse.1798

Brice, W. D., & Richardson, J. (2009). Culture in family business: A two-country empirical investigation. *European Business Review*, *21*(3), 246–262. doi:10.1108/09555340910956630

Brundtland, G. H. (1987). *Our Common Future. The World Commission on Environment and Development*. Oxford, UK: Oxford University Press.

Burke, K. (1985). Dramatism and logology. In D. L. Sills (Ed.), International encyclopedia of the social sciences (pp. 445–452). London, UK: Macmillan Publishers. doi:10.1080/01463378509369584

Cabrera-Suarez, K., Saa-Perez, P., & Garcia-Almeida, D. (2001). The succession process from a resource- and knowledge-based view of the family firm. *Family Business Review*, *14*(1), 37–48. doi:10.1111/j.1741-6248.2001.00037.x

Carlsen, J., Getz, D., & Ali-Knight, J. (2001). The environmental attitudes and practices of family businesses in the rural tourism and hospitality sectors. *Journal of Sustainable Tourism*, *9*(4), 281–297. doi:10.1080/09669580108667403

Chichilnisky, G. (1996). An axiomatic approach to sustainable development. *Social Choice and Welfare*, *13*(2), 231–257. doi:10.1007/BF00183353

Chrisman, J. J., Chua, J. H., Pearson, A. W., & Barnett, T. (2012). Family involvement, family influence, and family_centered non_economic goals in small firms. *Entrepreneurship Theory and Practice*, *36*(2), 267–293. doi:10.1111/j.1540-6520.2010.00407.x

Chrisman, J. J., & Patel, P. C. (2012). Variations in R&D investments of family and nonfamily firms: Behavioral agency and myopic loss aversion perspectives. *Academy of Management Journal*, *55*(4), 976–997. doi:10.5465/amj.2011.0211

Chua, J. H., Chrisman, J. J., & Sharma, P. (1999). Defining the family business by behavior. *Entrepreneurship Theory and Practice*, *23*(4), 19–40. doi:10.1177/104225879902300402

Craig, J., & Dibrell, C. (2006). The natural environment, innovation, and firm performance: A comparative study. *Family Business Review, 19*(4), 275–288. doi:10.1111/j.1741-6248.2006.00075.x

Czarniawska, B. (1999). *Writing management: Organization theory as a literary genre*. Oxford, UK: Oxford University Press. doi:10.1093/acprof:oso/9780198296140.001.0001

Dovers, S. R., & Handmer, J. W. (1992). Uncertainty, sustainability and change. *Global Environmental Change, 2*(4), 262–276. doi:10.1016/0959-3780(92)90044-8

Drexhage, J., & Murphy, D. (2010). *Sustainable Development: from Brundtland to Rio 2012*. United Nations, Background Paper. Retrieved June 2015 from http://www.un.org/wcm/webdav/site/ climatechange/shared/gsp/docs/GSP1-6_Background%20on%20Sustainable%20Devt.pdf

Duran, P., Kammerlander, N., Van Essen, M., & Zellweger, T. (2016). Doing more with less: Innovation input and output in family firms. *Academy of Management Journal, 59*(4), 1224–1264. doi:10.5465/amj.2014.0424

Eisenhardt, K. M. (1989). Building theories from case study research. *Academy of Management Review, 14*(4), 532–550. doi:10.5465/amr.1989.4308385

Eisenhardt, K. M., & Graebner, M. E. (2007). Theory building from cases: Opportunities and challenges. *Academy of Management Journal, 50*(1), 25–32. doi:10.5465/amj.2007.24160888

European Commission. (2001). *Green Paper: Promoting a European Framework for Corporate Social Responsibility. 28 Official Journal C 340*. Brussels: Treaty on European Union.

European Commission. (2002). Communication - Corporate Social Responsibility: A Business Contribution to Sustainable Development. COM(2002) 347 final, Brussels.

Fisher, D. R., & Freudenburg, W. R. (2001). Ecological modernization and its critics: Assessing the past and looking toward the future. *Society & Natural Resources, 14*(8), 701–709. doi:10.1080/08941920152524891

Garaud, R., Jain, S., & Tuertscher, P. (2008). Incomplete by design and designing for incompleteness. *Organization Studies, 29*(3), 351–371. doi:10.1177/0170840607088018

Getz, D., & Carlsen, J. (2000). Characteristics and goals of family and owner-operated businesses in the rural tourism and hospitality sectors. *Tourism Management, 21*(6), 547–560. doi:10.1016/S0261-5177(00)00004-2

Getz, D., & Carlsen, J. (2005). Family business in tourism: State of the art. *Annals of Tourism Research*, *32*(1), 237–258. doi:10.1016/j.annals.2004.07.006

Getz, D., & Petersen, T. (2005). Growth and profit-oriented entrepreneurship among family business owners in the tourism and hospitality industry. *International Journal of Hospitality Management*, *24*(2), 219–242. doi:10.1016/j.ijhm.2004.06.007

Gladwin, T. N., Kennelly, J. J., & Krause, T. S. (1995). Shifting paradigms for sustainable development: Implications for management theory and research. *Academy of Management Review*, *20*(4), 874–907. doi:10.5465/amr.1995.9512280024

Gomez-Mejıa, L. R., Haynes, K. T., Nunez-Nickel, M., Jacobson, K. J., & Moyano-Fuentes, J. (2007). Socioemotional wealth and business risks in family-controlled firms: Evidence from Spanish olive oil mills. *Administrative Science Quarterly*, *52*(1), 106–137. doi:10.2189/asqu.52.1.106

Hall, J., & Vredenburg, H. (2003). The Challenges of Innovating for Sustainable Development. *MIT Sloan Management Review*, *45*(1), 61–68.

Hall, J. K., Daneke, G. A., & Lenox, M. J. (2010). Sustainable development and entrepreneurship: Past contributions and future directions. *Journal of Business Venturing*, *25*(5), 439–448. doi:10.1016/j.jbusvent.2010.01.002

Hamilton, C., Gemenne, F., & Bonneuil, C. (2015). *The anthropocene and the global environmental crisis: rethinking modernity in a new epoch*. London: Routledge. doi:10.4324/9781315743424

Hart, S. L., & Milstein, M. B. (1999). Global Sustainability and the Creative Destruction of Industries. *MIT Sloan Management Review*, *41*(1), 23–33.

Heck, R. K., & Stafford, K. (2001). The vital institution of family business: Economic benefits hidden in plain sight. In G. K. McCann & N. Upton (Eds.), Destroying myths and creating value in family business (pp. 9–17). Deland, FL: Stetson University.

Heck, R. K., & Trent, E. S. (1999). The prevalence of family business from a household sample. *Family Business Review*, *12*(3), 209–219. doi:10.1111/j.1741-6248.1999.00209.x

Hernadi, P. (1987). Literary interpretation and the rhetoric of the human sciences. In J. S. Nelson, A. Megill, & D. N. McCloskey (Eds.), The rhetoric of the human sciences (pp. 263–275). Madison, WI: University of Wisconsin Press.

Hove, H. (2004). Critiquing sustainable development: A meaningful way of mediating the development impasse? *Undercurrent*, *1*(1), 48–54.

Howorth, C., Rose, M., Hamilton, E., & Westhead, P. (2010). Family firm diversity and development: An introduction. *International Small Business Journal*, *28*(5), 437–451. doi:10.1177/0266242610373685

James, H. S. Jr. (1999). What can the family contribute to business? Examining contractual relationships. *Family Business Review*, *12*(1), 61–71. doi:10.1111/j.1741-6248.1999.00061.x

Jick, T. D. (1979). Mixing qualitative and quantitative methods: Triangulation in action. *Administrative Science Quarterly*, *24*(4), 602–611. doi:10.2307/2392366

Kammerlander, N., Dessì, C., Bird, M., Floris, M., & Murru, A. (2015). The Impact of Shared Stories on Family Firm Innovation A Multicase Study. *Family Business Review*, *28*(4), 332–354. doi:10.1177/0894486515607777

Kelly, R., Sirr, L., & Ratcliffe, R. (2004). Futures thinking to achieve sustainable development at local level in Ireland. *Foresight*, *6*(2), 80–90. doi:10.1108/14636680410537547

König, A., Kammerlander, N., & Enders, A. (2013). The family innovator's dilemma: How family influence affects the adoption of discontinuous technologies by incumbent firms. *Academy of Management Review*, *38*(3), 418–441. doi:10.5465/amr.2011.0162

Kotlar, J., & De Massis, A. (2013). Goal setting in family firms: Goal diversity, social interactions, and collective commitment to family_centered goals. *Entrepreneurship Theory and Practice*, *37*(6), 1263–1288. doi:10.1111/etap.12065

Le Breton-Miller, I., & Miller, D. (2006). Why do some family businesses out_compete? Governance, long-term orientations, and sustainable capability. *Entrepreneurship Theory and Practice*, *30*(6), 731–746. doi:10.1111/j.1540-6520.2006.00147.x

Le Breton-Miller, I., & Miller, D. (2008). To grow or to harvest? Governance, strategy and performance in family and lone founder firms. *Journal of Strategy and Management*, *1*(1), 41–56. doi:10.1108/17554250810909419

Lumpkin, G. T., & Brigham, K. H. (2011). Long-term orientation and intertemporal choice in family firms. *Entrepreneurship Theory and Practice*, *35*(6), 1179–1197. doi:10.1111/j.1540-6520.2011.00495.x

Malesios, C., Skouloudis, A., Dey, P. K., Abdelaziz, F. B., Kantartzis, A., & Evangelinos, K. (2018). (in press). The impact of SME sustainability practices and performance on economic growth from a managerial perspective: Some modeling considerations and empirical analysis results. *Business Strategy and the Environment*. doi:10.1002/bse.2045

Maruffi, BL., & Petri, R., & Malindretos, J. (2013). Corporate Social Responsibility and the Competitive Advantage of Multinational Corporations: What is the Right Balance? *The Journal of Global Business Issues*, *7*(2), 69–81.

Miller, D., & Le Breton-Miller, I. (2005). Management insights from great and struggling family businesses. *Long Range Planning*, *38*(6), 517–530. doi:10.1016/j.lrp.2005.09.001

Nieto, M. J., Santamaria, L., & Fernandez, Z. (2015). Understanding the innovation behavior of family firms. *Journal of Small Business Management*, *53*(2), 382–399. doi:10.1111/jsbm.12075

Oertwig, N., Galeitzke, M., Schmieg, H. G., Kohl, H., Jochem, R., Orth, R., & Knothe, T. (2017). Integration of Sustainability into the Corpory Strategy. In R. Stark, G. Seliger & J. Bonvoisin (Eds.), Sustainable Manufacturing. Challenge Solution and Implementations Perspective (pp. 175–200). SpringerOpen.

Parr, A. (2014). *The wrath of capital: neoliberalism and climate change*. New York: Columbia University Press.

Patton, M. Q. (1990). *Qualitative evaluation and research methods*. Newbury Park, CA: SAGE Publications.

RAS. (2017). *Sardegna in cifre 2017. Cagliari: Servizio della Statistica Regionale.* Retrieved from http://www.sardegnastatistiche.it/documenti/12_103_20170727151245.pdf

Reganold, J. P., Glover, J. D., Andrews, P. K., & Hinman, H. R. (2001). Sustainability of three apple production system. *Nature*, *410*(6831), 926–930. doi:10.1038/35073574 PMID:11309616

Rimanoczy, I., & Pearson, T. (2010). Role of HR in the new world of sustainability. *Industrial and Commercial Training*, *42*(1), 11–17. doi:10.1108/00197851011013661

Rowe, B. R., Haynes, G. W., & Stafford, K. (1999). The contribution of home-based business income to rural and urban economies. *Economic Development Quarterly*, *13*(1), 66–77. doi:10.1177/089124249901300109

Sage, A. P. (1999). Sustainable Development: Issues in Information, Knowledge, and Systems Management. *Information, Knowledge, Systems Management*, *1*(3-4), 185–223.

Sava, T., Rizea, C., & Flood, I. (2010). Benefits of the implementation and certification for environmental management system SMEs in Romania. *Quality*, *2*, 248–254.

Shanker, M. C., & Astrachan, J. H. (1996). Myths and realities: Family businesses' contribution to the US economy—a framework for assessing family business statistics. *Family Business Review*, *9*(2), 107–123. doi:10.1111/j.1741-6248.1996.00107.x

Sharma, P., Chrisman, J. J., & Chua, J. H. (2003). Predictors of satisfaction with the succession process in family firms. *Journal of Business Venturing*, *18*(5), 667–687. doi:10.1016/S0883-9026(03)00015-6

Sharma, P., & Manikutty, S. (2005). Strategic divestments in family firms: Role of family structure and community culture. *Entrepreneurship Theory and Practice*, *29*(3), 293–311. doi:10.1111/j.1540-6520.2005.00084.x

Short, J. C., Payne, G. T., Brigham, K. H., Lumpkin, G. T., & Broberg, J. C. (2009). Family firms and entrepreneurial orientation in publicly traded firms: A comparative analysis of the S&P 500. *Family Business Review*, *22*(1), 9–24. doi:10.1177/0894486508327823

Shrivastava, P. (1995). The role of corporations in achieving ecological sustainability. *Academy of Management Review*, *20*(4), 936–960. doi:10.5465/amr.1995.9512280026

Spaargaren, G., & Mol, A. P. (1992). Sociology, environment, and modernity: Ecological modernization as a theory of social change. *Society & Natural Resources*, *5*(4), 323–344. doi:10.1080/08941929209380797

Spaargaren, G., & Van Vliet, B. (2000). Lifestyles, consumption and the environment: The ecological modernization of domestic consumption. *Environmental Pollution*, *9*(1), 50–76.

Springett, D. (2003). Business conceptions of sustainable development. A perspective from critical theory. *Business Strategy and the Environment*, *12*(2), 71–86. doi:10.1002/bse.353

Starik, M., & Rands, G. P. (1995). Weaving an integrated web: Multilevel and multisystem perspective of ecologically sustainable organizations. *Academy of Management Review*, *20*(4), 908–935. doi:10.5465/amr.1995.9512280025

Suddaby, R. (2006). From the editors: what grounded theory is not. Academy of Management Journal, 49(4), 633–642.

Sutton, P. W. (2007). *The environment: a sociological introduction*. London: Polity Press.

Upton, N., Teal, E. J., & Felan, J. T. (2001). Strategic and business planning practices of fast growth family firms. *Journal of Small Business Management*, *39*(1), 60–72. doi:10.1111/0447-2778.00006

Viederman, S. (1993). *The economics and economy of sustainability: Five capitals and three pillars, talk delivered to Delaware Estuary Program*. New York: Noyes Foundation.

Villalonga, B., & Amit, R. (2006). How do family ownership, control and management affect firm value? *Journal of Financial Economics*, *80*(2), 385–417. doi:10.1016/j.jfineco.2004.12.005

Welter, F., Baker, T., Audretsch, D. B., & Gartner, W. B. (2017). *Everyday Entrepreneurship—A Call for Entrepreneurship Research to Embrace Entrepreneurial Diversity*. Thousand Oaks, CA: SAGE Publications. doi:10.1111/etap.12258

World Business Council for Sustainable Development. (2000). Corporate Social Responsibility: Making Good Business Sense. Geneva, Switzerland: Author.

World Commission on Environment and Development. (1987). *Our common future*. Oxford, UK: Oxford University Press.

Yin, R. K. (2009). *Case Study Research. Design and methods*. Thousand Oaks, CA: SAGE.

Yin, R. K. (2012). *Applications of case study research* (3rd ed.). Thousand Oaks, CA: SAGE.

Zadek, S. (2006). Responsible Competitiveness: Reshaping Global Markets through Responsible Business Practices. *Corporate Governance*, *6*(4), 334–348. doi:10.1108/14720700610689469

Zahra, S. A., Hayton, J. C., Neubaum, D. O., Dibrell, C., & Craig, J. (2008). Culture of family commitment and strategic flexibility: The moderating effect of stewardship. *Entrepreneurship Theory and Practice*, *32*(6), 1035–1054. doi:10.1111/j.1540-6520.2008.00271.x

ADDITIONAL READING

Broccardo, L., Truant, E., & Zicari, A. (2019). Internal corporate sustainability drivers: What evidence from family firms? A literature review and research agenda. *Corporate Social Responsibility and Environmental Management*, *26*(1), 1–18. doi:10.1002/csr.1672

Dibrell, C., & Memili, E. (2019). A brief history and a look to the future of family business heterogeneity: an introduction. In The Palgrave Handbook of Heterogeneity among Family Firms (1–15). Palgrave Macmillan, Cham. doi:10.1007/978-3-319-77676-7_1

Hsueh, J. W. J. (2018). Governance structure and the credibility gap: Experimental evidence on family businesses' sustainability reporting. *Journal of Business Ethics*, *153*(2), 547–568. doi:10.100710551-016-3409-y

Kallmuenzer, A., Nikolakis, W., Peters, M., & Zanon, J. (2018). Trade-offs between dimensions of sustainability: Exploratory evidence from family firms in rural tourism regions. *Journal of Sustainable Tourism*, *26*(7), 1204–1221. doi:10.1080/09669582.2017.1374962

López-Pérez, M., Melero-Polo, I., Vázquez-Carrasco, R., & Cambra-Fierro, J. (2018). Sustainability and Business Outcomes in the Context of SMEs: Comparing Family Firms vs. Non-Family Firms. *Sustainability*, *10*(11), 4080. doi:10.3390u10114080

Memili, E., Fang, H. C., Koç, B., Yildirim-Öktem, Ö., & Sonmez, S. (2018). Sustainability practices of family firms: The interplay between family ownership and long-term orientation. *Journal of Sustainable Tourism*, *26*(1), 9–28. doi:10.1080/09669582.2017.1308371

Núñez-Cacho, P., Molina-Moreno, V., Corpas-Iglesias, F., & Cortés-García, F. (2018). Family businesses transitioning to a circular economy model: The case of "Mercadona". *Sustainability*, *10*(2), 538. doi:10.3390u10020538

Rondi, E., De Massis, A., & Kotlar, J. (2018). (in press). Unlocking innovation potential: A typology of family business innovation postures and the critical role of the family system. *Journal of Family Business Strategy*. doi:10.1016/j.jfbs.2017.12.001

KEY TERMS AND DEFINITIONS

Case Study: In-depth investigations of a single person, group, event, or community.

Development: The process of developing or being developed.

Family Business: Firm characterized by the pervasive role of the family as controlling and influencing family firms, thanks to the active participation in the management of two or more family members.

Innovativeness: Tending to innovate or to introduce something new or different characterized by innovation.

Narratives: A spoken or written account of connected events.

Sustainability: The quality of causing little or no damage to the environment and therefore able to continue for a long time.

Section 2

Renewable Energy and Energy-Efficient Technologies

Chapter 10
Economically Optimal Solar Power Generation

Sana Badruddin
University of Ottawa, Canada

Cameron Ryan Robertson-Gillis
University of Ottawa, Canada

Janice Ashworth
Ottawa Renewable Energy Cooperative, Canada

David J. Wright
University of Ottawa, Canada

EXECUTIVE SUMMARY

The Ottawa Renewable Energy Cooperative is considering installing solar modules on the roofs of two buildings while they stay connected to the public electricity grid. Solar power produced over their own needs would be sent to the public electricity grid for a credit on their electricity bill. When they need more power than they are generating, these buildings would purchase electricity from the grid. In addition to paying for the electricity they purchase, they would be subject to a "demand charge" that applies each month to the hour during which their consumption is at a peak for that month. Any electricity consumed during that peak hour would be charged at a rate about 100 times the rate for other hours. The case addresses three questions: (1) Is it profitable for these organizations to install solar on their roofs? (2) Can profitability be increased by adding a battery? and (3) How sensitive is profitability to uncertainty in future electricity prices? The case shows how the answers to these questions depend on the profile of hourly electricity consumption during the day, which is very different from one building to the other.

DOI: 10.4018/978-1-5225-8559-6.ch010

BACKGROUND INFORMATION

This section provides an introduction to solar power and how it is deployed.

Photovoltaic technology, also known as PV, is the conversion of incoming solar radiation into electricity using semiconductor materials, most commonly silicon. Most readers of this case study will be familiar with images of solar panels installed in rows on rooftops and on the ground. A "solar panel" is the popular terminology for a "PV module" and most PV modules are installed in this way. In the Northern hemisphere, solar panels typically are tilted at an angle approximately equal to the latitude and are oriented towards the South since the daily path of the sun is from East to West across the Southern sky. At the equator there is zero tilt, i.e. the modules are horizontal. If PV modules were installed at the North Pole they would be vertical, i.e. tilted at 90°. In Ottawa, the ideal tilt for solar panels is approximately the latitude, 45° (Tomosk et al., 2017).

PV modules generate more electricity when they are pointed directly at the sun and some solar installations use trackers to adjust the angle of the modules, tracking the path of the sun from morning to evening. Other installations avoid the cost of a tracker and accept the fact that they will generate less electricity from a fixed orientation of their PV modules. The use of trackers in PV modules generally implies that the panels will be ground-mounted, as trackers on rooftops cause a lot of strain to the structure of roofs during periods of high wind. Since most buildings are located in an urban setting with very little adjacent land, those looking to install solar typically use rooftops with non-tracked PV.

There are two situations in which PV modules are used:

1. Utility-Scale: A large-scale solar installation supplies electric power directly to the public electricity grid in much the same way that electricity is supplied by coal, natural gas or nuclear generating stations.
2. Behind-the-Meter: Residences and businesses install solar on their own premises and typically consume most of their solar electricity for their own use. They also stay connected to the public electricity grid which is their source of electricity when there is insufficient solar power, e.g. at night or when it is cloudy.

PV is being deployed on a global scale and is becoming an increasingly popular method of energy production due to its low maintenance and zero carbon emissions. However, power generated from solar energy is limited by the availability of sunlight, which by nature is intermittent and cannot be supplied at will. As such, there is much research interest in the integration of energy storage into a PV system, to increase the capability of providing power as needed to meet customer demand (McLaren et al., 2018). A solar system installed in a commercial building and connected to a battery and to the electrical system of the building is known as a "microgrid." Microgrids can also include other sources of electricity generation such as small-scale wind and hydro, however all microgrids referred to in this chapter are powered by solar energy. Microgrids incorporate a controller to schedule electric power flow to and from the battery so that sufficient power is available to supply the electrical loads in the building in an economically efficient manner. Such a microgrid can operate as a stand-alone system, not connected to the public electricity grid, e.g. at a mining site in a remote area. However, the case studies covered in this chapter are for buildings in an urban area where the microgrid is connected to the public electricity grid.

Storage technologies used in a solar microgrid include lithium-ion (Li-ion), lead-acid, saltwater, and sodium-sulfur. Batteries can be heavy and are typically installed at ground level even when the solar modules are on a rooftop. Li-ion batteries provide several advantages such as longer lifespans, lower weight, volume and maintenance, and better performance compared to other battery types (Hesse et al., 2017; Raszmann, Baker, Shi & Christensen, 2017) and the analysis in this case study uses Li-ion batteries.

Industry Trends

The case studies described in this Chapter are in Ontario, Canada and this section gives the relevant specifics of the solar industry in Canada.

In the province of Ontario, as in most jurisdictions internationally, the majority of solar projects were implemented under the Feed-in-Tariff (FIT) program, a policy intended to increase investment in novel and upcoming renewable energy technology. Under FIT, producers of renewable electricity are guaranteed through long-term contracts, a specified rate in $/kWh for every kilowatt hour of energy produced (Auditor General, 2011). In Canada, the FIT program has so far been implemented in the province of Ontario, and to a lesser extent in Nova Scotia and Prince Edward Island (International Energy Agency, 2019).

The FIT program in Ontario led to major investment in wind, solar and bio-based energy which were used to replace energy previously generated from coal (Auditor General, 2015). Energy producers were guaranteed set prices through FIT that covered the cost of solar modules at that time. Over time, module costs declined such that the FIT program became no longer necessary. Solar power in Ontario is now produced through Net-Metering programs (NM) under which customers producing their own electricity behind-the-meter receive a credit for excess power generated that is injected into the public grid (Ontario Energy Board, 2018). Rates are more flexible under NM than FIT as tariffs can vary depending on the time of day and level of electricity demand. For example, for medium sized business customers, extra charges are commonly applied when customer demand is greatest or at peak (also referred to as 'demand charges') in addition to charges for electricity. Solar projects under NM thus involve a more complex economic assessment than under FIT since many business customers are subject to demand charges for peak usage, something which was not relevant under FIT.

In commercial buildings, electricity demand may be high due to air conditioning in the summer or due to heating in the winter. When demand is high due to air conditioning, solar irradiance can also be expected to be high. This creates an opportunity for business customers to install solar and save money through reduced demand charges as well as reduced electricity charges. Commercial buildings also offer greater economies of scale because of larger roof areas compared to households installing rooftop solar. Moreover, the cost of lithium-ion batteries, which can be used to store power for use during periods when solar is not sufficient to reduce demand charges, is declining rapidly. Solar power is therefore increasingly being integrated with energy storage on customer premises to help reduce electricity bills.

SETTING THE STAGE

The following section provides details related to the specific case studies described in this chapter.

Focus

Recent research has focused on the potential added benefits that energy storage could offer in a solar PV system. A common strategy in this type of analysis is to use a linear optimization model to determine the optimal scheduling of battery charge and discharge cycles to achieve maximum savings in demand plus electricity charges. The results are varied as to whether battery storage is a financially viable option, with the analysis of some projects finding that despite decreasing costs in recent years, battery prices are still too high to be considered economical, e.g. Merei et al. (2016) and Zhang et al. (2017). By contrast batteries significantly increased financial returns in McLaren et al. (2018) and Khalipour et al. (2016). Many researchers assess their results by looking at the Net Present Value (NPV) and Lifetime Costs of Electricity (LCOE), e.g. Mariaud et al. (2017), Parra et al. (2017) and Makbul et al. (2015). However this requires assuming a value for the discount rate, reflecting the "time value of money". Discount rates vary from one organization to another, e.g. 5% in Bouloukat et al. (2016), 7.5% in Wouters et al. (2015) and 9% in Parra et al. (2017) and depend on what other investment opportunities an organization has as alternatives to installing solar power. In this chapter, the authors therefore measure the economic benefit of solar using the Internal Rate of Return (IRR), which avoids making any assumptions regarding the discount rate and is used, for example, by Mariaud et al. (2017) in UK, Tomosk et al. (2017) in USA and MacDougall et al. (2018) in Canada. The IRR is a metric used to estimate the profitability of a potential investment and one can use IRR as a means to evaluate and choose among multiple projects. Any project that has a positive IRR is profitable, and generally projects that have a higher IRR are more desirable to undertake.

The aim of this chapter is to determine the maximum IRR as a result of installing solar modules with lithium-ion batteries of various sizes under NM for two medium-sized business customers in Ottawa, Ontario subject to peak demand charges. Due to a multitude of factors involved, the analysis in this chapter was divided into four stages which will be discussed in more detail in the 'Methodology' section:

1. The first stage uses local information about the building to determine the desired orientation and tilt of the solar modules, after which an electric energy yield model calculates the amount of solar electricity that can be generated by the panels for each hour of the day and month of the year.
2. The second stage uses a linear optimization model (Wright et al., 2018), to match this solar electricity generation profile against the customer demand profile and the profile of electricity prices to calculate an optimum schedule of power flowing into and out of the battery. This calculation is repeated for a range of different battery sizes from 0-500 kWh and the results are used to calculate the total savings generated by the microgrid for each battery size.
3. The third stage calculates all expenses associated with the solar microgrid system, including the capital costs of the solar system and battery pack, and the operating costs including operations and maintenance and the cost of electricity drawn from the grid.
4. The fourth stage calculates the IRR by subtracting the expenses (step 3 above) from the total savings of the system (step 2 above). This is repeated for each battery size and the battery size that maximizes the IRR is selected.

Partnership

The Ontario Green Energy and Economy Act in 2009 introduced the FIT program and also established renewable energy co-operatives (RECs), which are for-profit organizations that develop renewable energy systems within their respective communities (Oxford Community Energy Cooperative, n.d.). Residents invest in solar projects through their local REC and receive a return on their investments. As with other Ontario co-operatives, they are subject to a number of internationally adopted guiding principles such as the "International Co-operative Principles" (Oxford Community Energy Cooperative, n.d.). However, they differ from most co-operatives as the business of these organizations is limited to generating and selling electricity, and all surplus is to be distributed among the members.

In order to ensure that both case studies were conducted using accurate information, the analysis was undertaken in partnership with the Ottawa Renewable Energy Cooperative (OREC). OREC develops, owns and operates solar systems while providing acceptable return on investment to community investors. It is therefore crucial to OREC's continued success to assess the economic viability of solar projects under the NM system. Like most renewable energy co-operatives in the province, OREC is currently considering medium-sized business customers in order to achieve economies of scale. This chapter presents an economic analysis of net-metered solar microgrids for two of OREC's potential customers. The analysis determines the optimum IRR that can be expected from these potential solar installations. OREC can then use these IRRs to assess whether they are sufficient to attract local investors whilst also covering ORECs administrative and other expenses.

This partnership provides access to empirical electricity consumption data from OREC's customers together with the practical expertise that OREC has gained from previous solar projects. OREC was consulted throughout the analysis and provided input whenever necessary.

CASE DESCRIPTION

This chapter evaluates the economic viability of solar power at two of OREC's potential customer sites. The analysis of these two sites will incorporate energy storage, connection to the public grid and net-metering. The sections below describe the significance of the case studies, the data sources and methodology used in the analysis, microgrid generation and scheduling, and present the results of this work. It is important to note that the models developed for this case study are based on real consumption data obtained from OREC's potential customers that has been anonymized. The battery price and electricity cost projections were extrapolated from historical values.

Significance of this Case

Social Significance

Local residents and businesses help to finance renewable energy projects developed and owned by OREC. Increased participation by communities in the local energy industry facilitates the adoption of newer technologies on a local scale and gives individuals a greater voice in the region's energy system. Under the NM system, it is essential that an economic analysis is conducted for solar projects which incorporate demand charges for peak usage. Doing so will help the continued success of OREC, which serves the needs of people in the Ottawa community who are interested in investing in solar power.

Environmental Significance

On a larger scale, many jurisdictions around the world are beginning to, or are in the process of, replacing FITs with NM. This chapter will provide jurisdictions around the world with solid research to build their economic analysis of customer-based solar projects, easing their energy transition towards renewable power. This will help to facilitate the worldwide diffusion of solar technology, thereby helping to reduce carbon emissions.

Economic Significance

With nations seeking to meet international and domestic carbon emission reduction targets, countries are working to transition away from fossil fuel-based economies to integrate greener sources of power. This requires a fundamental change in the electricity system away from large-scale, fossil-fuel plants towards small-scale, locally-based renewable energy systems. Under this setting, solar energy can be seen as an investment which provides a financial return to the customer, thereby incentivizing customers who may not be environmentally motivated to install solar.

Data Sources

This section describes the sources of data used to populate the linear optimization model that determines the IRR for the two case studies:

Average Workday Demand

An important benefit of the partnership with OREC was access to the anonymized data on the electricity demand and site configurations for two commercial buildings that OREC is considering as potential customers. The two case studies assessed have differing load profiles, ranging between 40 kW to 1,100 kW: case 1 closely resembles the load of an office building, peaking around noon, whereas case 2 has an overall flatter curve with several smaller peaks, most prominently in the morning and in the late evening hours. The analysis for workdays and weekends for both cases were conducted separately, as there were notable differences in the average workday versus weekend demand curves. Figure 1 (a) and Figure 1 (b) show the

Figure 1. Average annual workday demand (kW) for case 1 (a) for the months of January, May and July, and for case 2 (b) for February, March and May

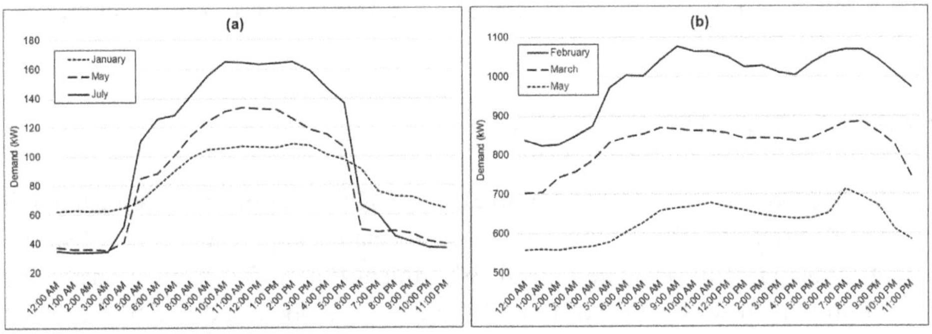

average workday (Monday to Friday) demand schedules respectively for case 1 and case 2. Three different months were chosen from each case to illustrate the range of variability of demand from one month to the other. In February, demand is high between 8am and 10am as electric fans are used to circulate hot air from the natural gas heating system.

Note that in both cases, demand begins to increase between the hours of 3:00AM and 5:00AM for most months throughout the year. This can in part be explained by early morning caretaking facilities requiring electricity consumption. However, sudden and intense surges of electricity demand occurring for example in the month of July for case 1, are likely the result of pre-cooling. In this case, building management has chosen to turn on air conditioning in the early hours of the morning with the intent of reducing peak demand occurring later in the day. Similar early morning peaks in the winter may be due to pre-heating the building. Although the heat is provided by natural gas (not electricity), the heating system uses fans to pump hot air around the building and those fans consume electric power.

Capacity of Solar System

The consumption patterns of case 1 and case 2 show a significant difference in the average workday demand, with case 2 having up to ten times the demand of case 1 at certain times of the day. Moreover, the rooftop area of case 2 is 4.8 times as large as that of case 1, meaning that case 2 has much more available space to install solar panels. Capacities of solar systems were sized according to demand but were limited by the available roof space as there is no adjacent land that can be used for additional solar modules. Based on the rooftop surface area, the appropriate capacity for the system was determined to be 100 kW AC for case 1 and 500 kW AC for case 2.

Capacity of Energy Storage

The range of battery sizes investigated was 0 – 100 kWh for case 1 and 0 – 500 kWh for case 2.

Retail Electricity Prices

Electricity prices are an important variable in the IRR calculation. In Ontario, the retail price of electricity for the two customers has two main components: Hourly Ontario Energy Price (HOEP) and the Global Adjustment (GA). Commercial customers face an additional constant fee of $0.01/kWh which cannot be reduced by the use of solar power. The HOEP represents the wholesale price of electricity (derived from the spot market price). However, some regulated and contracted generators have contracts under which they receive a different price for the electricity they generate. The GA accounts for the difference between the HOEP and the price paid to these generators. It can be positive or negative, but in recent years it has been positive and has gradually increased each year. Electricity operations in Ontario are managed by the Independent Electricity System Operator (IESO), and monthly historical rates used to project HOEP and GA values were obtained from IESO (2018).

Peak Demand Charges

A rate of CAD $8.67/kWh is charged to commercial customers in the City of Ottawa for the peak hour of demand each month, as set by Hydro Ottawa (2018), the city's electric distribution company. For instance, if the peak hour for the month of January occurred on January 14th from 2:00pm – 3:00pm, then all electricity consumed during that hour has a demand charge added to the electricity price (above). This process is repeated for each month. This demand charge is very high compared to the electricity charge of about CAD $0.10/kWh, but it is only charged for one hour per month and provides an incentive to the customer to flatten their load profile. When that flattening is achieved by a solar microgrid, the reduction in demand charge contributes significantly to the profitability of the microgrid. Calculations to determine IRR in this chapter were done in US$, and the average annual exchange rate of 1.299 (Bank of Canada 2018) was applied to convert retail electricity prices in CAD$ to US$.

PV Capital Costs

The capital cost for installing solar PV at the time of study was USD \$1.77/W_{DC} according to the National Renewable Energy Laboratory 2017 quarterly report (Fu, Feldman, Margolis, Woodhouse & Ardani, 2017).

Battery Pack Costs (\$/kWh)

These capital costs were based on information from BCG (Rubel, Pieper, Zenneck & Sunak, 2017) and Bloomberg (Sekine & Goldie-Scott, 2017) and are a key factor for determining the IRR for battery-integrated microgrids. The cost of installing a battery pack and management system is USD \$254/kWh at the time of this case study. This, however, is a lower-end estimate. Including factors such as project development, engineering, procurement and construction, grid connection and balance of system hardware, the 'all-in' battery price is estimated to be USD \$591/kWh. It should be noted that the actual battery cells that store the electricity cost only \$195/kWh and are thus a small proportion of the total cost, which includes the installation-related items listed above. Electric vehicle batteries are lower cost (Soulopoulos, 2018) since they are installed on a vehicle assembly line, whereas microgrid battery installation involves customization for each commercial building individually.

DC-Connected PV Generation Model

The model used for the two case studies is based on the classical model of Masters (2004), Chapter 7, extended by MacDougall et al. (2018). The model calculates the path of the sun across the sky and its angle of incidence on the solar modules. It uses data on two measures of solar irradiance. The first is the irradiance due to the direct beam of light from the sun, known as Direct Normal Incidence (DNI). The second includes, in addition, the diffuse light from the entire sky, known as Global Horizontal Irradiance (GHI). The model also calculates the light that is reflected from the ground or roof onto the solar modules. The third data set is the ambient temperature which affects the efficiency of the solar modules. The data is obtained from Natural Resources Canada (2018) and represents a typical meteorological year.

Methodology

In order to determine the internal rate of return in both case studies, it was necessary to follow a sequence of calculations, as illustrated in Figure 2. The section below gives an overall description of this methodology. Works by Wright et al. (2018) and MacDougall et al. (2018) provide further details on the mathematical models involved.

Figure 2. Basic methodology used for calculating the IRR. Italics indicates data sources. Bold indicates final results

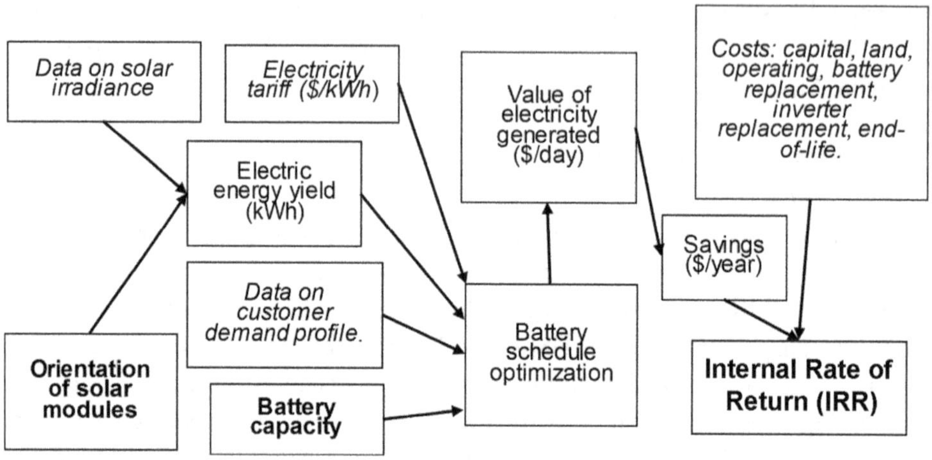

Step 1: Finding the Electric Energy Yield

The first step was to determine the amount of electricity generated at different times of day and months of the year. The three factors that must be acknowledged in order to calculate a realistic energy yield are the solar irradiance (the power per unit area received from the sun), optimal tilt and the optimal azimuth of solar modules, on the left side of Figure 2. Solar irradiance data is available from Natural Resources Canada (2018) and includes the irradiance due to the direct beam of light from the sun together with the diffuse irradiance coming from the whole sky.

Ottawa is located in a region prone to minor earthquakes caused by the land gradually rebounding after being depressed by the weight of ice during the last ice age. It is also subject to high winds during winter storms. Building codes require solar modules to be mounted on buildings in such a way as to withstand a storm during an earthquake. The buildings under consideration have flat roofs so that the modules cannot be tilted at a high angle which would result in high wind load. OREC's experience is that solar modules are stable when tilted at a maximum of 10° and held in place by ballast. A higher tilt angle (up to the latitude of 45° in Ottawa) would result in greater electric energy yield, however this would require the modules to be bolted to the roof, a procedure that requires penetration of the bolts through the roof with the consequent danger of leaking. Typically building managers are reluctant to risk damage to the roof and hence OREC decided to install modules with a tilt angle of 10° using ballast instead of bolts.

For the azimuth angle, the solar panels were planned according to the architectural design of the building roof. Although maximum energy yield can be obtained by orienting the modules due south, more modules can be installed if the rows of modules are aligned with the sides of the roof. Building orientation for case 1 was 20° west of south and was 29° east of south for case 2 and therefore these angles were set as the azimuths for the solar module installations.

Step 2: Calculating the Savings in Electricity Charges and Demand Charges

The second step was to calculate the value of this electric power in terms of savings in electricity charges and demand charges. This can be calculated using data on the customer demand profile, battery capacity and electric power generated. The demand profile represents the amount of electricity used by a building for every hour of every day. 24 demand profiles were used, one for each month of the year with weekends and workdays treated separately. Workday demands are shown previously in Figure 1. The buildings in the two cases have very different demand profiles. Case 1 analyzes an office building that consumes the majority of its energy between 7:00 am and 5:00 pm on workdays, peaking around 10:00 am to 3:00 pm, and consumes very little on weekends. Case 2 studies a building that is also used in the evening and at weekends. It consumes a large amount of energy between 6:00 am and 12:00 pm on both weekdays and weekends, peaking in the morning and in the late evening hours but with a much flatter profile than the office building in case 1.

Since solar energy is generally produced between the hours of 7:00 am and 7:00 pm, Figure 7, it can be used to decrease the amount of energy purchased from the grid in the middle of the day. This reduces the amount of energy needed to be purchased during the peak consumption hour in Case 1 as its peak falls between these hours, and hence reduces the demand charges.

In addition to the solar modules, a battery pack could be used to decrease the peak demand. Batteries can be charged using both solar and grid energy at any point throughout the 24 hours. This allows for the consumer to purchase energy from the grid overnight as it is generally at lowest cost at that time, and use it to decrease the amount purchased from the grid in the middle of the day.

Once all constraints and variables, including the customer demand profile, retail electricity prices, size of the battery pack, have been inserted into the model, it optimizes the schedule of power flow in and out of the battery so as to maximize the savings due to a reduction in both the demand charges and the electricity charges, see center box in Figure 2. The decrease in energy purchased from the grid during the peak consumption hour is multiplied by the demand charge to calculate the reduction in peak demand charge. This is added to the value of electricity generated to calculate the total savings from use of the solar modules plus the battery pack.

Step 3: Calculating the Expenses Associated With the Solar Microgrid

The third step is to calculate the expenses associated with the battery and solar modules over the lifespan of the solar microgrid system. Solar module warranties are typically 25 years based on conservative estimates from manufacturers, and we therefore estimate their life as slightly longer than this: 32 years. Evidently the life time of a solar module installed today is unknown, but the estimate does not need to be very accurate since the contribution to the net present value of revenues 30 years in the future is reduced by about 80% at the discount rates appropriate for this study. The solar modules represent the largest expense. According to data collected by NREL, PV modules on roofs of commercial buildings cost US $1.77/W including the cost of installation (Fu et al., 2017). However, solar modules are subject to several other expenses throughout their operational life. Operations and maintenance costs an average of 0.86% of the total cost of the solar modules every year (NREL, 2018). In addition, the inverter (which converts DC electric power from the battery and from the solar modules to AC, for use in the commercial building) needs replacing after 15 years. The battery cells need replacing when they reach their limit of charge/discharge cycles (Kittner et al., 2017). It also costs around USD $18.25/m^2 to recycle the system at the end of its useful life (Di Francia, 2013).

Step 4: Calculating the IRR

Once all savings and expenses have been determined, the IRR can be calculated. The IRR is the discount rate at which the net present value (NPV) of this sequence of cash flows is zero over the 32-year life of the solar modules:

$$NPV = \sum_{n=0}^{32} \frac{CF_n}{\left(1 + IRR\right)^n} = 0$$

where CF_n is the net cash flow (savings - expenses) in year n. The net cash flow n years into the future is discounted at a rate of the IRR to bring it back to its net present value and then all future cash flows are totaled. The IRR is the discount rate that gives a total NPV of zero. Once the IRR is calculated, it can be used by organizations that have their own individual discount rates. If the company owning and operating the solar microgrid uses a discount rate higher than the IRR, they will make a loss. If their discount rate is lower than the IRR, they will make a profit.

The larger the battery, the larger the savings calculated in Step 2 above. However, since batteries are priced per kWh of storage capacity, the largest battery pack will not necessarily be the most profitable. The optimal battery size was chosen by calculating the IRR for batteries between 0 kWh and 100 kWh for case 1 and between 0 kWh and 500 kWh for case 2 and determining which battery size gave the highest IRR.

Key components in the estimation of future cash flows, CF_n, are electricity prices which affects savings, and battery prices which affect cost of battery cell replacement. These are now described in detail.

Electricity Price Projections

The unit cost of electricity is an important variable in the linear optimization model. As stated previously, for medium-sized businesses in Ontario, this consists of the HOEP, the GA, and a constant of $0.01/kWh. Using historical values from 2005 to present found on the IESO website, it was possible to extrapolate future HOEP and GA prices (the constant value will not vary). It should be noted that calculations for the unit cost of electricity were done from 2005 to 2017 as at the time of writing, 2018 values for the full year were not available.

- HOEP: Ontario's wholesale market price varies every hour of every day of the year. It was found that since 2005, HOEP values have been declining and have largely leveled out since 2015. Annual figures from 2015 to 2017 were averaged for each hour of each month, treating workdays and weekends separately, and the resultant values were held constant for the 32-yr lifespan of the solar system.
- GA: Conversely to the HOEP, the GA has been increasing since 2005 to 2017. The GA only changes monthly, not hourly, therefore there was no distinction between workday and weekend values. Annual averages for the GA were plotted, and it was found that the GA increased at a relatively steady rate such that the relationship between the points could be described using a linear equation (shown in Figure 3). Using this equation, GA values were extrapolated until 2025 after which they were held constant as detailed below until the end of the 32-year analysis.

In general, electricity prices are difficult to forecast as they are influenced by a number of political and market factors. For example, changes in provincial government can result in different energy policies which could alter the trend in electricity prices. It was also the view of OREC that electricity prices cannot continue

Figure 3. Linear equation for GA, based on data from 2005 - 2017
Source: IESO, 2018

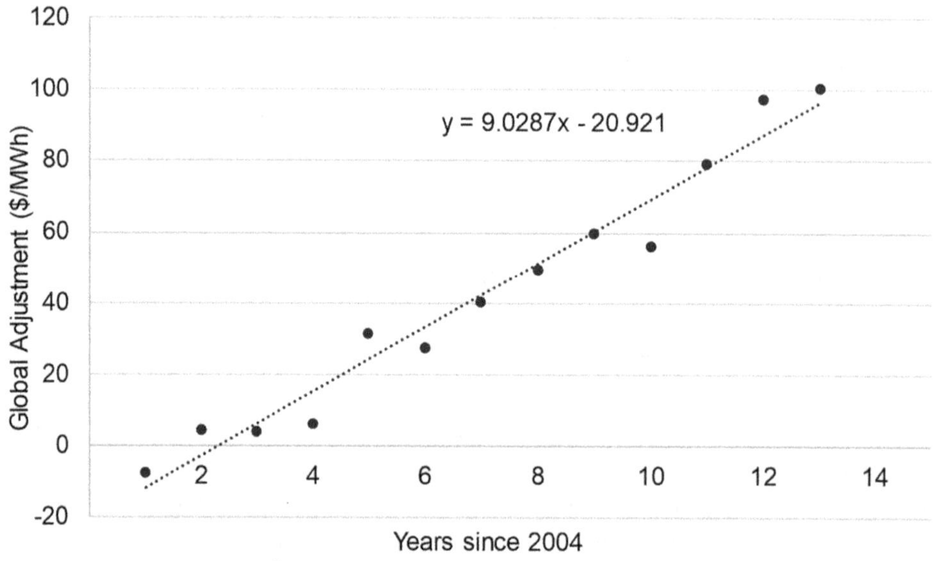

to increase indefinitely, as otherwise consumers would cease to buy power from the grid. This is why it was assumed in the model that the GA would increase until 2025 and then be held constant.

Battery Pack Price Projections

Li-ion battery packs consist of many Li-ion battery cells that store the electricity together with some associated wiring, electronics and casing. Distinguishing between the two is important since the battery cells eventually need replacing depending on how many charge/discharge cycles they go through, but the wiring and casing typically last as long as the solar panels. The analysis allows for 5000 charge/discharge cycles, which is typical of commercial Li-ion batteries (Saft, 2018). In both case studies, the batteries typically made one or two cycles per day (depending on the time of year and on whether it is a workday or weekend), as determined by the battery schedule. Under these circumstances, the cells would need to be replaced approximately every 10-13 years and they constitute 76% of the total battery pack cost.

Projections of the prices of battery packs from Bloomberg (until 2025) (Sekine & Goldie-Scott, 2017) and BCG (until 2030) (Rubel et al., 2017) are shown in Figure 4. The analysis extrapolated the Bloomberg forecast to 2030 and averaged the two forecasts to come up with a single forecast. Of this cost, 76% was taken and extrapolated to 2050 in order to obtain the cost of replacing battery cells at

Figure 4. Forecasted costs of battery pack (including battery management system) from Bloomberg and BCG to 2030, and their average. Forecasted battery cell costs to 2050, which are used to estimate the costs of replacing battery cells in the future.
Source: Sekine & Goldie-Scott, 2017; Rubel et al., 2017

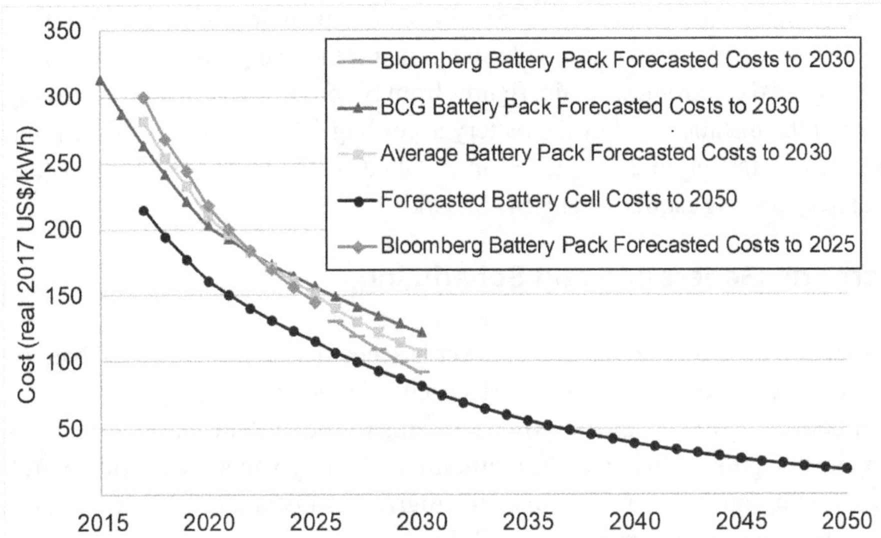

Figure 5. All-in battery costs from Bloomberg and BCG (including, overhead for the project developer, engineering and procurement costs, the energy management system and associated electronics)
Source: Sekine & Goldie-Scott, 2017; Rubel et al., 2017

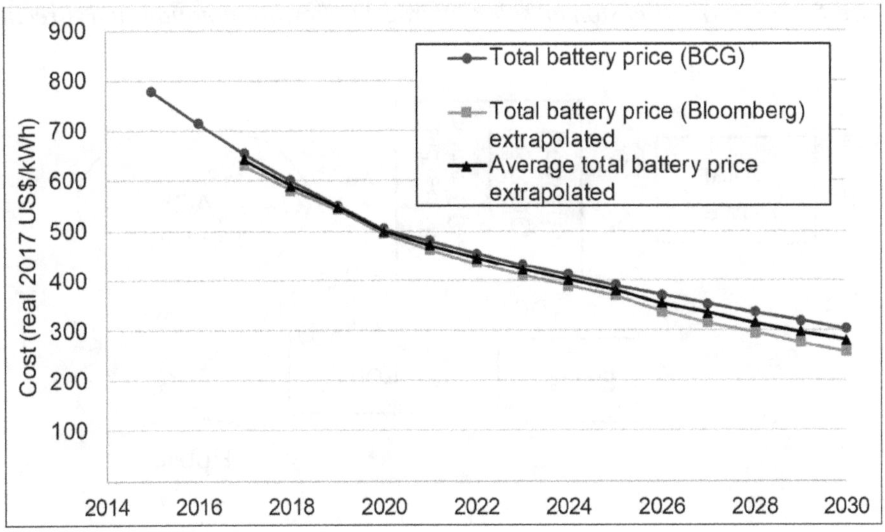

the appropriate year in the project. For instance, if battery cells need replacing after 12 years, for a 32-year project starting in 2018, battery cell costs in 2030 and 2042 were used in the IRR calculation. The installation of the battery pack adds considerably to the cost as shown in Figure 5, which has a very different vertical axis to Figure 4. In addition to the battery pack, it includes overhead costs for the project developer, engineering and procurement, the energy management system and associated electronics. Again, figures from Bloomberg and BCG were averaged to obtain the installed cost of the battery according to the start year of the project. For projects starting in 2018, the all-in installed cost is USD $591/kWh, whereas the battery pack is estimated at USD $254/kWh.

Microgrid Generation and Scheduling

This section describes how electric power is scheduled in and out of the battery in the solar microgrid to minimize the electricity charges plus demand charges.

In both case studies, the proposed solar microgrid systems have three primary components: photovoltaic panels, a lithium ion battery and a microgrid controller incorporating an inverter. The solar microgrid is illustrated in Figure 6 and the associated notation is summarized in Table 1.

An important component of the case studies was to determine the cost-minimizing hourly flow of energy in and out of the lithium ion battery. The model used generates a different schedule based on the type of day (workday or weekend) and the month of the year given that demand profiles and HOEP electricity charges vary based on

Figure 6. Commercial customer solar microgrid showing notation and direction of power flows

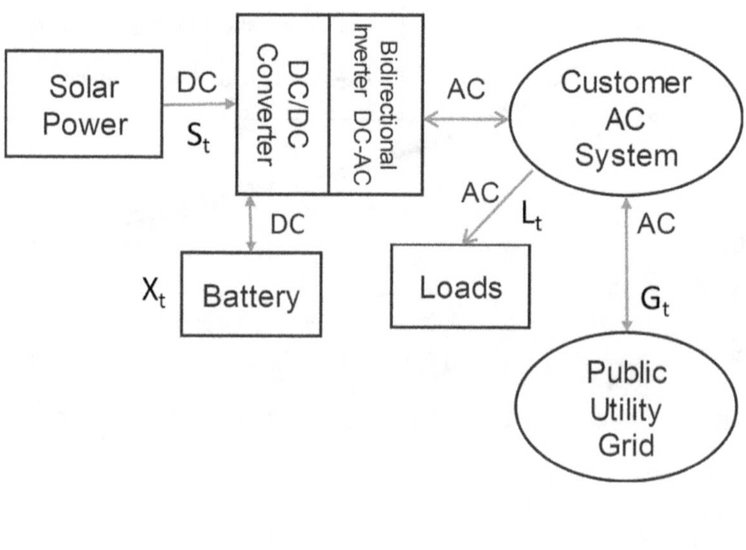

Table 1. Notation used in microgrid modeling. Since the time unit is one hour, energy (kWh) and power (kW) are equivalent.

Notation	Definition
t	Time (hours)
S_t	DC solar energy yield (kWh)
L_t	AC loads in customer premises (kW)
X_t	State of charge of the battery (kWh), i.e. the amount of electric energy stored in the battery
G_t	AC power purchased from grid (kW)

Figure 7. Case 1 battery schedule for a 50 kWh battery: (a) June (b) February (L_t=energy demanded, S_t=solar power generated, G_t= energy purchased from the grid and X_t=state of charge of the battery)

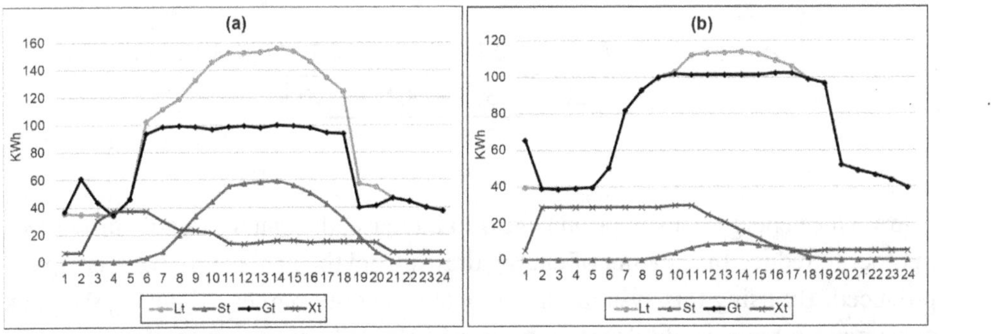

these two factors. It accounts for discretionary factors such as the inverter capacity, battery capacity and size of the solar system as well as independent factors such as peak demand charge, hourly electricity tariffs, building demand profile, inverter efficiency and battery efficiency. The model can be modified to suit the solar system, demand profile and cost of electricity for any building in any community.

An example of the results that can be generated by using the microgrid scheduling model with a 50 kWh battery is given in Figure 7.

Solar panels typically receive the most exposure to sunlight in the summer months because the sun rises early in the morning and sets late in the afternoon. This enables the customer to use the solar energy as it is being produced throughout the day and decrease the amount of electricity that must be purchased from the grid in order to satisfy loads. However, between 6:00am to 8:00am, very little energy is produced but the building's demand rapidly increases. To solve this problem, the battery is fully charged at night by purchasing energy from the grid, during hours when the HOEP

Figure 8. Case 2 battery schedule for a 400 kWh battery: June (L_t=energy demanded, S_t=solar power generated, G_t= energy purchased from the grid and X_t=state of charge of the battery)

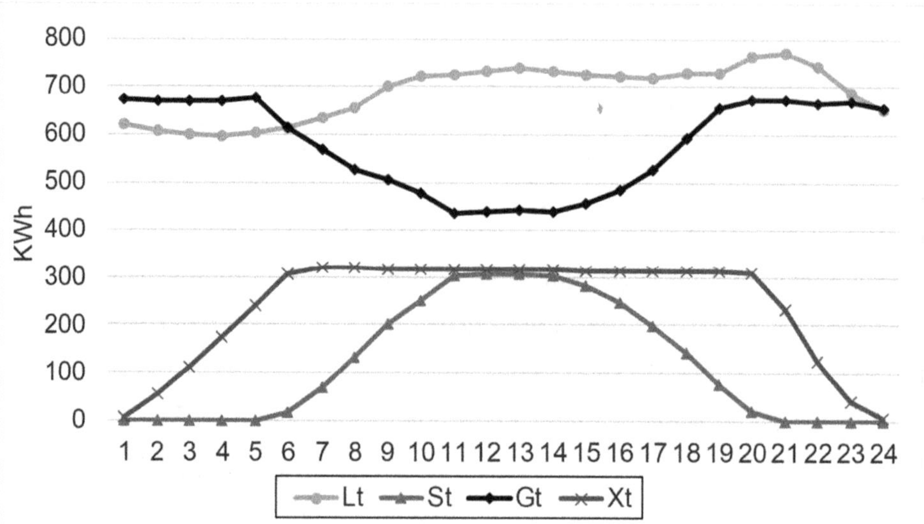

is low. Once energy consumption begins to increase, the battery rapidly discharges in order to reduce the amount of energy drawn from the grid before solar energy is produced. By using this system, the amount of energy purchased during the peak consumption hour would decrease by nearly 60 kWh, as shown by the difference between L_t and G_t in Figure 7 (a).

In the winter months, solar panels do not receive as much exposure to sunlight since days are short and atmospheric conditions, such as cloud, can prevent the panel from functioning at full capacity. Thus, during the month of February, the solar system is not able to significantly reduce the mid-day demand peak. This makes the battery all the more important as it is charged overnight and discharged gradually between 11:00am to 6:00pm. In this scenario, the building manages to reduce the amount of energy purchased during the peak hour by 15 kWh, see Figure 7 (b).

The demand curve L_t from case 2 (Figure 8) differs from that of case 1 as it is flatter and has several small peaks. The model accommodates for these differences by modifying the battery schedule for each month, as is made apparent when contrasting Figures 7 (a) and Figure 8. In June, the battery for case 2 is charged from 1:00 – 6:00 am, when energy from the grid is cheapest. It stores this energy throughout the day, using solar power to decrease the midday peak. Once solar power is no longer being generated, the energy from the battery is gradually released to decrease the peak between 8:00pm and 11:00pm.

The model determines the optimal flow of energy throughout the microgrid 24 times: for workdays and weekends for each month of the year. It then calculates an estimate of yearly savings. The customer can continue to run the model using a variety of battery sizes to determine that which yields the highest IRR.

Results

This section displays the results of the analysis of case 1 and case 2 by showing annual and peak demand charge savings and the IRR, as well as an overview of cash flow for case 1 for generating solar power over a 32-year time span.

A central component of this study was to determine the optimal battery size for a given solar system, such that it produces the greatest IRR. It was found that though a larger battery size increases annual savings for potential customers, it does so at a higher battery cost, which in turn impacts the IRR and financial benefits accrued to customers. With the higher all-in battery price of \$591/kWh, the results indicate that inclusion of a battery is currently uneconomical and that an alternative possibility would be to retrofit batteries at a later date, when their price has declined. Alternatively, if a battery can be obtained at the basic price of the battery pack plus management system at \$254/kWh, then a small battery results in a higher IRR than no battery at all.

Annual and Peak Demand Charge Savings as a Function of Battery Size

Each point on Figure 9 shows the result of optimizing the corresponding battery schedule for case 1 and case 2 for weekends and workdays of each month. It is clear

Figure 9. Total annual savings and savings due to peak demand charges (a) for Case 1, for battery sizes ranging from 0 kWh to 100 kWh, and (b) for case 2 for battery sizes ranging from 0 kWh to 500 kWh

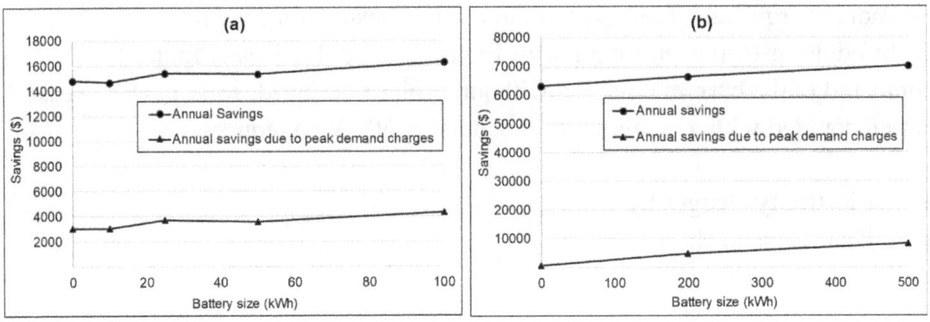

Figure 10. Annual savings from electricity charges versus peak demand charges for three battery capacities for (a) case 1 and (b) case 2

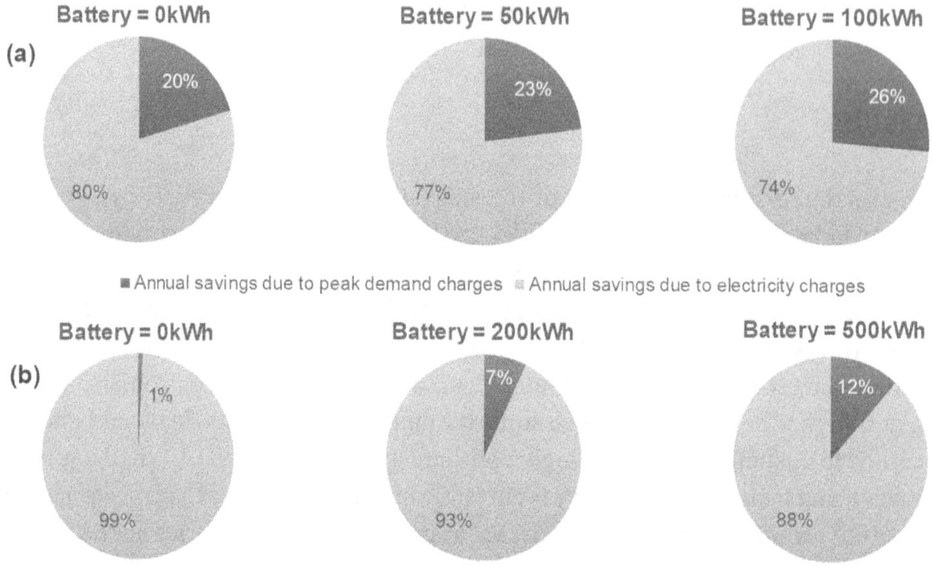

from Figure 9 that the use of a battery increases savings compared to no battery. The battery provides flexibility in when to use power generated from solar and also allows for power to be purchased at times when it is low cost and used at times when it would have cost more to purchase it from the grid. The larger the battery, the more flexibility in power scheduling and the greater the savings both in terms of total electricity charges and also in terms of demand charges.

The pie charts in Figure 10 compare savings due to peak demand charges and savings from electricity charges for three battery capacities for case 1 (a) and case 2 (b). Slight increases in peak demand charge savings were observed with greater battery capacity. This is because energy can be stored in the battery and released when it is most needed, thus decreasing the amount of energy drawn from the grid when there is a peak in the consumer demand profile. The higher the storage capacity, the more energy can be released to satisfy the consumer's demand.

In addition to the information in Figures 9 and 10, it was found that, when compared to the current situation without a solar microgrid, solar power in case 1 (when coupled with a medium-sized battery of 50 kWh) reduces:

- Electricity charges by 14%
- Peak demand charges by 25%

Similarly, when compared to the current situation without a solar microgrid, solar power in case 2 (when coupled with a medium-sized battery of 200 kWh) reduces:

- Electricity charges by 8%
- Peak demand charges by 5%

Overview of Cash Flow

Figure 11 shows increasing annual savings for the first 8 years of solar installation. This is due to estimates that retail electricity prices, in particular the global adjustment component, will continue to increase for several years, after which they become relatively stable. The slight linear decrease in annual savings after year 8 is due to reduced performance as a result of degradation of the solar modules over time. Operating costs occur every year for general operations and maintenance and also occur at specific years in the timeline to replace battery cells or inverters. At the end of the life cycle, additional costs are imposed on the customer to recycle all materials. In the bottom left, an arrow pointing downwards represents the enormous capital costs associated with the initial purchase of equipment and installation of solar, as compared to the other items illustrated in this Figure.

Figure 11. Cash flow for Case 1 installing solar power, over a 32-year period, i.e. the lifespan of a solar module

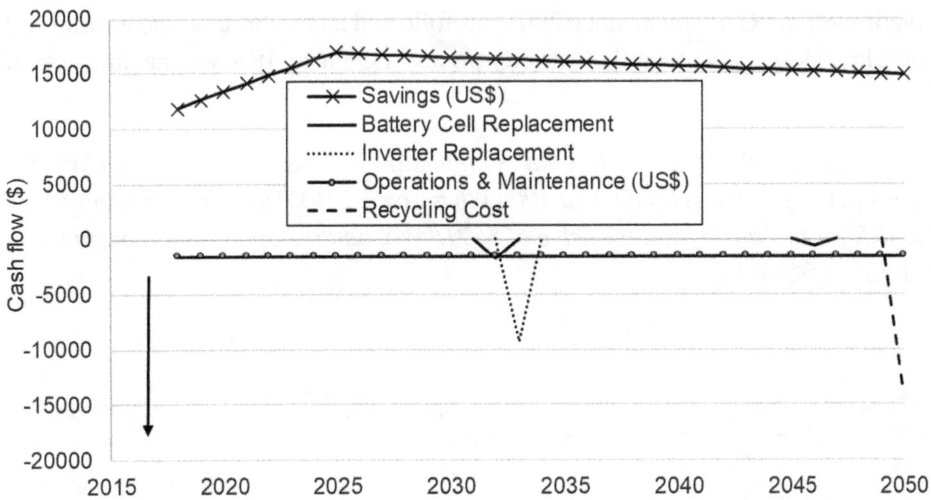

Internal Rate of Return

The calculations for IRR used upper and lower estimates for battery prices installed in 2018: the basic price including just the battery pack and management system ($254/kWh) and the 'all-in' battery price ($591/kWh), shown in Figure 12. In case 1 (a), an IRR of 6.30% with a 0 kWh battery decreased to 5.95% with a 100 kWh battery at a cost of $254/kWh and decreased to 4.60% at a battery cost of $591/kWh. For the lower battery price, the maximum IRR was 6.39% with a 25 kWh battery and for the higher battery price the maximum IRR was achieved without a battery. Case 2 (b) had an IRR of 4.17% with a 0 kWh battery which marginally increased to 4.18% with a 200 kWh battery but decreased to 4.02% with a 500 kWh battery at a cost of $254/kWh. For the higher battery price of $591/kWh, the maximum IRR was achieved without a battery and dropped to 2.86% with a 500 kWh battery.

These results indicate that currently large batteries are too costly to be used in a microgrid in a medium sized building. Smaller batteries improve profitability if they can be obtained at a low cost. Even though Figure 9 indicates that batteries are useful in achieving savings in electricity and demand charges, they do so at a cost that eats into those savings to a very large extent.

Effect of Electricity Price Uncertainty

Uncertainty regarding electricity prices is expressed in Figure 13. It accounts for the volatility of the GA which, since 2005 has followed a more or less linear trend and has a large impact on overall electricity prices in Ontario. IRR was again assessed

Figure 12. The IRRs for various battery capacities between 0 kWh and 100 kWh for case 1 (a) and between 0 kWh and 500 kWh for case 2 (b). The lower line represents the IRR for an 'all-in' battery price of $591/kWh, and the upper line represents an IRR for $254/kWh.

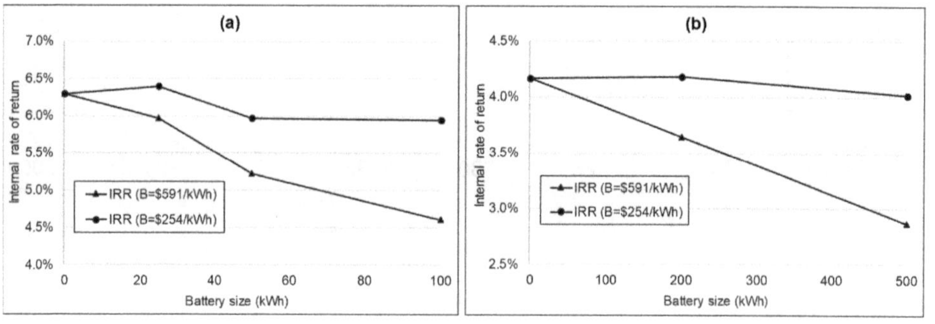

Figure 13. Prediction interval in GA extrapolation

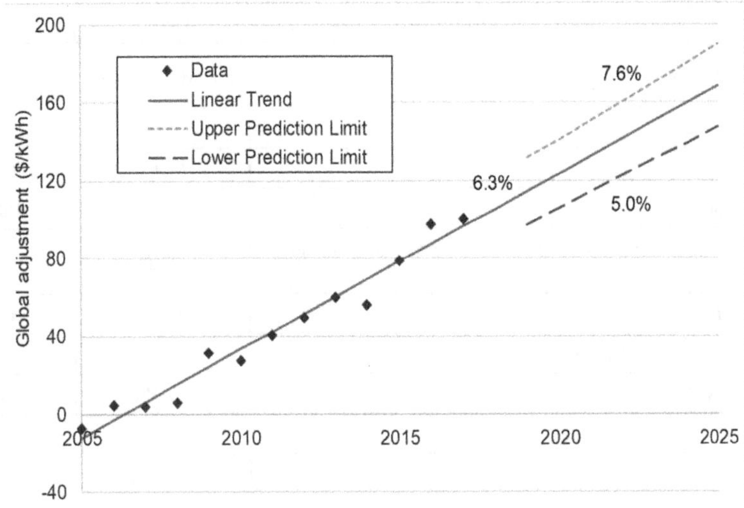

using upper and lower 95% prediction intervals for GA extrapolation. Case 1, which had an IRR of 6.30% with no battery, was found to have upper and lower IRRs of 7.6% and 5.0% respectively.

ANALYSIS AND DISCUSSION

The results of the above analysis are summarized in Table 2.

Annual Savings

The graphs in Figure 9 and pie charts in Figure 10 show that peak demand charges are significantly more reduced relative to total annual savings in case 1 than in case 2. This is mainly due to differences in the shape of their respective average demand curves (Figure 1). In case 1, demand is highest during midday when solar power is generating well, hence solar helps to lower the peak and associated demand charges incurred. Solar reduces demand charges in case 1 by nearly 25% and these demand charges can account for 20% and 26% of total annual savings depending on the size of the battery. The average demand curve in case 2 however, is much flatter than what is seen in case 1, with peaks generally occurring early morning and in the evening. This results in only a 5% reduction in demand charges which range between 1% and 12% of total annual savings with various battery capacities.

Table 2. Summary of main features of Case 1 and Case 2

	Case 1	Case 2
Azimuth of PV modules.	20° west of south	29° east of south
Tilt of PV modules	10°	10°
PV system size	100 kW	500 kW
Capital cost for solar	$177 K	$885 K
Percentage of annual savings due to demand charge reduction (no battery)	20%	1%
IRR (no battery)	6.3%	4.17%
All-in battery price $591/kWh		
Optimal battery size	0 kWh	0 kWh
Battery pack price $254/kWh		
Optimal battery size	25 kWh	200 kWh
Capital cost of optimal battery at $254/kWh	$6.35 K	$50.8 K
Percentage of annual savings due to demand charge reduction (optimal battery)	24%	7%
IRR with optimal battery	6.39%	4.18%

Internal Rate of Return

One of the primary objectives of this study was to determine the IRR which, as seen in Figure 11, is shown to decrease when a battery is incorporated into the microgrid despite slight increases in annual savings. The optimal battery size for case 1 and case 2 is therefore 0 kWh with all-in battery prices.

Two key variables in determining the IRR were battery prices and electricity prices. The model showed the IRR to decrease for the all-in battery prices making batteries an uneconomical investment for commercial customers paying those prices. However, when using the lower battery pack price, Figure 11 (a) shows a slight increase in IRR for a 25 kWh battery for Case 1, and Figure 11 (b) shows a slight increase for a 200 kWh battery for Case 2. Figure 13 shows the effect of uncertainty regarding GA extrapolation, and future electricity prices may also be impacted by other factors in the future, such as government policy.

Energy Management

The average demand curves for case 1 and case 2 show that these potential customers pre-cool their buildings with the intention of reducing peak demand later in the day, an action which creates very early morning peaks at a time when solar is not generating well. This essentially negates financial benefits that would otherwise be gained through solar for lowering peak demand at midday. Therefore, if solar is to be installed in commercial buildings, discussions with building management about heating and cooling cycles is imperative in order to ensure better efficiency of the use of solar to reduce peak demand charges.

The analysis above has been conducted prior to the installation of the microgrid and has been conducted using *average* electricity demand and *average* solar power generation. During the actual operation of an installed system, it is likely that with daily weather forecasting available, solar generation and battery scheduling will be done with greater accuracy, thus achieving higher savings for electricity and demand charges. For example, a battery schedule could be planned for higher solar generation on a sunnier day, or optimized to reduce peak demand charges if the manager anticipates increased demand, for example due to forecast of an exceptionally hot day with higher than usual air conditioning demand.

CURRENT CHALLENGES AND SOLUTIONS

Challenges Facing the Organization

The above analysis was provided to OREC to assess whether investment in PV modules on the roof tops of case 1 and case 2 would be profitable, and whether the inclusion of a battery would help to increase the IRR. None of OREC's previous projects used batteries, therefore the inclusion of batteries introduces an added complexity as a schedule has to be determined for power flow into and out of the battery. In making its decision, OREC would also need to consider the dividends previously paid to investors (3%-4% under FIT) and OREC's administrative costs. This means that in order for the project to be profitable, it is not sufficient for the IRR to simply be positive; it would need to be greater than the percentage required for dividends and administrative costs.

OREC's previous projects have been low-risk due to the 20-year contracts and guaranteed pricing offered under the FIT program. Now that the FIT program has been discontinued, OREC faces risks in their future investments due to profitability being dependent on future electricity prices which are uncertain, as well as the pros and cons associated with battery storage, including ever-changing technology and price fluctuations. A quantification of the extent of some of the risks associated with solar power in Ontario is provided by Tomosk et al. (2017).

Solutions and Recommendations

OREC is staffed by dedicated, environmentally motivated people, supported by many volunteers who are knowledgeable professional engineers and by investors in the community keen to support the renewable energy industry. Although the environmental benefits of solar energy are clear in their reduction of carbon dioxide emissions, the economics of rooftop solar in some cases remain marginal. In order to address this concern and challenges mentioned above, continued dedication as well as an open mind to adapt to changes in the market is of utmost importance. One intuitive solution could be for OREC to install solar today and retrofit batteries at a later date when prices have reduced enough to increase the IRR sufficiently.

Challenges Facing the Electricity Industry

As seen in the results, case 1 and case 2 were only able to derive marginal profits from installing solar. However, as the prices of solar and batteries gradually decrease over time and as electricity prices go up, the profits obtained from installing solar behind-the-meter will continue to improve, driving more organizations to implement solar. This trend would reduce revenue for the electricity industry including generating companies, transmission companies (that operate inter-city transmission lines) and distribution companies (that operate the distribution network within a city).

Generating companies that operate hydro, natural gas and nuclear generators provide power and sell it on the wholesale market, which, in Ontario, is the basis for the HOEP. Although they generate power 24 hours a day, they make most of their profits during the day when demand, and hence wholesale prices, are high. Solar generation occurs at precisely this time, reducing the profits of these generating companies. Transmission companies have invested heavily in high voltage transmission lines that bring power from the generating stations, to the cities where it is needed. They get paid per kWh transmitted, and if solar power is generated behind-the-meter, less power needs to be sent over the long-distance transmission lines, thus reducing profits of transmission companies. Distribution companies use their investment in the distribution grid within a city to take power from the transmission company and distribute it to the residences and businesses within a city. They incur operating costs to maintain the distribution network and any behind-the-meter power generated reduces their revenues and hence their profits.

In response to the reduction in the profits of these three industry players as a result of competition from solar, they could simply increase electricity prices so as to keep their profits stable and continue to maintain their equipment, which society depends on for its electricity needs. However, as seen in this case study, any increase in electricity prices will improve the profits from solar and result in yet more solar installations. This is sometimes referred to as the "death spiral" of the electricity industry.

The irony of this situation is that even organizations with behind-the-meter solar still need the public electricity grid for those times at which they do not have enough power from their solar microgrid. All that equipment in the public grid needs to be maintained because customers are relying on the public grid to be available if they need it, even though they are generating much of their power themselves.

Solutions and Recommendations

At present, there is little certainty as to how quickly electricity systems will undergo this change, and whether it would be as drastic as to create the sort of utility death spirals mentioned above. What is important though, is that utilities and distribution companies, instead of resisting change, direct their resources towards adapting to the electrification of the economy. The "death spiral" described above happens if overall demand for electricity remains reasonably constant. In a world where electric vehicles constitute an increasing part of the market, expanding services offered by distribution companies would help meet this new stream of electricity demand. Also, the roll-out of electric heating (for instance using heat pumps) creates an additional demand for electric power. If the rate at which behind-the-meter solar is installed matches the rate at which electricity demand increases due to demand from electric vehicles and electric heating, then electric distribution companies should be able to avoid any "death spiral".

Government intervention may become a necessary force to stabilize the electricity market. Policies to reduce use of fossil fuels by incentivizing electric vehicles and electric heating will need to be continuously fine-tuned so that the demand for electricity can continue to be supplied by renewable energy microgrids such as those presented in this chapter.

ACKNOWLEDGMENT

This case study was funded by the Social Sciences and Humanities Research Council of Canada, grant #892-2017-2060 and the Natural Sciences and Engineering Research Council of Canada grant #CREATE 497981-2017. The authors also acknowledge the contribution of Emma Mildren of the University of Ottawa for providing the analysis of battery prices.

REFERENCES

Bank of Canada. (2018). *Annual Exchange Rates*. Retrieved from: https://www.bankofcanada.ca/rates/exchange/annual-average-exchange-rates/

Boloukat, M., & Foroud, A. (2016). Stochastic-based resource expansion planning for a grid-connected microgrid using interval linear programming. *Energy, 113*, 776–787. doi:10.1016/j.energy.2016.07.099

Di Francia, G. (2013). The impact of recycling policies on the photovoltaic Levelized Cost of the Electricity. In *Proceedings of the 2013 International Conference on Renewable Energy Research and Applications (ICRERA)*. Piscataway, NJ: IEEE 10.1109/ICRERA.2013.6749894

Fu, R., Feldman, D., Margolis, R., Woodhouse, M., & Ardani, K. (2017). U.S. Solar Photovoltaic System Cost Benchmark: Q1 2017. *National Renewable Energy Laboratory*. Retrieved from: https://www.nrel.gov/docs/fy17osti/68925.pdf

Hesse, H. C., Martins, R., Musilek, P., Naumann, M., Truong, C. N., & Jossen, A. (2017). Economic optimization of component sizing for residential battery storage systems. *Energies*, *10*(7), 835. doi:10.3390/en10070835

Hydro Ottawa. (2018). *Business Rates*. Retrieved from: https://static.hydroottawa.com/documents/business/rates_brochures/rates_business_e.pdf

IESO, Independent Electricity System Operator. (2018). *Data Directory*. Retrieved from: http://ieso.ca/en/Power-Data/Data-Directory

International Energy Agency. (2019). *IEA/IRENA Global renewable energy policies and measures database*. Retrieved from https://www.iea.org/policiesandmeasures/renewableenergy/?country=Canada

Khalilpour, R., & Vassallo, A. (2016). Planning and operation scheduling of PV-battery systems: A novel methodology. *Renewable & Sustainable Energy Reviews*, *53*, 194–208. doi:10.1016/j.rser.2015.08.015

Kittner, N., Lill, F., Kammen, D. M. (n.d.). Energy storage deployment and innovation for the clean energy transition. *Nature Energy, 2*. Doi:10.1038/nenergy.2017.125

MacDougall, H., Tomosk, S., & Wright, D. J. (2018). Geographic Maps of the Impact of Government Incentives on the Economic Viability of Solar Power. *Renewable Energy, 122*, 497–506. doi:10.1016/j.renene.2017.12.108

Mariaud, A., Acha, S., Ekins-Daukes, N., Shah, N., & Markides, C. N. (2017). Integrated optimisation of photovoltaic and battery storage systems for UK commercial buildings. *Applied Energy*, *199*, 466–478. doi:10.1016/j.apenergy.2017.04.067

Masters, G. (2004). *Renewable and efficient electric power systems*. Hoboken, NJ: John Wiley & Sons, Ltd. doi:10.1002/0471668826

McLaren, J., Anderson, K., Laws, N., Gagnon, P., DiOrio, N., & Li, X. (2018). Identifying critical factors in the cost-effectiveness of solar and battery storage in commercial buildings. *National Renewable Energy Laboratory*. Retrieved from: https://www.nrel.gov/docs/fy18osti/70813.pdf

Merei, G., Moshovel, J., Magnor, D., & Sauer, D. U. (2016). Optimization of self-consumption and techno-economic analysis of PV-battery systems in commercial applications. *Applied Energy*, *168*, 171–178. doi:10.1016/j.apenergy.2016.01.083

Natural Resources Canada. (2018). *Canadian Weather Year for Energy Calculation*. Retrieved from: http://climate.weather.gc.ca/prods_servs/engineering_e.html

NREL, National Renewable Energy Lab. (2018). *Distributed Generation Energy Technology Operations and Maintenance Costs*. Retrieved from: https://www.nrel.gov/analysis/tech-cost-om-dg.html

Office of the Auditor General of Ontario. (2011). *Annual Report: Chapter 3: Reports on value-for-money audits and reviews*. Retrieved from: http://www.auditor.on.ca/en/content/annualreports/arbyyear/ar2011.html

Office of the Auditor General of Ontario. (2015). *Annual Report: Chapter 3: Reports on value-for-money audits and reviews*. Retrieved from: http://www.auditor.on.ca/en/content/annualreports/arbyyear/ar2015.html

Ontario Energy Board. (2018). *What initiatives are available?* Retrieved from: https://www.oeb.ca/industry/tools-resources-and-links/information-renewable-generators/what-initiatives-are-available

Oxford Community Energy Cooperative. (n.d.). *What is a renewable energy co-operative?* Retrieved from http://www.oxford-cec.ca/page-1741085

Parra, D., Norman, S. A., Walker, G. S., & Gillott, M. (2017). Optimum community energy storage for renewable energy and demand load management. *Applied Energy*, *200*, 358–369. doi:10.1016/j.apenergy.2017.05.048

Ramli, A. M., Hiendro, A., & Ssennoga, T. (2015, June). Economic analysis of PV/diesel hybrid system with flywheel energy storage. *Renewable Energy*, *78*, 398–405. doi:10.1016/j.renene.2015.01.026

Raszmann, E., Baker, K., Shi, Y., & Christensen, D. (2017). Modeling stationary lithium-ion batteries for optimization and predictive control. *National Renewable Energy Laboratory*. Retrieved from: https://www.nrel.gov/docs/fy17osti/67809.pdf

Rubel, H., Pieper, C., Zenneck, J., & Sunak, Y. (2017). *How batteries and solar power are disrupting electricity markets*. Retrieved March 3, 2018, from https://www.bcg.com/publications/2017/energyenvironment-how-batteries-and-solar-power-are-disruptingelectricity-markets.aspx

Saft. (2018). *Lithium-ion battery life*. Retrieved from www.saftbatteries.com

Sekine, Y., & Goldie-Scott, L. (2017). *2017 Global energy storage forecast*. Retrieved December 22, 2017, from Bloomberg database.

Soulopoulos, N. (2018). Cost Projections for Batteries, Vehicles and Total Cost of Operation. *Bloomberg New Energy Finance*. Retrieved from: https://www.iea.org/media/workshops/2018/aces/NikolasSOULOPOULOSBNEF.pdf

Tomosk, S., Haysom, J. E., Hinzer, K., Schriemer, H., & Wright, D. J. (2017). Mapping the Geographic Distribution of the Economic Viability of Photovoltaic Load Displacement Projects in SW USA. *Renewable Energy*, *107*, 101–112. doi:10.1016/j.renene.2017.01.049

Tomosk, S., Haysom, J. E., & Wright, D. J. (2017). Quantifying Economic Risk in Photovoltaic Power Projects. *Renewable Energy*, *109*, 422–433. doi:10.1016/j.renene.2017.03.031

Wouters, C., Fraga, E., & James, A. (2015). An energy integrated, multi-microgrid, MILP (mixed-integer linear programming) approach for residential distributed energy system – planning – A south Australian case-study. *Energy*, *85*, 30–44. doi:10.1016/j.energy.2015.03.051

Wright, D. J., Badruddin, S., & Robertson-Gillis, C. (2018). Micro-Tracked CPV Can Be Cost Competitive With PV in Behind-The-Meter Applications With Demand Charges. *Frontiers in Energy Research*, *6*(97). doi:10.3389/fenrg.2018.00097

Zhang, Y., Lundblad, A., Campana, P. E., Benavente, F., & Yan, J. (2017). Battery sizing and rule-based operation of grid-connected photovoltaic-battery system: A case study in Sweden. *Energy Conversion and Management*, *133*, 249–263. doi:10.1016/j.enconman.2016.11.060

ADDITIONAL READING

Apostoleris, H., Stefancich, M., & Chiesa, M. (2018). *Concentrating Photovoltaics (CPV): The Path Ahead*. Springer. doi:10.1007/978-3-319-62980-3

Cole, W., Frew, B., Gagnon, P., Richards, J., Sun, Y., Zuboy, J., ... Margolis, R. (2017). *SunShot 2030 for Photovoltaics (PV): Envisioning a Low-cost PV Future*. National Renewable Energy Laboratory. doi:10.2172/1392206

Masters, G. (2004). *Renewable and efficient electric power systems*. Hoboken, NJ: John Wiley & Sons, Ltd. doi:10.1002/0471668826

Tomosk, S., Wright, D. J., Hinzer, K., & Haysom, J. E. (2015). Analysis of Present and Future Financial Viability of High Concentrating Photovoltaic Projects. In E. F. Fernandez (Ed.), *High Concentrator PhotoVoltaics*. Boston, MA: Springer. doi:10.1007/978-3-319-15039-0_14

KEY TERMS AND DEFINITIONS

Azimuth: Orientation, measured as an angle east or west of south in the northern hemisphere and vice versa in southern hemisphere.

Demand Charge: Extra charge applied during the hour when customer demand is greatest or at peak for each month. Also called "peak demand charge." This type of charge applies to the mid-sized commercial buildings analyzed in this case study. By contrast small commercial buildings are subject to a time-of-use charge in which the price of electricity depends on time of day, independent of the customer demand profile.

Feed-in-Tariff: Program guaranteeing certain electricity generators a set rate per unit of energy produced ($/kWh) over a future time horizon of 10 – 25 years.

Global Adjustment: Difference between HOEP and price guaranteed to certain generators in Ontario. The global adjustment is added as a separate line item to the electricity charges of mid-sized commercial customers.

Hourly Ontario Energy Price: Wholesale price of electricity in Ontario.

Internal Rate of Return: Metric used to estimate profitability of a potential investment. It is the discount rate at which the net present value is zero.

Microgrid: Small, local energy grid at a customer premises consisting of solar modules, a battery, an inverter, and a controller with a connection to the customer's electric loads.

Net Present Value: The value of a future stream of cash flows discounted back to their value today.

Net-Metering: Program where behind-the-meter energy producers may receive a credit on their electricity bill for excess power generated and injected into the public grid.

Photovoltaics: Conversion of solar radiation into electricity using semiconductor materials.

State of Charge: The amount of electric energy stored in the battery (kWh).

Tilt: Angle at which modules are tilted from the horizontal plane.

APPENDIX 1: ACRONYMS

AC: Alternating Current
CAD: Canadian Dollars
DC: Direct Current
FIT: Feed-in-Tariff
GA: Global Adjustment in Ontario
HOEP: Hourly Ontario Energy Price in Ontario
IESO: Independent Electricity System Operator in Ontario
IRR: Internal Rate of Return
kWh: Kilowatt Hour
LCOE: Lifetime Cost of Electricity
NM: Net-Metering
NPV: Net-Present Value
OREC: Ottawa Renewable Energy Cooperative
PV: Photovoltaics

APPENDIX 2: ANALYSIS AND DISCUSSION QUESTIONS

1. Discuss reasons for and against OREC going ahead with Case 1 and/or Case 2 solar installations ***without*** a battery. (see Table 3)
2. Discuss reasons why or why not OREC would go ahead with Case 1 and/or Case 2 solar installations ***with*** a battery. (see Table 4)
3. Why do the results show a significantly higher IRR for Case 1 than for Case 2?
   The main difference is due to the average demand profiles as shown in Figure 1. Case 2 has a very flat demand profile so it is hard to reduce the demand charges by very much. This is also evident in Figure 10 (b). Demand in Case 2 is higher in the winter when solar is not generating well, whereas in Case 1 demand is higher in the summer and can be supplied by solar, see Figure 7 (a), thus reducing demand charges significantly, see Figure 10 (a).
4. What are the characteristics that OREC should look for in other buildings that would make them profitable for solar installations?
   The building should have an electricity demand profile with a single peak during summer days, as in Case 1, Figure 1 (a). Ideally the building should be facing in the direction where the sun is at the time of the peak demand so that

Table 4.

Reasons for going ahead with a battery	Reasons against going ahead with a battery
OREC could look for ways to reduce the 'all-in' battery cost, e.g. buy refurbished batteries from electric vehicles, government incentives, discounts from suppliers on the grounds that OREC is a coop.	Increasing the IRR for Case 1 from 6.3% to 6.39% is only a minor benefit compared to the statistical uncertainty in the electricity price forecast which could cause the IRR to swing between 5.0% and 7.6%. The increase in IRR from a battery for Case 2 is even smaller.
A battery allows more flexibility in when power is bought and when it is used. An optimal scheduling system taking into account weather forecasts and planned changes in building loads could increase the IRR beyond 6.39%	OREC has no experience with batteries and batteries require optimal scheduling.
OREC should use Case 1 as an opportunity to learn how to use batteries, since battery prices are declining fast and will probably be more important in the future.	The battery is only marginally profitable even at the basic battery pack price. OREC should install solar for Case 1, wait for battery prices to decline and then add a battery to Case 1.

Table 3.

Reasons for going ahead	Reasons against going ahead
Case 1 has an IRR of 6.3% which covers the 3% - 4% needed by the investors.	Case 2 has an IRR of 4.2% which is probably insufficient to cover the needs of investors plus OREC's administrative expenses.
Even with the uncertainty in future electricity prices in Figure 13, the IRR for Case 1 is between 5.0% and 7.6%, which should be high enough.	The uncertainty in Figure 13 only covers the statistical uncertainty of extrapolating a past trend. Politics can also influence electricity prices.
OREC is surrounded by people who support environmental projects.	

solar modules can face in that direction, thus increasing solar generation and reducing the size of any battery that may be needed. A building with a sloping roof would be ideal so that snow slides off the solar modules during the winter.

5. Why are peak demand charges more reduced relative to total annual savings in case 1 than in case 2, see Figures 9 and 10?

This is mainly due to differences in the shape of their respective average demand curves (Figure 1). In case 1, demand is highest during midday when solar power is generating well, hence solar helps to lower the peak and associated demand charges incurred. The average demand curve in case 2 is much flatter, with peaks generally occurring in the morning and evening.

6. In what ways does the installation of solar impact existing energy management of the building? What about solar with battery?

Discussions with building management is imperative in order to ensure efficiency in the use of solar to reduce peak demand charges. For example, buildings used to pre-cooling and pre-heating in early morning would need to adapt to the installation of solar modules, for instance by allowing solar to reduce demand charges when it is generating well and by not creating early morning peaks. When batteries are installed with solar, they can be used to purchase energy when it is cheapest. Battery schedules could also be optimized to accommodate for irregular weather (such as rain or snow) which reduce the efficiency of solar installations, or special events which may increase energy demand.

7. Discuss government policies that could impact the feasibility of installing solar power on commercial buildings.
    - **Feed-in-Tariff:** Guarantees a price for solar electricity over 10-25 years, protecting solar investors.
    - **Investment tax credit:** Can be used to reduce the effective cost of a solar installation.
    - **A capital cost subsidy:** Can be offered to reduce the cost of a solar installation.
    - **Policies regarding electricity prices:** Can affect the profitability of solar, with higher prices making solar more profitable.
    - **Incentives to electrify transportation and heating:** Helps stabilize the market and avoid utilities falling into a 'death spiral'.

Epilogue

Feed-in-tariffs have encouraged the development of solar projects throughout the world and the vast majority of installations have used conventional silicon solar modules. As this case has shown, batteries may compensate for the phase-out of feed-in-tariffs if their price is low enough, and new battery technologies such as flow-through batteries may be commercialized at lower prices than lithium-ion. New solar technologies are also being commercialized; bifacial solar modules which generate electricity from light on the back as well as the front are particularly suitable in Ottawa with its snowy (but sunny) winters. Snow reflects almost 100% of sunlight onto the back of solar panels. Concentrating photovoltaics use exotic semiconductor materials to double the efficiency of silicon modules, making them particularly suitable for urban use where space is limited.

As OREC was deciding whether to go ahead with these projects, a tornado hit Ottawa, knocking out a major transformer station bringing power into the city. Tens of thousands of homes and businesses were without electricity for days. Only one solar panel on just one of OREC's many solar installations was damaged. If more businesses had installed solar they may not have been without power since damage from tornados is very localized. The same is true of residences, but solar installations on residences are costly. Small solar farms in residential neighborhoods (community solar) would have more returns to scale. Solar generation that is distributed around a city can result in more reliability than centralized grid equipment during a storm. It is tough to quantify the economic benefit of this reliability, but it can be concluded that the benefits of solar power are threefold: environmental, economic and reliability.

Lessons Learned

There were five primary lessons learned from this case.

1. The economics of solar power are as important as its environmental benefits. The capital cost is high and investors expect a return on their investment even if they have an environmental motivation.
2. Batteries are currently too expensive to have a major impact on the profitability of solar in Ottawa, but their prices are declining rapidly.
3. A mid-day peak in the summer electricity demand profile of a commercial building can improve the profitability of solar, since it enables solar to reduce the peak demand charges as well as the electricity charges.
4. Solar power, if planned appropriately with building managers and if proper economic analysis is done, can be profitable and thus desirable even for those who are not environmentally motivated.
5. Widespread diffusion of solar power may reduce the profitability of existing electric power industry players, resulting in a fear of a "death spiral" if electricity demand remains constant. Government, industry and other actors need to work together to plan energy systems in parallel with the growth of new electricity demand from electric vehicles and electric heating so as to ensure reliability of an electrical grid in which supply from renewables matches growth in demand.

Chapter 11
Hydrogen Fuel Cells as Green Energy

Padmavathi Rajangam
https://orcid.org/0000-0002-1647-5436
Sree Sastha Institute of Engineering and Technology, Chennai, India

EXECUTIVE SUMMARY

To reduce reliance on fossil fuels and increase demands for clean energy technology worldwide, there is currently a growing interest in the use of fuel cells as energy-efficient and environmentally-friendly power generators. With this inevitable depletion, fossil fuels will not be able to respond to energy demand for future. Among all major types of fuel cells, hydrogen fuel cells (HFCs) are in the forefront stage and have gained substantial attention for vehicle and portable applications, which is composed of a cathode, an anode, and a PEM. The heart of the fuel cells is membrane electrode assembly (MEA). An electro-deposition technique for preparing the nano-catalyst layer in PEMFCs has been designed, which may enable an increase in the level of Pt utilization currently achieved in these systems. Functionalization process has been done using a mixture of concentrated nitric acid and sulfuric acid in refluxing condition. The hydrocarbon-based polymer membrane has been used as electrolyte part.

BACKGROUND INFORMATION

Today, when many legal restrictions are applied for environmental pollution and human health, while other technologies are increasing the cost too much, the environmental friendliness of this system is a valuable alternative.

DOI: 10.4018/978-1-5225-8559-6.ch011

Figure 1. Alkaline fuel cells

Anode reaction: $2H_2 + 4OH^- \longrightarrow 4H_2O + e^-$ (1)

Cathode reaction: $O_2 + 4e^- + 2H_2O \longrightarrow 4OH^-$ (2)

Overall reaction: $2H_2 + O_2 \longrightarrow 2H_2O$ (3)

Fuel cells, as highly efficient and environmentally friendly energy conversion devices, have been in the spotlight of energy research in the last few decades. Their origin can be dated back to the 19th century, when Christian F. Schonbein first discovered in 1838 that, once connected by electrodes, hydrogen and oxygen or chlorine could react to generate electricity, which is named as the polarization effect. Shortly afterwards in the 1800's Sir William Grove discovered by accident during an electrolysis experiment. When Sir William disconnected the battery from the electrolyzer and connected the two electrodes together, he observed a current flowing in the opposite direction, consuming the gases of hydrogen and oxygen. He called this device a 'gas voltaic battery'. His gas battery consisted of platinum electrodes placed in test tubes of hydrogen and oxygen, immersed in a bath of dilute sulphuric acid. It generated voltages of about one volt. In 1842 Grove connected a number of gas batteries together in series to form a 'gas chain'. He used the electricity produced from the gas chain to power an electrolyzer, splitting water into hydrogen and oxygen. However, due to problems of corrosion of the electrodes and instability of the materials, Grove's fuel cell was not practical. As a result, there was little research and further development of fuel cells for many years. Significant work on fuel cells began again in the 1930s, by Francis Bacon, a chemical engineer at Cambridge University, England. In the 1950s Bacon successfully produced the first practical fuel cell, which was an alkaline version. It used an alkaline electrolyte (molten KOH) instead of dilute sulphuric acid (Figure 1).

In the early 1960s, General Electric (GE) also made a significant breakthrough in fuel cell technology. Through the work of Thomas Grubb and Leonard Niedrach, they invented and developed the first polymer electrolyte membrane (PEM) fuel cell. It was initially developed under a program with the US Navy's Bureau of Ships and U.S. Army Signal Corps to supply portable power for personnel in the field. In 1983, Geoffrey Ballard a Canadian geophysicist, chemist Keith Prater and engineer Paul Howard established the company, Ballard Power. Ballard took the abandoned GE fuel cell, whose patents were running out and searched for ways to improve its power and build it out of cheaper materials (Brian, 2001).

Figure 2. Schematic representation of fuel cells

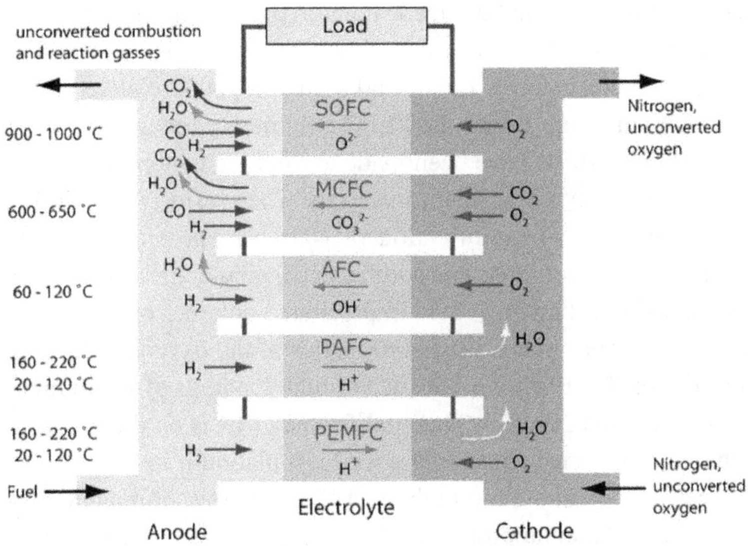

Proton exchange membrane fuel cells (PEMFCs) were first used by NASA in the 1960's as part of the Gemini space program, and were used on seven missions. Those fuel cells used pure oxygen and hydrogen as the reactant gases and were small-scale, expensive and not commercially viable. NASA's interest pushed further development, as did the energy crisis in 1973. Since then, fuel cell research has continued unabated and fuel cells have been used successfully in a wide variety of applications.

Based on the types of the electrolytes and charge carriers, fuel cells can be categorized into five major types: proton exchange membrane fuel cells (PEMFCs), phosphoric acid fuel cells (PAFCs), alkaline fuel cells (AFCs), molten carbonate fuel cells (MCFCs) and solid oxide fuel cells (SOFCs) (Figure 2).

Each type of fuel cell has its own advantages and disadvantages. For example, alkaline fuel cells allow the use of non precious metal catalysts because of facile oxygen reduction kinetics at high pH conditions, but suffer from the problem of liquid electrolyte management and electrolyte degradation. Similarly, molten carbonate fuel cells can tolerate high concentrations of carbon monoxide in the fuel stream (CO is a fuel for such fuel cells), but their high operating temperature precludes rapid start-up and sealing remains an issue. Solid oxide fuel cells offer high performance, but issues such as slow start-up and interfacial thermal conductivity mismatches must be addressed. High cost is an issue that affects each type of fuel cell.

There has been enhanced interest in recent times to operate PEFCs at temperatures >100°C for a multitude of reasons: (i) to minimize the poisoning effect of carbon monoxide impurities in the reformate hydrogen feed stream; (ii) to enable rapid heat rejection with a reasonably sized radiator (for transportation applications); and (iii) to enhance reaction rates and simplify water management issues. It is difficult to operate much above 100°C and maintain 100% relative humidity. This is one important challenge facing the PEFC community today. In response to this challenge, extensive efforts have been made to improve the properties of Nafion® and to identify alternate replacement materials. These efforts can be divided into three broad categories: (i) modification of Nafion® and similar perfluorinated membranes by inclusion of inorganic additives; (ii) membranes based on sulfonated hydrocarbon backbones; and (iii) acid impregnated polymer matrices.

A second important issue is the poor kinetics of the oxygen reduction reaction (ORR) in acidic media. Even with the use of high loadings of platinum catalyst in the electrode, the activation over potential for the ORR is on the order of 500 mV at acceptable current densities. Using such high platinum loadings (0.4 mg/cm^2) lead to fuel cell costs that are too high by at least an order of magnitude to permit commercialization. Three approaches are being pursued to resolve this issue: (i) efforts are ongoing to enhance the activity of the noble metal catalyst through alloying. By enhancing activity, it is anticipated that the amount of catalyst required can be lowered; (ii) a variety of less expensive, non-noble metal catalysts such as metal porphyrins are being investigated in acidic media; and (iii) in a more indirect approach, attempts are being made to engineer the electrode in a manner designed to ensure that ohmic and mass transport losses are kept to an absolute minimum and that the reaction is activation controlled.

The third challenge relates to durability. Typically, lifetimes of 40,000 h of continuous operation (for stationary applications) and 5,000-10,000 h of cyclic operation (for transportation applications) are targeted as a precursor to commercialization (Vijay, 2006).

SETTING THE STAGE

A first version of a PEMFC battery that had a power of 1 kW was built in the early 1960s by GE for the Gemini spacecraft. A sulfonated polystyrene ion-exchange membrane was used as the electrolyte in these cells. The electrodes contained about 4 mg/cm^2 of platinum catalyst. Because of the marked ohmic resistance of the membrane, the current density was below 100 mA/cm^2, with a voltage of about 0.6 V for an individual element. This corresponds to a specific power of the cells of about 60 mW/cm^2. Due to insufficient chemical stability of the membrane used, the total lifetime of the battery was below 2000 hours.

After a dormant period lasting almost three decades, striking improvements in PEMFC properties were achieved after 1990. The specific power of current PEMFCs went up to 600 to 800 mW/cm^2 while less than 0.4 mg/cm^2 of platinum catalysts are used, and the lifetime is now several tens of thousands of hours (Vladimir, 2012).

A fuel cell may be one of a variety of electrochemical power sources (EPSs), but is more precisely a device designed to convert the energy of a chemical reaction directly to electrical energy. Fuel cells differ from other EPSs: the primary galvanic cells called batteries and the secondary galvanic cells called accumulators or storage batteries, 1) in that they use a supply of gaseous or liquid reactants for the reactions rather than the solid reactants (metals and metal oxides) built into the units; 2) in that a continuous supply of the reactants and continuous elimination of the reaction products are provided, so that a fuel cell may be operated for a rather extended time without periodic replacement or recharging.

Possible reactants or fuels for the current-producing reaction are natural types of fuel (e.g., natural gas, petroleum products) or products derived by fuel processing, such as hydrogen produced by the reforming of hydrocarbon fuels or water gas (syngas) produced by treating coal with steam. This gave rise to their name: fuel cells.

Fuel cells directly convert the chemical energy in hydrogen to electricity, with pure water and potentially useful heat as the only byproducts. Hydrogen-powered fuel cells are not only pollution-free, but they can also have more than two times the efficiency of traditional combustion technologies. A conventional combustion-based power plant typically generates electricity at efficiencies of 33-35%, while fuel cell systems can generate electricity at efficiencies up to 60% (and even higher with cogeneration).

The gasoline engine in a conventional car is less than 20% efficient in converting the chemical energy in gasoline into power that moves the vehicle, under normal driving conditions. Hydrogen fuel cell vehicles, which use electric motors, are much more energy efficient and use 40-60% of the fuel's energy - corresponding to more than a 50% reduction in fuel consumption, compared to a conventional vehicle with a gasoline internal combustion engine.

Energy is released whenever a fuel reacts chemically with the oxygen in air. In an internal combustion engine, the reaction occurs combustively and the energy is released in the form of heat, some of which can be used to do useful work by pushing a piston. In a fuel cell, the reaction occurs electrochemically and the energy is released as a combination of low-voltage dc electrical energy and heat. The electrical energy can be used to do useful work directly while the heat is either wasted or used for other purposes.

In the first half of the 20th century, there were isolated attempts to develop FCs, such as by Francis T. Bacon, who started his alkaline fuel cell (AFC) development in

1932 and presented a practical 5 kW system in 1959. In the same year, Harry K. Ihrig (Allis-Chalmers) demonstrated the first FC vehicle, a 15 kW AFC powered tractor.

In the 1960s and 1970s, FCs found application in the space program (AFC-Apollo, and polymer electrolyte fuel cell (PEFC)-Gemini). However, this development occurred without substantial impact in the civil sector. Abundant availability of energy further stood against FC commercialization. Only with the first oil crisis in the beginning of 1970s, was energy efficiency again addressed, causing an increase of FC development activities in the years to follow. However, in spite of all these developmental activities, no commercial market was found for FCs. In the beginning of the 1990s, global environmental and resource problems as well as related legislation, such as the Clean Air Act and Zero Emission Mandates in California, drove the automotive industry to develop electric vehicles (EVs), also powered by FCs. In 1997, Daimler-Benz announced the commercial market introduction of FC-EVs for 2004. Although this date was considered to be about 10 years too early, its announcement triggered a renaissance in FC development. Nearly all car manufacturers worldwide started FC-EV development programs after 1997, leading to a huge boost in the fundamental understanding and a concomitant lowering in cost. Today, FC costs are still considered too high for unsubsidized commercialization. Promising developments have taken place aside from light-duty road vehicle application; examples include FCs for residential combined heat and power (CHP), propulsion of forklifts, generation of backup power, and off grid and portable power. Of the total fuel cell megawatts for 2014, the distribution mainly revolves around PEFC (~70 MW), MCFC (~70 MW), and SOFC (~32 MW). About 80% of the power was delivered by Fuel Cell Energy (FCE) and Bloom Energy for the stationary industrial market, Panasonic and Toshiba for the residential CHP market, and Plug Power for the material handling market. In 2013, worldwide fuel cell industry sales surpassed $1 billion for the first time, reaching $1.3 billion. As of 2015 more than 100,000 residential CHP-installations are operative in Japan.

Whereas the FC and H_2 technology development was driven mainly by industry, these technologies are now included in national energy and environmental programs. Problems with the storage and transport of hydrogen arise primarily when using electrolytic hydrogen. Technical hydrogen is produced by the reforming or gasification of products (natural gas, LPG, coal, and carbon products) for which distribution and storage infrastructures have long been established and will reach even very remote areas. Practically in all applications, FCs is competing with well-established technologies (heat engines, batteries, etc.). Because the costs of these competing technologies are determining the market accepted price, FCs initially will substitute the most expensive rival technologies (Jürgen & Ludwig, 2015).

Because of their attractive properties fuel cells have already been developed and demonstrated in the following applications.

1. Automobile: almost every car manufacturer has developed and demonstrated at least one prototype vehicle, and many have already gone through several generations of fuel cell vehicles. Some car manufacturers are working on their own fuel cell technology (gnerral motors, Toyota, Honda) and some buy fuel cell stacks and systems from fuel cell developers such as Ballard, UTC fuel cells, and DeNora (DaimlerChrysler, Ford, Nissan, Mazda, Hyundai, Fiat, Volkswagen).
2. Scooters and bicycles: several companies (Palcan, Asian Pacific, Manhattan Scientific) have demonstrated fuel cell powered scooters and bicycles using either hydrogen stored in metal hydrides or methanol in direct methanol fuel cells.
3. Portable power: Many companies (MTI, Motorola, NEC, Fuji, Matsuhita, Medis, Manhattan Scientific, Polyfuel) are developing miniature fuel cells as battery replacements for various consumer and military electronic devise.
4. Boats: MTU Fridrichschaffen demonstrated a sailboat on lake Constanze (2004) powered by a 20kW fuel cell, developed jointly with Ballard.
5. Space: Fuel cells continue to be used in the U.S. Space Program, providing power on the space orbiters. Although this proven technology is of the alkaline type, NASA announced plans to use PEM fuel cells in the future (Fanco, 2005).

CASE DESCRIPTION

Hydrogen is a versatile energy carrier that can be used to power nearly every end use energy need. The fuel cell is an energy conversion device that can efficiently capture and use the power of hydrogen is the key to making it happen.

Stationary fuel cells can be used for backup power, power for remote locations, distributed power generation, and cogeneration (in which excess heat released during electricity generation is used for other applications). In general, all fuel cells have the same basic configuration-an electrolyte and two electrodes. But there are different types of fuel cells, classified primarily by the kind of electrolyte used. The electrolyte determines the kind of chemical reactions that take place in the fuel cell, the temperature range of operation, and other factors that determine its most suitable applications (Jurgen & Ludwig, 2015).

Technology Concerns and Components

Fuel cells can be grouped by the type of electrolyte they use, namely:

- Polymer electrolyte membrane or proton exchange membrane fuel cells (PEMFC) use a thin proton conductive polymer membrane (such as perfluorosulfonated acid polymer) as the electrolyte. The catalyst is typically platinum supported on carbon with loadings of about 0.3 mg/cm^2, or, if the hydrogen feed contains minute amounts of CO, Pt-Ru alloys are used. Operating temperature is typically between 60 and 80°C. Low-temperature operation allows them to start quickly (less warm-up time) and results in less wear on system components, resulting in better durability. However, it requires that a noble-metal catalyst (typically platinum) be used to separate the hydrogen's electrons and protons, adding to system cost. The platinum catalyst is also extremely sensitive to carbon monoxide poisoning, making it necessary to employ an additional reactor to reduce carbon monoxide in the fuel gas if the hydrogen is derived from a hydrocarbon fuel. This reactor also adds cost.
- Alkaline fuel cells (AFCs) were one of the first fuel cell technologies developed, and they were the first type widely used in the U.S. space program to produce electrical energy and water on-board spacecraft since the 1960s. These fuel cells use a solution of potassium hydroxide in water as the electrolyte and can use a variety of non-precious metals as a catalyst at the anode and cathode. The high performance of AFCs is due to the rate at which electro-chemical reactions take place in the cell. They have also demonstrated efficiencies above 60% in space applications. A key challenge for this fuel cell type is that it is susceptible to poisoning by carbon dioxide (CO_2). In fact, even the small amount of CO_2 in the air can dramatically affect cell performance and durability due to carbonate formation. Alkaline membrane fuel cells (AMFCs) address these concerns and have lower susceptibility to CO_2 poisoning than liquid-electrolyte AFCs do. However, CO_2 still affects performance, and performance and durability of the AMFCs still lag that of PEMFCs. AMFCs are being considered for applications in the W to kW scale. Challenges for AMFCs include tolerance to carbon dioxide, membrane conductivity and durability, higher temperature operation, water management, power density, and anode electrocatalysis.
- PEM fuel cells are used primarily for transportation applications and some stationary applications. Due to their fast startup time and favorable power-to-weight ratio, PEM fuel cells are particularly suitable for use in passenger vehicles, such as cars and buses. PEM fuel cells are a serious candidate for

automotive applications, but also for small-scale distributed stationary power generation, and for portable power applications as well.

- Phosphoric acid fuel cells (PAFC) use concentrated phosphoric acid (-100%) as the electrolyte. The matrix used to retain the acid is usually SiC, and the electrocatalyst in both the anode and the cathode is platinum. Operating temperature is typically between 150 and 220°C. Phosphoric acid fuel cells are already semi commercially available in container packages (200 kW) for stationary electricity generation. Hundreds of units have been installed all over the world. PAFCs are more tolerant of impurities in fossil fuels that have been reformed into hydrogen than PEM cells, which are easily "poisoned" by carbon monoxide because carbon monoxide binds to the platinum catalyst at the anode, decreasing the fuel cell's efficiency. PAFCs are more than 85% efficient when used for the co-generation of electricity and heat but they are less efficient at generating electricity alone (37% – 42%). PAFC efficiency is only slightly more than that of combustion-based power plants, which typically operate at around 33% efficiency. PAFCs are also less powerful than other fuel cells, given the same weight and volume. As a result, these fuel cells are typically large and heavy. PAFCs are also expensive. They require much higher loadings of expensive platinum catalyst than other types of fuel cells do, which raises the cost. The PAFC is considered the "first generation" of modern fuel cells. It is one of the most mature cell types and the first to be used commercially. This type of fuel cell is typically used for stationary power generation, but some PAFCs have been used to power large vehicles such as city buses.
- Molten carbonate fuel cells (MCFC) have the electrolyte composed of a combination of alkali (Li, Na, K) carbonates, which is retained in a ceramic matrix of LiAlOi. Operating temperatures are between 600 and 700°C where the carbonates form a highly conductive molten salt, with carbonate ions providing ionic conduction. At such high operating temperatures, noble metal catalysts are typically not required. Improved efficiency is another reason MCFCs offer significant cost reductions over phosphoric acid fuel cells. Molten carbonate fuel cells, when coupled with a turbine, can reach efficiencies approaching 65%, considerably higher than the 37%–42% efficiencies of a phosphoric acid fuel cell plant. When the waste heat is captured and used, overall fuel efficiencies can be over 85%. Unlike alkaline, phosphoric acid, and PEM fuel cells, MCFCs do not require an external reformer to convert fuels such as natural gas and biogas to hydrogen. At the high temperatures at which MCFCs operate, methane and other light hydrocarbons in these fuels are converted to hydrogen within the fuel cell itself by a process called internal reforming, which also reduces cost. The primary disadvantage of

current MCFC technology is durability. The high temperatures at which these cells operate and the corrosive electrolyte used accelerate component breakdown and corrosion, decreasing cell life. Scientists are currently exploring corrosion-resistant materials for components as well as fuel cell designs that double cell life from the current 40,000 hours (~5 years) without decreasing performance.

- Solid oxide fuel cells (SOFC) use a solid, nonporous metal oxide, usually Y203-stabilized ZrOi (YSZ) as the electrolyte. These cells operate at 800 to 1000°C where ionic conduction by oxygen ions takes place. SOFCs use a hard, non-porous ceramic compound as the electrolyte. SOFCs are around 60% efficient at converting fuel to electricity. In applications designed to capture and utilize the systems waste heat, overall fuel use efficiencies could top 85%. SOFCs are also the most sulfur-resistant fuel cell type; they can tolerate several orders of magnitude more sulfur than other cell types can. In addition, they are not poisoned by carbon monoxide, which can even be used as fuel. This property allows SOFCs to use natural gas, biogas, and gases made from coal. High-temperature operation has disadvantages. It results in a slow start up and requires significant thermal shielding to retain heat and protect personnel, which may be acceptable for utility applications but not for transportation. The high operating temperatures also place stringent durability requirements on materials. The development of low-cost materials with high durability at cell operating temperatures is the key technical challenge facing this technology. Scientists are currently exploring the potential for developing lower-temperature SOFCs operating at or below 700°C that have fewer durability problems and cost less. Lower-temperature SOFCs have not yet matched the performance of the higher temperature systems, however, and stack materials that will function in this lower temperature range are still under development.
- Sometimes, a direct methanol fuel cell (DMFC) is categorized as yet another type of fuel cell; however, according to the previous categorization (based on electrolyte), it is essentially a polymer membrane fuel cell that uses methanol instead of hydrogen as a fuel. Direct methanol fuel cells do not have many of the fuel storage problems typical of some fuel cell systems because methanol has a higher energy density than hydrogen though less than gasoline or diesel fuel. Methanol is also easier to transport and supply to the public using our current infrastructure because it is a liquid, like gasoline. DMFCs are often used to provide power for portable fuel cell applications such as cell phones or laptop computers.

Figure 3. Hydrogen- oxygen fuel cell reaction

$$\text{Anode reaction:} \quad 2H_2 \longrightarrow 2H^+ + 2e^- \quad (4)$$

$$\text{Cathode reaction:} \quad 1/2O_2 + 2H^+ + 2e^- \longrightarrow H_2O \quad (5)$$

$$\text{Overall reaction:} \quad H_2 + 1/_2O_2 \longrightarrow H_2O \quad (6)$$

The membrane electrode assembly (MEA) is the heart of the single cell. It is the critical component of the PEMFC, since it is the site of the fuel cell reactions. The MEA is less than a millimetre thick and consists of a solid polymer, proton conducting membrane electrolyte, with a layer of platinum-based catalyst and a gas-porous electrode support material on both sides of it, forming the anode and cathode of the cell. The membrane electrolyte is typically bonded to the electrodes by hot pressing at a temperature above the glass transition temperature of the membrane (Ralph, 1997).

ELECTRODE PART

The electrodes must be porous because the reactant gases are fed from the back and must reach the interface between the electrode and the membrane, where the electrochemical reactions take place in the so-called catalyst layers. A fuel cell electrode is essentially a thin catalyst layer pressed between the ionomer membrane and porous, electrically conductive substrate. It is the layer where the electrochemical reactions take place. More precisely, the electrochemical reactions take place on the catalyst surface. Because there are three kinds of species that participate in the electrochemical reactions, namely gases, electrons and protons.

Electrons travel through electrically conductive solids, including the catalyst itself, but it is important that the catalyst particles are somehow electrically connected to the substrate. Protons travel through ionomer, therefore the catalyst must be in intimate contact with the ionomer. And finally, the reactant gases travel only through the porous electrode. At the same time, product water must be effectively removed; otherwise the electrode would flood and prevent oxygen access (Figure 3).

The following processes take place inside the fuel cell

1. Gas flow through the channels, some convective flows may be induced in the porous layers.
2. Gas diffusion through porous media.
3. Electrochemical reactions, including all the intermediary steps.
4. Proton transport through proton conductive polymer membrane.
5. Electron conduction through electrically conductive cell components.
6. Water transport through polymer membrane including both electrochemical drag and back diffusion.
7. Water transport through porous catalyst layer and gas diffusion layers.

Carbon fiber paper is typically used for this purpose since it is porous, hydrophobic, conductive (electrons) and non-corrosive. A catalyst is added to the surface of each electrode where it contacts the electrolyte in order to increase the rate at which the chemical reaction occurs. A catalyst promotes a chemical reaction by providing ready reaction sites but is not consumed in the process. Platinum is typically used for this purpose due to its high electro-catalytic activity, stability and electrical conductivity. Platinum is very expensive, so the amount used (known as the catalyst loading) is a significant factor in the cost of a fuel cell. Fuel cell designers strive to minimize the amount of platinum used while maintaining good cell performance.

The fuel cell seems to be a very simple device, numerous processes takes place simultaneously. It is therefore important to understand those processes, their mutual interdependence, and dependence on components design and materials properties (Fanco, 2005).

However, one of the major challenges for PEMFC commercialization is the production of low-cost electrocatalysts with high catalytic performance and long-term durability under their working conditions. The most commonly used electrocatalysts currently are nanosized platinum (Pt) particles supported on carbon black (CB), commercially called Vulcan XC-72 due to its relatively large surface area and excellent chemical stability in the fuel cell environment. However, approximately 47% of carbon black is comprised of micropores, some of which are too small (less than 1 nm in diameter) to be accessible to the electrolyte polymer, resulting in the entrapped Pt NPs not contributing to the electrode reactions due to the absence of the triple phase boundaries (i.e., gas–electrolyte–electrode) (Baizeng et al., 2011).

In the late 1980s and early 1990s Los Alamos National Laboratory and Texas A & M University also made significant developments to the PEM fuel cell. They also found ways to significantly reduce the amount of platinum required and developed a method to limit catalyst poisoning due to the presence of trace impurities in the hydrogen fuel (Los Alamos National Laboratory).

Some novel nanostructured carbon materials have recently been investigated as possible catalyst supports for PEMFCs, including carbon nanotubes (CNTs), carbon nanofibers (CNFs) and graphene, owing primarily to the unique electrical and structural properties. However, since the surface of these is highly inert because of high graphitization. Pt nanoparticles are very difficult to deposit directly and evenly onto such surfaces without active functional groups. Therefore, how to evenly deposit Pt nanoparticles on the surfaces of these catalyst supports and to reinforce the binding strength between Pt and catalyst supports become a big challenge. Thus far, the most established protocols for catalytic metal immobilization on catalyst supports include generating functional groups on the external walls, mostly through harsh oxidative treatment, such as refluxing in HNO_3 followed by metal deposition on activated catalyst supports. Such surface functionalization introduces an avenue for metal precursors to correlate with catalyst supports and prompts the deposition of metals on the anchoring sites (Daping et al., 2011).

Carbon nanotubes (CNTs) consists of large number of concentric cylinders of graphene of length several micrometers and distance between neighboring shells equal to interlayer spacing of graphene sheets i.e. 0.34 nm. Several methods are employed for the synthesis of MWCNTs, which include carbon arc process, chemical vapour deposition (CVD) and pulsed laser deposition technique. It has high surface area, chemical stability, excellent mechanical and electrical properties and particularly abundant CNTs are commercially available at a relatively low cost, making them potential competitive candidates for use as catalyst supports. Although the CNT-supported Pt electrode revealed a higher power density than the Vulcan XC-72 supported one, the realistic application of CNTs as cathode catalyst supports has been greatly hindered by several difficulties associated with processes for their purification and dispersion, particularly in supporting high loading of noble metals (i.e., Pt) with a uniform distribution of Pt NPs. there is great difficulty in preparing CNT supported Pt catalysts with high metal loading because pristine CNTs are chemically inert, which make it difficult to anchor and deposit Pt NPs without the activation of the graphitic surface. So far, much effort has been made toward creating defect sites in CNTs and improving the hydrophilicity of CNTs in order for CNTs to support high loading of the metal catalyst (i.e., Pt) (Datsyuk et al., 2008).

Post synthesis treatments of CNTs, such as chemical oxidation by strong acids and sonochemical treatment have been commonly used to introduce defect sites into CNTs for improved deposition of precious metals on the CNT supports. However, the introduction of a large number of defects may reduce the electrical conductivity and corrosion resistance of CNTs, resulting in a loss in the electrochemical surface area (ECSA) of the supported Pt NPs and reduced durability during fuel cell operation. MWCNT has been functionalized by chemical method using the mixtures of HNO_3 (69%), H_2SO_4 (96.2%) (Figure 4) (Baizeng et al., 2011; Padmavathi et al., 2015).

Figure 4. Schematic diagram of functionalization process of MWCNTs

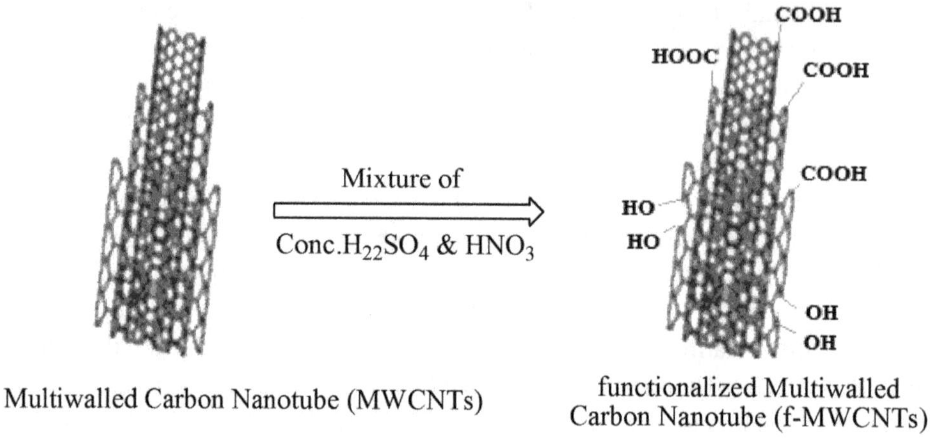

Figure 5. Schematic representation of electrospinning setup

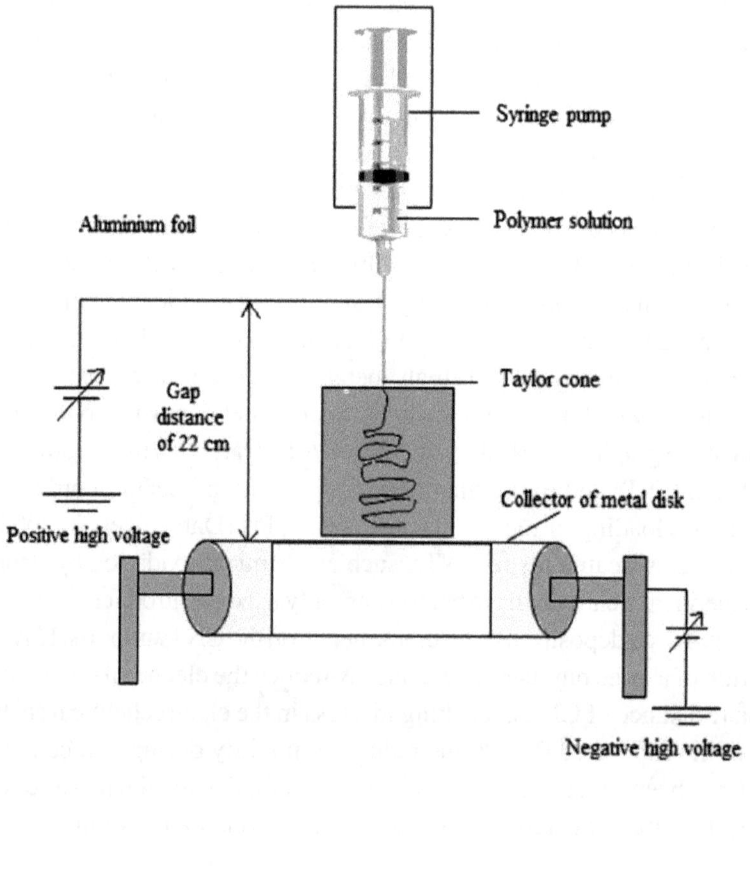

Figure 6. Diagram of the molecular changes occurring during the chemical process of stabilization and carbonization of PAN (a) the stabilization step occurs at 200–300oC in airenvironment, the carbonization step, shown in (b) is the mechanism of intermolecular cross-linking at 800–1000oC through oxygen-containing groups, and dehydrogenationis a possible mechanism as well. In (c), cyclized sequences are cross linked and nitrogen atoms are released.

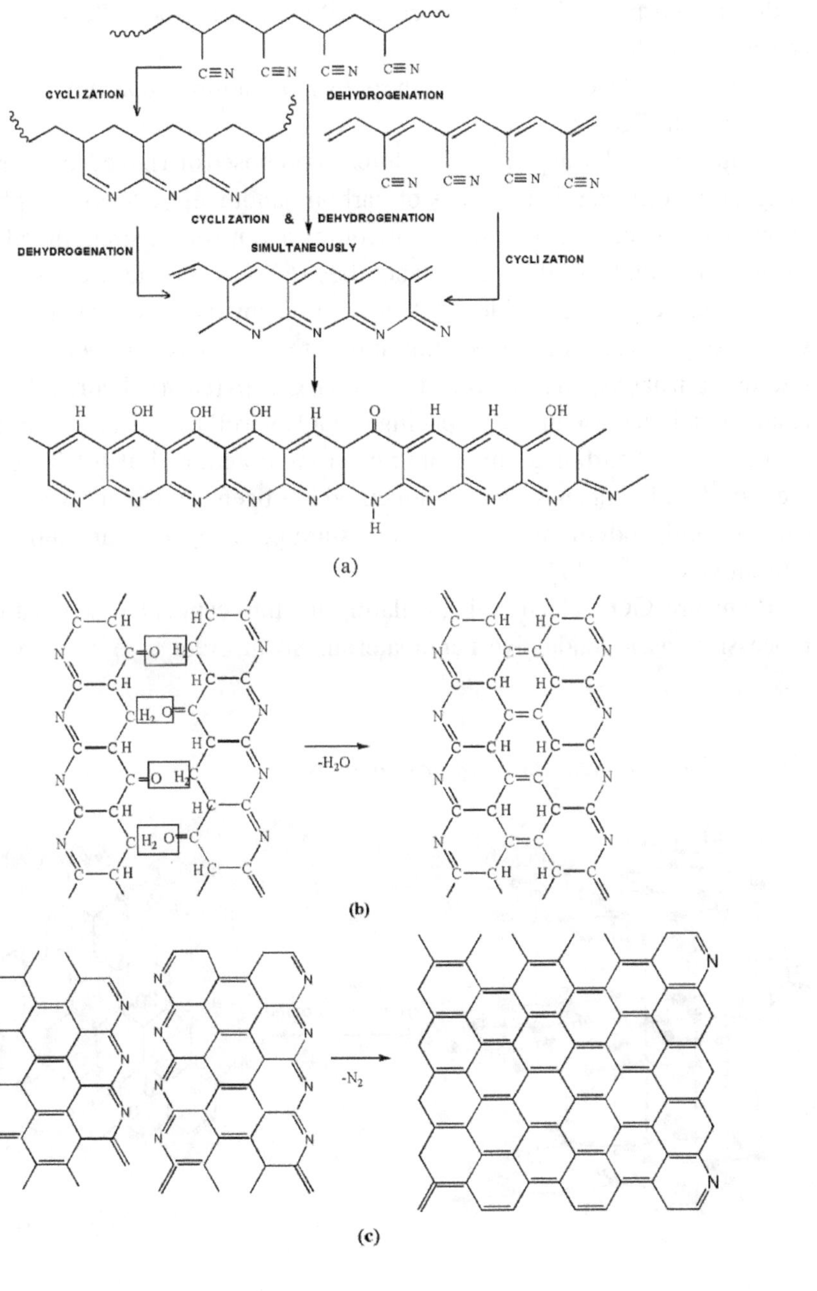

By electrospinning process are shown in Figure 5, the polyacrylonitrile polymer nanofibers has been prepared. Using this process it is possible to generate fibers with diameters ranging from the sub-micrometer size down to the nanometer range by applying a high voltage to produce an electrically charged jet of polymer solution or melt, which solidifies during its travel to the collector. Due to the small size of fiber diameters, electrospun fibers possess a large specific surface area and are expected to display morphologies and properties different from the bulk state (Padmavathi & Sangeetha, 2013).

The various stages of processing the prepared nanofibers mat into carbon nanofiber are shown in Figure 6.

Graphene, the basal plane of graphite, is composed of a honeycomb arrangement of carbon atoms and is the basis of carbon nanotubes (CNTs). Graphene single sheets are expected to have tensile modulus and ultimate strength values similar to those of single wall carbon nanotubes (SWCNTs) and are also electrically conducting. Graphene oxide (GO) formation involves the reaction of graphite with strong oxidizers such as sodium nitrate and potassium permanganate. The introduction of oxygen containing functional groups (such as hydroxyl and epoxide) results in an increase in the d-spacing of GO as well as a change of hybridization of the oxidized carbon atoms from planar sp^2 to tetrahedral sp^3. To prepare GO, the modified Hummer's oxidation method has been carried out with a mixture of sulfuric acid, sodium nitrate, and potassium permanganate are shown in Figure 7 (Sehkyu et al., 2011).

By nature, GO is electrically insulating and thus cannot be used, without further processing, as a conductive nanomaterial. So in order to obtain an electrically

Figure 7. Schematic diagram of graphene oxide

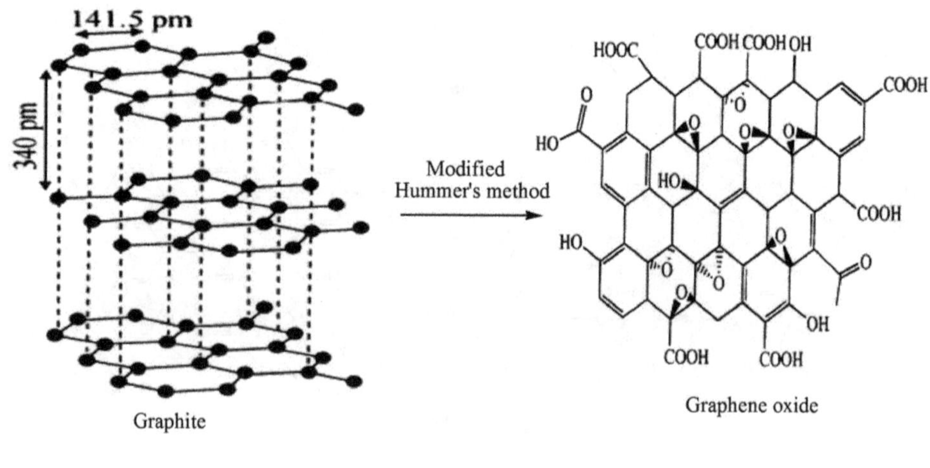

Figure 8. Schematic diagram for epoxide with hydrazine

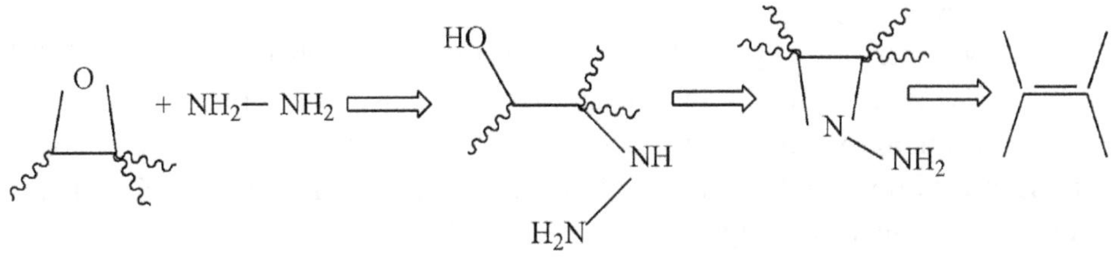

conducting material, GO can be reduced either by heating or by reducing agents. The reduction of GO to graphene has been performed by chemical reduction method using hydrazine hydrate as a reducing agent (Figure 8) (Peiying et al, 2011; Min, Gyeong & Rodney, 2009).

CORROSION OF CATALYST SUPPORT

In addition to the cost and performance, long term stability of the catalyst is also important. Sintering of Pt particles on the carbon support decreases the catalytically active surface areas. The sintering of the catalyst may be reduced by strengthening the metal-support interaction, like functionalization of the carbon support. Carbon support corrosion of cathode catalysts in PEMFCs is a major contributor to the catalyst degradation. In the state of the art PEMFC, carbon black is normally used as catalyst support material for PEMFCs. Despite its widespread use, high surface-area carbon support in PEMFC electrodes is susceptible to corrosive conditions, which include high water content, low pH, high temperature, high oxidative potential, and high oxygen concentration. Oxygen atoms are being generated by the catalyst particles, and at elevated temperature, carbon atoms are able to react with oxygen atoms and/or water to generate gaseous products such as CO and CO_2 which leave the cell. Furthermore, during the start-up and shutdown of a fuel cell, local cathode potential can reach as high as 1.5V, which significantly speeds up the carbon corrosion. In addition, Pt catalysts have been suspected to accelerate the rate of carbon corrosion. This loss mechanism removes carbon from the cell, leading to a reduction of the carbon content in the catalyst layer with time. As carbon is corroded away, noble metal nanoparticles will be lost from the electrode or aggregate to larger particles, which may cause a loss of catalyst activity, and, in the extreme, a structural collapse of the electrode. Oxidation of carbon support can also lead to changes in surface hydrophobicity that can cause gas transport difficulties (Jiajun Wang et al 2008).

ELECTROLYTE PART

The solid polymer electrolyte is the ultimate distinguishing characteristic of a PEM fuel cell. The electrolyte is a thin membrane of a plastic like film that ranges in thickness from 50 to 175 microns. Commercially these membranes are composed of perfluorosulfonic acids, which are fluorocarbon polymers that have side chains ending in sulfonic acid groups.

The SO_3H group is ionically bonded, and so the end of the side chain is actually an SO_3^- ion with H^+ ion. This is why such structure is called ionomer. Because of their ionic nature, the ends of the side chains tend to cluster within the overall structure of the membrane (Fanco, 2005).

All acidic solid polymer electrolytes require the presence of water molecules for hydrogen ion conductivity since hydrogen ions move together with water molecules during the ion exchange reaction. The ratio of water to hydrogen ions for effective conductivity is typically about 3:1. For this reason, the gases in contact with the membrane must be saturated with water for effective fuel cell operation.

All electrolytes must perform the fundamental functions of being a proton conductor, an electron insulator and a gas separator. In addition, the electrolyte should have reasonable mechanical strength, dimensional stability (resistance to swelling), high ionic conductivity, low equivalent weight, and that are easily manufacturable. To some extent, mechanical and dimensional stability of the polymer is provided through its integration into a membrane electrode assembly which adds a supporting structure.

A number of commercial membranes are available such as Nafion®, produced by Dupont (Figure 9), and others by the Dow Chemical Company. In addition, fuel cell manufacturers like Ballard Power Systems have developed their own proprietary membranes.

Figure 9. Molecular structure of Nafion

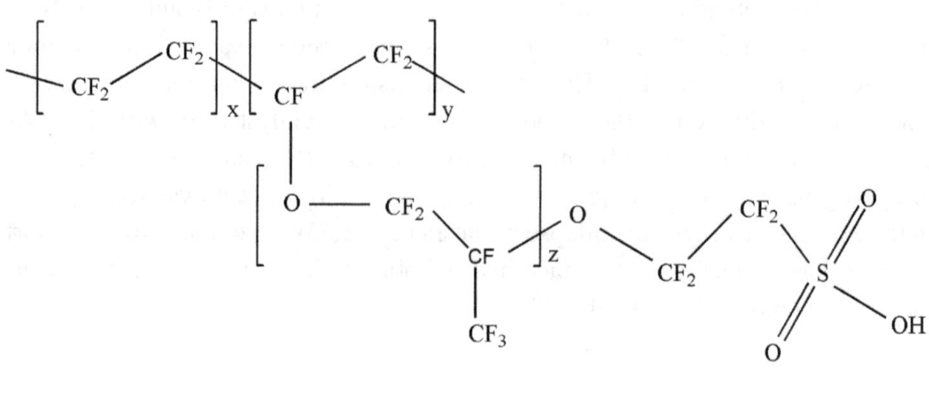

Currently, the best known commercialized membrane materials for PEMs are perfluorosulfonic acid polymers (represented by DuPont's Nafion series) that exhibit good thermal and chemical stabilities, and excellent proton conductivity under wet conditions. Nevertheless, Nafion series membranes suffer from some drawbacks, such as severe fuel permeability and dramatically decreased proton conductivity at low hydration levels, which significantly affect the efficiency and performance of fuel cells. Moreover, the complex preparation process and high cost of Nafion also bring adverse effects on the large scale commercial application of PEMFCs and DMFCs. Accordingly numerous studies have been performed for the development of alternative PEMs with adequate properties and low cost (Ying et al., 2018).

In this regard, much work has been devoted to the development of fluorine free and inexpensive alternative hydrocarbon membranes in order to meet high requirements for sustainability and environmental friendliness. The currently popular alternative polymer membranes under investigation for use in fuel cells, are prepared from thermostable aromatic polymers such as sulfonated polyether sulfones, sulfonated polysulfone (Adnan et al 2017), sulfonated polyether ether ketone (SPEEK), sulfonated polyether ether ketone/ zirconium oxide (Silva et al 2005), sulfonated polyimide, sulfonated polybenzimidazole (PBI) and acid doped PBI. The most frequent approach used for the synthesis of such polymer electrolytes, i.e. polymers with sulfonic acid groups in the backbone, consists of either a post sulfonation method performed on available bare polymers, or sulfonation of monomers and subsequent polymerization of the sulfonated monomers are shown in Figure 10 (Padmavathi & Sangeetha, 2013).

Figure 10. Sulfonation of polyether ether ketone

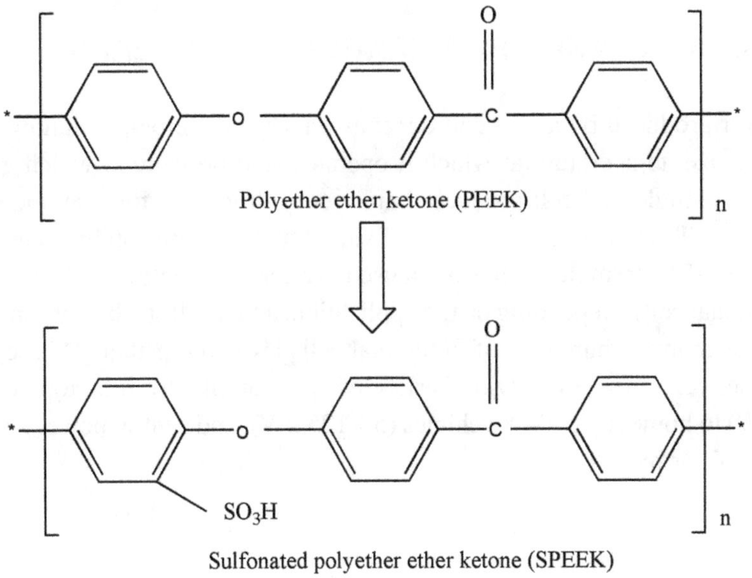

Figure 11. Sulfonation of PSEBS using CSA

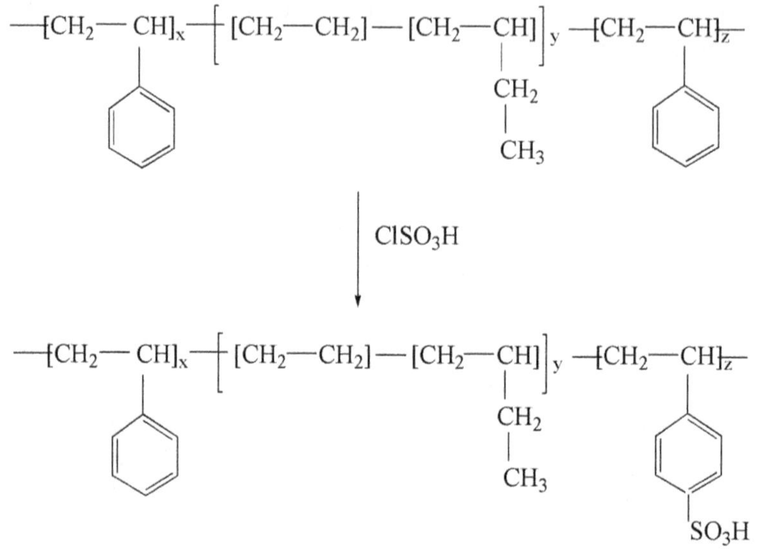

One of the alternatives for Nafion is sulfonated form of polystyrene-ethylene-butylene-polystyrene tri-block polymer (S-PSEBS). This ionomer, which is based on a hydrocarbon polymer, polystyrene-block-(ethylene-ran-butylene)- block-polystyrene (PSEBS), is a thermostable polymer with an aromatic non-fluorinated backbone. It is sulfonated using chlorosulfonic acid (CSA) in the medium of chloroform are shown in Figure 11 (Elamathi et al., 2008; Tomoya, Kazuya & Mitsuru, 2009).

MANAGEMENT AND ORGANIZATIONAL CONCERNS

The power produced by a fuel cell depends on several factors, including the fuel cell type, size, temperature at which it operates, and pressure at which gases are supplied. A single fuel cell produces barely enough voltage for even the smallest applications. To increase the voltage, individual fuel cells are combined in series to form a stack (The term fuel cell is often used to refer to the entire stack, as well as to the individual cell). Depending on the application, a fuel cell stack may contain only a few or as many as hundreds of individual cells layered together. This scalability makes fuel cells ideal for a wide variety of applications, from laptop computers (20-50 W) to homes (1-5 kW), vehicles (50-125 kW), and central power generation (1- 200 MW or more).

CURRENT CHALLENGES AND OPPORTUNITIES

Reducing cost and improving durability are the two most significant challenges to fuel cell commercialization. Fuel cell systems must be cost competitive and perform better than, traditional power technologies over the life of the system. Ongoing research is focused on identifying and developing new materials that will reduce the cost and extend the life of fuel cell stack components including membranes, catalysts, bipolar plates, and membrane electrode assemblies. Low-cost, high volume manufacturing processes will also help to make fuel cell systems cost competitive with traditional technologies.

At present there are many uncertainties to the success of fuel cells and the development of a hydrogen economy:

- Fuel cells must obtain mass-market acceptance to succeed. This acceptance depends largely on price, reliability, longevity of fuel cells and the accessibility and cost of fuel. Compared to the price of present day alternatives e.g. diesel-engine generators and batteries, fuel cells are comparatively expensive. In order to be competitive, fuel cells need to be mass produced less expensive materials developed.
- An infrastructure for the mass-market availability of hydrogen, or methanol fuel initially, must also develop. At present there is no infrastructure in place for either of these fuels. As it is we must rely on the activities of the oil and gas companies to introduce them. Unless motorists are able to obtain fuel conveniently and affordably, a mass market for motive applications will not develop.
- At present a large portion of the investment in fuel cells and hydrogen technology has come from auto manufacturers. However, if fuel cells prove unsuitable for automobiles, new sources of investment for fuel cells and the hydrogen industry will be needed.
- Changes in government policy could also derail fuel cell and hydrogen technology development. At present stringent environmental laws and regulations, such as the California Low Emission Vehicle Program have been great encouragements to these fields. Deregulation laws in the utility industry have been a large impetus for the development of distributed stationary power generators. Should these laws change it could create adverse effects on further development.
- At present platinum is a key component to fuel cells. Platinum is a scarce natural resource; the largest supplies to the world platinum market are from South Africa, Russia and Canada. Shortages of platinum are not anticipated; however changes in government policies could affect the supply (Brian 2001).

Phosphoric acid doped proton conducting polymers are also used as polymer electrolytes in hydrogen fuel cells. But the main problems of such electrolyte are leaching out of the phosphoric acid, flooding, low mechanical stability and fuel cell crossover (Ghosh et al., 2018).

Looking broadly at the fuel cell energy based technologies for converting fuels such as hydrogen into electric power. There will be losses in these processes, but as ways of handling the fluctuating production from solar or wind power devices, there is no technology more efficient for matching supply and demand. Of course, only electricity that cannot be used when produced will have to be sent through the loss-prone conversions.

The structure of the energy market may be influenced in a positive way by the hydrogen technologies. This is at least true for the PEM fuel cells, because they are modular and can be installed at nearly the same price per kW at any size, from family installations to central power plants. This means a large freedom for independent power producers who may avoid the restraints of conventional energy companies in regard to inertia in technologies used and services offered. Everyone can install a PEM fuel cell to produce power and associated heat for the building or for portable equipment on the basis of piped or locally stored hydrogen. Primary, intermittent power may come from rooftop photovoltaic panels or from a wind farm in which the user may possess shares.

In terms of environmental impacts, the hydrogen conversion and infrastructure technologies are generally benign. Some of the high temperature fuel cells may produce nasty wastes that have to be dealt with delicately, but generally, from hydrogen through storage and converters to end use energy, the negative impacts are small. Due to high cost, it is an advantage that fuel cells can be introduced gradually through hybrids.

SOLUTIONS AND RECOMMENDATIONS

In hydrogen fuel cells, the most considerable disadvantage is the cost, due to the requirement for materials with specific qualities. For instance, platinum and Nafion membrane are the most common used materials for catalyst layers and electrolytic membrane, respectively. However, there is an ongoing need to develop lower cost alternatives to these materials for commercialization. Another big disadvantage could be the preference of using hydrogen as a fuel for which there is no production and distribution systems currently in place. Moreover, catalyst degradation and electrolyte poisoning will occur over time if another fuel besides hydrogen is fed into the fuel cell system (or CO contamination in the hydrogen) which by degrees results in the overall efficiency decrease. Another drawback refers to the desirable lifetime for fuel cell systems. For commercial portable and mobile devices a lifetime of around 5000 hrs and for stationary and power plant applications up to 40,000 hrs operations is required.

The ORR is a key electrochemical reaction that enables the conversion of oxygen and hydrogen to water in PEFCs. Platinum-based catalysts are normally used in PEFCs but high cost, scarcity and CO poisoning remains a risk hindering their commercialization. Intensive research efforts have been directed towards designing efficient alternatives to Pt based catalysts, primarily to reduce cost (improve the performance/cost ratio) and to achieve higher reactant selectivity (e.g. cathode catalysts that are tolerant to fuel cross-over from the anode).

Electrode Part

The catalyst performance and stability in the electrochemical environment during operation has been and is still among the most frequently addressed issues in terms of fuel cell optimization. Catalyst diffusion, agglomeration, dissolution and Ostwald ripening are the most commonly known phenomena of catalyst degradation, but also CO tolerance of the anode catalyst of PEMFCs has been the focus of many research activities. Two types of free Pt site adsorption mechanisms are proposed: a linear and a bridge-type adsorption. Both mechanisms result in a positive shift of the reversible potential so the desorption process of CO from the surface of the Pt nanoparticles becomes difficult. The poisoning effect of CO can be overcome by either reducing the CO adsorption or enhancing the CO oxidation or reduction. To accomplish these requirements, several binary, ternary or even quaternary electrocatalysts are under investigation (Hengge et al., 2017).

To achieve easily accessible active sites, the following features are recognized as desirable: i) a high surface area to accommodate an adequate number of active sites; ii) a tailored morphology that enhances the exposure of catalytic centers to maximize utilization efficiency; iii) a hierarchical porous structure to facilitate the diffusion of O_2 molecules. Therefore, finding a strategy that combines abundant active sites with desirable structure is essential for obtaining an efficient electrocatalyst with the optimal mass transfer of ORR-relevant species. The incorporation of nitrogen atoms within a carbonaceous skeleton creates a net positive charge on the adjacent carbon atom and modulates O_2 chemisorption energy, weakens the O-O bond strength, and facilitates the direct oxygen reduction via the desired four electron pathway (minimizes generation of reactive peroxide species), leading to a significant enhancement in ORR performance (Yaxiang et al., 2017).

The oxidation treatment using microwave assisted polyol of CNTs introduced primarily -OH and -COOH groups, thereby enhancing the reduction of Pt ionic species, resulting in smaller Pt particles with improved dispersion and attachment properties. The Pt particles supported on oxidized CNTs displayed superior durability to those on pristine CNTs or commercially available Pt/C. These improvements are most likely associated with the percentage of metallic Pt in the particles. After 400 cycles, the losses of electrochemical surface area in Pt nanoparticle supported on oxidized CNTs and pristine CNTs catalysts were 66 and 84%, respectively, of that associated with commercial Pt/C. A single proton exchange membrane fuel cell using Pt supported on oxidized CNTs at the cathode with a total catalytic loading of 0.6 Pt mg cm^{-2} exhibited the highest power density of 890 mW cm^{-2} and displayed a lower mass transfer loss, compared to Pt/C (Weimin et al., 2010; Yu & Jhao, 2011).

Mei et al., synthesized the Pt catalysts supported on graphites and were developed for oxygen reduction reaction in PEMFCs. Catalytic activity and carbon corrosion of the developed catalysts were evaluated using rotating disc electrode techniques and results were compared with those of a state-of-the-art commercial Pt catalyst supported on Vulcan XC-72. The results showed that the electrochemical active surface area losses after 1500 cycles were 46.92% and 62.2% for the developed catalyst and the commercial catalyst, respectively, while mass activity losses were 45.3% and 84.2%, respectively. The developed Pt catalysts had similar catalytic performance to the commercial catalyst; however, the developed catalysts had much better corrosion resistance than the commercial catalyst. They concluded that the overall nanoscale graphite can be a promising electrocatalyst support to replace the currently used Vulcan XC-72 carbon black (Mei et al., 2011).

Nanostructured platinum dispersed on functionalized graphene and functionalized multiwalled carbon nanotube [Pt/f-G-f-MWNT] hybrid nanomaterials used as an electrocatalyst for ORR in PEMFC. Electrochemical studies performed on Pt/f-G-f-MWNT composite materials by varying the ratio of the composition of f-G and f-MWNT for the investigation of the electrochemical active surface area (ECSA) have resulted in an ECSA as high as 108 m^2/g for the Pt dispersed on nanocomposite containing equal proportions of f-G and f-MWNT. Polarization graphs for the ORR reaction in PEMFC with Pt/f-G-f-MWNT as an electrocatalyst resulted in the best performance of 540 mW/cm^2 for the Pt/(50wt% f-G + 50wt% f-MWNT) cathode catalyst, agreeing with the electrochemical active surface area of Pt, due to good accessibility and uniform dispersion of the nanostructured Pt catalyst dispersed on the f-G-f-MWNT catalyst support, making them a suitable electrocatalyst for advanced PEMFC (Imran et al., 2010).

ELECTROLYTE PART

Abundant and inexpensive natural polymers such as chitosan, alginate and cellulose have been widely applied in the field of membrane technologies in order to meet high requirements for sustainability and environmental friendliness due to its low toxicity and cost. But it has relatively poor mechanical properties and low intrinsic proton conductivity of pristine chitosan limits its application as PEMs. Hence creation of organic-inorganic composite nanostructures to improve the mechanical strength and construct facile channels for H^+ migration is a straight forward and effective strategy to solve the above problems (Ying et al., 2018).

Incorporation of hygroscopic inorganic nanomaterials such as zirconia (ZrO_2), silica (SiO_2) titania (TiO_2), phosphorous pentoxide (P_2O_5), Zeolite, dialuminum dioxide, zirconium dioxide, and zirconium hydrogen phosphate onto the backbone of PFSA has shown to be a potential replacement for Nafion at high temperatures, due to their good water retention and effect on the membrane stability enhancement (Jing et al., 2018; Nattinee et al., 2018). The proton conductivity of these hybrid membranes is still inferior when compared to recast Nafion membranes, which is believed to derive mainly from the increased barrier properties of the membranes, due to the incorporation of inorganic fillers (an increase in inorganic particle agglomeration at higher concentrations of inorganic additives) (Chen et al., 2006). In order to increase the proton conductivity of these hybrid membranes, functionalized nanoparticles have been employed. After functionalization, the nanoparticles can display improved proton conductivity, the ability to hold more water and more homogeneous dispersion among the hydrophilic sites of the Nafion, compared with non-functionalized nanoparticles (Chalkova et al., 2009).

Incorporation of nanosized inorganic filler functionalized graphene oxide (f-GO), functionalized carbon nanotubes (f-CNTs) and functionalized carbon nanofibers (f-CNFs) materials into the hydrocarbon based polymer matrix can influence the properties of the parent polymer, such as proton conductivity, mechanical and thermal stability (Ravi, Mohamed & Keith 2014) (Minghui et al., 2018).

Phosphosilicate gel is one of the most attractive solid state highly proton conductive material. For low temperature range PEMFC it is composited with sulfonated poly(ether ether ketone) (SPEEK). The fuel cell performance of the membranes was taken at different operating temperatures. Best performance obtained for SPEEK at temperature 70 °C. Performance enhances with phosphosilicate gel loading. Around 1.83 times enhancement in the peak power is achieved on going from 50 °C to 70 °C operating temperature for SPEEK with phosphoslicate gel of less than 10 nm particle sizes (Ghosh et al., 2018).

Functionalization of GO with aspartic acid, amino acid family considering the Zwitter ion interaction for better ionic transport through the additive in SPEEK matrix. Amino acids consist both amino (-NH_2) and carboxylic acid (-COOH) groups and play significant role in biosynthesis involving proton transport. Amino acids generally exist as zwitter ions through intra molecular proton exchange, as a result amino acids can act as both proton donor and acceptor and hence mediating proton transport. Among the various amino acids, aspartic acid is preferred to functionalize GO due to the presence of more carboxylic acid groups which can favor ionic transport. Aspartic acid functionalized GO prepared by the condensation reaction between amino group of aspartic acid and carboxylic acid group of GO. Addition of aspartic acid functionalized graphene oxide (ASPGO) improves the membrane dynamics by forming interfacial interactions with the host matrix in turn enhancing the proton transport and mitigating the methanol permeability through the membrane (Gutru & Santoshkumar, 2018).

The promising candidates for the breakthrough in PEM application are categorized into the following four materials; (1) sulfonated/non-sulfonated multiblock copolymers which contain hydrophobic and hydrophilic oligomers,(2) branched polymers in which the inner hydrophilic domains are surrounded by outer hydrophobic domains, (3) locally and densely sulfonated polymers with multiple sulfonic acid moieties along the main chains, and (4) high-IEC polymers (IEC >3 meq./g) with a highly hydrophobic main chain skeleton. All methods aim at the generation of clear and well-connected proton channels while maintaining the water stability for high proton conduction even at low RH and high operating temperature. By further pursuing new and/or optimal polymer structures following (1)–(4), extraordinarily useful materials possessing well-balanced properties must emerge for fuel cell applications in the near future (Tomoya, Kazuya & Mitsuru, 2009).

CONCLUSION

As our demand for electrical power grows, it becomes increasingly urgent to find new ways of meeting it both responsibly and safely. Energy is one of the driving factors that determine the overall sustainability of humanity. PEM fuel cell is regarded as an alternative because it gives near zero emission, high efficiency and near zero noise pollution. Numerous studies of potential catalyst supports have been performed, mainly focusing on carbon materials. Although numerous effective electrocatalysts have been studied, there were only a few that exhibited good performance in acidic media. To overcome this limitation, recent studies have focused more on discovering other modifications of carbon material supports to grant enhanced capability and stability in acid conditions, better thermal stability, strong interactions with the catalyst that can enhance the electrocatalytic activity and extended fuel cell lifetimes. Despite all the notable achievement in the development of the PEM fuel cell and it excellent feature as a power source, commercialization is still a major concern. It faces serious challenges regarding cost, durability, and performance. The success of fuel cell industry depends greatly on the membrane technology. Since the biggest market of fuel cells will lie in the electric vehicles, most of the effort on the development of proton exchange membranes has been and will be devoted to high-temperature and low-humidity operable membranes. Composite membranes are suitable for use in fuel cells for the automobile industry due to their wider operating temperature range above 95°C.This indicates that investment in such research is essential to achieve reductions in cost.

REFERENCES

Adnan, O., Mustafa, E., Yilser, D., Ozgur, C., & Feridun, H. (2017). Evaluation of sulfonated polysulfone/zirconium hydrogen phosphate composite membranes for direct methanol fuel cells. *Electrochimica Acta*, *256*, 196–210. doi:10.1016/j.electacta.2017.10.002

Baizeng, F., Min, S. K., Jung, H. K., Min, Y. S., Yan, J. W., Haijiang, W., ... Jong, S. Y. (2011). High Pt loading on functionalized multiwall carbon nanotubes as a highly efficient cathode electrocatalyst for proton exchange membrane fuel cells. *Journal of Materials Chemistry*, *21*(22), 8066–8073. doi:10.1039/c1jm10847f

Brian, C. (2001). An introduction to fuel cells and hydrogen technology. Vancouver, Canada: Academic Press.

Chalkova, E., Wang, C., Komarneni, S., Lee, J., Fedkin, M., & Lvov, S. (2009). Composite proton conductive membranes for elevated temperature and reduced relative humidity PEMFC. *Electrochemical society. Transactions*, *25*, 1141–1150.

Chen, Z., Holmberg, B., Li, W., Wang, X., Deng, W., Munoz, R., & Yan, Y. (2006). Nafion/zeolite nanocomposite membrane by in situ crystallization for a direct methanol fuel cell. *Chemistry of Materials*, *18*(24), 5669–5675. doi:10.1021/cm060841q

Daping, H., Chao, Z., Cheng, X., Niancai, C., Huaiguang, L., Shichun, M., & Mu, P. (2011). Polyaniline-Functionalized Carbon Nanotube Supported Platinum Catalysts. *Langmuir*, *27*(9), 5582–5588. doi:10.1021/la2003589 PMID:21476530

Datsyuk, V., Kalyva, M., Papagelis, K., Parthenios, J., Tasis, D., Siokou, A., ... Galiotis, C. (2008). Chemical oxidation of multiwalled carbon nanotubes. *Carbon*, *46*(6), 833–840. doi:10.1016/j.carbon.2008.02.012

Elamathi, S., Nithyakalyani, G., Sangeetha, D., & Ravichandran, S. (2008). Preparation and evaluation of ionomeric membranes based on sulfonated-poly (styrene_isobutylene_styrene) membranes for proton exchange membrane fuel cells (PEMFC). *Ionics*, *14*(5), 377–385. doi:10.100711581-007-0163-2

Fanco, B. (2005). *PEM Fuel Cells: Theory and Practices, Academic press sustainable world series*. Elsevier Academic Press.

Ghosh, P., Dhole, C. K., Ganguly, S., Banerjee, D., & Kargupta, K. (2018). Phosphosilicate gel-sulfonated poly(ether ether ketone) nanocomposite membrane for polymer electrolyte membrane fuel cell. *Materials Today: Proceedings*, *5*, 2186–2192.

Gutru, R., & Santoshkumar, D. B. (2018). Amino acid functionalized graphene oxide based nanocomposite membrane electrolytes for direct methanol fuel cells. *Journal of Membrane Science*, *551*, 1–11. doi:10.1016/j.memsci.2018.01.026

Hengge, K., Heinzl, C., Perchthaler, M., Varley, D., Lochner, T., & Scheu, C. (2017). Unraveling micro- and nanoscale degradation processes during operation of high-temperature polymer-electrolyte-membrane fuel cells. *Journal of Power Sources*, *364*, 437–448. doi:10.1016/j.jpowsour.2017.08.042

Imran, R. J., Arockiados, T., Rajalakshmi, N., & Ramaprabhu, S. (2010). Nanostructured Pt dispersed on graphene multiwalled carbon nanotube hybrid nanomaterials as electrocatalyst for PEMFC. *Journal of the Electrochemical Society*, *157*(6), B874–B879. doi:10.1149/1.3374353

Jiajun, W., Geping, Y., Yuyan, S., Zhenbo, W., & Yunzhi, G. (2008). Investigation of further improvement of platinum catalyst durability with highly graphitized carbon nanotubes support. *The Journal of Physical Chemistry C*, *112*(15), 5784–5789. doi:10.1021/jp800186p

Jing, L., Guoxiao, X., Xingying, L., Jie, X., Zhao, L., & Weiwei, C. (2018). Effect of nano-size of functionalized silica on overall performance of swelling-filling modified Nafion membrane for direct methanol fuel cell application. *Applied Energy*, *213*, 408–414. doi:10.1016/j.apenergy.2018.01.052

Jurgen, G., & Ludwig, J. (2015). Applications of Fuel Cell Technology: Status and Perspectives. *The Electrochemical Society Interface*, 39–43.

Mei, X. W., Fan, X., Hong, F. S., Qi, L., Kateryna, A., Eric, A. S., & Jian, X. (2011). Nanoscale graphite-supported Pt catalysts for oxygen reduction reactions in fuel cells. *Electrochimica Acta*, *56*(5), 2566–2573. doi:10.1016/j.electacta.2010.11.019

Min, C. K., Gyeong, S. H., & Rodney, S. R. (2009). Epoxide reduction with hydrazine on graphene: A first principles study. *The Journal of Chemical Physics*, *131*(6), 064704–064709. doi:10.1063/1.3197007 PMID:19691400

Minghui, W., Ge, L., Xuejun, C., Yuxiang, F., Huan, Z., Gaoxu, W., ... Yunqing, L. (2018). Self-crosslinked organic-inorganic nanocomposite membranes with good methanol barrier for direct methanol fuel cell applications. *Solid State Ionics*, *315*, 71–76. doi:10.1016/j.ssi.2017.12.001

Nattinee, K., Kodchakorn, V., Thirayu, C., Sairung, C., Katesara, P., Anuvat, S., & Karnthidaporn, W. (2018). Preparation of sulfonated zeolite ZSM-5/sulfonated polysulfone composite membranes as PEM for direct methanol fuel cell application. *Solid State Ionics*, *319*, 278–284. doi:10.1016/j.ssi.2018.02.019

Padmavathi, R., Sandhya, D. A., Saranya, N., Gnanasundaram, P., & Sangeetha, D. (2015). Synthesis and characterization of Pt supported on multiwalled carbon nanotubes for improved catalytic performance in fuel cell Applications. *Journal of Porous Materials*, *22*(3), 647–658. doi:10.100710934-015-9937-5

Padmavathi, R., & Sangeetha, D. (2013). Design of novel SPEEK-based proton exchange membranes by self-assembly method for fuel cells. *Ionics*, *19*(10), 1423–1436. doi:10.100711581-013-0867-4

Padmavathi, R., & Sangeetha, D. (2013). Synthesis and characterization of electrospun carbon nanofibersupported Pt catalyst for fuel cells. *Electrochimica Acta*, *112*, 1–13. doi:10.1016/j.electacta.2013.08.078

Peiying, Z., Ming, S., Shuhua, X., & Dong, Z. (2011). Experimental study on the reducibility of graphene oxide by hydrazine hydrate. *Physica B, Condensed Matter*, *406*(3), 498–502. doi:10.1016/j.physb.2010.11.022

Ralph, T. R. (1997). Proton Exchange Membrane Fuel Cells: Progress in cost reduction of the key components. *Platinum Metals Review*, *41*(3), 102–113.

Ravi, K., Mohamed, M., & Keith, S. (2014). Sulfonated polyether ether ketone-sulfonated graphene oxide composite membranes for polymer electrolyte fuel cells. *Royal Society of Chemistry Advances*, *4*, 617–621.

Sehkyu, P., Yuyan, S., Haiying, W., Peter, C. R., Vilayanur, V. V., Silas, A. T., ... Yong, W. (2011). Design of graphene sheets-supported Pt catalyst layer in PEM fuel cells. *Electrochemistry Communications*, *13*(3), 258–261. doi:10.1016/j.elecom.2010.12.028

Tomoya, H., Kazuya, M., & Mitsuru, U. (2009). Sulfonated aromatic hydrocarbon polymers as proton exchange membranes for fuel cells. *Polymer*, *50*(23), 5341–5357. doi:10.1016/j.polymer.2009.09.001

Vijay, R. (2006, Spring). Fuel Cells. *The Electrochemical Society Interface.*

Vladimir, S. B. (2012). Fuel cells Problems and solutions (2nd ed.). Moscow, Russia: Electrochemical Society.

Yaxiang, L., Lianqin, W., Kathrin, P., Mo, Q., Maria, M. T., John, V., & Qiong, C. (2017). Halloysite-derived nitrogen doped carbon electrocatalysts for anion exchange membrane fuel cells. *Journal of Power Sources*, *372*, 82–90. doi:10.1016/j.jpowsour.2017.10.037

Ying, O., Wen, C. T., Shin, C. J., Fu, S. C., Jie, W., Hai, L., ... Chunli, G. (2018). Novel composite polymer electrolyte membrane using solid superacidic sulfated zirconia - Functionalized carbon nanotubes modified chitosan. *Electrochimica Acta*, *264*, 251–259. doi:10.1016/j.electacta.2018.01.131

Yu, C. C., & Jhao, R. C. (2011). Effects of surface chemical states of carbon nanotubes supported Pt nanoparticles on performance of proton exchange membrane fuel cells. *International Journal of Hydrogen Energy*, *36*(11), 6826–6831. doi:10.1016/j.ijhydene.2011.02.114

KEY TERMS AND DEFINITIONS

Bipolar Plate: A bipolar plate is a multi-functional component within the PEM fuel cell stack. It connects and separates the individual fuel cells in series to form a fuel cell stack.

Carbon Corrosion: Carbon corrosion is a major concern for long-term durability of PEM fuel cells. Carbon corrosion can also be active during operation if there are localized regions that are temporarily starved of hydrogen, e.g., flooding of the anode catalyst.

Catalyst Poison: If even minute quantities of carbon monoxide are present in that gas, it can poison the platinum catalysts that are key to driving the fuel cells.

Electrocatalysts: Electrocatalysts are a specific form of catalysts that function at electrode surfaces or may be the electrode surface itself.

End Plate: End plates or otherwise called clamp plates are needed at either end of the stack to apply pressure on the cells to maintain the structure as well as to prevent the gases from escaping from between the plates.

Functionalization: It is the process of adding new functions, features, capabilities, or properties to a material by changing the surface chemistry of the material.

Hydrophobic: The hydrophobic effect is the observed tendency of nonpolar substances to aggregate in an aqueous solution and exclude water molecules.

Ionomer: After sulfonation the polymer will have SO_3H group on its backbone.

Membrane Electrode Assembly: It is an assembled stack of proton exchange membranes (PEM) or alkali anion exchange membrane (AAEM), catalyst and flat plate electrode used in fuel cells and electrolyzers.

Stack: A single fuel cell consists of a membrane electrode assembly (MEA) and two flow-field plates delivering about 0.5 and 1V voltage (too low for most applications). Just like batteries, individual cells are stacked to achieve a higher voltage and power. This assembly of cells is called a fuel cell stack, or just a stack.

Vulcan XC-72R: It is a trade name for carbon powder used as supporting material in the anode and cathode electrodes of PEMFC, DMFC, etc.

Zwitter Ion: A zwitter ion is a molecule with two or more functional groups, of which at least one has a positive and one has a negative electrical charge and the net charge of the entire molecule is zero.

APPENDIX 1: ACRONYMS

AFC: Alkaline Fuel Cells
CB: Carbon Black
CHP: Combined Heat And Power
CNT: Carbon Nano Tube
CNF: Carbon Nano Fiber
CO: Carbon Monoxide
CVD: Chemical Vapor Deposition
DMFC: Direct Methanol Fuel Cells
ECSA: Electrochemical Surface Area
GE: General Electric
GO: Graphene Oxide
HFC: Hydrogen Fuel Cells
LPG: Liquefied Petroleum Gas
ORR: Oxygen Reduction Reaction
PAFC: Phosphoric Acid Fuel Cells
PEMFC: Proton Exchange Membrane Fuel Cells
YSZ: Yttria-Stabilized Zirconia

APPENDIX 2: ANALYSIS AND DISCUSSION QUESTIONS

1. What Are the Latest Developments in Fuel Cell Catalyst Technology?

   Zero-emissions energy enthusiasts remember the early demonstrations of hydrogen fuel cell technology. Hydrogen fuel cell solutions are in use today, in real world applications ranging from industrial-scale backup power systems to mobile power for electric trains, buses, material handling systems and heavy-duty transport trucks. Until recent years, widespread adoption has been constrained by two hurdles: i) manufacturing costs & ii) durability of some fuel cell designs.

2. What markets are hydrogen fuel cells viable in today?

   Several major automakers offer fuel cell vehicles on a limited basis and fuel cell buses are in service in several states. There are thousands of fuel cell-powered forklifts working around the clock in America's warehouses and factories, and fuel cells are powering some data centers, communications networks, retails sites, and municipal facilities across the country.

3. Why hydrogen could be the future of green energy?

From the way the power and heat our homes to the fuel we use in our vehicles, the energy sources on which we depend release harmful carbon dioxide into the atmosphere. Given the scale of the decarbonisation challenge, we need to use many technological solutions in tandem. But one element has so far been forgotten: hydrogen. Hydrogen has the potential to decarbonise electricity generation, transport and heat. That's because when produced by electrolysis - using electricity to split water (H_2O) into hydrogen and oxygen - hydrogen does not produce any pollutants.

Chapter 12
Smart Grid:
An Overview

Maheswari M.
Nalla Malla Reddy Engineering College, India

Gunasekharan S.
Lords Institute of Engineering and Technology, India

EXECUTIVE SUMMARY

The electric grid that has the tendency to communicate two-way and can sense various parameters in the transmission line is termed as smart grid. This chapter deals about the overview of smart grid evolution, characteristics, and operation. There are various benefits in smart grid like improvement in efficiency, adaptive, self-healing, and optimized than conventional grid. The smart grid composition is complex and defined based on standards adaption, technical components perspective, technical perspective, and conceptual reference model perspective. In the architecture of smart grid, the role of advanced metering infrastructure (AMI) plays a vital role to sense, measure, record, and communicate the data from load centre to data centre. AMI consists of smart meter, communication network, data reception, and management system. This chapter also covers the IEEE and IEC standards defined for smart grid operation. It also envisages the barriers in the implementation of smart grids.

INTRODUCTION TO THE EXISTING GRID

The word grid refers to the electric grid which consists of network of transmission lines, substations and transformers. The current electric grid was built in the 1890s and improved according to the development of the technology. Today, it consists

DOI: 10.4018/978-1-5225-8559-6.ch012

of more than 9,200 electric generating units with more than one million megawatts generating capacity is connected to more than 3,00,000 miles of transmission lines. The expected global demand of energy is about 25 percent by 2040. Hence the performance of the grid has to be improved to meet the future demand. At the same, few factors such as

- Increasing demand of electricity
- Supply shortfalls of electricity
- Need of reducing losses
- Peak demand management
- Integration of renewable energy generation systems
- Solution to global warming
- Better customer satisfaction
- Ageing assets and lack of circuit capacity
- Security of supply

are affecting the performance of the existing grid. So it is necessary to move forward in terms of digital and computerized equipment and technologies in grids. It can automate and manage the complexity and needs of electricity in this era.

INTRODUCTION TO SMART GRID

The term Smart grid means the digital technology that allows for two-way communication between the utility and the customers and also the sensing along the transmission line. As per US department of Energy (DOE) the smart grid is defined as "Grid 2030 envisions a fully automated power delivery network that monitors and controls every customer and node, ensuring two-way flow of information and electricity between the power plant and the appliance and all points in between". Similarly, International Electro technical Commission (IEC) defined "The smart grid is a developing network of transmission lines, equipment, controls and new technologies working together to respond immediately to our 21st Century demand for electricity".

The smart grid characteristics are identified based on (i) functionality approach and (ii) broad approach as discussed by Gharavi et al., in the year 2011. The functionality approach of smart grid has the following seven important characteristics,

- Optimization of asset utilization and operating efficiency
- Accommodate generation and storage options
- Provide power quality for the range of needs in a digital economy.
- Respond to system disturbance in a self-healing manner.

- Operate robustly for physical and cyber-attacks and natural disasters.
- Enable active participation by consumers.
- Enable new products, services and markets.

Similarly, the characteristics of smart grid based on broad approach as said by Miller and Hamilton in the year 2009 and 2010 respectively are given below,

- *Adaptive and Self – healing:* The smart grid is self-healing means it has the capability of repair or removes faulty equipment automatically from service before it gets damaged. Also it has the ability to reconfigure the system such a way to ensure the continuity of the energy to all customers.
- *Flexible:* It has the ability to interconnect distributed generation and energy storage devices at any place of the system at any instant.
- *Predictive:* Smart grid has the ability to apply the operational data for the maintenance and can identify potential outages before it occurs. It is achieved through weather learning, machine learning and stochastic analysis. This will help to reconfigure the system before happening of the next worst event.
- *Integrated:* It is very much useful in terms of control functions and real time communications.
- *Interactive:* It should have the capability to provide information about the status of the system to both the operators and the customers. It will lead to an active participation of operators and customers to participate actively in optimal management of contingencies.
- *Optimized:* This is achieved by understanding the status of every major component in real time and controlling it. It provides alternate routing paths for autonomous optimization of the flow of electricity throughout the system to maximize availability, reliability, economic performance and efficiency.
- *Secure:* Two-way communication is the fundamental and basic requirement of the smart grid, hence the physical and cyber – security of all critical assets is essential to secure from attack and naturally occurring disruptions.

BENEFITS OF SMART GRID

The benefits obtained from smart grid includes technical, environmental and electricity marketing benefits,

1. Technical Benefits:

   Full operation of smart grid would result in the following technical benefits like,

Energy Efficiency Improvement: It is obtained by peak shaving, reduction in loss, automated energy system operation and implementation of AMI.
Grid reliability Improvement: It is achieved by reducing the duration and frequency of power interruptions.
Operational Efficiency Improvement: It is achieved through implementing automation, active control and management services in distribution grids. Also encouraging the customers to use smart appliances and home automation.
Security and safety Improvement: Security of power system can be improved by using sensors, automated operations and proper coordination with the transmission and distribution operation. Similar way safety can be improved by reducing the vulnerability of the grid to unexpected hazards.
Quality of Supply: Supply quality is maintained by means of maintaining voltage magnitude within the permissible limit through sensors, two-way communication embedded in smart grid technologies.
Improved connection and access of the grid: The access of the grid and connection has to be improved for distributed energy sources (DERs), like renewable energy sources and plug in electric vehicles.

2. Environmental Benefits:
The smart grid deployment offers reduction in carbon emissions and climate change benefits to the environment. The integration of renewable, distributed generation, plug in electric vehicles and reduction in grid losses provides reduction in carbon emissions. This in turn will improve the climatic condition also.
3. Electricity Marketing Benefits:
The price for electricity can be reduced than conventional grid through dynamic interaction between the consumer and the supplier compared to the conventional grid in smart grid environment. The information about the electricity price for different suppliers will be made available and based on that the consumers can choose the least electricity price suppliers. This will create healthy electricity market competition which will be beneficial to the consumers.

SMART GRID INITIATIVES/PROJECTS

The smart grid concept has recently promoted in many countries leads to initiation of several projects/initiatives to realize this concept. Such projects planned /executed in the USA, Europe and China are listed below,

USA: EPRI's Intelli Grid SM

In USA, EPRI's Intelli Grid SM initiative and DOE's Grid wise vision are the two important examples of projects carried out in Smart Grid. The EPRI has initiated a research program called "Intelli Grid SM" which is aimed to establish the best way to implement smart grid and incorporating into the operation of individual electrical utilities. The IEC has recognized EPRI's Intelli grid methodology as a standard in the year 2007. The intelli Grid initiative addresses the following key industrial issues,

- Understanding the function of smart grid for a particular utility
- Developing an architecture that enables interoperable systems and components.
- Conducting technology assessment for the potential components that can make up a smart grid.

The Grid Wise vision aimed to work in the assumption of conventional power system planning and operation revolutionized by the information technology such as happened in business, education and entertainment. The vision Grid is expected to

- Permit electric devices, enterprise systems and their owners to communicate and involve in the grid operation.
- Connect all automation components such as buildings, HVAC systems, distribution, transmission and bulk generation of the electric system for intelligent and interoperability.
- Use of new and extensible technologies which are compatible with the existing infrastructure and address the issues of universal grid access, grid communication and information exchange.

European Smart Grid Efforts

Europe smart grid project expressed as a series of three documents. The first document highlights the need for the future European electricity networks entitled as "Vision and strategy for Europe's Electricity Networks of the Future". "Strategic Research Agenda" is the second document which emphasis on the views of stakeholders on the research priorities necessary to deliver these networks. The deployment of new network technologies and the delivery of smart grid is concluded in "The smart Grid strategic deployment document for Europe's Electricity networks of the future" as third document.

China's Smart Grid Efforts

The recent hot topic in China is Smart Grid and it is viewed as a sophisticated control system to efficiently manage resources and consumptions. It also emphasis (i) grid reliability (ii) supply capacity and (iii) reduce grid losses. The source of China's smart grid project is State Grid Corporation of China (SGCC), which is the largest transmission company in China. SGCC has undertaken the smart grid project and invested a total of US$601 billion into a nationwide transmission network out of which US$101 billion is decided to develop smart grid technology. Few specific Smart Grid projects in China are

- Long-distance, large-capacity, low-loss **UHV core technology**, and localization of power equipment
- ± 800 kV UHV DC converter station
- Power system digital real-time simulation device
- Electric vehicle charging stations

SMART GRID VS. CONVENTIONAL ELECTRICAL NETWORKS

Conventional Electric Networks

A conventional electric network consists of generating stations where the electric power is generated and passed to the transmission system which is then transferred to distribution network to feed the loads as discussed by Salman in the year 2017. Figure 1 shows the vertical structure of the conventional power system which is vertical in nature. The conventional system has the following characteristics,

- It has vertical structure
- Power flow is unidirectional
- The price of electricity is dictated by the utility and the customer has no option to buy their electricity

Smart Grid Concept

In recent years, the electrical networks are subjected to the changes which are unable to meet the future challenges. It will make the need of urgent requirement of modernizing the electrical networks which in turn the reason for the birth of smart grid concept. The smart grid concept is mainly developed to solve issues related to growing energy consumption, integration of distributed generation, energy efficiency,

Figure 1. Vertical structure of conventional power system

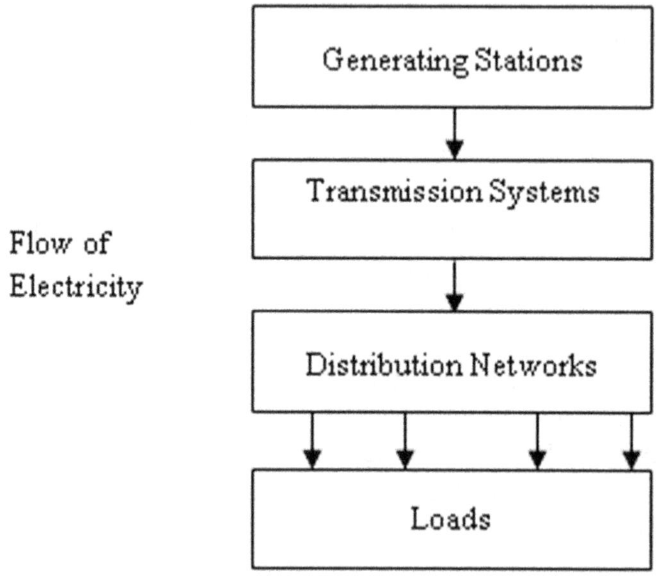

power supply reliability and power quality. The aspects that led to the development of smart grid concept are,

Aging of conventional electrical networks-In many countries, the Conventional electrical networks were designed in 1950s and 1960s. This is well before the microprocessor era where huge advancements in communication and automatic control happened. These advancements and new applications will affect the management and operation of electrical networks. Hence it is recognized that the electrical networks are old and out-of-date and need to be modernized.

Political and environmental factors–The oil price has been raised steeply due to the conflict between Arab – Israeli in 1973. Due to this many nations decided to harvest their energy requirement through renewable energy sources. Also the conventional generating stations are omitting 25% of the greenhouse gas emissions. The integration of renewable energy sources, made the passive electrical network into active and power flow in bidirectional. This will affect the operation of protection relays and voltage control devices.

Liberalization of electricity market – The reliability and quality of electric supply are necessary to liberalize the electricity market. Also the efficiency of transmission and distribution system needs to be improved to limit the grid tariffs.

Motivation and inclusion of customers as players to support the grid- Customers are perceived as passive users in conventional system where as it can be changed into active participants.

Comparison Between Smart Grid and Conventional Electrical Networks

There are several expectations in smart grid compared to conventional grid and the fundamental differences are listed in Table 1.

The principal characteristics of smart grid as defined by the USA are,

- Efficient transmission of electricity
- Fast restoration of electricity after disturbance
- Reduction in operation and management costs for utilities and intern lower power costs for consumers.
- Reduction in peak demand will help in low electricity rates.
- Increase in integration of renewable energy systems
- Improvement in security

SMART GRID ARCHITECTURE

In general, smart grid consists of electrical power system, communication and information system, automation, distributed control system, intelligent protection and marketing systemas discussed by Salman in the year 2017. The composition of smart grid is complex and many efforts made to define it. Based on the efforts, the smart grid compositions are,

Table 1. Comparison between smart grid and conventional electrical networks

Parameters	Existing Grid	Smart Grid
Metering	Electromechanical	Digital
Communication	One way Communication	Two way communication
Customer Interaction	Limited	Extensive
Generation	Generalized Generation	Distributed Generation
Use of Sensors	Limited sensors used	Sensors will be used throughout
Monitoring System	Manual Monitoring	Self-Monitoring
Control	Limited control	Pervasive control
Reliability	Cascaded Outages	Automated, Proactive protection and Prevent outages
Restoration	Manual restoration	Self-Healing

- Standards Adaption
- Technical Components' perspective
- Technical Perspective
- Conceptual reference model perspective

The following sections will discuss about the composition of the smart grid.

Composition of Smart Grid Based on Standards Adaptation

To ensure the interoperability and security, it is necessary to develop and adapt open standards for the smart grid. In this perspective, the smart grid is assumed to comprise of (i) "Utility Electric System" consists of many individual systems, including generation, transmission, distribution and customer systems within the utility (ii) Another entity system consists of the many customer systems, service provider systems and resource supplier systems (iii) An Overall macro system like wide area control system and RTO/ISO systems. The interfacing of utility systems, other entity systems and macro system together results in the smart grid as shown in Figure 2. Inter system interfaces and intra system interfaces are two different types of interfaces are used to link the above said systems. The intersystem interfaces are the interfaces between the boundary of the utility's transmission, distribution and customer systems. For example, this may be an interface between a customer device and a meter or between the RTO system and the Utility system. In the similar way, the interfaces are within the boundary of the utility systems of generation, transmission, distribution and customer systems then it is known as intra system interfaces. It may be the interface between the utility fault detector and distribution management system as discussed by Salman in the year 2017.

Composition of Smart Grid Based on Technical Components' Perspective

As per technical components' perspective of smart grid, it is viewed as a highly complex combination and integration of multiple digital and non-digital technologies and systems as shown in Figure 3. The major components are

- New and advanced grid components
- Smart Devices and smart metering
- Integrated communication technologies
- Software programs for decision support and human interfaces and
- Advanced control systems

Figure 2. Composition of smart grid based on standards adaptation

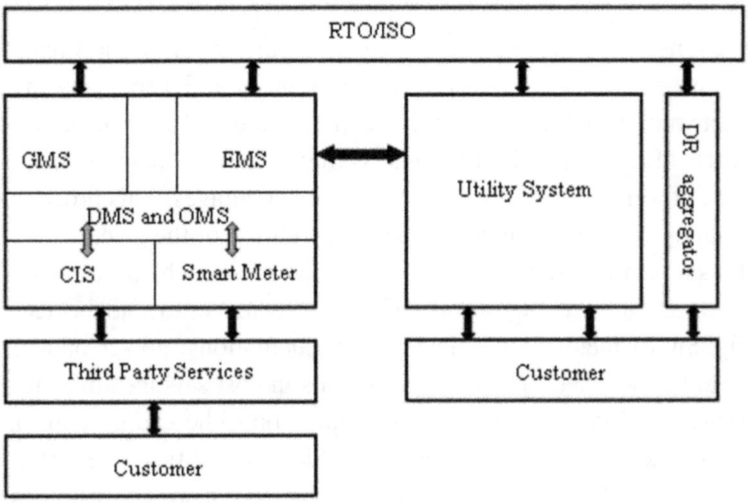

Figure 3. Smart Grid Major Components

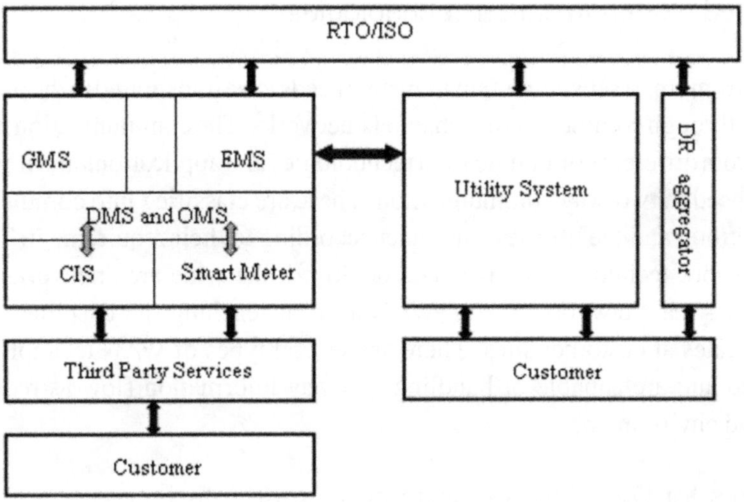

New and Advanced Grid Components

It includes the advanced storage devices, super conductors, advanced power electronic devices and distributed energy generation. These advanced components would improve the efficiency of the supply and increase the availability and the reliability of power.

Smart Devices and Smart Metering

Sensors and sensor networks are the main parts of smart devices and smart metering. Generally sensors are located at different parts of the grid. Sensors play an important role in remote monitoring and demand side management. As the sensors are located at all the parts of the grid, it is able to measure/monitor temperatures, operation of grid devices, power quality disturbances and detect outages. This process helps the control centre to receive accurate and actual condition of the grid.

Smart meters are fixed at the customer locations which provide real-time determination and information storage of energy. It enables to read the data remotely and locally. Smart meters can detect power fluctuations, power outages, permit the meters to be switched off and it will results in cost saving. Smart meters help the electricity providers to have a clear picture about the energy consumption of different consumers at different points. It enables the utilities to establish demand side management and develop new pricing mechanisms also. These processes make the customers more interactive with the suppliers and make them to concentrate on energy consumption habits.

Integrated Communication Technologies

The data from smart sensors and smart meters needs to be transmitted to the processing locations through communication channels/networks. The communication networks formed from different communication technologies and applications and it should be of high speed in two-way communication. These are classified into communication services group and the utilities can select according to their requirement. There are lot of network technologies have been deployed like wide area networks (WAN) and Local Area Networks (LAN). WAN aims at reaching the customer whereas LAN operates at customer sites. There are several types of WAN technologies are developed and are capable of handling two-way information flow as required by smart grid environment.

Programs for Decision Support and Human Interfaces

It is considered as the major component of the smart grid to handle huge volume of data generated under the smart grid environment. There are two main tasks to be performed by these components such as, (i) properly handles the integration and management of the generated data and (ii) present the data to grid operators in a user friendly manner to support their decisions. These should have the capability of handling the data from the grid about the status of grid, power quality, instabilities and outages. It should capable of handling the data from geographical information system about geographic, spatial and location of sources.

Advanced Control Systems

To monitor and control the major components of the smart grid, advanced control systems are essential. It includes, distributed intelligent control systems, substation automation, distributed automation and integration systems.

Composition of Smart Grid Based on Technical Perspective

The smart grid should consist of the following major systems based on technical perspective,

- Smart Infrastructure System
- Smart Management and Control system
- Smart Protection System

Smart Infrastructure System

Smart infrastructure system consists of three subsystems namely electrical subsystem, information subsystem and communication infrastructure subsystem. The function of electrical subsystem is to generate the electricity, transmit it and distribute to the customers. The support to information metering, monitoring and management systems comes under the function of smart information subsystem. The communication and information transmission between various subsystems, devices and application of smart grid are managed by communication infrastructure system. The smart infrastructure system has to support two-way flow of electricity and information. In smart grid, the electrical energy may be generated at the customer side through renewable energy sources and it may be supplied to the grid.

Smart Management and Control System

The advanced management and control services are provided by the smart management and control system and it aims to improve energy efficiency, ensure reliability of supply, greenhouse gas emission control, and reduction in operation cost and maximize the profit.

Smart Protection System

The smart protection system provides advanced grid reliability analysis, failure protection, and security and privacy protection services. It protects the grid from unintentional faults due to users, errors, equipment failures and natural disasters. Apart from this, it protects the grid from planned cyber-attacks also.

Composition of Smart Grid Based on Conceptual Reference Model Perspective

The National Institute of Standard and Technology (NIST) in the USA proposed the smart grid conceptual reference model and it provides a high level, overall view of the smart grid. This model is a useful tool for identifying actors and possible communication paths between various players and various interfaces in the smart grid. According to this model smart grid is divided into several domains as given in Table 2.The graphical representation of the smart grid conceptual model is shown in Figure 4 and is used to analyze, define solution and implementation. The conceptual model is descriptive and not prescriptive. It will help to understand the smart grid operational details and complexities.

Figure 4. Conceptual model of smart grid

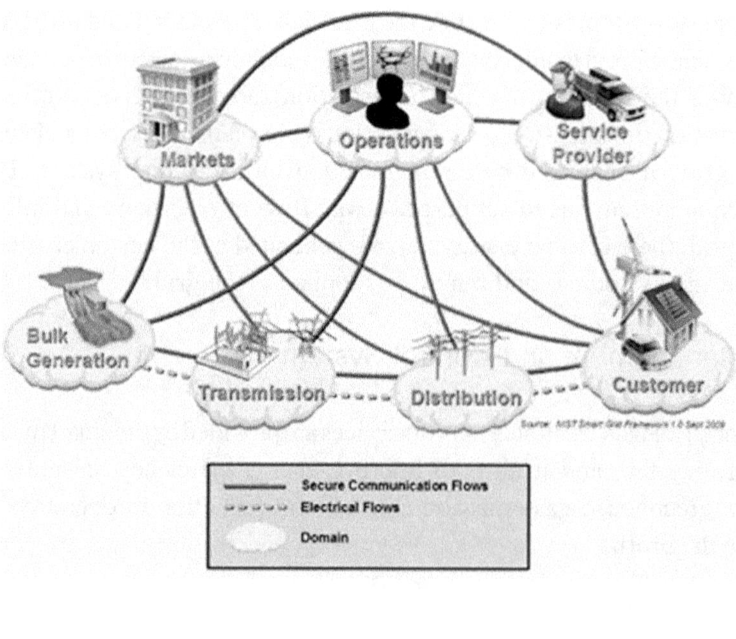

Table 2. Different domain of smart grid

S.No.	Domain	Actors in the domain
1	Customers	The end-users of electricity. May also generate, store, and manage the use of energy. Three customer types are considered, each with its own domain: residential, commercial, and industrial.
2	Markets	The operators and participants in electricity markets.
3	Service Providers	The organizations providing services to electrical customers and utilities.
4	Operations	The managers who control the flow of electricity.
5	Generation	Generators of electricity may also store energy for later distribution. This domain includes traditional generation sources (traditionally referred to as generation) and distributed energy resources (DER). At a logical level, ''generation'' includes coal, nuclear, and large scale hydro generation usually attached to transmission. DER (at a logical level) is associated with customer- and distribution domain- provided generation and storage, and with service provider- aggregated energy resources.
6	Transmission	The carriers of bulk electricity over long distances may also store and generate electricity.
7	Distribution	The distributors of electricity to and from customers may also store and generate electricity.

In the above Table, each domain comprises smart grid actors and applications. A device, a system or a program that makes decisions and exchange information necessary for performing applications is called as an actor such as smart meters, solar generators and control systems. Each applicationis performed by one or more actors within a domain. The actors in the same domain will have the same objective and they often communicate with the actors in the other domain for smooth operation of the smart grid.

Basically, the conceptual model is a legal and regulatory framework which gives polices and requirements that apply to various actors and applications and to their interactions. They have to ensure that the reliability of the supply, security, safety, fair and privacy are met. The objective of conceptual model development is to use it as useful and effective tool for regulators in all levels to achieve the public policy goals, maintain business objectives, motivate investments in modernizing conventional electric power infrastructure and building a clean economy.

BASIC COMPONENTS OF SMART GRID AND ITS TECHNICAL INFRASTRUCTURE

In the previous section, the different attempts have been discussed about the composition of the smart grid. In this section, it is essential to identify the basic components of the smart grid. The basic components of smart grid are,

- ***Generating Stations:*** It include a combination of bulk generation plants, renewable energy plants like solar, wind and hydro.
- ***Transmission System:*** It includes transmission lines and transmission substations.
- ***Distribution System:*** It consists of distribution substation and others
- ***Consumer Load network***
- ***Information and Communication Technologies:*** It is required to have communication technologies to transmit data safely and reliable. It consists of communication networks, media and protocols. The main duty of communication technologies is to communicate between systems, devices and applications through proper connectivity.
- ***Advanced Metering Infrastructure (AMI)***: AMI is defined by Bian et al., in the year 2014 as " a measurement and collection system that includes smart meters, communication networks and data management systems that make the information available to the service provider". It is the key component of the smart grid and helps to implement residential demand response and mechanism for dynamic pricing.
- ***Intelligent Electronic Devices (IEDs)***
- Computer control, control devices and distributed control systems
- ***Smart Grid*** - Enabled Home Appliances: The home appliances such as smart grid enabled washing machine, are enabled by the smart grid for electricity price and other demand signals. It will help the customers to save the energy. Generally these devices can plug-in and register with their respective service providers through a web portal or phone call. This arrangement will be much helpful to the customers who are not having any expertise, not ready to spend money and time. It is called as plug and play and auto configuring by Salman et al., in the year 1994.
- ***Smart Interfacing Devices:*** It is used to interface between the smart grid and the customers and is visible to the customers. Interfacing devices consists of two elements such as a meter and an Energy Services Interface (ESI). The meter does (i)measure, record and communicate energy usage (ii) Communicate information for outage management (iii) enable automated provisioning and maintenance functions. In case if the customers have

distributed generations then the meters should record the power flow from the premises to the grid also. The ESI is the information management gateway through which the interaction happens between the customer and the energy service provider. The functions of ESI are (i) demand response signaling i.e., the communication about the price of energy at peak load period (ii) provide information about the energy usage.

- ***Internet Protocol (IP):*** IP networks are the basic element for the smart grid information system networks. It acts as a bridge between the communication medium and the application.
- ***Standard Models and Protocols:*** To ensure interconnection and interoperability, standard models and protocols are essential as discussed by Miller et al., in the year 2008. The important process in the smart grid deployment is to identify the existing standards and protocol documents or develop a new one. For safe and secure smart grid operation, it requires many standards. Hence the sincere efforts are required to select the standards.
- ***Software and programs*** for decision making and interfacing are essential.

ADVANCED METERING INFRASTRUCTURE (AMI)

The milestone of modernization of conventional electrical power system is the development of AMI. It supplies information about the electricity pricing and helps the utilities to achieve load reductions. An AMI is an integration of many technologies to provide intelligent connection between consumers and the system operators. It makes easy to supply the energy to the consumers with proper information with lot of choices which they don't have in the conventional electrical network. Based on the information given by AMI, it facilitates the system operators to improve customer service. AMI provides the link between the grid, consumers and their loads, generation and storage devices and it is the initial step for the modernization of the existing conventional electrical networks.

Main Components of AMI

AMI system consist of many technologies and applications that are integrated together to perform as a single system. The major components of AMI system are shown in Figure 5 and as follows

- Smart Meters
- Communication Network
- Data reception and management system

Figure 5. Major components of AMI

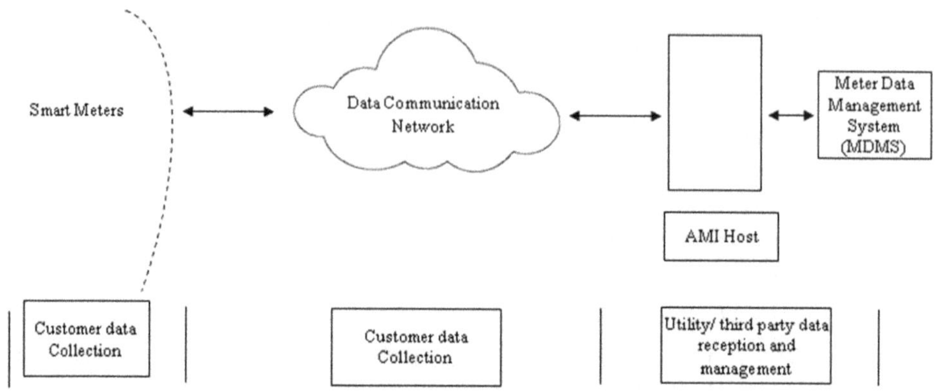

Smart Meters

Smart meters are digital programmable devices which record the consumption of electrical energy in the customer side and communicate to the energy suppliers periodically to monitor and billing purpose. The other responsibilities of smart meters are

- Net metering
- Loss of power and restoration notification
- Pricing based on time
- Turn on/off remotely
- Power quality monitoring
- Prepayment for energy
- Detection of energy theft
- Interactions with other intelligent devices

Smart Meters also can be called as green meters, since it provides energy consumption information to the supplier and emphasis on energy conservation.

Communication Network

The second important component of the AMI is communication network. The objectives of these networks are to continuously provide support for the interaction between the supplier, consumer and the controllable load. It is possible only by offering open bidirectional communication with high security. Due to bidirectional communication utilities can monitor real time consumption by the end-users and also make the active participation of end-user in the system operation.

Concentrator is the key element for enabling the bidirectional communication. It is classified into two types such as local concentrator and backbone concentrator. The local concentrator collects information form smart meters and forward to the backbone network, also distribute commands received from a backbone concentrator to meters. Backbone concentrators collect information from local concentrators and communicate the commands received from the utility side. There is no need of local concentrators when the system has less number of customers. Under such situations the smart meters are configured to communicate directly with the backbone concentrators. The following communication networks are employed in the smart grid,

- Power Line Carrier Communication (PLC)
- Broadband over power lines (BPL)
- Copper or optical fiber communication
- Wireless communication
- Internet
- Combination of all the above

Data Reception and Management System

The data transferred by the smart meter through communication network is received at utility/third party site by the AMI host system and then sent to meter data management system (MDMS) as shown in Figure 5. MDMS plays a major role in AMI and can perform full potential functions of AMI as discussed by Moore in the year 2008. The functions of MDMS are

- Organizing and atomize the data collected from the multiple meters.
- Evaluate the quality of collected data and calculate the errors and gaps exist.
- Delivering the data in the form that suits for billing systems.

Hence in AMI system, smart meters are located at the customer location and MDMS are located at the utility side as shown in Figure 5. So the smart meters and the MDMS communicate with each other through neighborhood area network (NAN) as discussed by Kuzlu and Bian et.al, in the year 2013 and 2014 respectively.

In addition to the above three components, Home area networks (HANs) and operational gateways are also considered as the essential parts of AMI. In smart grid, it is required to have relatively low bandwidth with regular communications.

HAN is used to interface with the costumer portal to connect the smart meters with the controllable loads. HANs role of energy management functions include

- Home displays to aware about the energy and its cost used by the consumer
- Responses to price signals based on consumer entered preferences
- Set points to limit utility
- Automatic control of loads
- Consumer override capability

HAN provides smart interface to the market so it acts as consumer's agent. It is also support new services like security monitoring and it can be implemented in different ways.

AMI COMMUNICATION INFRASTRUCTURE

AMI communication infrastructure consists of two layers namely, lower layer and upper layer as shown in Figure 6. Lower layer connects all smart meters and local concentrators. It is of mesh type network, whose function is to collect energy consumption data from end users and upload it to the backhaul network. And also

Figure 6. AMI communication infrastructure

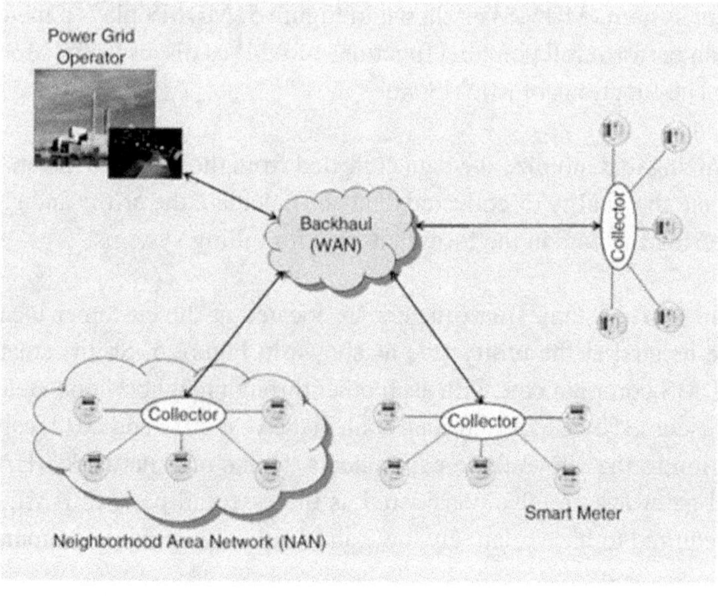

communicates the price/command signals among the smart meters. Upper layer connects backbone concentrators and MDMS. It collects data from local concentrators and also sends commands from control centre. It is required to maintain high security and reliability in these networks.

Communication Technologies Used in AMI

In AMI deployment, either wired or wireless communication technologies can be used. Generally, wired communication technologies used for AMI deployment are power line communication (PLC) and fiber optic. Similarly, the wireless communication technologies used in AMI deployment are cellular, Long-term Evolution (LTE), WiMAX, WLAN, Zig-Bee and RF mesh. All these technologies have their own merits and demerits as discussed below,

Wired Communication Technologies

PLC–It uses existing power lines as a medium to transmit the data and it is a promising technology for smart grid applications because of the availability of power lines. It is well suited for applications like smart metering, home automation and others. It is well suited for rural areas which do not have the communication infrastructure. Low bandwidth, noise, hard to transmit the signal through power distribution devices and security are the demerits of PLC.

Fiber optic communications –Fiber optic communication has high data rate and high immune to noise. Hence it is more suited for backbone communications for various smart grid applications and long distance network. The high installation cost becomes the major drawback of this technology.

Wireless Communication Technologies

Cellular Communication (4G, LTE and WiMAX)–A utility can save the cost and time required for communication network setup if a cellular network infrastructure exists already. Hence an existing cellular network can be used to setup an AMI system to support communication between the concentrator and the control centre. The strong security system becomes its highlight whereas network congestion, unavailability of service under abnormal conditions are disadvantages. WiMAX is the most promising wireless technology among all based on IEEE 802.16 series of standards. It has high data rate of up to 75 Mbps and distance up to 50 km and low communication latency. All these qualities make the WiMAX as the good choice for smart grid deployment. WiMAX consumes more power and installation cost also more becomes its drawbacks.

ZigBee–ZigBee is the most considered industry mesh network to connect sensors, loads and control devices as said by Tomar in the year 2011. Its standards are defined by IEEE 802.15.4. The distance covered by ZigBee is up to 100 m whereas ZigBee pro covers up to 1600 m. ZigBee pro is capable of transmitting data from 20 Kbps to 250 Kbps. Low cost, less power consumption and more secure are the characteristics of ZigBee. It has the drawbacks of severe interference problems and low processing capabilities. Hence it should be implemented with well-organized network and communication traffic and is suited for large scale applications.

WLAN –It is known as Wi-Fi and is a high speed wireless technology. IEEE 802.11 defines the standards and operate at 2.4 GHz, 3.6 GHz and 5 GHz bands. Reliability, security and high speed communication are the merits of WLAN. However, high cost and high power consumptions are the major demerits of these technology. WLAN is also a mesh network like ZigBee.

900 MHz band –It is an unlicensed scientific, industrial and medical RF band. 900 MHz has a range two times that at 2.4 GHz. It is also a type of mesh network and has properties like self-healing, highly reliable and cost effective for wide range. It is most suited to deploy in urban and suburban areas. High bandwidth consumption, less security and lack of interoperability are the drawbacks of this technology.

SMART GRID INTEROPERABILITY STANDARDS

Smart grid is the complete transformation of conventional electrical power system into fully automated with intelligent control devices and communication among them. To accomplish the communication between the various components it is essential to follow international recognized standards for interoperability. Interoperability defined by the IEEE as said by De Blasio and Tom in the year 2008 " the ability of two or more systems or components to exchange information and to use the information that has been exchanged." As per the above definition, the digital based devices should have the capability to exchange the information and also they should understand the information for effective usage.

Benefits of the Interoperability

The benefits of interoperability are

- Easy to smart devices regardless of location & provider.
- Protect privacy
- Facilitate future upgrade and future update
- Provide for backward compatibility
- Ensure the security and enhance the reliability of grid.

Standard Development Organization (SDOs)

The standards for electrical power industry are developed by several SDOs worldwide. In addition to SDOs, the Utility Standard Board (USB) also providing lot of inputs and guidance for the development and implementation of the standards as discussed by Narayana Prasad Padhy and Premalatha Jena in the year 2017. The important SDOs are

- National Institute of Standards and Technology (NIST)
- American National Standards Institute (ANSI)
- International Electro technical Commission (IEC)
- Institute of Electrical and Electronics Engineers (IEEE)
- International Organization for standardization (ISO)
- International Telecommunication Union (ITU)

A brief discussion about the SDOs as well as users group and consortia in smart grid is listed in the following sections. The broad classification of smart grid standards is given in Figure 7.

Interconnection of Distributed Energy Resources (DERs)

IEEE 1547: IEEE 1547 is a series of standards to provide criteria and requirements for the interconnection of distributed generation resources into the power grid, published since 2003.

Figure 7. Broad classifications of smart grid standards

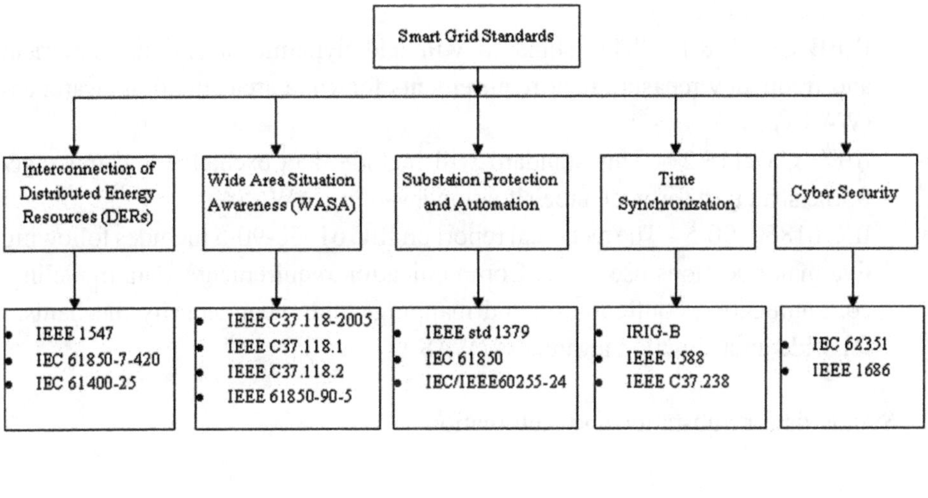

IEEE 1547 has several parts:

- IEEE standard 1547.1 "IEEE standard conformance Test Procedure for Equipment Inter connecting Distributed Resources with Electric Power Systems"
- IEEE standard 1547.2 "IEEE Application Guide for IEEE std 1547, IEEE standard for interconnecting distributed resources with electric power systems".
- IEEE standard 1547.3 "IEEE guide for monitoring, information exchange and control of distributed resources interconnected with electric power systems".
- IEEE standard 1547.4 "IEEE Guide for design, operation, and integration of distributed resource Island systems with electric power systems".
- IEEE standard 1547.5 has not been issued, yet. Its intended scope is to address issues when interconnecting electric power sources of more than 10 MVA to the power grid.
- IEEE standard 1547.6 "IEEE Recommended practice for interconnecting distributed resources with electric power systems distribution secondary networks".
- IEC 61850-7-420 – defines the communication and control interfaces for all DER devices.
- IEC 61400-25 – Provides uniform information exchange for monitoring and control of wind power plants.

Standards for Wide Area Situation Awareness (WASA)

- IEEE C37.118-2005 – It proposed a method of evaluating a PMU measurement and requirements for steady state measurement in 2005 for wide area situation awareness (WASA).
- IEEE C37.118.1 – This standard will add dynamic phasor measurement and frequency measurement requirements for wide area situation awareness (WASA)..
- IEEE C37.118.2 – This standard will include data exchange and message/ frame structure for wide area situation awareness (WASA).
- IEC 61850 -90-5 – The technical report on IEC 61850-90-5 includes following five major sections use cases, Communication requirements, data modeling, communication configuration and mappings and cyber security mechanism for wide area situation awareness (WASA).

Standards for Substation and Automation

- IEEE std 1379 – It recommends practice for data communications between remote terminal units (RTUs) and intelligent Electronic Devices (IEDs) published in 2000.
- IEC 61850 – Defines communication protocol with various system IEDs
- IEC/IEEE 60255-24 – It proposes common format for transient data exchange (COMTRADE). It provides a common format for the data files and exchange medium needed for the interchange of various types of fault, test or simulation data.

Standards for Cyber Security

- IEC 62351 – It proposes data communication security for power system management published in May 2007.
- IEEE 1686 – It states which safeguards, audit mechanism and alarm indications shall be provided by the developer of an IED with regard to all activities associated with access, operation, configuration, firmware revision and data retrieval from an IED.

BARRIERS TO SMART GRID TECHNOLOGIES

There are lots of advancements and advantages are there in smart grid deployment. At the same time there are some barriers to implement these technologies in the developing countries, they are

- It requires huge amount of investment hence the lack of financial resources
- Uncertainty in market
- Lack of regulatory framework
- Low public awareness and engagement
- Lack of innovativeness in the industry
- Lack of infrastructure
- Technology Immaturity
- Integration of the grid with large scale renewable generation
- Need of advanced bidirectional communication systems, cyber security and data privacy.

CONCLUSION

This chapter discussed about the need for smart grid technologies and difference between the conventional and smart grid technologies. It also emphasized on the smart grid architecture based on Standards Adaption, Technical Components' perspective, Technical Perspective and Conceptual reference model perspective. Also it deals about the major components of the smart grid and AMI infrastructure. The standards required for smooth operation of the smart grid are listed as per IEEE, IEC standards.

REFERENCES

ABB. (2009). *Towards a Smarter Grid—ABB's vision for the power system of the future*. Retrieved from http://www02.abb.com/db/db0003/db002698.nsf /0/e30fc9 d5f79d4ae8c12579e2002a4209/%24file/Toward_a_smarter_grid_Julþ09.pdf

Bian, B., Kuzlu, M., Pipattanasomporn, M., & Rahman, S. (2014). *Analysis of communication schemes for advanced metering infrastructure (AMI)*. National Harbor, MD: IEEE PES General Meeting Conference and Exposition. doi:10.1109/ PESGM.2014.6939562

Bian, D., Kuzlu, M., Pipattanasomporn, M., & Rahman, S. (2014). Assessment of communication technologies for a home energy management system. *Proceedings of the IEEE Innovative Smart Grid Technologies (ISGT) PES Conference*, 1–5. 10.1109/ISGT.2014.6816449

DeBlasio, R., & Tom, C. (2008). Standards for the Smart Grid. IEEE Energy 2030 Conference, 1–7.

Gharavi, H., & Ghafurian, R. (2011). Smart grid: The electric energy system of the future. *Proceedings of the IEEE*, *99*(6), 917–921. doi:10.1109/JPROC.2011.2124210

Hamilton, B. A., Pullins, S., Miller, J., Renz, B., & Hanley, M. (2010). Smart grid principal characteristic enables new products, services, and markets. *US DOE/ NETL*. Retrieved from http://www.smartgridinformation. info/pdf/1267_doc_1.pdf

ISA. (2014). *Information sharing environment-information interoperability framework (I2F)* (Version 0.5). Retrieved from http://www. ise.gov/sites/default/ files/FINAL%20-%20ISE_I2F_v0%205.pdf

Kuzlu, M., & Pipattanasomporn, M. (2013). Assessment of communication technologies and network requirements for different smart grid applications. *Proceedings of the IEEE Innovative Smart Grid Technologies (ISGT) PES Conference*, *1*(6), 24–7. 10.1109/ISGT.2013.6497873

Miller, J. (2009). Structuring the smart grid framework: application of complex systems engineering. *US DOE/NETL Modern Grid Team*. Retrieved from http://www.smartgrid.gov/sites/default/files/pdfs/structuring_smart_grid_framework_05-2009.pdf

Miller, J., Pullins, S., & Bossart, S. (2008). (NETL). *The modern grid*. Retrieved from http://wpui.wisc.edu/programs/Institute%20Lunches/Smart_Grid/Presentations/Miller.pdf

Moore, S. (2008). *Key features of meter data management systems*. Retrieved from https://www.itron.com/na/PublishedContent/Key%20MDM%20Features%20Whitepaper_FINAL.pdf

Padhy & Jena. (n.d.). Introduction to Smart Grid. *NPTEL online videos*. Retrieved from https://onlinecourses.nptel.ac.in/noc18_ee42/

Salman, K. (2017). Introduction to the Smart Grid – Concepts, Technologies and Evolution. The Institution of Engineering and Technology.

Salman, S. K., Jiang, F., & Rogers, W. J. S. (1994). Effects of wind power generators on the voltage control of utility's distribution networks. *Wind Engineering*, *18*(4), 181–187.

Tomar, A. (2011). *Introduction to Zig Bee Technology*. Retrieved from https://www.element14.com/community/servlet/JiveServlet/previewBody/37177-102-1-219424/Introduction%20to%20Zigbee%20Technology.pdf

Chapter 13
Alternative Energy Source of Auxiliary Systems of the Pumping and Hydroelectric Power Stations Using Jet Pumps

Vladimir Aleksandrovich Khokhlov
https://orcid.org/0000-0003-0964-5094
National Research University "MPEI", Russia

Aleksandr Vasilevich Khokhlov
Research-and-Production Enterprise "Vodopodyomnik", Uzbekistan

Janna Olegovna Titova
Research-and-Production Enterprise "Vodopodyomnik", Uzbekistan & Tashkent State Technical University, Uzbekistan

Tatiana Aleksandrovna Shestopalova
National Research University "MPEI", Russia

EXECUTIVE SUMMARY

This chapter describes technology that ensures reliable pumping of drainage and sewage water during electromechanical and hydro-mechanical transients from blocks of the hydroelectric power stations and pumping stations. As an alternative source of energy, it is proposed to use the energy of the liquid column of the pressure penstock of the stations, and as an auxiliary, to use jet pumps. Transmission of energy to the suction stream is carried out without direct usage of electrical and mechanical energy. During total shutdown of electric power, reliable evacuation

DOI: 10.4018/978-1-5225-8559-6.ch013

of drainage and seepage water and reduction the influence of electromechanical and hydro-mechanical transients on power equipment and pipelines can be ensured with the use of self-regulating jet pumps over a period of several days; this cannot be accomplished by any other pump. The scientific results of the research are recommended to allow efficient use of water and energy resources and to ensure reliable operation of the power equipment of stations, especially in the events of sudden power outages.

INTRODUCTION

To provide reliable and trouble-free operation of the main equipment and structures of pumping and hydroelectric power stations, auxiliary drainage systems are provided. As an alternative source of energy, it is proposed to use the energy of the liquid column of the pressure penstock of the stations, and as an auxiliary, to use jet pumps. At the same time, taking into account the special significance and importance of the work of the hydroelectric power stations and pumping stations drainage systems, a technology is implemented that ensures reliable pumping of drainage and sewage water from all blocks of the stations and from canals, with efficient hydraulic automation in any combination of hydraulic units' work.

About 1500 pumping stations and about 50 hydroelectric power stations are operated in the Republic of Uzbekistan, many of which are related to the first and second class of hydraulic facilities, i.e. they are the objects of state importance, and any accidents can lead to catastrophic consequences or significant economic damage.

The pumping stations are among the main electric power consumers in the Republic of Uzbekistan. From about 50,7 billion kWh/year the electricity supplied to consumers in Uzbekistan (http://uzbekenergo.uz/ru/activities/indicators, 2019) approximately 7,8 billion kWh/year or 16% are spent for covering the demands of water management organizations containing more than 1500 state irrigative pumping stations (http://agro.uz/ru/information/about_agriculture/420/5856, 2019). All of them are now under the jurisdiction of the Ministry of Water Resources of the Republic of Uzbekistan. At these pumping stations about 5000 pump units are established. Their total power is about 3.8 million kW and the general water delivery is about 7000 m^3/s.

In Uzbekistan 2.4 million hectares out of total 4.3 million hectares of irrigated lands now are irrigated by means of pumping stations. Annually about 55 billion m^3 of irrigating water are pumped. This volume consists of 24 billion m^3 pumped by the first lifting, and 31 billion m^3 by the subsequent liftings of pumping stations stages.

Such a significant consumption of electricity by pumping stations, gives the problem of energy conservation to them a status of special importance and relevance.

It is possible to judge the large scale of machine irrigation by the following examples: the Karshi pumping stations stage – total delivery is 200 m^3/s; height of lifting is 132 m; the area of irrigation is 350 thousand hectares; the Amu-Bukhara pumping stations stage – total delivery is 263 m^3/s, height of lifting is 69 m, the area of irrigation is 285 thousand hectares; the Djizak pumping stations stage – total delivery is 138 m^3/s; height of lifting is 112 m, the area of irrigation is 110 thousand hectares.

Among the most important auxiliary systems there are drainage systems. The in-depth position of station buildings, the presence of technological leaks, as well as possible damages and accidents, cause the accumulation in stations buildings the amounts of filtration, drainage and sewage water.

As the world experience of operation shows, the overwhelming number of accidents at pumping stations occurs during electromechanical or hydro-mechanical unsteady processes, when the system experiences the greatest dynamic loads.

Most of the stations were built in the 1960s-1980s and the wearing of the auxiliary equipment of the stations is considerable. As practice has shown, centrifugal pumps do not always meet reliability requirements, especially with sudden power outages. A number of flooding of pumping station buildings is known because at that time all auxiliary systems are paralyzed (Khokhlov, Khokhlov & Titova, 2011, 2018).

In rural brunch of economy very actual are ensuring of reliability, continuity, high energy efficiency and flexibility of pumping stations work, as well as their performance under transient conditions. In order to achieve all of the abovementioned parameters in different countries of the world the certain progress has been made where modern pumping equipment is constantly updating and improving. In this regard, the use of high-tech specialized basic and auxiliary power equipment and new efficient and safety operating modes of pumping stations, prolonging the life of the large systems is necessary.

RELATED WORKS

In the world, the questions of reliability and uninterrupted operation of pumping stations during steady and transient conditions, the creation of new techniques and technologies of protection of the power equipment and pipelines of pumping stations against sudden transients are becoming particularly important.

In this area, the implementation of targeted research in developing the methods of control the steady and transient modes of pumping stations operation are the priority problems. Among them the creation of the main power equipment and pipelines protection devices against water hammer, the creation of technical solutions and improving the reliability of the power equipment. They include following: principles of smoothing of water hammer pulsation at pumping stations and new air valves for pressure normalization are developed (Val-Matic Valve & Manufacturing, 2016); drives to ensure different closing speed for different parts of the valve stroke in the case of transients at pumping stations are designed (Sipos Aktorik, 2016); automatic control valves, in which the output speed adjustment is performed separately for normal and emergency operation modes of the pumping stations are developed (GA Valves, 2016) and many others.

LITERATURE REVIEW

Researches aimed at the creation of algorithms and methods of regulation of pumping stations modes protecting from negative effects of electromechanical and hydro-mechanical transients are conducted in the leading research centers and higher educational institutions of the world.

The development of intelligent algorithms and optimal control modes of large energy complexes and hydroelectric objects in a number of priority areas of research are carried out, including: the theory and methods of study on renewable energy sest transients (Ilyin, Shestopalova, Vaskov & Ko, 2019), (Kharchenko, Panchenko, Tikhonov & Vasant, 2018); the creation of algorithms and methods of regulation of large distribution systems transients (Anderson, Lowry, & Thomte, 2001); the development of operational and energy conservation modes of pumping stations operating by means of adjustable speed drives (Cookson, Lang & Thornton, 2008); computational tools for simulating the interaction of the different pumping facilities (Garcia-Hernandez, Wilcox, & Moore, 2010); the creation of basis of the simulation of water hammer frequency control laws with scalar and vector control of speed of asynchronous motors at the pumping stations (Larsen, 2012); the development of energy-efficient, optimal modes of pumping stations based on computer control technology (Li, Ma, Liu, Wu, Gao & Yang, 2016).

In the Republic of Uzbekistan, also the large-scale activities on the effective organization of measures for the creation of energy-saving techniques and the control system of pumping stations operating modes are carried out. In Tashkent State Technical University, Uzbekistan, the mathematical models of the pumping stations transients at the electric motor after a short break, self-catering are developed (Khashimov & Abidov, 2012); the formalization of the relationship of electromechanical and hydro-mechanical transients at pumping stations is elaborated (Allayev, Khokhlov, Sytdykov & Titova, 2013). In LLc "scientific and technical centre", Uzbekistan, rational structures of irrigation systems pumping stations control through variable frequency drives were built (Kamalov, 2014). In the Research-and-production Enterprise "Vodopodyomnik" ("Water lifting") with Turin

Polytechnic University in Tashkent, Uzbekistan, the notable progress has been made in developing of effective methods, techniques and devices to reduce the influence of electromechanical and hydro-mechanical transients on power equipment and pipelines of pumping stations (Khokhlov, Khokhlov & Titova, 2011) and ensuring of the reliability of the main and auxiliary pumping equipment of large irrigative pumping stations (Khokhlov, Khokhlov & Titova, 2014, 2015).

Literature review shows that number of research results is received in the world to improve the algorithms and methods of control of pumping and hydroelectric power stations operating modes and protection against negative effects of electromechanical and hydro-mechanical transients.

BACKGROUND

For pumping out drainage water, a rather complicated system of auxiliary drainage facilities is created (Figure 1).

Figure 1. The diagram of the system of drainage and pumping of water from the buildings of large hydropower plants: I - set with vertical centrifugal pumps or radial-axial turbines; II - set with axial pumps or rotary vane turbines; 1 - drainage gallery; 2, 12, 14 - valves, respectively, discharge, receiving, bypass; 3, 7 - drain pipes, respectively, for emptying pressure pipelines and supplying water from the downstream to compensate for leaks through the intake valves; 4 - pumping system pumps; 5,6 - drainage pumps with electric drives and with an internal combustion engines; 8, 15 - wells, respectively, for drainage and for pumping systems; 9 - switching on the second drainage pump and giving the signal to the duty personnel; 10 - the inclusion of the first drainage pump; 11 - shutdown of all pumps; 13 - electrode sensors.

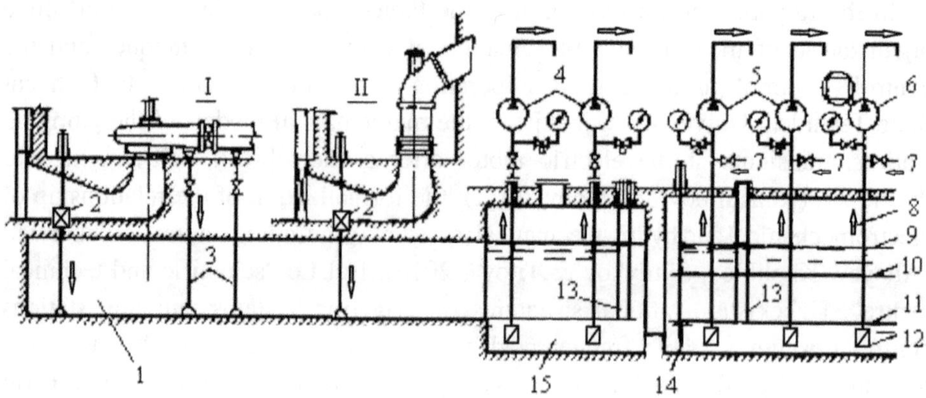

Most often, drainage sets use self-priming centrifugal pumps filled with water, as well as submersible pumps.

The drainage system should have at least two pumps. The second pump is a backup one. It automatically turns on if the first pump fails and during an emergency overflow of the drainage well. Drainage systems at small and medium-sized stations are usually equipped with horizontal centrifugal pumps, and at large ones with vertical artesian pumps.

Drainage pumps are installed on non-flooding marks or on high foundations, not lower than 0.7 m above the floor level. They must be energized throughout the year. Emergency shutdowns of electricity can lead to flooding of the station, which in turn will lead to flooding of the drainage pump electric motor and to complete loss of control over the station.

Drainage flow is defined as the sum of the drainage flow of water through the glands, covers and pump housings and filtration flow of water through the walls and bottom of the station building, located below the maximum water level.

The drainage well is most often arranged in the end part of the station building or under the installation site. The working capacity of the drainage wells is calculated for a 20–30 minute inflow of water.

As an example, is cited the flooding of a pumping plant in Uzbekistan. The pumping plant is located near the Syrdarya River, and is distinguished by a high influx of seepage water. Emergency disconnection of electric power for 10 h and for 8 h resulted in complete flooding of the pumping plant and its shutdown for virtually one month in each case. All twelve electric motors installed at the pumping plant were flooded with water. Some of them had to be dried out and were returned to normal service over an extended period, while others were discarded and removed from service. Use of a jet pump in this case would have made it possible to evacuate the seepage and drainage water by means of the discharge of three pressure conduits each 3300 m long in eight days.

A case of serious flooding of the building of the Jizzak Head pumping station in 1985 is known, when a hatch cover on the pressure pipe broke out during the start-up of the 2400V - 25/40 pumping unit. (Subsequent calculations and experiments showed that the cause of the accident was insufficient fastening of the cover at the factory.) As a result, the inflow of water from the hatch exceeded the drainage pumping units' possibilities. The water level in the station building began to grow continuously and soon reached the level of drainage pumping electric motors, which were the first to be flooded. The station building was flooded completely. The most unpleasant thing was that the electric motors of the 1600V - 10/40 pumping units came under flooding and were out of operation during almost a year. It took two weeks to pump the water out of the pumping station, and vegetative watering was late for the whole of Jizzak steppe.

Thus, serious difficulties arose in the system of operation of irrigative pumping stations and hydropower stations of Uzbekistan. In the absence of the required pumps at some stations, they try to solve the pumping problem using borehole pump systems. These pumps are designed for pumping water at the depths of more than 8-10 m. Their use in the drainage systems of buildings of stations where the head does not exceed 1.0-1.5 m leads to non-observance of the temperature mode of the electric motors and the pumps getting into the cavitation mode and, as a result, the rapid failure of the units.

The tragic accident at the Sayano-Shushenskaya HPP (Russia), which occurred in August 2009 (Belash, 2010), once again pointed out with particular urgency the need for an additional source of energy for pumping drainage water. As a result of emergency flooding and during power outages, drainage pumps, in fact, are the first to be damaged. This is due to the fact that centrifugal pumps used in auxiliary systems are electrically driven and must be constantly supplied with electricity.

It is undeniable that for pumping it is necessary to have an alternative source of energy. As such, diesel pumping systems are most often used at large stations, but as the experience of operating of hydropower plants have shown, diesel systems cannot provide timely results.

According to all directives for auxiliary station systems, it is necessary to have an alternative source of energy.

MAIN FOCUS OF THE CHAPTER

Authors have substantiated the method of using an alternative source of energy in the auxiliary systems of pumping and hydroelectric stations - the potential energy of a liquid column in a pressure head conduit of stations, using jet pumps. For this technology the invention Patent number UZ IAP 03670 is taken out.

Jet pumps have come into widespread use in many branches of engineering. This has been promoted by their high structural and service indicators. Jet pumps are simple in terms of design, easy to fabricate and repair, perform reliably, do not require preliminary priming prior to start up, and permit the pumping of contaminated water and a water-air mixture. It is proposed to use jet pumps to evacuate drainage and seepage water from buildings housing pumping and hydroelectric power stations. They are more reliable than the centrifugal pumps traditionally used for these purposes, since they operate on the water pressure of a pressure conduit, and make it possible to evacuate drainage and seepage water even when the entire station has lost electric power.

Jet pumps have high performance, they are reliable in operation, do not require pre-fill before starting. The advantage of jet pumps is that they do not have rotating

Figure 2. Drainage system of a pumping or hydroelectric station with a jet pump: 1 - building of a pumping or hydroelectric station; 2 - main units (pumps or turbines); 3 - pressure pipe; 4 - drainage well; 5 - jet pump; 6 - supply pressure pipe; 7 - butterfly valve.

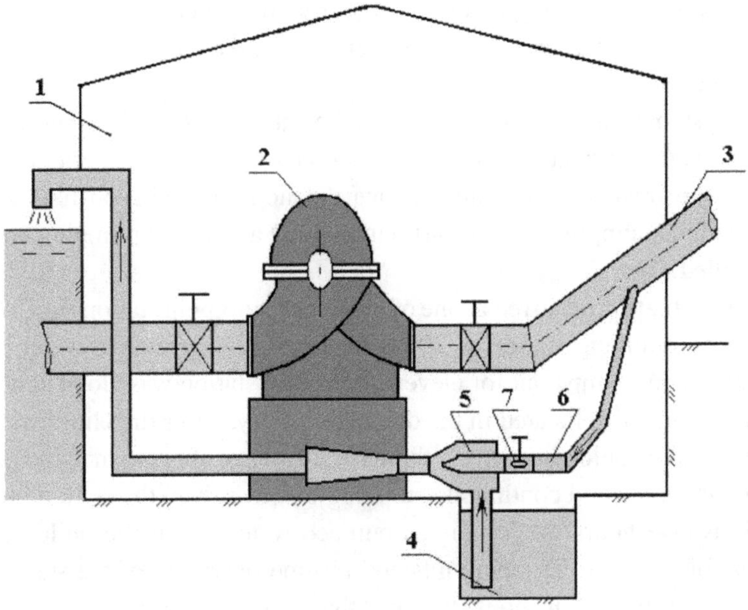

or moving parts, and the actuating fluid of the pumps is water from the station's pressure penstock. Transmission of energy to the suction stream is carried out without direct usage of electrical and mechanical energy.

Water from the pressure conduit is fed through the supply pipe into the cavity of the jet pump. Under the action of the discharge created in the suction chamber of the jet pump, the drainage water enters the pump cavity, mixes with the working fluid and is removed from the drainage well to the lower reach of the station.

The system with the jet pump is shown schematically in Figure 2.

Filtration and drainage waters accumulate in the drainage well. Water is gradually drained from the station's pressure pipe through the supply pipe and enters the jet pump nozzle. In the nozzle, the potential energy of the station's pressure conduit is converted into kinetic energy, and water flows at high velocity into the chamber of the jet pump.

Under the action of a vacuum created in the suction chamber of the jet pump, the filtration and drainage waters are removed from the drainage well to the downstream of the station. At hydroelectric power plants, jet pumps can be a significant source of energy savings, because jet pump does not consume electricity.

The jet pumps, originally intended as backup, due to their high performance, in practice began to be used as the main, and centrifugal pumps are displayed in the reserve.

At pumping stations equipped with long pressure pipelines, when electricity is switched off, the water will gradually merge from the pipeline, and the work of the jet pumps will continue for 3-18 days depending on the length of the pipeline and the pumped out flow rate.

If such systems with jet pumps were installed at the Sayano-Shushenskaya HPP, then immediately after the removal of the source of flooding, i.e. after closing the valves on the upstream side and shutting down the suction chamber on the downstream side, continuous pumping would start without any additional pumping equipment being installed.

A similar situation occurred at one of the largest pumping stations in Uzbekistan - Jizzak Head pumping station in 2009. At this station, drainage water has been pumped by the jet pumps out for eleven years. The station was flooded due to the failure of a drain valve located in the drainage gallery. After the elimination of the source of flooding, automatically, without the influence of personnel, the jet pumps began an immediate and continuous pumping of water from the station building.

Within twelve hours the jet pumps pumped water out of the building so that it was possible to start the main units and resume operation of the stage of three Jizzak pumping stations. It should be noted that a similar mentioned above accident that occurred in 1985 at the same station before the installation of jet pumps, was eliminated within two weeks.

It should be pointed out that jet pumps operate effectively only under the strictly defined conditions for which they are designed. It is known that the geometric height of the liquid column will be lowered when the reservoir at hydroelectric plant is drawn down, and, consequently, the pressure of the working fluid diminished. During the winter months, some of the working pumps at pumping plants are disconnected for repair and preventive maintenance, and the pressure of the working fluid is also diminished for this reason. The reduction in pressure may reach 20% at hydroelectric plants, and 15% at pumping plants. When the pressure of the working fluid is reduced, the flow velocity of the fluid from the nozzle of the jet pump is reduced, and the efficiency of the pump will fall off sharply. The inability of a jet pump to evacuate drainage and seepage water in a sufficient volume comes about under certain conditions. The probability that the plant building will be flooded increases as a result.

These inadequacies can be eliminated by use of self-regulating jet pumps. A self-regulating jet pump (Figure. 3) is built with regulating needle connected by pusher to moving axle on which piston is mounted and spring-loaded relative to the housing. The invention Patent number UZ N° IDP 04411 is taken out. When the pressure of

Figure 3. Self-regulating jet pump with a needle: 1 - nozzle; 2 - moving axis; 3 - the piston; 4 - spring; 5 - discharge port; 6 - suction nozzle; 7 - needle; 8 - mixing chamber.

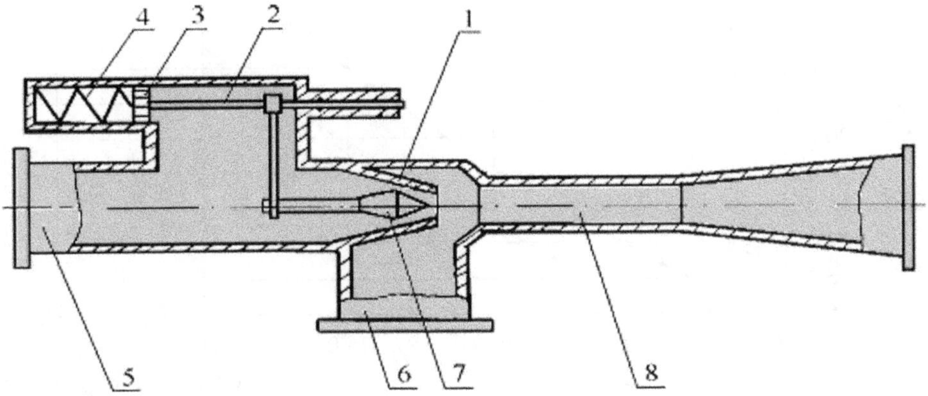

the working fluid drops in the pressure conduit, the pressure of the liquid in jet pipe falls off, and, consequently, the force of the pressure on the piston from the liquid side deceases. Under the action of spring, the piston, together with the axle and needle, is displaced to the right in the axial direction. The needle partially covers nozzle, reducing its cross-section area. The change in the cross-sectional area of the nozzle leads to an increase in the discharge velocity of the working fluid from the nozzle, and to restoration of the jet-pump's suction capacity. A pump with a spring-loaded piston makes it possible to vary the section of the nozzle automatically as a function of the pressure of the working fluid. Use of self-regulating jet pumps makes it possible to maintain a rather high pump efficiency, and increase the reliability of the flood-protection system.

An increase in the head of the jet pump when regulated by the needle has a natural limit associated with in increase in hydraulic resistances in the nozzle. To determine the range of regulation, was developed a procedure to calculate the variation in the characteristics of a jet pump as external working parameters are varied.

Figure 4 shows the performance characteristics of a self-regulating jet pump with a needle. As in the case of a self-regulating jet pump with an elastic nozzle, the operational characteristics are dependences of the change in intake flow on the working head when the nozzle is closed from $\overline{S} = 0$ to 0.2. The pressure of the jet pump is constant 9 m.

With a working head equal to 21.7 m and a head at the exit of 9 m, the unregulated pump will pump out 10 l/s. If the working head decreases to 19.1 m, the intake flow rate will decrease to 5.5 l/s, and if the working head decreases to 17.5 m, the pump will be able to pump out only 3.0 l/s.

Figure 4. Performance characteristics of a self-regulating jet pump with a needle

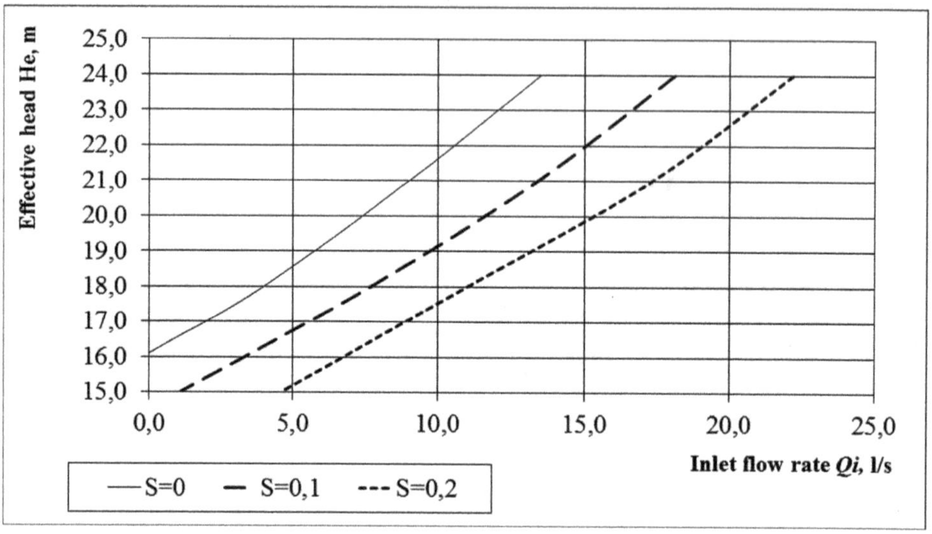

When self-regulating the jet pump with a needle, it can create the necessary pressure at the exit without loss of suction ability. A self-regulating jet pump with a needle, when the working head decreases from 21.7 to 17.5 m, will be able to pump out a flow equal to 10 l/s, and the nozzle section of the jet pump will decrease by 10% at $H_e = 21$ m and by 20% at $H_e = 18.5$ m.

Another design of a self-regulating jet pump is design with an elastic nozzle. The invention Patent number UZ Nº IDP 04305 is taken out.

To ensure effective drainage water pumping at a decrease in the pressure of the working fluid, was proposed the design of a self-regulating jet pump, the nozzle of which is made of two parts: external metal and internal in the form of an elastic sleeve made of an elastic material.

This embodiment of the pump allows you to automatically change the nozzle cross section depending on the pressure of the working fluid and thereby create the necessary flow rate of the working fluid through the nozzle.

The proposed construction is shown schematically in Figure 5.

Self-regulating jet pump with an elastic nozzle contains a nozzle, nozzle nozzle, suction nozzle, mixing chamber, confuser and diffuser. The nozzle is made of mating cones with a common base of the outer metal part and the inner of elastic material, divided by an annular slot into the outer and made into the mating petals of triangular cross section and the body divided by a transverse partition into cavities, in which plungers move with a common stock. In the cavity on the side of the elastic part

Figure 5. Self-regulating jet pump with an elastic nozzle: 1 - metal part of the nozzle; 2 - elastic part of the nozzle; 3 - plungers with common stem; 4 - nozzle pipe; 5 - suction nozzle; 6 - spring; 7 - mixing chamber.

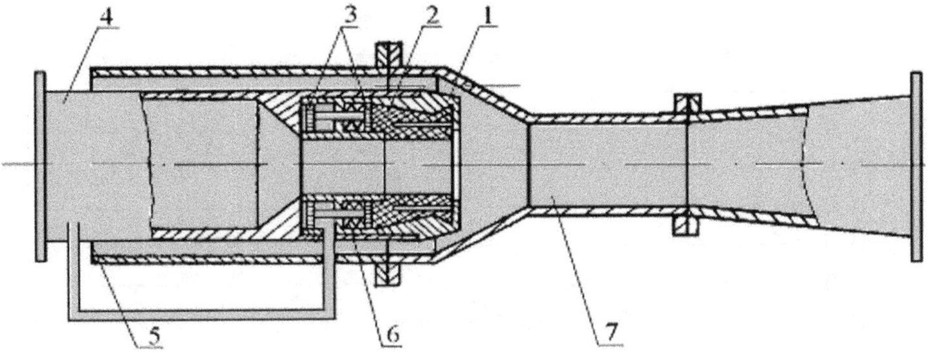

there are springs, and the other cavity is connected with a nozzle pipe by a pipeline. To support the elastic part is a ledge.

The pump works as follows. When a hydropower plant, pumping station or pumping unit is in operation for pumping water from wells in the design mode, when the pressure of the working fluid is normal, the nozzle is in the position shown in Figure 5. The working fluid from the discharge pipe enters the nozzle pipe and the nozzle. Coming out of the nozzle at high speed, the working fluid creates a vacuum in the confuser, thanks to which the pumped water from the well is sucked in and together with the working fluid enters the mixing chamber, diffuser and discharge line. Under optimal conditions, the force of the springs balances the pressure in the cavity.

When the pressure of the working fluid in the pressure pipeline is reduced, the pressure of the fluid in the cavity between the plunger and the bulkhead drops. Under the action of springs, the elastic part of the plunger is shifted to the right. The outer part, moving along the cones, forms a protrusion acting on the petals of the inner part. The petals are shifted to the center, reducing the nozzle section. Changing the diameter of the joint is automatically, without an additional drive. Changing the diameter of the nozzle depending on the pressure of the working fluid allows you to maintain a sufficiently high efficiency of the pump, reduce the non-production costs of the working fluid, and increase the reliability of the protection system against flooding.

Figure 6 shows the performance characteristics of a self-regulating jet pump with an elastic nozzle. Operational characteristics are dependences of the change in intake flow rate on the working head with a relative closure of the nozzle $\overline{S}$ from

Figure 6. The performance characteristics of a self-regulating jet pump with an elastic nozzle

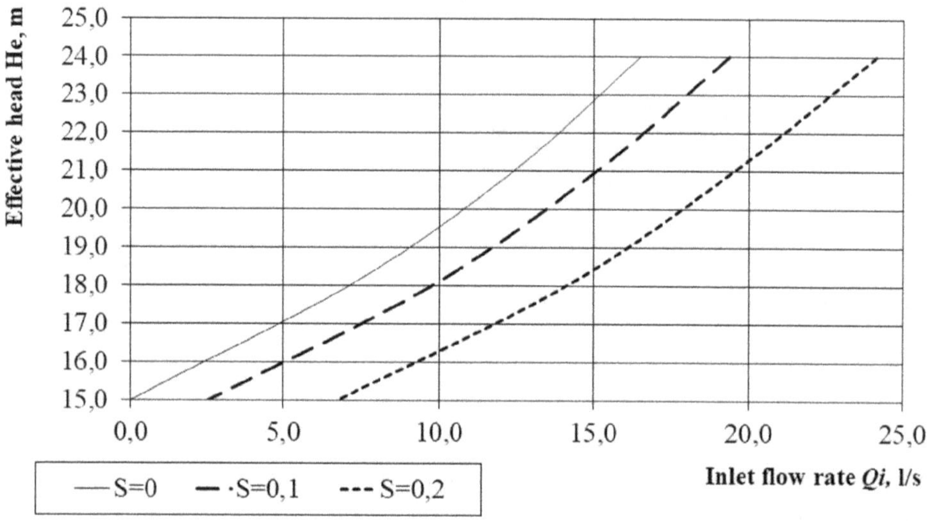

0 to 0.2, located in the zone of effective action of self-regulating jet pumps. At the same time, the operational characteristics are built on a specific pressure, injected by an H_e jet pump, which is assumed to be unchanged and depends on the maximum difference in water levels in the drainage well and downstream of the station, as well as on the pressure loss in the discharge line. In particular, the pressure of the jet pump in Figure 6 amounted to 9 m.

The resulting performance characteristics are confirmed by multiple tests and implementations.

When considering the upper operational characteristics of the corresponding $\overline{S}$ = 0, i.e. in fact, the characteristics of an unregulated jet pump show that when the working head decreases, the value of the suction inlet flow rate *Qi* correspondingly decreases. During self-regulation of the jet pump, the mode point will decrease "down" down to the "$\overline{S}$ = 0.2" characteristic when the working head decreases. So with a working head equal to 22.9 m and a head at the exit of 9 m, the unregulated pump will pump out 15 l/s. If the working head decreases to 21 m, it will be able to pump out about 12 l/s, and if the working head decreases to 18.5 m, the intake flow will decrease to 8 l/s.

Self-regulating jet pump with an elastic nozzle in the same conditions, i.e. when the working head decreases from 22.9 to 18.5 m, it will be able to pump out a flow equal to 15 l/s, while the cross-sectional area of the jet pump nozzle will decrease accordingly at H_e = 21 m by 10% and at H_e = 18.5 m at 20%

Figure 7. Schematic diagram showing operation of self-regulating jet pump; Q_w -flow of working fluid; Q_i - inlet flow of fluid; $Q_p = Q_w + Q_i$ - mixed pump flow; F_n - sectional area of pump nozzle; F_c - sectional area of mixing chamber; V_d -discharge velocity of liquid from nozzle; V_i - velocity of liquid sucked into mixing chamber; V_t - velocity of liquid at terminus of mixing chamber.

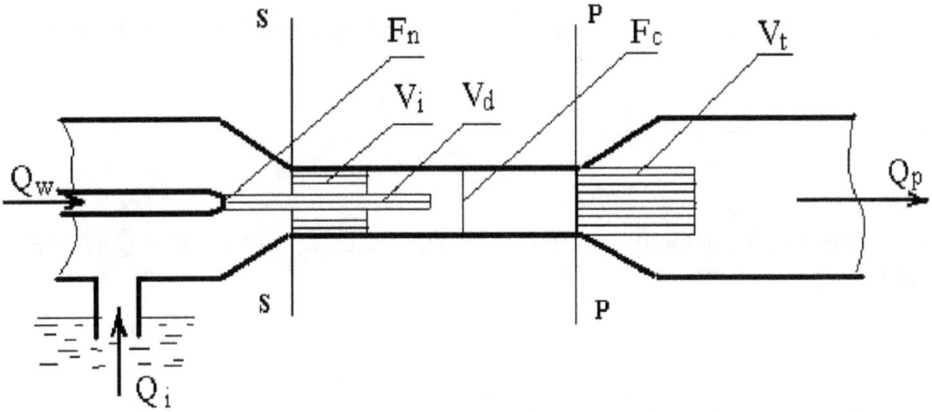

Authors have established hydraulic systems for jet pumps regulation. Mathematical models and algorithms of calculation and optimum power-hydraulic operating modes are developed.

The schematic diagram showing operation of self-regulating jet pump is presented on Figure 7.

Authors determine the head H_p of a self-regulating jet pump. It is equal to the difference between the total specific energies of the flow in the pressure E_p and suction E_s lines

$$E_p = \frac{P_p}{\rho g} + \frac{V_t^2}{2g}$$

$$E_s = \frac{P_s}{\rho g} + \frac{V_i^2}{2g}$$

$$H_p = E_p - E_s = \left(\frac{P_p}{\rho g} - \frac{P_s}{\rho g}\right) + \left(\frac{V_t^2}{2g} - \frac{V_i^2}{2g}\right) - \Delta h - \Delta h_n \qquad (1)$$

where P_p and P_s are, respectively, the pressure in the pressure and suction pipes of the jet pump, Δh is the head loss in the setting of the jet pump, and Δh_n is the bead loss in the regulating needle.

To determine the potential component of the head $(P_p - P_s)/\rho g$, the equation is formulated of the amount of motion for sections S-S and P-P

$$\rho \cdot Q_w \cdot V_d + \rho \cdot Q_i \cdot V_i - \rho \cdot (Q_i + Q_w) \cdot V_t = (P_p - P_s) \cdot F_c \tag{2}$$

Dividing both sides of the equation by the acceleration of free fall and density ρ,

$$\frac{Q_w \cdot V_d + Q_i \cdot V_i - (Q_i + Q_w) \cdot V_t}{g} = \frac{(P_p - P_s) \cdot F_c}{\rho \cdot g}$$

and expressing the velocity in terms of the flow rates $V_d = Q_i/F_n$, $V_i = Q_i/(F_c - F_n)$, and $V_t = (Q_w + Q_i)/F_c$

$$\frac{P_p - P_s}{\rho \cdot g} = \frac{\dfrac{Q_w^2}{F_n} + \dfrac{Q_i^2}{F_c - F_n} - \dfrac{(Q_i + Q_w)^2}{F_c}}{g \cdot F_c}$$

Introducing the known coefficients $q = Q_i/Q_w$ - the relative suction factor, and $f = F_c/F_n$ - the relative geometric factor of the jet pump, and considering that $H_e = \dfrac{Q_w^{\ 2}}{F_n^2 \cdot 2g}$ - the effective head of the jet pump,

$$\frac{P_p - P_s}{\rho \cdot g} = \frac{2H_e}{f}\left(1 + \frac{q^2}{f-1} - \frac{(q+1)^2}{f}\right) \tag{3}$$

The kinetic component of the head of a self-regulating jet pump

$$\left(\frac{V_t^2}{2g} - \frac{V_s^2}{2g}\right) = \frac{(Q_i + Q_w)^2}{2g \cdot F_c^2} - \frac{Q_i^2}{2g(F_c - F_n)^2} =$$

$$\frac{Q_w^2}{2g \cdot F_n^2}\left[\frac{F_n^2}{F_c^2} \cdot \left(\frac{Q_t}{Q_w} + 1\right)^2 - \left(\frac{Q_t}{Q_w}\right)^2 \cdot \frac{1}{\left(\dfrac{F_n}{F_c} - 1\right)^2}\right] = H_e\left(\frac{(q+1)^2}{f^2} - \frac{q^2}{(f-1)^2}\right) \tag{4}$$

can be transformed in a similar manner.

The head loss Δh in the setting of the jet pump are summed from the loss in the inlet confuser, nozzle, mixing chamber, and diffuser, and are determined from hydraulics formulas as a function of the design of the pump components. The local-resistance factors in the components for the analyzed and fabricated jet pump are constants because their design does not change. The local-resistance factor depends on the design of the local resistance. And the head loss, in turn, depends on this factor and the fluid velocity. Head loss is not constant. The head loss Δh_n at the regulating needle are variable: they depend on the relative closing S of the nozzle.

After substituting relationships (3) and (4) in Eq. (1), and performing transformations, was obtained the basic relationship for determination of the performance curves of a self-regulating jet pump

$$H_p = H_e \left[\frac{2}{f} - \frac{(q+1)^2}{f^2} + \frac{q^2 (f-2)}{(f-1)^2 \cdot f} \right] - \Delta h - \Delta h_n \tag{5}$$

A jet pump with the following basic parameters was adopted as the basis of the calculation performed in accordance with relationship (5): a sectional area F_c = *397.4* mm^2, and a sectional nozzle area F_n = 176.6 mm^2.

Figure 8. Curves of head H_p of self-regulating jet pump versus relative nozzle closing for various effective heads: H_e = 24m; H_e = 21m; H_e = 18m; H_e = 15m

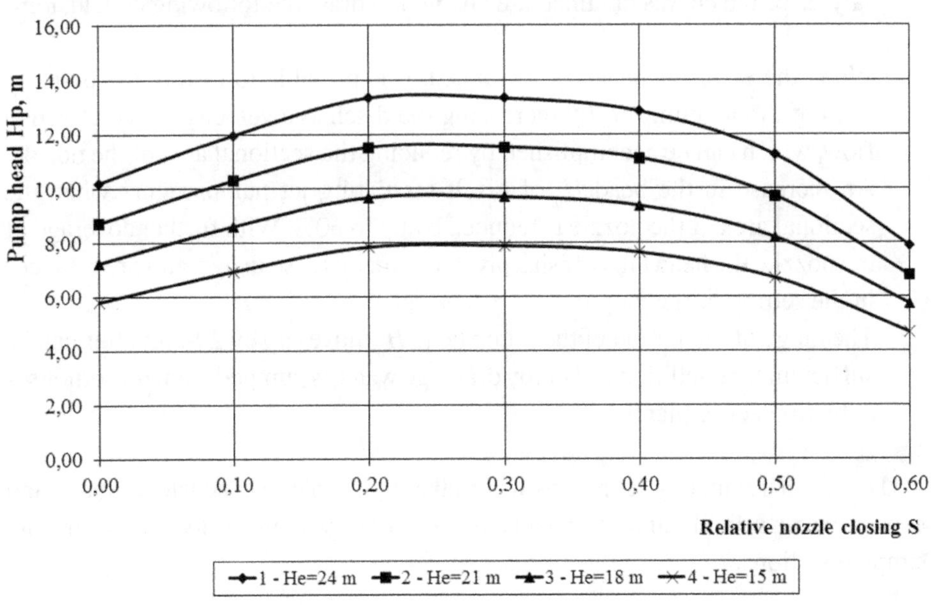

Figure 9. Curves of head H_p of self-regulating jet pump versus with an elastic nozzle closing for various effective heads: H_e = 24m; H_e = 21m; H_e = 18m; H_e = 15m

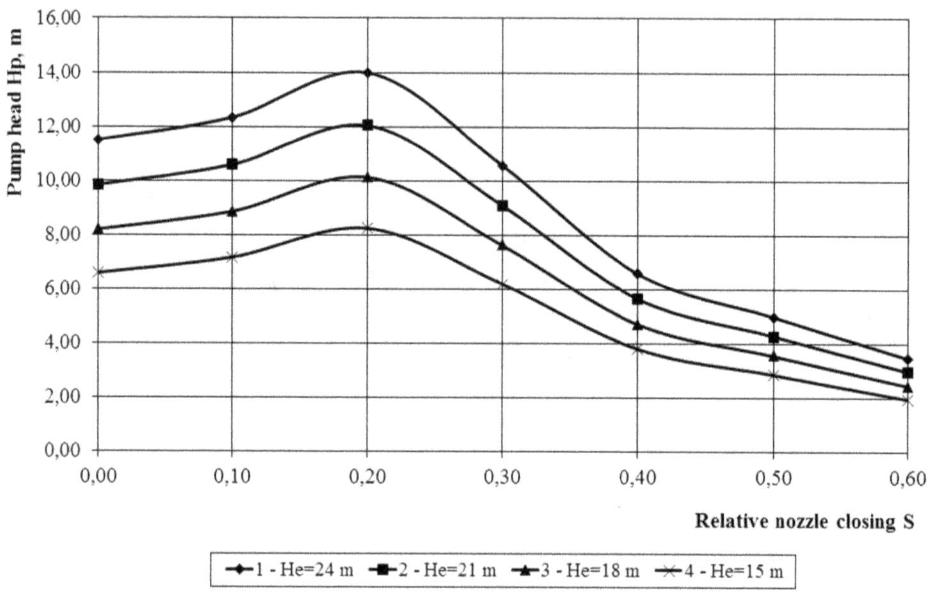

Basic results of the investigations are presented in Figure 8 and Figure 9 in the forms of the relationship between the pump head H_p and the relative nozzle closing S during its self-regulation. The curves are plotted for the condition that the suction flow Q_i is constant and amounts to 1 l/s.

Analyses of the curves obtained are enabled to draw the following conclusions:

1. When the effective head H_e is reduced, it is possible to restore the delivery head of the jet pump H_p by increasing the discharge velocity of the effective flow, which can be accomplished by reducing the sectional area of the nozzle.
2. An increase in the bead H_p of a self-regulating jet pump occurs when the sectional area of the nozzle is reduced by up to 60%. With further coverage of the nozzle, the head falls off sharply due to an increase in hydraulic resistances in the setting of the jet pump.
3. The range of regulation of the pump head H_p moves up by 20-25%; this can be sufficient for conditions whereby drainage water is pumped from powerhouses at hydroelectric plants.

Use of self-regulating jet pumps will make it possible to evacuate drainage and seepage water reliably and effectively from buildings housing hydrostations and pumping stations.

Analyses performed for pumping plants equipped with long pressure conduits indicated that during total shutdown of electric power, reliable evacuation of drainage and seepage water can be ensured with the use of self-regulating jet pumps over a period of several days; this cannot be accomplished by any other pump.

Due to the constant increase of sewage water from the territory associated with duration of stations and pressure pipelines exploitation the leakage water through the walls of the station building drains inside the building via channels with electrical cables, which may cause accidents and floods.

At the research-and-production enterprise "Vodopodyomnik" ("Water lifting") new device for the simultaneous pumping of filtration, drainage and sewage waters from buildings and territories of pumping and hydroelectric stations is developed. The utility model Patent number UZ FAP 00592 is taken out. The device comprises a drainage hole in the territory, drainage well in the station building where the jet pump with a suction pipe, butterfly valve, connected via a lever with a load-float are located.

The proposed system is shown schematically in Figure 10. At the average value of the water flowing into the drainage well the load-float is positioned so that cinematically associated with it butterfly valve is in the half-closed state. In this

Figure 10. A system for the simultaneous removal of filtration, drainage and sewage water from adjacent territory and the building of pumping or hydroelectric station: 1 - system with a jet pump; 2 - drainage well in the building station; 3 - drainage hole at the station; 4 - additional pipeline; 5 - additional valve.

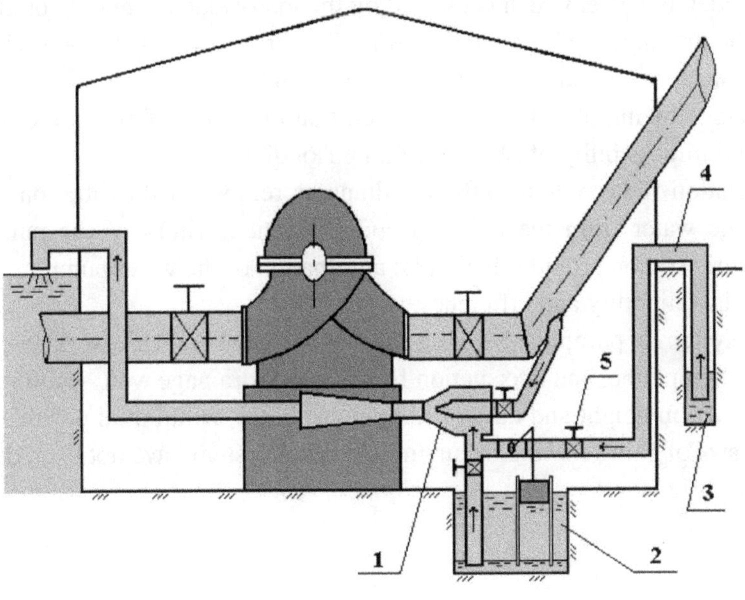

Figure 11. Calculation scheme of regulated pumping system of filtration, drainage and sewerage water from the building and adjacent territory of pumping station

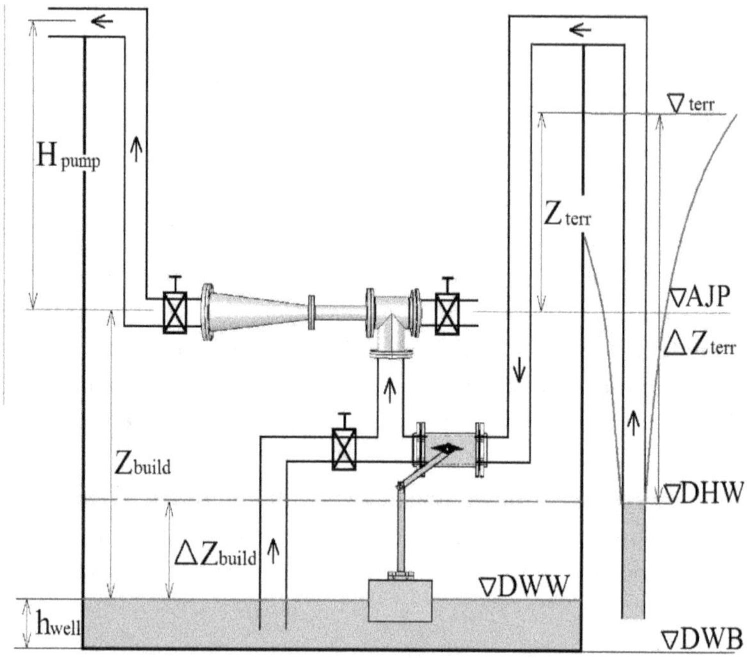

case the jet pump removes all of the water flowing into the drainage well and pumps water out of the drainage hole, lowering the water level to a certain value.

The butterfly valve, which is exerted by the load-float, depending on the water level in the drainage well produces a redistribution of the water flow sucked from out of drainage well and drainage hole. The mandatory condition is pumping all of the water flowing into the drainage well that provides safety work of drainage system and impossibility of station building flooding.

Application of the system for the simultaneous removal of the filtration, drainage and sewage water from the building and adjacent territory of the pumping or hydroelectric station provides hydraulic automation of the water pumping process, increases its reliability and efficiency.

In the system of pumping there is a mixture of three flows: the workflow – out of the pressure pipeline, and two suction flows: out of drainage well – with the initial negative suction height and out of the drainage hole - with initial positive suction height. Laws of flows mixing occurring in such a system have not been described mathematically.

The processes occurring in the jet pump at mixing of three flows as well as the processes of regulating of the pumping system by means of a load-float are formalized. Calculation scheme of regulated pumping system is shown in Figure 11.

Suction from drainage well in the station building is:

$$H_{\mathrm{S\ build}} = \nabla DWW - \nabla AJP - \Delta h_{build} = \nabla DWB + h_{well} - \nabla AJP - \Delta h_{build} \tag{6}$$

where: ∇DWW – watermark in the drainage well; ∇AJP – mark of the axis of the jet pump; ∇DWB – mark of the drainage well bottom; h_{well} – depth of water in the well, a variable that depends on the flow of water into the drainage well $Q_{\mathrm{S\ build}}$; Δh_{build} – head loss in the suction pipeline located in the building.

Suction height from the drainage hole in the territory is:

$$H_{\mathrm{S\ terr}} = \nabla DHW - \nabla AJP - \Delta h_{terr} - \Delta h_{valve} = \nabla DWB + H_0 - S_c - \nabla AJP - \Delta h_{terr} - \Delta h_{valve} \tag{7}$$

where: ∇DHW – watermark in the drainage hole; Δh_{ter} – head loss in the suction pipeline located in the territory, a variable that depends on the flow of water into the drainage hole $Q_{\mathrm{S\ terr}}$; Δh_{valve} – head loss in the butterfly valve.

Head loss are determined by the formulas of hydraulics:

$$\Delta h_{build} = \lambda \frac{8 l_{build} \left(Q_{Sbuild}\right)^2}{\pi^2 g d_{build}^5}, \quad \Delta h_{terr} = \lambda \frac{8 l_{terr} \left(Q_{Sterr}\right)^2}{\pi^2 g d_{terr}^5}, \quad \Delta h_{valve} = \xi_{valve} \frac{8 \left(Q_{Sterr}\right)^2}{\pi^2 g d_{terr}^4} \tag{8}$$

The value of the minor loss coefficient ξ_{valve} is variable and depends on the angle α of the closing disk of butterfly valve.

At angles α from 0 to 50^0 value of the minor loss coefficient is approximated by a polynomial of the sixth degree:

$$\xi_{valve} = 9.4\cdot10^{-11}\alpha^6 + 3.56\cdot10^{-7}\alpha^5 - 2.47\cdot10^{-5}\alpha^4 + 6.81\cdot10^{-4}\alpha^3 - 5.0\cdot10^{-3}\alpha^2 + 4.24\cdot10^{-2}\alpha + 0.097, \text{(9-a)}$$

at angles α from 50^0 to 90^0 – by a polynomial of fourth degree:

$$\xi_{valve} = 0.0155\alpha^4 - 3.502\alpha^3 + 296.965\alpha^2 - 11181.76\alpha + 157675.6 \text{ (9-b)}$$

The value of the reliability of the approximation was $R^2 = 0.99$.

In turn, the angle of the closing disk of butterfly valve depends on the water depth in the well and the geometric parameters of the lever.

A mathematical model of the hydraulic regulation of jet pump during pumping of flows with different suction heights has the form:

$$z_{terr} + \frac{8Q_{Sterr}^2}{g\pi^2 d_{terr}^4}\left(1 - \frac{\lambda_{terr} l_{terr}}{d_{terr}} - \xi_{valve}\right) = z_{build} + \frac{8Q_{Sbuild}^2}{g\pi^2 d_{build}^4}\left(1 - \frac{\lambda_{build} l_{build}}{d_{build}}\right) \tag{10}$$

It determines the amount of jet pump suction flow rates from the building and territory of hydropower station.

$$Q_{Sterr} = \sqrt{\frac{z_{build} - z_{terr} + \frac{8Q_{Sbuild}^2}{g\pi^2 d_{build}^4}\left(1 - \frac{\lambda_{build} l_{build}}{d_{build}}\right)}{\frac{8}{g\pi^2 d_{terr}^4}\left(1 - \frac{\lambda_{terr} l_{terr}}{d_{terr}} - \xi_{valve}\right)}} \tag{11}$$

The results of calculation of the system parameters made according to the developed mathematical model, conducted by the example of the pumping station "M-2-2" of the Kashkadarya region are shown in a graphic based on Figure 12.

Figure 12. The pressure dependence of the jet pump on suction flow rate from the building and from the territory of the pumping station

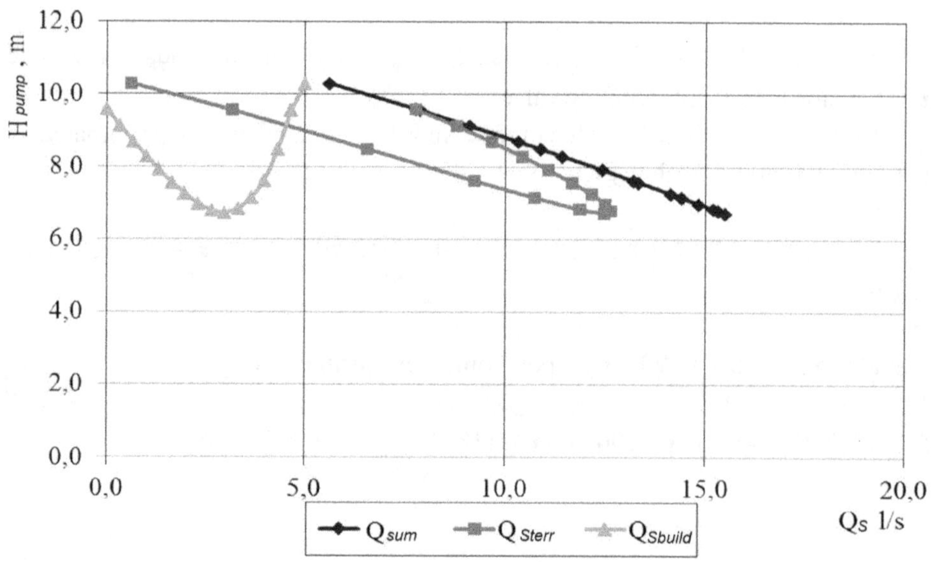

As can be seen from Figure 12 pressure dependence of the total suction rate, which has the name of the operating characteristics of the jet pump, has a slightly downward shape facing convexity upwards. This shape is typical of the performance of jet pumps and was obtained in previous studies.

However, when considering separately dependences of the jet pump head on suction flow rate from the pumping station building and dependences of the jet pump head on suction flow rate from the pumping station territory, it can be noted that the characteristics have different contours. Characteristics face convexity downwards and have minimum points. Modes of operation of the jet pump, corresponding to these minima, have the lowest heads and therefore are the most energy efficient.

The new devices were implemented at the pumping stations "M-2-2", "Pakhtakor", "Guvalak" of the Kashkadarya region, PS2 - PS6 of the Karshi Main Canal, "Karayantag" and "DM-1" of the Jizzak region, "Turkistan 1-1" and "Turkistan 1-2" of the Namangan region as well as at the Charvak and Farhad hydroelectric power stations.

The results of field tests of the new devices carried out at the pumping station "M-2-2" are presented.

According to the results of tests of the pumping system of drainage, drainage and waste water, the following conclusions were made:

1. Tests with a constant head are allowed to build a full-scale characteristic of the jet pump system at a stationary operating head $H_e = 27$ m. The average intake flow rate of the system with the jet pump with one main pump unit running during the entire observation period was $Q_i = 15.2$ l/s.
2. Tests at variable modes are allowed to build field dependencies of the injected head of the system with a jet pump on various values of intake flow from the territory and from the building of the pumping station. With the values of intake flow rates $Q_i = 12.6$ l/s and $Q_i = 2.6$ l/s, the head pressure is minimal, i.e. the most economical mode of operation of the jet pump for simultaneous pumping is achieved. With other ratios of intake flow rates, the pressure value increases due to an increase in energy loss in the inlet and outlet pipelines of the system and in the jet pump.

The developed jet pump test procedure and the employed means and methods of measurement allowed caring out the measurements of the head with an error not exceeding 3%, the jet pump delivery and the working liquid flow rate with an error not exceeding 3.5%. Statistical analysis of the calculation data and carried out field tests showed that the maximum difference of values of theoretical and experimental studies did not exceed 6% at the critical points. Comparison of the calculation results and field test results indicates the accuracy of the research.

Authors have installed more than 50 systems with jet pumps at pumping stations in Uzbekistan. At present, authors have begun to implement them for pumping out drainage water from hydroelectric power stations. At hydroelectric power stations, jet pumps become a significant source of energy savings, as they do not consume electricity. An experimental jet pumps, installed at the Charvak and Farhad hydroelectric power stations, showed their good work and efficiency. The annual economic effect from their introduction allows the payback period about 1 year. Jet pumps, installed as a backup in the event of power outage, have become, in practice, the main pumping equipment due to their high operational parameters and the centrifugal pumps are taken to the reserve.

An experimental system with jet pumps, was implemented in 2015-2016 at the largest in Uzbekistan Charvak hydroelectric power station with a capacity of 620 MW (four identical radial-axial hydro turbines RO-170/638a-VM-410, each of them has capacity of 155 MW). Charvak hydroelectric power station is located 70 km from Tashkent, the capital of Uzbekistan. The hydroelectric power station is located in the headwaters of the Chirchik river - the right tributary of the Syrdarya river. The average annual water flow rate of the river at the site of the hydrosystem is 208 m^3/s; the maximum measured flow rate is 1600 m^3/s. The Charvak hydroelectric complex is of great importance for the economy of Uzbekistan and, above all, for irrigation of fertile lands.

The inflow of water into the drainage well of the Charvak hydroelectric power station is formed of three sources with different suction heights: the filtration water from the pressure side of the building, emerging through the seam of the wall near the hydro unit No. 2 with a maximum flow rate of 20 l/s; filtration waters from the pressure side of the building, emerging through the seam of the wall near hydroelectric unit No. 4 with a maximum flow rate of 10 l/s and drainage water formed directly in the building with a maximum flow rate of 10 l/s.

It should be taken into account that significant fluctuations of the water level in the reservoir occur in hydroelectric power stations and, accordingly, fluctuations in the inflow of drainage waters into the building occur. At the same time, taking into account the special significance and importance of hydroelectric power stations, it is necessary to develop a system for automatic control, signaling and monitoring of the drainage system equipped with jet pumps.

An experimental jet pump with a total flow rate of 40 l/s, installed in 2015-2016 at the Charvak hydroelectric power station, showed its good work and efficiency. The annual economic effect from its introduction amounted to 28.3 million Uzbek soums, the payback period is about 1 year.

Jet pumps, installed by our specialists as a backup in the event of power outage, have become, in practice, the main pumping equipment due to their high operational parameters and the centrifugal pumps are taken to the reserve.

The same happened at the Charvak hydroelectric power station, where during the last two years the pumping of drainage water from the station's building is carried out exclusively by jet pumps. However, this does not mean that the jet pumps must completely replace the electric centrifugal pumps installed by previous design.

Nevertheless, the drainage well as the place of installation of the jet pump, dictated by the lowest gathering point of all flows of filtration and drainage water, is not the most optimal from the energy point of view. For the operation of the jet pump, where possible, it is economically more advantageous to work with the head, i.e. with a positive flow suction height than pumping water from the drainage well with a negative suction height. And besides, the higher the drainage water is caught, the less height it will have to lift.

Therefore, to ensure economical pumping of the filtration and drainage waters of the Charvak hydroelectric power station, it was decided to install three more jet pumps: near the unit No. 2 at a flow rate of 20 l/s, with the pumping of water into the suction tube of the unit No. 2, near unit 4 at a flow rate of 10 l/s with the pumping of water into the suction tube of the unit No. 4 and inside the drainage well at flow rate of 10 l/s with the pumping of water into the same pipeline which is used by the first jet pump installed in 2015-2016.

The installation of two pumps at higher elevations than the mark of the drainage well will significantly save the consumption of actuating fluid necessary for the operation of the jet pump. The installation of the last pump inside the well will allow it to operate in a non-cavitation mode, which will increase its durability and reliability. The jet pump, installed in 2015-2016, will be taken to emergency reserve. It will be equipped with electrified valves on the side of the actuating fluid and on the side of the discharge of water into the tail water. These valves will be opened from the control panel of the hydroelectric power station at any sudden increase of the water level in the drainage well. Thus, for safety reasons, two-times pumping will be created.

The second largest hydroelectric power station in Uzbekistan is the Farhad hydroelectric power station. Initially, the appointment of the Farhad reservoir on the Syrdarya River was energy, irrigation and technical water supply. However, as a result of the silting of the reservoir during the past 80 years, the area of its water surface has decreased by more than 20%, and, accordingly, it is now working to create a water head, performing only daily regulation.

The installed capacity of Farhad hydroelectric power station is 126 MW, the available capacity is 114 MW and the average annual power generation is 825 million kWh. The average long-term consumption is 538 m^3/s, the maximum measured flow is 4270 m^3/s

At Farhad hydroelectric power station, unlike the Charvak hydroelectric power station, the station's building is divided into two separate blocks, which should be drained. In each block there are two hydro turbines, which are of different types: two rotary-blade PL-587-VM-400 and two radial-axial RO/45-VM-406,4.

Drainage system proposed by Research and Production Enterprise "Vodopodyomnik" may be used at Farhad hydroelectric power station. There is a need in installation of four jet pumps at Farhad hydroelectric power station: two pumps in drainage wells of each block and two pumps in two shafts of the second block where hydro turbines are situated.

In 2016-2017, the installation of two jet pumps was started. For Farhad hydroelectric power station, with its multi-typed hydro units and independent blocks and non-connected drainage wells, a scheme for connecting the supply, pumping, discharge lines and shut-off valves will be introduced.

SOLUTIONS AND RECOMMENDATIONS

1. Analysis of the state of removal of seepage, drainage and wastewater from adjacent territories and buildings of hydropower and pumping stations in Uzbekistan and a literature review showed that the main hydraulic structures of the stations are not fully provided with measures of reliability and safety. The constant increase in the flow of wastewater associated with the aging of hydraulic structures and pressure pipes of water stations, leads to waterlogging of the territory, suffusion and failure of the soil under the concrete hydraulic structures, destruction of foundations and concrete blind areas and other negative consequences.
2. A mathematical model of the hydraulic control of a jet pump is developed for pumping streams with different suction heights from a building and the adjacent territory of hydroelectric power plants. As a result of mathematical modeling, the law of regulation of the hydraulic float system has been revealed. The optimal ratio of intake flows has been found for the most efficient operation of the simultaneous pumping system, which is achieved in the area of the shutter shut-off of the regulating float device at an angle of about 40^0–45^0. With further shut-off of the flop valve of the float system, the intake flow decreases due to increasing energy loss.
3. A method has been developed for calculating the geometric parameters of a jet pump and a system of hydromechanical control for the simultaneous removal of seepage, drainage and wastewater. It was determined that, in terms of its geometrical parameters, the float regulating device will be able to perform both of its functions - the function of the load and the function of the lifting mechanism without flooding the device it will be operational throughout the entire range of changes in intake costs from the building and from the territory of the pumping station.

4. Conducted field tests of the system with a jet pump for simultaneous removal of filtration, drainage and wastewater from the building and adjacent areas of the pumping station confirmed that jet pumps in auxiliary systems of pumping stations can also be used to drain the adjacent territory.
5. The scientific results of the research are recommended to allow efficient use of water and energy resources and to ensure reliable operation of the power equipment of stations, especially in the events of sudden power outages. The practical significance of the results of the work is in developing the devices for energy saving and protection of the main and auxiliary power equipment from accidents.

FUTURE RESEARCH DIRECTIONS

As further researches can be proposed the following:

- to establish a relationship of electromechanical and hydro-mechanical transients of pumping stations, taking into account the life and ageing of the pumping stations power equipment;
- to develop the algorithms with a law of voltage regulation and a law of control frequency change in function of time to protect the main power equipment of the pumping stations from water hammer;
- to conduct field tests of operating modes of the proportional closing of pump butterfly valve for the purpose of protection against water hammer and to compare the field tests results with the results obtained in the developed software package.

CONCLUSION

Energy efficient technology for pumping stations and hydroelectric power stations reliable pumping of drainage and sewage water, effective systems of reduction of transients in the stations pipelines, adjustable system for simultaneous pumping of filtration, drainage and sewage water are implemented at a number of large pumping and hydroelectric power stations of agricultural and water systems. Among them Uzbekistan largest stations such as Charvak hydroelectric power station, Farhad hydroelectric power station, pumping stations No 1÷5 of Karshi stage of Directorate of Exploitation of Karshi Main Canal, pumping stations DPS-1, DPS-2, DPS-3, "Khavast-Gallakor" of Djizak stage of Inter-provincial Directorate of Exploitation of Pumping Stations "Mirzachul" and a number of medium pumping stations including the pumping stations "Turkistan 1-1", "Turkistan 1-2" of Namangan Directorate of Pumping Stations Energy and Communications, pumping stations "Pakhtakor", "Guvalak" of Kashkadarya Directorate of Pumping Stations Energy and Communications, pumping stations "Karayantag", "DM-1" of Djizak Directorate of Pumping Stations Energy and Communications, pumping station "M-2-2" of Karshi Directorate of Exploitation of Small Pumping Stations (Certificate given by the Ministry of Agriculture and Water Resources of the Republic of Uzbekistan N° 04/31-700 on June 28, 2016).

REFERENCES

Allayev, K.R., Khokhlov, V.A., Sytdykov, R.A., & Titova, J.O. (2013). *Elektromekhanicheskiye i gidromekhanicheskiye protsessy v gidroenergeticheskikh ustanovkakh* [Electromechanical and hydro-mechanical processes in hydropower plants]. Tashkent: Fan va texnologiya.

Anderson, J. L., Lowry, M. V., & Thomte, J. C. (2001). Hydraulic and water quality modeling of distribution systems: What are the trends in the U.S. and Canada? In *Proceedings of the AWWA Annual Conference*. Denver, CO: American Water Works Association (AWWA).

Belash, I. G. (2010). Causes of the accident of the hydraulic unit No. 2 at the Sayano-Shushenskaya HPP. *Hydrotechnical construction*, *3*, 25–31.

Cookson, T., Lang, N., & Thornton, E. (2008). Adjustable speed drives applied to large AC induction motor and pump systems. In *Proceedings of the Twenty-fourth International Pump Users Symposium*, (pp. 75-80). Houston, TX: GA Valves. Retrieved January 20, 2016, from http://www.gavalves.co.uk

Garcia-Hernandez, A., Wilcox, M., & Moore, T. (2010) Hydraulic modelling and simulation of pumping systems. *Proceedings of the Twenty-sixth International Pump Users Symposium*, 81-88.

Ilyin, D., Shestopalova, T., Vaskov, A., & Ko, A. (2019). Strategies for Protection Systems of Wind Turbines with Doubly Fed Induction Generator: Control Strategies and Techniques for Fault Ride Through of Doubly Fed Induction Wind Generator. In Renewable Energy and Power Supply Challenges for Rural Regions (pp. 267-288). Hershey, PA: IGI Global.

Kamalov, T.S. (2014). *Chastotno-reguliruyemyy elektroprivod nasosnykh stantsiy sistem mashinnogo orosheniya* [Frequency-regulated electric drive of pumping stations of machine irrigation systems]. Tashkent: Fan.

Kharchenko, V., Panchenko, V., Tikhonov, P. V., & Vasant, P. (2018) Cogenerative PV thermal modules of different design for autonomous heat and electricity supply. In Handbook of Research on Renewable Energy and Electric Resources for Sustainable Rural Development (pp. 86-119). Hershey, PA: IGI Global. doi:10.4018/978-1-5225-3867-7.ch004

Khashimov, A.A. & Abidov, K.G. (2012) *Energoeffektivnyye sposoby samozapuska elektroprivodov nasosnykh stantsiy* [Energy-efficient ways to self-start electric drives of pumping stations]. Tashkent: Fan va texnologiya.

Khokhlov, A.V., Khokhlov, V.A. & Titova, J.O. (2014) *Rezhimy raboty nasosnykh stantsiy Dzhizakskogo kaskada* [Operating modes of Djizak pumping stations stage. Monograph.]. Tashkent: Fan va texnologiya.

Khokhlov, A. V., Khokhlov, V. A., & Titova, J. O. (2015). *Rezhimy raboty nasosnykh stantsiy Karshinskogo kaskada* [Operating modes of Karshi pumping stations stage]. Tashkent: Navruz.

Khokhlov, A. V., Khokhlov, V. A., & Titova, J. O. (2018) Energy Saving and Safe Operating Modes of the Large Irrigative Pumping Stations. In Handbook of Research on Renewable Energy and Electric Resources for Sustainable Rural Development (pp. 176-203). Hershey, PA: IGI Global. doi:10.4018/978-1-5225-3867-7.ch008

Khokhlov, V.A., Khokhlov, A.V. & Titova, J.O. (2011) *Regulirovaniye rezhimov raboty struynykh nasosov* [Regulation of operating modes of jet pumps. Monograph.]. Tashkent: Fan va texnologiya.

Larsen, T. (2012). *Water hammer in pumped sewer mains*. Aalborg, Denmark: University of Aalborg, Department of Civil Engineering.

Leznov, B. S. (2006). *Energosberezhenie i reguliruemyy privod v nasosnykh i vozdushnykh ustanovkakh* [*Energy saving and adjustable drive for pump and blower sets*]. Moscow: Energoatomizdat.

Li, Ch., Ma, Sh., Liu, H., Wu, H., Gao, P., & Yang, F. (2016). *Research on scheme of pumping energy-saving based on computer control technology*. MATEC Web of Conferences. Retrieved March 21, 2017, from http://www.matec-conferences.org

Sipos Aktorik. (n.d.). Retrieved January 20, 2016, from http://www.sipos.de

Technical Report of the Main Directorate of Pumping Stations Energy and Communications of the Ministry of Ministry of Agriculture and Water Resources of the Republic of Uzbekistan. (2006). Tashkent.

Val-Matic Valve & Manufacturing Corp. (n.d.). Retrieved January 20, 2016, from http://www//valmatic.com

Vishnevskii, K. P. (n.d.). *Transient processes of pumping stations. Calculations of hydraulic shock in pressure pipelines*. Retrieved January 29, 2016, from http://library.fsetan.ru/doc/perehodnyie-protsessyi-nasosnyih-stantsij-raschetyi-gidravlicheskogo-udara-v-napornyih-truboprovodah-posobie-k-vsn-33-2212-87-meliorativnyie-sistemyi-i-sooruzheniya-nasosnyie-stantsii-normyi-proektirovaniya

Chapter 14
Self-Assistive Controller Using Voltage Droop Method for DC Distributed Generators and Storages

Ranjit Singh Sarban Singh
Universiti Teknikal Malaysia Melaka, Malaysia

Maysam Abbod
https://orcid.org/0000-0002-8515-7933
Brunel University London, UK

EXECUTIVE SUMMARY

With the rapid growth of distributed generation currently, DC microgrids energy system structure is being deployed in parallel with, or independently from, the main power grid network. The DC microgrids energy system structure is designed to provide an effective coordination with the aggregating distributed generators, energy storage, and connected loads. In this sense, the DC microgrids energy system structure can be connected to the grid network or can be off-grid network. In the mode of grid network connected, DC microgrids energy system structure is presented as a controllable entity. When it is necessary, DC microgrids energy system is connected in islanded mode to deliver reliable power to the grid network during the interrupted power supply from the grid network system. Having said that, the DC microgrids energy system structure is encompassed of renewable energy sources, energy storages and loads, and not excluding the grid network transmission. Hence, this chapter proposes to focus on designing and modelling a self-assistive controller using voltage droop method for DC distributed generators and storages which is a part of the DC microgrids energy system structure.

DOI: 10.4018/978-1-5225-8559-6.ch014

INTRODUCTION

With the improving of community economy and expansion of power utilization demand, the generation of electric energy has to be increased. Nevertheless, large-scale power systems which are developed to bring great benefits to the society are potentially dangerous. Looking at the potential dangerous of large-scale power system, the first micro-grid concept was established by the Consortium for Electric Reliability Technology Solutions (CERTS)(Sun & Liu, 2018). The micro-grid system comprises of an AC microgrids energy system structure, DC microgrids energy system structure and AC / DC microgrids energy system structure. As the concept of microgrids energy system structure was introduced, AC microgrid energy structure system was easier to realized due to AC microgrids network system maturity. But due to the complexity and inefficiency of AC microgrids energy system structure as mentioned in (Sun & Liu, 2018), DC microgrids energy system structure are considered more efficient and flexible utilization of distributed energy resources (Davis & Costyk, 2005).

The advantage of the DC microgrids energy system structure has driven to capability of integrating different kinds of DC distributed energy resources and generators, such as renewable energy resources (RERs) and energy storage system and micro turbines and improve the overall system sustainability and efficiency. DC microgrids energy system structure is a small size DC distributed system which connecting multiple renewable energy resources and energy storage system. As mentioned in Xu et al. (2018), the rapid development of the semiconductor devices recently has made the DC conversion to AC much flexible. This has led to more complex design of DC microgrids energy system structure that poses greater challenges to organize, control and manage these systems. Therefore, this chapter describes the designing and modelling of a DC microgrids energy system structure based on self-assistive controller using voltage droop method for the connected DC distributed generators.

BACKGROUND INFORMATION

This chapter proposes a complete design and model of a self-assistive controller which uses the voltage droop method to sense, measure, coordinate and control the connected DC distributed generators. The self-assistive controller sense and measure the output voltages from the DC distributed generators and effectively coordinate and control the power flow from the DC distributed generators to the connected loads. The self-assistive controller for DC distributed generators is divided into two (2) stages. In Stage I, the methodological design of the self-assistive controller using voltage droop method for DC Distributed Generators and Storages is presented and fundamental operation of the proposed controller is described. In the Stage II, SIMULINK – MATLAB software is used to model the self-assistive controller in terms of its electronic components is presented. During the modelling of each part in Stage II, each developed part is referred with the methodological design in Stage I. Completion of each part in Stage II is simulated and results obtained is recorded as preliminary finding which is used as reference results to model the next part in Stage II. The previously obtained results from previously modelled part is used to analyze the electronically developed circuit using the SIMULINK - MATLAB software. This modelling process is continued till the proposed self-assistive controller for DC distributed generators for the connected loads is modelled successfully. The difference in the obtained results in Stage II are studied and analyzed to make any necessary changes to the modelled system. After studying, analyzing and making necessary changes to the modelled system, final results are presented which demonstrate the sensing, measuring, coordination and control of proposed self-assistive controller for DC distributed generators using the voltage droop method. The obtained final results also demonstrated the effectiveness of energy management and optimize the available renewable energy resources as primary energy source for the connected loads.

MOTIVATION AND LITERATURE REVIEW

To deal with the increasing number of penetration of DC distributed generators as a system, a proper coordination scheme is required to manage all the DC distributed generators and actively operate as a complete system. DC distributed generators is built using small-scale generation sources such as solar photovoltaic, wind energy turbine, biogas and etc. Hence, integration of these DC distributed generators into microgrid system requires a proper control technique. In the following paragraph, available control techniques from previously developed research work is discussed.

The microgrid concept has been introduced in Lasseter et al. (2002) and Lasseter and Paigi (2004). Microgrid is also known for its operation as islanded and grid connected mode (Katiraei, Iravani, & Lehn, 2005) and address economical, technical and concerns on the environmental in modern power systems (Vaccaro, Popov, Villacci, & Terzija, 2011). In Rey, Marti, Velasco, Miret, and Castilla (2017), microgrid has been disconnected from the main grid network, therefore, multi schemes are required to be implemented to control the distributed generators and distributed energy storages.

Among the schemes as mentioned in Iovine, Damm, De Santis, and Di Benedetto (2017), great effort has been put into developing dedicated DC microgrid energy system structure and network control system which employs different level of control which ineffectively to be implemented on the power flow control. As mentioned in Iovine et al. (2017), a dynamic power flow model or controller such as in Dou, Zhang, Yue, and Zheng (2017), Zhang and Ukil (2019), Worku, Hassan, and Abido (2019) would be more efficient and able to schedule the system operation based on real-time meteo forecasts (Bracco, Delfino, Pampararo, Robba, & Rossi, 2015).

Many schemes has been investigated such as Palizban and Kauhaniemi (2015), Zaheeruddin and Manas (2015), Arikiez, Grasso, and Zito (2016), Bhoye et al. (2016), Javaid et al. (2017), Jayachandran and Ravi (2017), Hussain et al. (2018), and Hemmati (2018). This research proposes flexible microgrid systems which are capable of intelligently operate in islanded and grid network connected. Besides that, the developed research work also focuses on the overall system stabilization and operation smoothness. As the research extended, the proposed schemes were also improved to overcome relevant conflicts in the previously developed microgrids.

However, mostly research presented in all the previous research work is focusing on the DC/AC hybrid based power system. After studying the importance of the DC microgrid energy system structure and network system, there is a need to also strategies the DC microgrid energy system structure for efficient power delivery. With that, this chapter study on the designing and modelling of a self-assistive controller using voltage droop method for DC distributed generators and energy storages. The self-assistive controller designed and modelled at some intermittent forms intelligently managed and handles uncertainties of the system operation.

RESEARCH DESCRIPTION

This chapter presents the designing and modelling of self-assistive controller using voltage droop methods for DC distributed generators and energy storage system in the SIMULINK – MATLAB software. Therefore, in presenting the designing and modelling, a comprehensive self-assistive controller using voltage droop method for DC distributed generators and energy storage system composed of the voltage sensing and measurement electronic circuit, voltage base conditioning chart and electronic conditioning circuits are presented in this section.

Voltage droop is known as accidental loss in output voltage of a device which drives the load. Hence, to be able to capture the accidental loss in output voltage from connected DC distributed generators it is important to sense and measure the amount of available output voltages from each DC distributed generators which drives the load. This to allow the DC distributed generator to switch to other high output voltage DC distributed generator to continuously drive the connected load. Because of all the aforementioned, the voltage droop method which uses the voltage sensing and measurement electronic circuit to sense the amount of output voltages from available DC distributed generators along with the self-assistive controller which operates based on the voltage droop method is discussed in the following section.

DC DISTRIBUTION GENERATORS AND ENERGY STORAGE SYSTEM

To sense, measure and switch between the DC distributed generators for continuous energy source supply to the load effective system design and control application is necessary. In this section, the block design of complete DC Distributed Generators and Energy Storage System with Self-Assistive Controller is illustrated and system operation methodologically is described.

Figure 1 shows the complete control and structure of DC Distributed Generators and Energy Storage System with Self-Assistive Controller. The methodological design is divided into four (4) sections which is interconnected to perform optimum operation among the DC distributed generators and connected loads. A detailed technical illustration of each section of DC Distributed Generators and Energy Storage System with Self-Assistive Controller is presented in Figure 1. Section A explains about the DC Renewable Energy Resources Energy and Voltage Base Conditioning Chart. The regulated DC voltages from the solar and wind is sent into the voltage sensing and measurement electronic circuit. These voltages are converted into digital values before sending the digital values into the voltage base conditioning chart to perform the logical self-assistive operation. Section B contains six (6) different types of configurations of electronic conditioning circuits. These six (6) configurations are able to identify the possibility of when both or either one of the solar or wind energy resource is available. Section C is about voltage switching and group switching. This section contains the control switches between the sources to load or BESS. Last but not least, section D is grid connected system. The grid connected system will only be activated if the DC Distributed Generators and Energy Storage System is unable to support the load demand. To further understand the operation of each sub-system in each section, a detail description is presented in the next section.

Figure 1. Block diagram of complete DC distributed generators and energy storage system with self-assistive controller

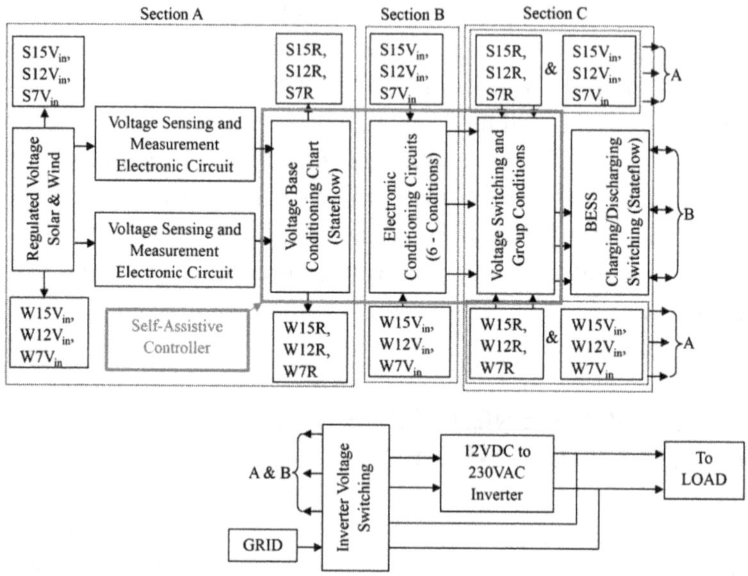

Figure 2. Block diagram of DC renewable energy resources and voltage base conditioning chart model

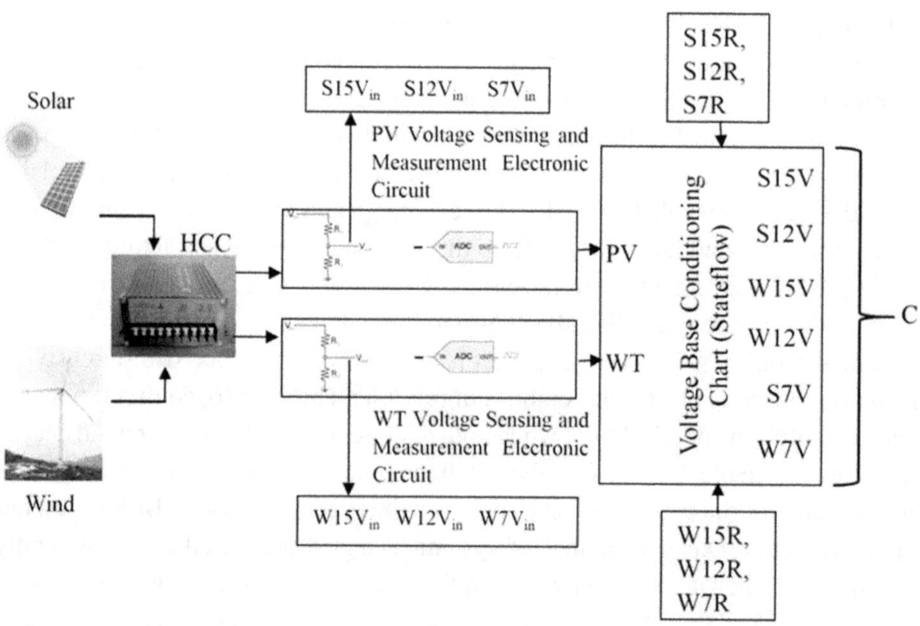

DC DISTRIBUTION GENERATORS: CIRCUIT SYSTEM COMPONENTS

The DC Renewable Energy Resources Energy and Voltage Base Conditioning Chart model is presented in Figure 2. Referring to block diagram in Figure 2, the solar energy and wind energy produces regulated voltage via the Hybrid Charge Controller (HCC). The HCC regulates the fluctuating output voltages from the solar and wind energy systems as stable output voltages before sending these voltages into the voltage sensing and measurement electronic circuit. In the proposed system, the solar and wind energy are used as primary power supply sources while Energy Storage System (ESS) is used as secondary power supply source.

Shown in Figure 2, the output voltages from HCC are connected to the input of the PV and WT voltage sensing and measurement electronic circuit. The voltage sensing and measurement electronic circuit is known as voltage divider circuit. This circuit sense and measure the amount of voltage at the respective resistor and send the amount of voltage to the analogue to digital (ADC) converter to be converted into digital voltage. The digital output voltage is sent to the voltage base conditioning chart to perform the logical self-assistive which is performing decision to connect to load or BESS.

The concept of voltage sensing and measurement electronic circuit which is based on voltage divider is used to sense and measure the increment and decrement of output voltages from solar and wind renewable energy resources. To understand the operation of voltage sensing and measurement electronic circuit, the following explains the concept of voltage sensing and measurement during voltage increment and decrement.

Let's assume voltage source at solar renewable energy output voltage is 17.5 Volt, Resistor 1 (R_1) = 10 kΩ, Resistor 2 (R_2) = 15 kΩ, hence output voltage at R_1 needs to be calculated.

Hence,

$$\text{Output Voltage at Resistor 1 } (VR_1) = \left(\frac{R_1}{R_1 + R_2}\right) V_S \qquad (1)$$

$$= \left(\frac{10k\Omega}{10k\Omega + 1.5k\Omega}\right)(17.5Volt)$$

$$= 15Volt$$

Table 1. Output voltage (analogue to digital) quantification from solar and wind renewable energy resources

No.	Sources	Conditions (Analogue Voltage)	Conditions (Digital Voltage)
1.	Solar Photovoltaic (PV)	12 Volt < PV ≤ 15 Volt	1404 < PV ≤ 1755
	Wind Turbine (WT)	12 Volt < WT ≤ 15 Volt	1404 < WT ≤ 1755
2.	Solar Photovoltaic (PV)	7 Volt < PV ≤ 12 Volt	819 < PV ≤ 1404
	Wind Turbine (WT)	7 Volt < WT ≤ 12 Volt	819 < WT ≤ 1404
3.	Solar Photovoltaic (PV)	0 Volt < PV ≤ 7 Volt	0 < PV ≤ 819
	Wind Turbine (WT)	0 Volt < WT ≤ 7 Volt	0 < WT ≤ 819

The output voltage VR_1 from the solar renewable energy resource is equal to 15 Volt. The configuration in Figure 2 and Equation 1 is used to calculate the output voltages from the wind renewable energy source and ESS. Sensing and measurement of output voltages from solar and wind renewable energy resources and ESS are important to allow to perform the self-assistive among the resources which effectively helps to optimize the resources for connected load or ESS charging.

Therefore, before discussing the self-assistive controller's methodology the output voltages from the solar and wind renewable energy resources and ESS are quantified into three sections. Table 1 shows the quantification of output voltages of solar and wind renewable energy resources. The output voltage quantifications in Table 1 is used to design and develop voltage base conditioning chart shown in Figure 1. The digital output voltages quantification shown in Table 1 is used to develop the decision making algorithm in the voltage base conditioning chart shown in Figure 1.

The following Equation 2 explains about the analogue output voltage to digital voltage conversion calculation and this is presented in Table 1.

Hence,

$$\frac{\mathrm{Re}\,solution}{System\,Voltage} = \frac{Digital Output\,Voltage}{Ana\log ue\,Voltage Measured} \tag{2}$$

Therefore, Equation 2 is used to calculate the digital voltage output for analogue output voltages of 15 Volt, 12 Volt and 7 Volt.

For 15 Volt

$$\frac{2048}{17.5 Volt} = \frac{x}{15 Volt}$$

$$x = \frac{(2048)(15Volt)}{17.5Volt}$$

= 1755 Volt Digital Voltage

For 12 Volt

$$\frac{2048}{17.5Volt} = \frac{x}{12Volt}$$

$$x = \frac{(2048)(12Volt)}{17.5Volt}$$

= 1404 Volt Digital Voltage

For 7 Volt

$$\frac{2048}{17.5Volt} = \frac{x}{7Volt}$$

$$x = \frac{(2048)(7Volt)}{17.5Volt}$$

= 819 Volt Digital Voltage

The decision making algorithm is developed for self-assistive controller of solar and wind renewable energy resources is presented in Figure 3. The digital output voltages presented in Table 1 is used to develop the self-assistive decision making among the solar and wind renewable energy resources. Hence, Figure 3 depicts the decision making algorithm methodology for solar and wind renewable energy resources which is illustrated based on output voltages quantification in Table 1.

Figure 3 shows nine different conditions which allows the solar and wind renewable energy resources to perform self-assistive based on the output digital voltages quantifications shown in Table 1. Each condition that helps the self-assistive controller to perform decision making between solar and wind renewable energy resources are described in the following.

Figure 3. Decision making algorithm for voltage base conditioning chart

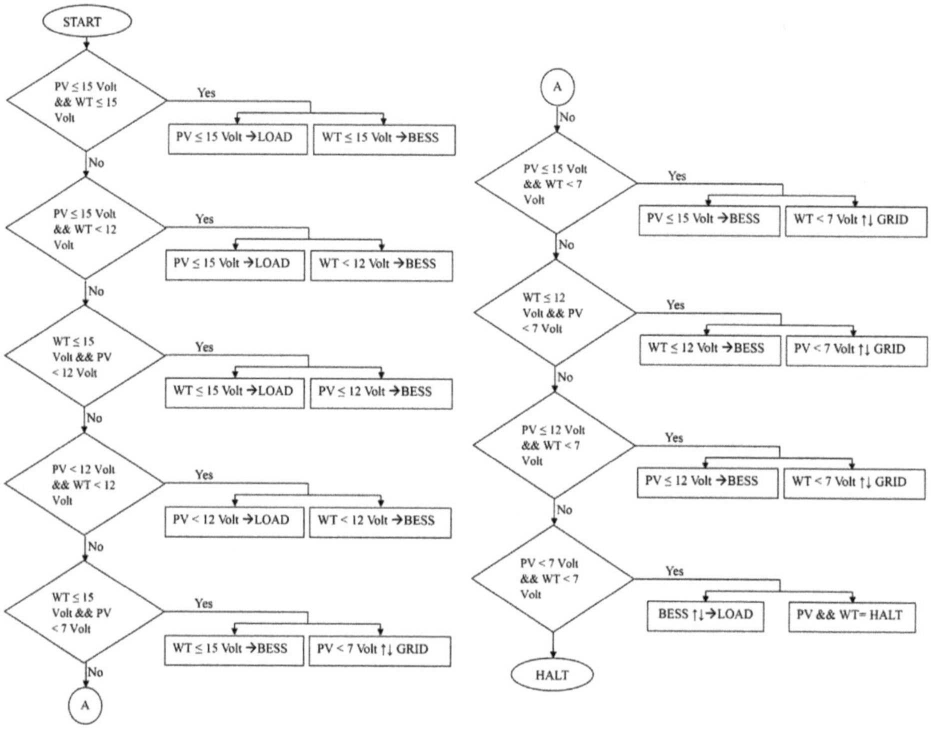

Self-Assistive Condition 1: 12 Volt < Solar and Wind ≤ 15 Volt

When the solar and wind renewable energy resources outputs 15 Volt voltage, the decision making algorithm in the voltage base conditioning chart produces high signals (S15V = 1) to activate the connectivity to load and (W15V = 1) to activate the connectivity of ESS for charging.

Self-Assistive Condition 2: 7 Volt < Solar and Wind < 12 Volt

When the solar and wind renewable energy resources outputs less than 12 Volt, the solar and wind renewable energy resources are connected to a DC to DC Boost Converter before supplying the output voltage to connected load. The decision making algorithm developed for voltage base conditioning chart produces high signals (S12V = 1) for solar renewable energy source and (W12V = 1) for wind renewable energy source. The high signal S12V = 1 activates the connectivity to load and W12V = 1 activates connectivity to ESS.

Self-Assistive Condition 3: 12 Volt < Solar ≤ 15 Volt and 7 Volt < Wind ≤ 12 Volt

When the solar renewable energy source produces output voltage of 15 Volt and wind renewable energy source produces output voltage of 12 Volt, the decision making algorithm in the voltage base conditioning chart produces high signal (S15V = 1) for solar renewable energy source and high signal (W12V = 1) for wind renewable energy source. The high signal S15V = 1 activates the connectivity to load and W12V = 1 activates connectivity to ESS. This condition also applied if 12 Volt <wind ≤ 15 Volt & 7 Volt < solar ≤ 12 Volt.

Self-Assistive Condition 4:12 Volt < Solar ≤ 15 Volt and 0 Volt < Wind ≤ 7 Volt

This condition is activated only when solar renewable energy source produces output voltage of 15 Volt and wind renewable energy source produce less than 7 Volt. During this condition, the decision making algorithm in voltage base conditioning chart produces high signal (S15V = 1) which signal is used to activate the connectivity to ESS for charging process if required. This condition also applied for 12 Volt < wind ≤ 15 Volt & 0 Volt < solar ≤ 7 Volt.

Self-Assistive Condition 5: 7 Volt < Solar ≤ 12 Volt and 0 Volt < Wind ≤ 7 Volt

This condition is activated only when solar renewable energy source produces output voltage of 12 Volt and wind renewable energy source produce less than 7 Volt. During this condition, the decision making algorithm in voltage base conditioning chart produces high signal (S12V = 1) which activates the connectivity to ESS for charging process if required. Prior to that, the 12 Volt output voltage will be boosted through DC to DC Boost Converter to charge the ESS. This condition also applied for 7 Volt < wind ≤ 12 Volt & 0 Volt < solar ≤ 7 Volt.

Self-Assistive Condition 6: 0 Volt < Solar and Wind ≤ 7 Volt

When the solar and wind renewable energy resources output voltages is less than 7 Volts, the decision making algorithm in voltage base conditioning chart sends a high signal (S7V = 1 and W7V = 1) to the system to inform that solar and wind renewable energy resources are not producing any output voltages. During this condition, the load is connected to the ESS or GRID network for power source supply.

Figure 4. Input output connectivity diagram

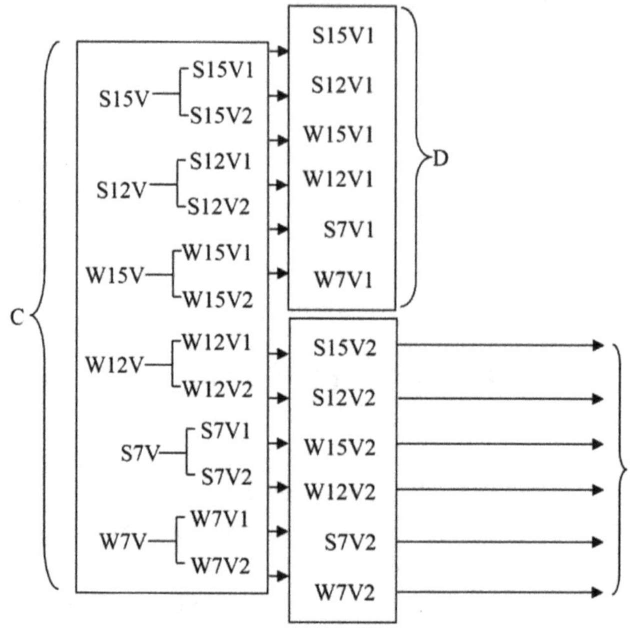

Figure 4 shows the input output connectivity block diagram between the voltage base conditioning chart and the electronic conditioning circuit's subsystem. As shown in Figure 2, the voltage base conditioning chart produces output logics HIGH or LOW base on the voltage quantifications in Table 1. Therefore, the input part C in Figure 4 is connected to the output part at Figure 2. Each HIGH and LOW output logic signal at the voltage base conditioning chart is divided into two output logics as shown in Figure 4. This section will explain the output logic group E signals as control signals for the voltage switching and group conditions shown in Figure 1.

The input-output connectivity shown in Figure 4 is the subsystem that provides an effective control between other connected subsystems. Figure 5 shows voltage switching and group conditions, relays are used as control switches between the individual solar and wind renewable energy resources as well as to allow self-assistive control operation between the resources. Figure 5 depicts six (6) types of different configurations for solar and wind renewable energy resources. The six (6) configurations are identified as the possible configurations when both or either one of the sources is available.

The voltage switching and group conditions shown in Figure 5 is controlled by the circuit breaker setup shown in Figure 6. The configuration shown in Figure 6 is used for four different pairs of solar and wind renewable energy resources condition shown in Table 2.

Figure 5. Voltage switching and group conditions

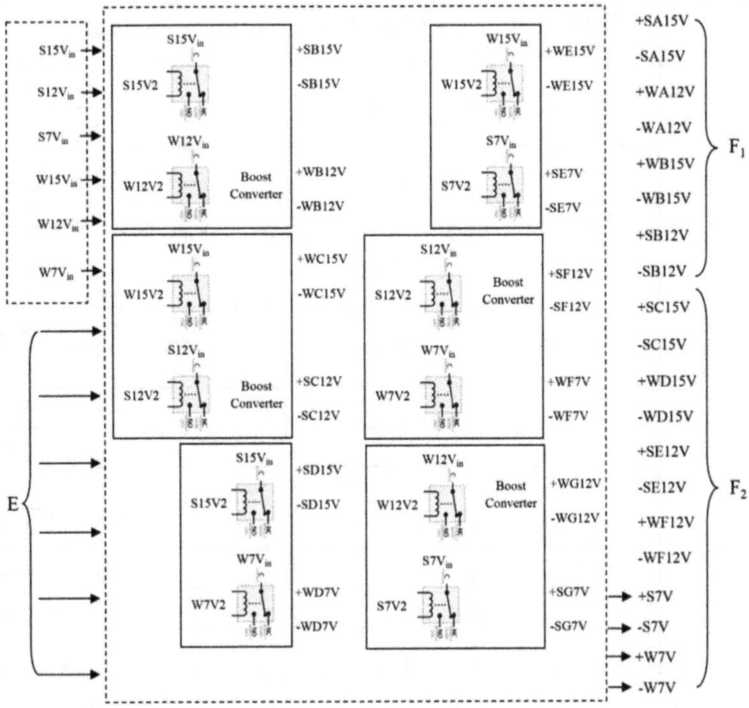

Figure 6. Part 1: Circuit breaker setup block diagram for voltage switching and group conditioning

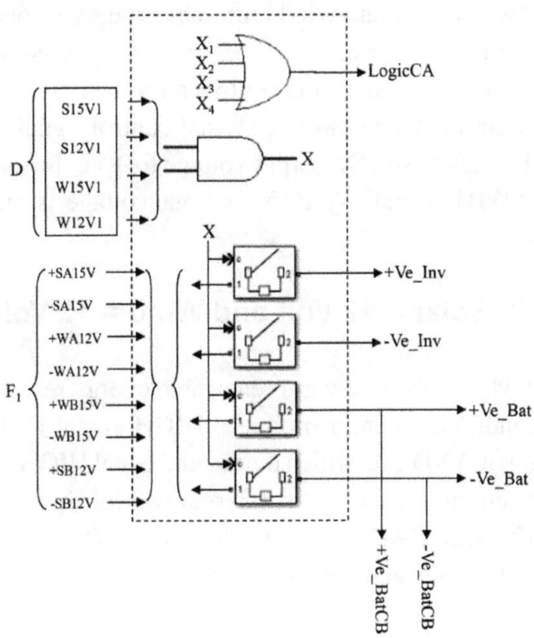

Table 2. Part 1: Pairing solar and wind renewable energy resources

Pairing No.	Voltage Base Controller Output	Analogue Input Voltage	Connectivity	
1.	S15V1	15 Volt	F1	+SA15V
				-SA15V
	W15V1	15 Volt	F1	+WA15V
				-WA15V
2.	S12V1	12 Volt	F1	+SB12V
				-SB12V
	W12V1	12 Volt	F1	+WA12V
				-WA12V
3.	S15V1	15 Volt	F1	+SA15V
				-SA15V
	W12V1	12 Volt	F1	+WA12V
				-WA12V
4.	W15V1	15 Volt	F1	+WA15V
				-WA15V
	S12V1	12 Volt	F1	+SB12V
				-SB12V

Configuration 1: Solar = 15 Volt and Wind = 15 Volt

Table 2 shows that when solar and wind renewable energy resources are 15 Volts, the voltage base conditioning chart produces HIGH signal logic for S15V1 and W15V1 to the 2 – inputs AND gate which output an equal HIGH signal. This HIGH signal is send to the circuit breaker as an external control signal (X) to activate the connectivity +SA15V and –SA15V output voltage to +Ve_Inv and –Ve_Inv ports. The connectivity +WB15V and –WB15V output voltage is output to +Ve_Bat and –Ve_Bat ports.

Configuration 2: Solar = 12 Volt and Wind = 12 Volt

Table 2 shows that when solar and wind renewable energy resources are 12 Volts, the voltage base conditioning chart produces HIGH signal logic for S12V1 and W12V1 to the 2 – input AND gate which output and equal HIGH signal. This HIGH signal is send to the circuit breaker as an external control signal (X) to activate the connectivity +SA12V and –SA12V output voltage to the DC to DC Boost Converter before supplying to +Ve_inv and –Ve_Inv ports.

Configuration 3: Solar = 15 Volt and Wind = 12 Volt

Table 2 shows that when Solar = 15 Volt and Wind = 12 Volt, the voltage conditioning chart produces HIGH signal for S15V1 and W12V1 to the 2 – inputs AND gate shown in Figure 6. The AND gate output HIGH signal is send into the circuit breaker as an external control signal (X) to activate the connectivity +SA15V and –SA15V output voltage to +Ve_Inv and –Ve_Inv ports. Whereas, the +WA12V and –WA12V HIGH signal activates the connectivity to the DC to DC Boost Converter prior to +Ve_Bat and –Ve_Bat ports.

Configuration 4: Wind = 15 Volt and Solar = 12 Volt

Table 2 shows that when Wind = 15 Volt and Solar = 12 Volt, the voltage conditioning chart produces HIGH signal for W15V1 and S12V1 to the 2 – input AND gate shown in Figure 6. The AND gate output HIGH signal is send into the circuit breaker as an external control signal (X) to activate +WB15V and –WB15V output voltage to +Ve_Inv and –Ve_Inv ports. Whereas, the +SB12V and –SB12V output voltage is connected to the DC to DC Boost Converter prior to +Ve_bat and –Ve_Bat ports.

Concluding the part 1 configuration, each pair of the setup only operate when the output logic from the voltage conditioning chart matches the condition part 1 in Table 2.

There are four (4) other configurations despite the one presented in part 1 Table 2. These configurations are presented in Table 3 and these configurations are mainly based on single availability of source from renewable energy resource. This single available source will mainly be used to charge the ESS only. Figure 7 shows the part 2 Circuit breaker setup block diagram for voltage switching and group conditioning. In part 2, the circuit breaker is setup only to activate one renewable energy source. The configuration is explained further in the following.

Configuration 5: Solar = 15 Volt and Wind = 7 Volt

Table 3 shows when Solar = 15 Volt and Wind = 7 Volt, the voltage conditioning chart produces HIGH logic for S15V1 and W7V1 to the 2 – inputs AND gate in Figure 7. The AND gate output HIGH signal is send to the circuit breaker as an external control signal (X) to activate the +SC15V and -SC15V output voltage connectivity to the +Ve_BatCB and –Ve_BatCB ports. The +W7V and –W7V output voltage is connected to the GROUND ports.

Configuration 6: Wind = 15 Volt and Solar = 7 Volt

Table 3 shows when Wind = 15 Volt and Solar = 7 Volt, the voltage conditioning chart produces HIGH logic for W15V1 and S7V1 to the 2–inputs AND gate in Figure 7. The AND gate output HIGH signal is send to the circuit breaker as an external control signal (X) to activate +WD15V and –WD15V output voltage connectivity to the +Ve-BatCB and –Ve_BatCB ports. The +S7V and –S7V output voltage is connected to the GROUND ports.

Figure 7. Part 2: Circuit breaker setup block diagram for voltage switching and group conditioning

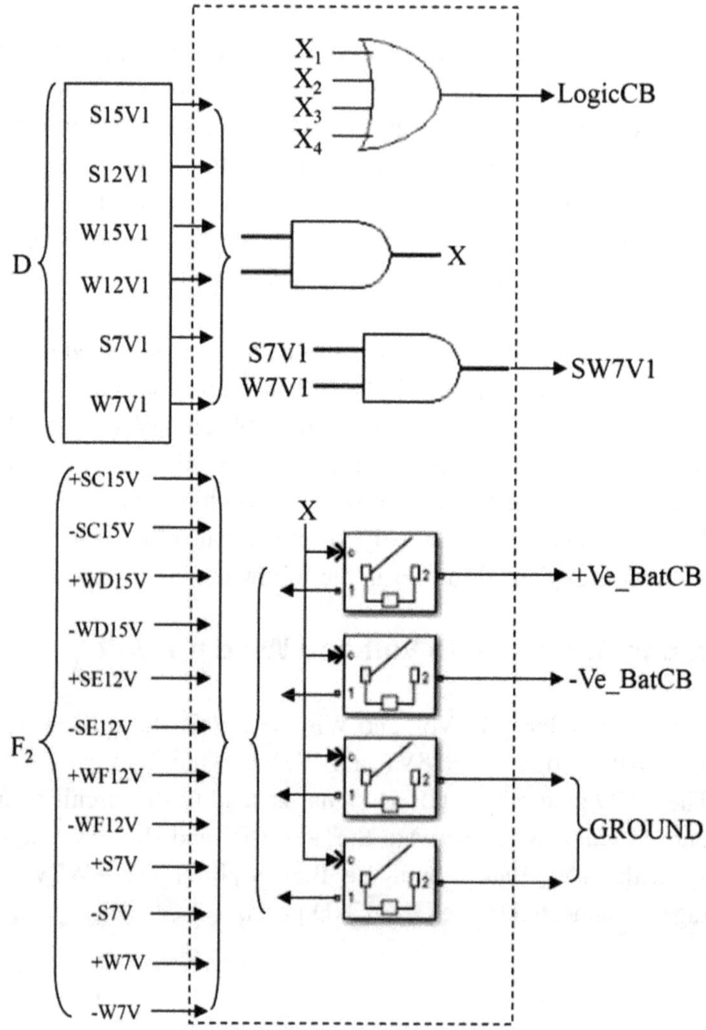

Configuration 7: Solar = 12 Volt and Wind = 7 Volt

Table 3 shows when Solar = 12 Volt and Wind = 7 Volt, the voltage conditioning chart produces HIGH logic for S12V1 and W7V1 to the 2 – inputs AND gate in Figure 7. The AND gate output HIGH signal is send to the circuit breaker as an external control signal (X) to activate +SE12V and –SE12V output voltage connectivity to the DC to DC Boost Converter before connecting to +Ve_BatCB and –Ve_BatCB ports. The +W7V and –W7V output voltage is connected to the GROUND ports.

Configuration 8: Wind = 12 Volt and Solar = 7 Volt

Table 3 shows when Wind = 12 Volt and Solar = 7 Volt, the voltage conditioning chart produces HIGH logic for W12V1 and S7V1 to the 2 – inputs AND gate in Figure 7. The AND gate output HIGH signal is send to the circuit breaker as an external control signal (X) to activate +WF12V and –WF12V output voltage connectivity to the DC to DC Boost Converter before connecting to +Ve_BatCB and –Ve_BatCB ports. The +S7V and –S7V output voltage is connected to the GROUND ports.

Table 3. Part 2: Pairing solar & wind renewable energy sources

Pairing No.	Voltage Base Controller Output	Analogue Input Voltage	Connectivity	
5.	S15V1	15 Volt	F2	+SC15V
				-SC15V
	W7V1	7 Volt	F2	+W7V
				-W7V
6.	W15V1	15 Volt	F2	+WD15V
				-WD15V
	S7V1	7 Volt	F2	+S7V
				-S7V
7.	S12V1	12 Volt	F2	+SE12V
				-SE12V
	W7V1	7 Volt	F2	+W7V
				-W7V
8.	W12V1	12 Volt	F2	+WF12V
				-WF12V
	S7V1	7 Volt	F2	+S7V
				-S7V

Concluding the four configurations part 2 presented in Table 3, each pair only operate for one available renewable energy source. During the operation of any one condition, the available output voltage is use to charge the BESS which is connected as secondary power source supply for load.

RESULTS AND DISCUSSION

After completing the designing and modelling of each subsystem of the self-assistive controller using voltage droop method for distributed generators and storages in Simulink/Stateflow MATLAB environment, all the subsystem is connected together to demonstrate the system's aims and objectives. The following section presents the system's simulation results for selected situation and condition presented in the methodology section. Based on this selected situation and condition the proposed system's operation is validated.

Self-Assistive Condition 1: 12 Volt < Solar and Wind ≤ 15 Volt

Figure 8 (a) and (b) presents the sensed and measured output voltages form solar and wind renewable energy resources.

Voltage conditioning chart shown in Figure 2 output a HIGH signal at S15V port when PV15 = 1 and at W15V port when WT15 = 1.

Referring to Figure 4, activated HIGH signals are received at S15V2 and W15V2 ports of Figure 5. The connected relays are switched ON. Hence, the 15 Volt output voltage result is presented in Figure 9.

The SA15V and WB15V output voltages are connected to Ve_Inv ports to supply power source to load and Ve_Bat ports to charge ESS as shown in Figure 6. The 12~15 Volt solar output voltage is send to Ve_Inv ports for inversion to AC voltage. The 230 VAC output voltage at the output Ve_Inv ports are shown in Figure 10. The 12~15 Volt wind output voltage is connected to Ve_Bat ports shown in Figure 6 is used to charge ESS.

Self-Assistive Condition 3: 12 Volt < Solar ≤ 15 Volt and 7 Volt < Wind ≤ 12 Volt

Figure 11 (a) and (b) presents the sensed and measured output voltages form solar and wind renewable energy sources. Voltage conditioning chart shown in Figure 2 output a HIGH signal at S15V port when PV15 = 1 and at W12V port when WT12 = 1.

Figure 8. Solar and wind renewable energy resources output voltages – voltage conditioning chart

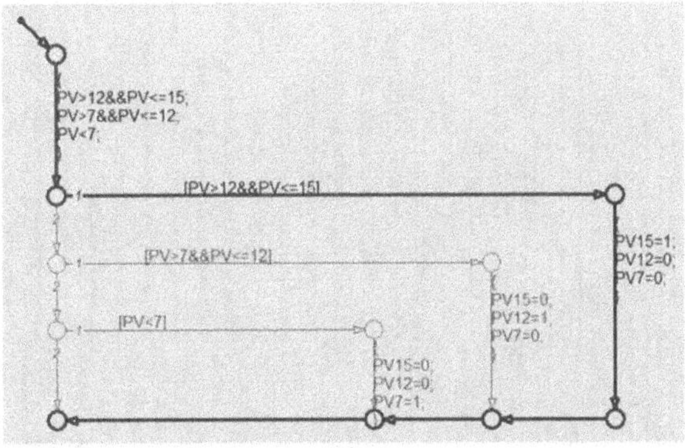

(a)

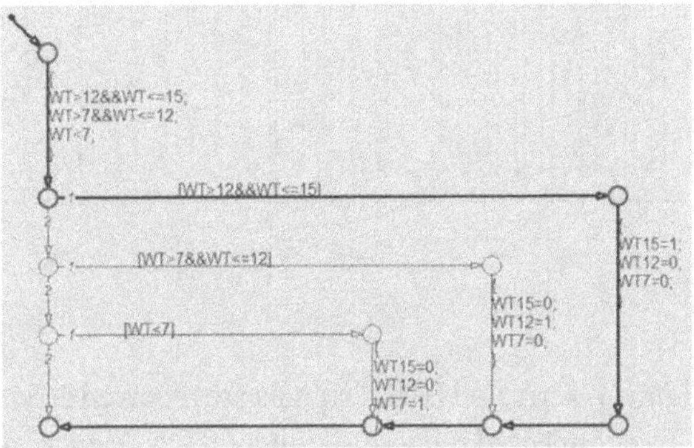

(b)

Figure 9. 15 Volt output voltage – Solar and Wing renewable energy resources

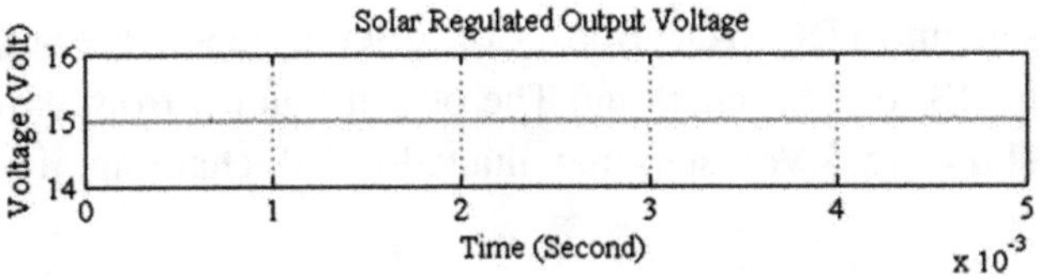

(a)

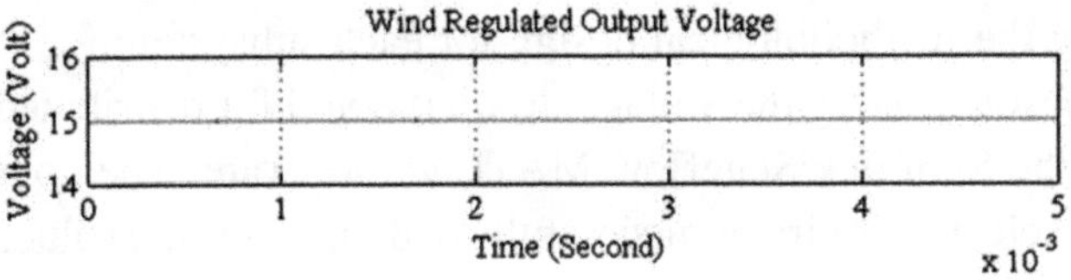

(b)

Figure 10. 240 VAC Inverter output voltage

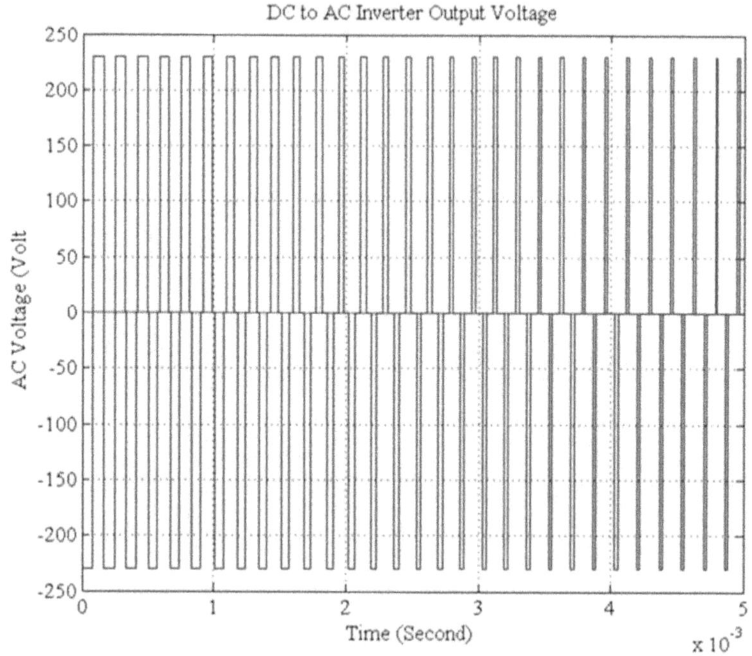

Referring to Figure 4, activated HIGH signals are received at S15V2 and W12V2 ports of Figure 5. The connected relays are switched ON. Hence, the 15 Volt output voltage result is presented in Figure 12.

The 12~15 Volt solar output voltage is connected at Ve_Inv ports and inverted into an AC voltage. Whereas the 7~12 Volt wind output voltage is connected at Ve_Bat ports is used to charge ESS, if required. Prior to that, the 7~12 Volt wind output voltage is sent into a DC to DC Boost Converter to step-up the output voltage as shown in Figure 13 for ESS charging. The output voltage from the DC to DC Boost Converter shows 15.5 Volt step-up voltage for ESS charging, if required.

CONCLUSION

After completing the methodological design for each subsystem for the proposed self-assistive controller using the voltage droop method for distributed generators and storages in the Simulink/Stateflow MATLAB software. The voltage sensing and measurement electronic circuit successfully demonstrated the voltage increment and decrement for solar and wind renewable energy sources. Also, the self-assistive decision making algorithm in the voltage conditioning chart also successfully

Figure 11. Solar and wind renewable energy resources output voltages – voltage conditioning chart

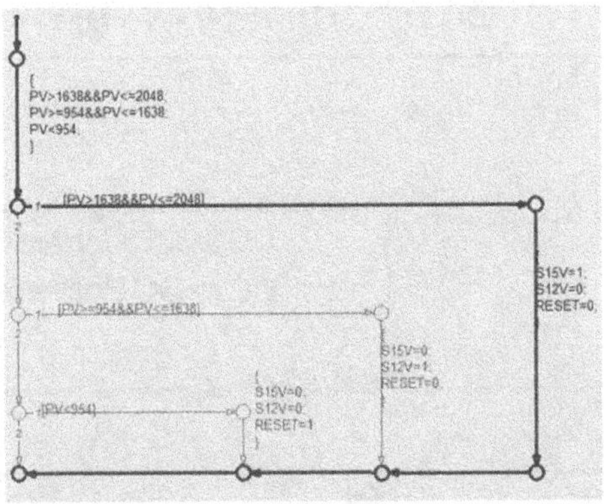

(a)

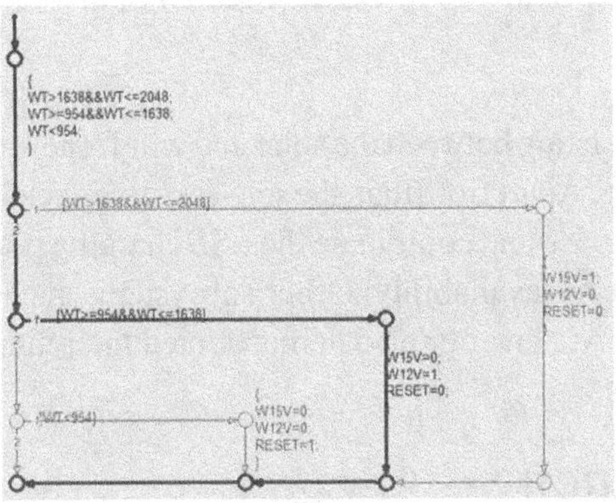

(b)

Figure 12. 15 Volt output voltage – solar renewable energy resource and 12 Volt output voltage = wind renewable energy source

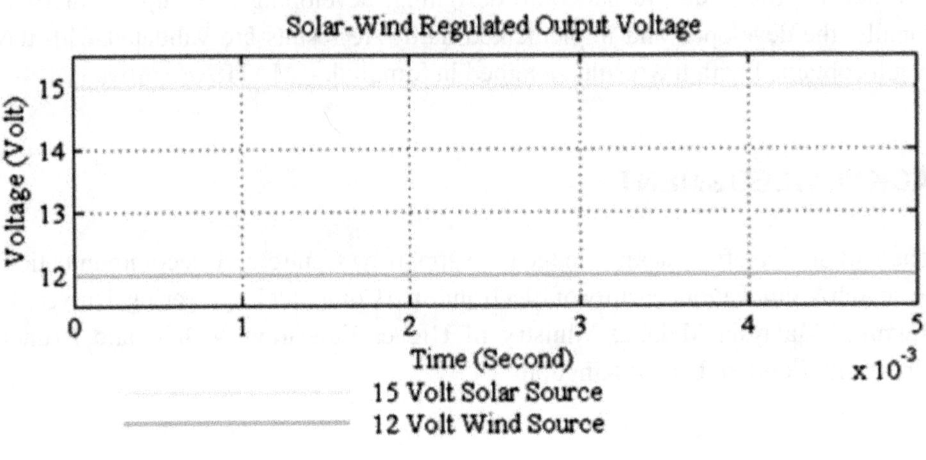

Figure 13. 12 Volt wind input voltage – 15.5 Volt DC to DC Boost Converter output voltage

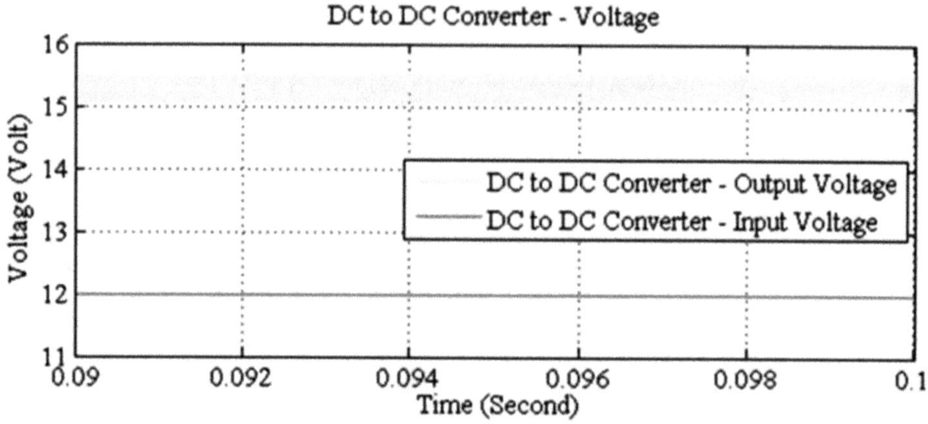

performed the self-assisting between the solar and wind renewable energy sources to load and ESS charging. Apart from that, the self-assistive decision making algorithm also managed to supervise and coordinate the ESS charging process during only one renewable energy source availability. Other subsystems such as DC to DC Boost Converter and DC to AC Inverter also demonstrated their aims and objectives.

FUTURE RESEARCH

It is important to note that, in this research work, a self-assistive controller design and modelling is described for DC distributed generators and energy storage system. Based on the designing and modelling obtained results, the current progress of this research is progressing to hardware designing, developing and implementation. Finally, the developed and implemented hardware results are validated with the results obtained with the results obtained in Simulink – MATLAB software.

ACKNOWLEDGMENT

The authors gratefully acknowledge the support by Centre of Telecommunication Research & Innovation, Faculty of Electronic and Computer Engineering, Universiti Teknikal Malaysia Melaka, Ministry of Higher Education MOHE and Brunel University London, United Kingdom.

REFERENCES

Arikiez, M., Grasso, F., & Zito, M. (2016). Heuristic algorithm for coordinating smart houses in MicroGrid. *2015 IEEE International Conference on Smart Grid Communications, SmartGridComm 2015*, 49–54. 10.1109/SmartGridComm.2015.7436275

Bhoye, M., Purohit, S. N., Trivedi, I. N., Pandya, M. H., Jangir, P., & Jangir, N. (2016). Energy Management of Renewable Energy Sources in a Microgrid using Cuckoo Search Algorithm. *IEEE Students' Conference on Electrical, Electronics and Computer Science.*

Bracco, S., Delfino, F., Pampararo, F., Robba, M., & Rossi, M. (2015). A Dynamic Optimization-based Architecture for Polygeneration Microgrids with Tri-Generation, Renewables, Storage Systems and Electrical Vehicles. *Energy Conversion and Management*, *96*, 511–520. doi:10.1016/j.enconman.2015.03.013

Davis, M., & Costyk, D. (2005). *IEEE Guide for Design, Operation, and Integration of Distributed Resource Island Systems with Electric Power Systems*. IEEE.

Dou, C., Zhang, Z., Yue, D., & Zheng, Y. (2017). MAS-Based Hierarchical Distributed Coordinate Control Strategy of Virtual Power Source Voltage in Low-Voltage Microgrid. *IEEE Access: Practical Innovations, Open Solutions*, *5*, 11381–11390. doi:10.1109/ACCESS.2017.2717493

Hemmati, R. (2018). Optimal cogeneration and scheduling of hybrid hydro-Thermal-wind-solar system incorporating energy storage systems. *Journal of Renewable and Sustainable Energy*, *10*(1), 014102. doi:10.1063/1.5017124

Hussain, H., Javaid, N., Iqbal, S., Hasan, Q., Aurangzeb, K., & Alhussein, M. (2018). An Efficient Demand Side Management System with a New Optimized Home Energy Management Controller in Smart Grid. *Energies*, *11*(1), 190. doi:10.3390/en11010190

Iovine, A., Damm, G., De Santis, E., & Di Benedetto, M. D. (2017). Management Controller for a DC MicroGrid integrating Renewables and Storages. *IFAC-PapersOnLine*, *50*(1), 90–95. doi:10.1016/j.ifacol.2017.08.016

Javaid, N., Naseem, M., Rasheed, M. B., Mahmood, D., Khan, S. A., Alrajeh, N., & Iqbal, Z. (2017). A new heuristically optimized Home Energy Management controller for smart grid. *Sustainable Cities and Society*, *34*, 211–227. doi:10.1016/j.scs.2017.06.009

Jayachandran, M., & Ravi, G. (2017). Design and Optimization of Hybrid Micro-Grid System. *Energy Procedia, 117*, 95–103. doi:10.1016/j.egypro.2017.05.111

Katiraei, F., Iravani, M. R., & Lehn, P. W. (2005). Micro-grid autonomous operation during and subsequent to islanding process. *IEEE Transactions on Power Delivery, 20*(1), 248–257. doi:10.1109/TPWRD.2004.835051

Lasseter, R., Akhil, A., Marnay, C., Stevens, J., Dagle, J., Guttromson, R., … Eto, J. (2002). White Paper on Integration of Distributed Energy Resources - The MicroGrid Concept. *Consortium for Electric Reliability Technology Solutions (CERTS) April*, 1–27.

Lasseter, R. H., & Paigi, P. (2004). Microgrid: A Conceptual Solution. *PESC Record - IEEE Annual Power Electronics Specialists Conference, 6*, 4285–4290. 10.1109/PESC.2004.1354758

Palizban, O., & Kauhaniemi, K. (2015). Hierarchical control structure in microgrids with distributed generation: Island and grid-connected mode. *Renewable & Sustainable Energy Reviews, 44*, 797–813. doi:10.1016/j.rser.2015.01.008

Rey, J. M., Marti, P., Velasco, M., Miret, J., & Castilla, M. (2017). Secondary Switched Control with no Communications for Islanded Microgrids. *IEEE Transactions on Industrial Electronics, 64*(11), 8534–8545. doi:10.1109/TIE.2017.2703669

Sun, S., & Liu, J. (2018). Research on DC Boost Type of Photovoltaic Power Generation System. *IOP Conference Series. Materials Science and Engineering, 382*(3). doi:10.1088/1757-899X/382/3/032045

Vaccaro, B. A., Popov, M., Villacci, D., & Terzija, V. (2011). An Integrated Framework for Smart Microgrids Modeling, Communication, and Verification. *Proceedings of the IEEE, 99*(1), 119–132. doi:10.1109/JPROC.2010.2081651

Worku, M. Y., Hassan, M. A., & Abido, M. A. (2019). Real Time Energy Management and Control of Renewable Energy based Microgrid in Grid Connected and Island Modes. *Energies, 12*(2), 276. doi:10.3390/en12020276

Xu, J., Khan, M., Mumtaz, M., Shahid, M., Habib, S., Hashmi, K., & Tang, H. (2018). A Hierarchical Control Methodology for Renewable DC Microgrids Supporting a Variable Communication Network Health. *Electronics (Basel), 7*(12), 418. doi:10.3390/electronics7120418

Zaheeruddin, & Manas, M. (2015). Renewable energy management through microgrid central controller design: An approach to integrate solar, wind and biomass with battery. *Energy Reports, 1*, 156–163. doi:10.1016/j.egyr.2015.06.003

Zhang, X., & Ukil, A. (2019). Enhanced Hierarchical Control of Hybrid Energy Storage System in Microgrids. *IECON 2018 - 44th Annual Conference of the IEEE Industrial Electronics Society, 1*, 1801–1806. 10.1109/iecon.2018.8591343

ADDITIONAL READING

Chen, N., Qu, X., Weng, W., & Xu, X. (2014). Design of Wind-solar Complementary Power System Based on Progressive Fuzzy Control. *Journal of Computers, 9*(6), 1378–1384. doi:10.4304/jcp.9.6.1378-1384

Fathima, A. H., & Palanisamy, K. (2015). Optimization in Microgrids with Hybrid Energy Systems – A Review. *Renewable & Sustainable Energy Reviews, 45*, 431–446. doi:10.1016/j.rser.2015.01.059

Hangaragi, G. B. (2015). Novel Integration of a PV-Wind Energy System with Enhanced Efficiency. *IEEE Transactions on Power Electronics, 30*(7), 3638–3649. doi:10.1109/TPEL.2014.2345766

Kamal, N., & Kumar, S. S. (2016). Hybrid Power Generation for Distributed Grid Applications. *Middle East Journal of Scientific Research, 24*(1), 51–58. doi:10.5829/idosi.mejsr.2016.24.01.22854

Khalid, M., Savkin, A. V., & Agelidis, V. G. (2014). Optimal Hybrid Wind-Solar System for Matching Renewable Power Generation With Demand. *11th IEEE International Conference on Control & Automation*, 1322–1326. 10.1109/ICCA.2014.6871115

Kumar, R. N., & Baskaran, J. (2014). Energy Management system for Hybrid RES with Hybrid Cascaded Multilevel Inverter. [IJECE]. *Iranian Journal of Electrical and Computer Engineering, 4*(1), 24–30.

Mahesh, A., & Sandhu, K. S. (2015). Hybrid wind/photovoltaic energy system developments: Critical review and findings. *Renewable & Sustainable Energy Reviews, 52*, 1135–1147. doi:10.1016/j.rser.2015.08.008

Sateesh, R., & Reddy, B. V. (2015). Design and Implementation of Seven Level Inverter with Solar Energy Generation System. *International Journal of Emerging Trends in Electrical and Electronics, 11*(June).

Schlager, R., Gawlik, W., Begluk, S., & Trster, E. (2013). Optimal distributed hybrid-storage and voltage support of photovoltaic systems. *22nd International Conference and Exhibition on Electricity Distribution (CIRED 2013)*, (1320), 1320–1320. 10.1049/cp.2013.1180

Tech, K. H. M. (2014). Battery Energy Management System for DC Micro Grids with Fuzzy Controller. *International Journal of Innovation Research in Science. Engineering and Technology*, *3*(1), 1486–1493.

Wijewardana, S., & Member-IESL. (2014). Real-Time Monitoring System for a Utility-Scale Photovoltaic Power Plant. *International Journal of Emerging Technology and Advanced Engineering*, *4*(2), 48–79.

KEY TERMS AND DEFINITIONS

Assistive Controller Voltage Droop: The assistive controller is a controller assisted using some external characteristic such as voltages in this study. Hence, the voltage droop technique is used to assist the controller to make decisions.

DC Distributed Generators and Storages: DC distributed generators and storages are known as only to produce direct current (DC) source of supply.

Hybrid Microgrid System: Hybrid is known having two (2) or more resources connected to generate energy source supply. In this study, solar, wind, and battery as energy storage system is used. And microgrid system is referred to a small power system.

Modelling Hybrid Renewable Energy System: Modelling a hybrid renewable energy system is to model a solar, wind, and battery as energy storage system as a power system.

Off-Grid Network: A power system which is not connected to any kind of grid network.

Renewable Energy Sources: Renewable energy sources are known as solar, wind, biomass, tidal, and many more.

Stand-Alone Hybrid Renewable Energy System: Stand-alone hybrid renewable energy system is referred to a power system which is not connected to any grid network.

APPENDIX: ANALYSIS AND DISCUSSION QUESTIONS

Based on the presented work, designing and modelling techniques were used to present the Self-Assistive Controller Using Voltage Droop Method for DC Distributed Generators and Storages. To initiate an idea which is still not be able to deliver, methods such as designing and modelling are the most suitable to start a meaningful experiment, project or research.

To further initiate this research work, one can start to think of developing this idea into hardware development. Hence, some considerations are required such as, technical understanding and designing understanding. Also, one should also think about types of hardware required to develop this system and will the hardware operate as it is desired also need to be considered.

Chapter 15
Hybrid Electric Energy Storage and Its Dynamic Performance

Syed Abid Ali Shah Bukhari
Aston University, UK

Wenping Cao
Aston University, UK

Xiangping Chen
Guizhou University, China

Fayyaz Jandan
Quaid-Awam University of Engineering Science and Technology, Pakistan

Debjani Goswami
Aston University, UK

EXECUTIVE SUMMARY

This chapter concerns energy storage technologies. It firstly outlines two popular storage technologies, batteries and supercapacitors, while their working principles are revealed. The key issues of these two technologies, such as costs, key types, capacities, etc., are also discussed. Afterwards, a hybrid electrical energy storage (HEES) system consisting of both technologies are demonstrated where the electrical circuit is illustrated. The design of the system aims to demonstrate different characteristics of these two technologies via their charging and discharging process. A test rig is explained in detail while other components, including a load bank, an inverter, a data acquisition subsystem (both the hardware and the software) are also clarified. The experimental results are illustrated and analyzed thereafter. Also, this chapter presents several other promising technologies where their key features, pros and cons, and core applications are pointedly reviewed. The concerned storage technologies include photovoltaic (PV) systems, pumped hydro-energy storage (PHES), superconducting

DOI: 10.4018/978-1-5225-8559-6.ch015

magnetic energy storage (SMES), gas, and other alternatives sources. The authors provide the readers with a brief insight of various energy storage technologies and the inspiration of developing a low-cost, accessible energy storage system for the reader's own purposes.

INTRODUCTION

Electrical energy is an incredible type of energy that can be efficiently changed into different types of energy with high effectiveness and, much more essentially, it is able to regulate lower class of energy easily. Currently, energy storage devices are playing very pivotal role and have a huge impact on harvesting technologies especially on the electrical grid. Storage of energy during off peak times and without the need of additional generation and then reuse the same energy whenever it is required is the biggest ability of the energy storage devices and it is called electrical supply matching with respect to the load. Though, it is a challenging job to build an economically cost effective electrical energy storage (EES) framework regardless of consistent advances in the plan and assembling of EES components including advances in different technologies in battery and supercapacitor. This actually assistances the generation of power and with the help of this the cost of distribution of electrical energy to the consumer is efficiently reduced. Starting today, no single sort of EES component satisfies high power conveyance limit, high energy density, minimal effort per unit of capacity, minimum leakage and long cycle life (Pedram, Chang, Kim, & Wang, 2010). The hybrid electrical energy storage(HEES) administration issues can be broken into charge portion into various banks of EES components, charge substitution (i.e., discharge) from various banks of EES components, and charge movement starting with one bank then onto the next bank of EES components. Regardless of the ideal charge portion and substitution, charge movement is required to use the EES framework productivity (Wang et al., 2011). Generally, there are two basic classes of energy storage devices e.g. batteries and supercapacitors which are considered as electrical energy storage devices. The other types of storage devices are non-electrical storage devices in which kinetic and thermal energy is exchanged into electrical energy e.g. hydro storage system, pumped hydro and pumped air and flywheels (Olabi, 2017). The actual attention of this book chapter is to classify battery and supercapacitor with their applications of energy storage systems and their advantages and disadvantages are presented with respect to electrical vehicle and electric grid applications. We propose a HEES framework that comprises of at least two heterogeneous EES components, along these lines understanding the upsides of each EES component while concealing their shortcomings. Further a potential available on the photovoltaic, pumped hydro and other alternative sources in the context of power system applications have been described in the book chapter.

OUTLINE OF STORAGE SYSTEM FOR ELECTRICAL ENERGY

Energy storage technology is a promising research branch in the research field of energy systems, as energy storage frameworks can be outlined with an expansive arrangement of innovations, for example, an extensive group of batteries, superconducting magnetic energy storage (SMES), pumped hydro, flywheels, and compacted air energy storage (CAES). According to IEA (International Energy Agency), since 1970 up to now the overall energy requirement in the world has been increased 100% due to growth in population and prosperity (Lefebvre & Tezel, 2017) and at the same time the major cause of atmospheric pollution in the environment is fossil fuels. Every innovation has its own particular execution attributes that makes it ideally appropriate for certain systems and less so for other applications of the grid. The capability of a storage system to meet the performance requirement of the grid also permits the similar storage system to deliver numerous services. Currently it perceived as a key component in the present supply chain of modern energy network. This is principally since it can upgrade stability of the grid, increment infiltration of sustainable power source assets, enhance the proficiency of energy frameworks, preserve fossil energy assets and decrease the environmental effects during the

Table 1. Comparison between different energy storage devices

Device	Power	MWh	Discharge duration	Efficiency	Cost $/kWh	Life Time (yr)	Class
Batteries	< 20MW	<200 MWh	Min/Hours	70-90%	85-4800	2-10	Long-term
Pumped Hydro	<2GW	<24GWh	Days	87%	45-85	40	Real long
Fuel Cells of Hydrogen	< 20MW	<200 MWh	Min/Hours	70-90%	-----	2-10	Long term
SMES	0.3-3 MW	<250 kWh	Sec/Min	90%	240-600	40	Short term
Flywheel	<100kW	<100 kWh	Sec/Min	90%	170-720	20-30	Short term
Super-capacitors	<250kW	<3MWh	Sec/Min	95%	85-480	30-40	Short term
Compressed Air	100-300 MW	0.4-7 GWh	Days	80%	12-85	30	Real long
Flywheel	<100kW	<100 kWh	Sec/Min	90%	170-720	20-30	Short term
CSP	0.1-200 MW	<2 GWh	Hours	< 60%	3500-7000	----	Long term

production of energy. Numerous development technologies of storage in addition to enhancement in the current storage expertise have been recently technologically advanced. There are multiple types of energy storage technologies which are present comprising chemical, magnetic, mechanical thermal and biological and the extensively used technologies will be reviewed in this section. The comparison table of different devices for storage of energy is shown in Table 1, (Masaud et al., 2010).

Currently, the 17% of the world's total energy is being derived from renewable sources, and has the main contribution by hydro and biomass whereas the solar and wind are at developing stage (Elliott, 2013).

Battery

Battery is a typical energy storage technology by using electrochemical way. In a battery, the storage of charge takes place by the exchange of electrons which creates a reaction in the electro-receptive medium. The features of battery operation with the requirement of electron and ion conduction is shown in Figure 1 (Winter & Brodd, 2004). The battery energy storage is a promising solution which is capable to contribute to the collective need of energy in the efficient environment. Currently, all the automakers are encouraging and announcing commitment planes for the

Figure 1. The features of battery operation with the requirements on electron and ion conduction

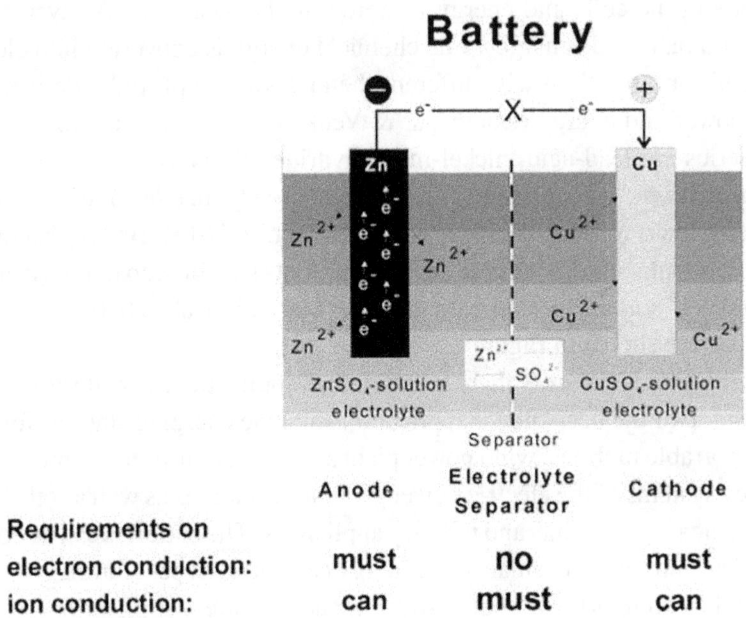

Figure 2. Applications of battery as an energy storage device
(Wang et al., 2016)

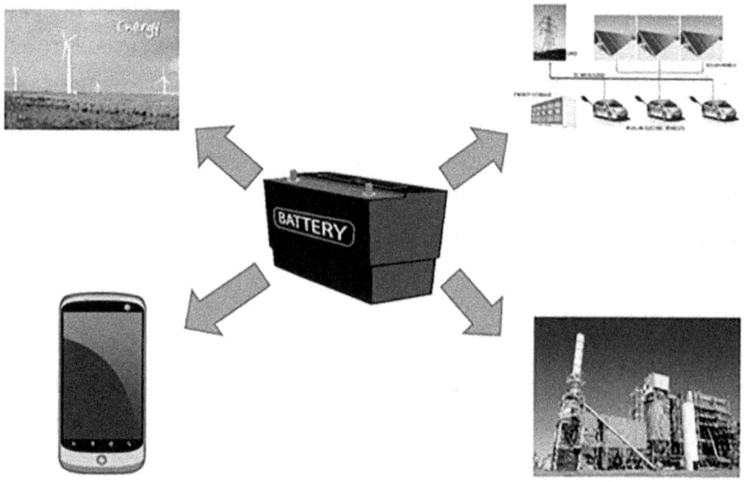

electrification of vehicle and certainly these assurances have caused an urgency to the supply of the batteries and the cars of upcoming future would need (Fletcher, 2011). A device in which energy is deposited in the arrangement of chemical energy is known as battery. Figure 1 shows the conduction process of electrons and ions (Winter & Brodd, 2004).

When the voltage is applied across the battery terminals, a chemical reaction takes place in the cells and energy is stored in chemical way. A reverse process reacts over a battery discharge while chemical energy is converted into electricity. Diverse cell configurations with different chemicals are employed to deliver specific voltage, power and energy (Bhatnagar & Venkatesh, 2012). The most commonly used batteries are lead-acid, nickel-metal hydride (Ni-MH), lithium-ion (Li-ion), vanadium redox, nickel-cadmium (NiCad), Sodium-sulphur (Na/S) and Zinc-bromine (Zn-Br) (Barnhart & Benson, 2013; Borden & Schill, 2013). Battery becomes one of the most popular energy storage devices due to its flexible capacities (from 100W up to 20MW), portability and easy usages (Masaud et al., 2010). Some battery applications are shown in Figure 2.

As can be seen in Figure 2, there are numerous applications of battery as an energy storage. Some of the best known applications include charging stations for electric vehicles, portable mobiles, wind power plant and excitation systems, protection and emergency systems of the electric power generating stations as well as other usages, such as medical instruments and military appliances. The batteries can be combined in parallel or/and series to enhance the energy capacities as per required application. This flexible feature brings about positive impact in battery applications, especially in coordination with other inexhaustible power sources required for the overall improvement of a supportable society. In such manner, to analyse the performance of electrochemical storage devices, for example, lead-acid batteries introduced on hybrid energy system and microgrids regarding their lifetime also, financial benefit is a critical research point. Lead–acid battery has been commercially utilized as an electric power supply or capacity framework for over 100 years and is still the most broadly utilized rechargeable electrochemical gadget (Ding et al., 2012; Shapira et al., 2013). A large portion of the customary valve-regulated lead–acid (VRLA) batteries are used for starting, lighting and ignition (SLI) batteries in cars and other vehicles, which typically work in shallow charge/release cycles. Recently, lead–acid batteries have drawn massive attentions for hybrid electric vehicles (HEVs) and vitality storage system applications as a result of low cost, simple design, good quality and safety and high reusing productivity (Albers, Meissner, & Shirazi, 2011; Pech, Brousse, Bélanger, & Guay, 2009). Lead acid batteries, an embedded innovation that is open to the world, risen as a promising technology for sustainable power

Table 2. Characteristics of commonly used rechargeable batteries

Specification	Lead Acid	NiCd	NiMH	Li-ion		
				Cobalt	Manganese	Phosphate
Specific energy (Wh/kg)	30-50	45-80	60-120	150-250	100-150	90-120
Internal resistance	Very Low	Very Low	Low	Moderate	Low	Very Low
Life cycle[2]	200-300	1000[3]	300-500[3]	500-1000	500-1000	1000-2000
Charge time[4]	8-16h	1-2h	2-4h	2-4h	1-2h	1-2h
Overcharge tolerance	High	Moderate	Low	Low. No trickle charge		
Self-discharge/month (room temp)	5%	20%	30%	<5% Protection circuit consume 3%/month		
Cell voltage	2V	1.2V	1.2V	3.6V	3.7V	3.2-3.3V
Maintenance requirement	3-6 months	Full discharge every 90 days when in full use		Maintenance-free		
In use since	Late 1800s	1950	1990	1991	1996	1999
Toxicity	Very high	Very high	Low	Low		
Cost	Low	Moderate		High		
Safety Requirement	Thermally stable	Thermally-stable, fuse- protection		Protection circuit mandatory		
Discharge cut-off voltage	1.75V	1.00V	2.50V - 3.00V	2.50V		

(Park et al., 2012; Pollet et al., 2012)

sources and applications of energy storage (Spiers & Rasinkoski, 1995). They are generally low costs (£150/kWh). These batteries are sensitive to faulty operation. Around at 25 °C temperatures, discharging are generally vitality, the performance deteriorate rapidly at high temperatures, which diminishes battery life by as much as half for each 8–10K temperature increment (Jossen et al., 2004). The characteristics of commonly used rechargeable batteries is shown in Table 2.

In the recent years, due to the double burden of the emergency in the energy crisis and environmental issues, utilizing renewable energy rather than oil as the power resources for electrical vehicle to minimise the energy utilization and emission has turned into the advancement bearing for the car industry (Eyer & Corey, 2010). Battery driven electric vehicles have great natural security for environment since no pollution is generated over driving.

Supercapacitors

The capacitor can be used to charge and discharge energy immediately. In an electric double layer capacitor, which is also called as a supercapacitor, no electron transfer happens over the big surface electrode interface. This makes an extraordinary high capacitance with a very quick charge cycle. The electrical energy stored by the capacitor as electrostatic charge and have the capability to do long haul storing the energy by means of slight losses. The electric double layers at the electrode (supercapacitor) is shown in Figure 2, (Winter & Brodd, 2004). In electrochemical capacitors (supercapacitors), energy may not be conveyed by means of redox responses and subsequently the utilization of the anode and cathode may not be fitting. By introduction of electrolyte particles at the electrolyte/electrolyte interface, known as electrical double layers (EDLs) are shaped and discharged, which brings about a parallel development of electrons in the outside wire, that is during the energy conveying process. Different capacitor types comprising capacitors with double layer, super capacitors, and ultra-capacitors. The super capacitors utilized as a part of strongly correlated electron system (SCES) are fundamentally two-fold layer capacitors that stores the energy within a field of magnet. They have quick reaction in the electric double layers at the electrode (Winter & Brodd, 2004).

operation that require short time span, for example, to build up voltage. They have a power capacity up to 250kW (Masaud et al., 2010), and higher power/energy density. An example of parallel plate capacitor is similar to the double layer capacitor, although the terminal region in double layer capacitor is expanded by utilizing enacted carbon. This has a surface region of 1000 m^2/gram of material (Spyker & Nelms, 1997). The expanded surface has high capacitance because of a littler plate division delivers the bigger capacitances in two-fold layer capacitors.

Figure 3. An electric double layer supercapacitor

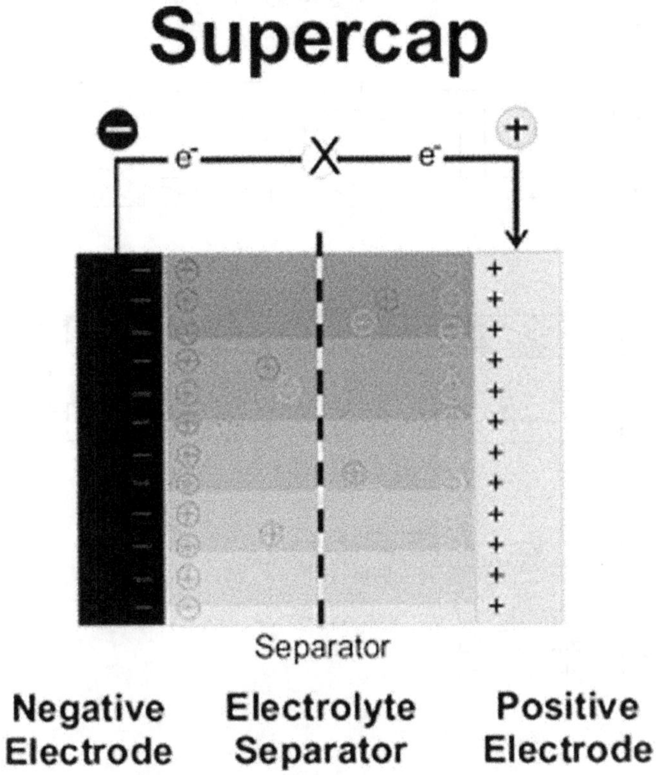

HYBRID ENERGY STORAGE SYSTEM

There are five topologies for a hybrid electrical energy storage consisting of batteries and supercapacitors, as shown in Figure 4. Topology 1 is the simplest among all where the supercapacitors connect with the batteries at the DC bus. In contrast, topology 4 employs two DC/DC converters to maintain the same output voltage at the DC Bus while allowing the output voltages of the batteries and super-capacitors to fluctuate. This is an advantage to store more energy. In Topology 5, DC power from batteries and super-capacitors are transformed into AC and then connect at the AC bus.

Different topologies have their own pros and cons. Topology 1 has the simplest structure and the least power loss among those topologies. However, the DC bus voltage is dictated by the battery output and thus the supercapacitors have a limited voltage variation (i.e. low energy storage). Topology 4 has two more DC/DC converters than Topology 1. In this case, batteries and supercapacitors can have different voltage levels while the output voltage of DC/DC converters can keep at

Figure 4. Different topologies of hybrid energy systems consisting of batteries and super-capacitors (a) Topology 1, (b) Topology 2, (c) Topology 3, (d) Topology 4, (e) Topology 5

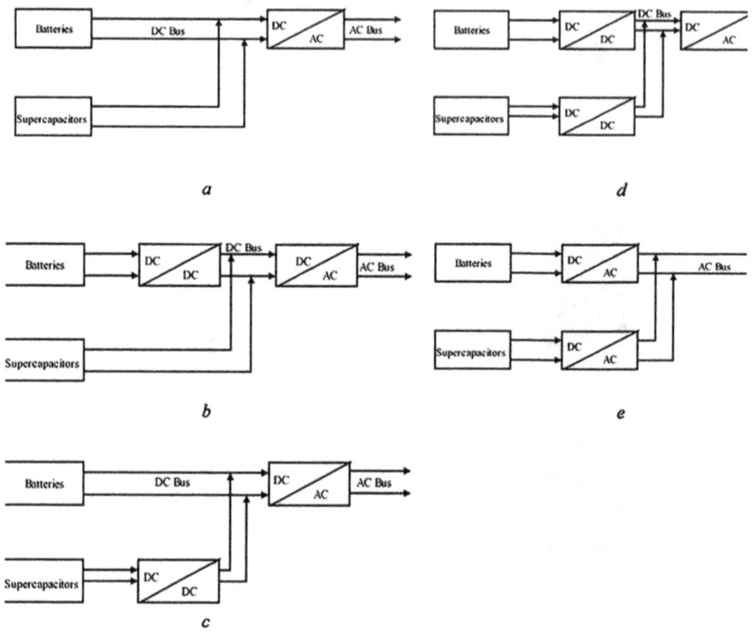

the same level. Therefore, maximum energy can be stored in the supercapacitors. However, power loss is the highest due to the use of three converters. In this work, topology 2 is selected to strike a balance between maximising the power range of the supercapacitors and complexity of the topology.

Batteries

Three units of batteries were selected as another main DC sources and linked together in serial. Each battery was 12V/12Ah lead-acid that used to be as common electrical energy storage device by storing/releasing DC power for backup power supply. The layout of the batteries in serial is illustrated in Figure 5.

The batteries are the main component to store electrical energy in the system to balance the supply and demand. Their performance is dependent on polarisation, which are expressed as follows.

$$\eta = E_{OCV} - E_T \quad (1)$$

Figure 5. The batteries

$$\text{ZONE 1 } \eta = a - blog\left(I / I_0\right) \quad (2)$$

$$\text{ZONE 2 } \eta = IR \quad (3)$$

$$\text{ZONE 3 } \eta = \left(RT / n\right) ln\left(C / C_0\right) \quad (4)$$

where, E_{OCV}, E_T stand for the polarisation, the open circuit voltage and the terminal voltage of the battery cell, respectively. Eq. 1 shows the voltage decrease during discharge of lead-acid batteries, the closer E_T is to E_{OCV}, the less energy is wasted. Eqs. 2-4 represent battery polarisation over three discharging zones. Parameters a, b, I, I0, R, T, C, C0 donates the Tafei equation constants, current flow, exchange current, concentration at the electrode surface, and the concentration in the solution, respectively. Figure 6 presents the curve developed by these equations. Three polarisation zones (zone 1-3) illustrate the severe degree as the discharging current increases. On the one hand, the batteries suffer from power loss (heat) during the discharge process. The severe degree depends on the voltage and current of the discharge. One the other hand, their equivalent resistance in this three zones shows that zone 1 and zone 3 are worse than zone 2 due to either high voltage or high current. The voltage will decrease proportionally to the increase of the discharge current in zone 2.

Figure 6. Polarizations of the batteries

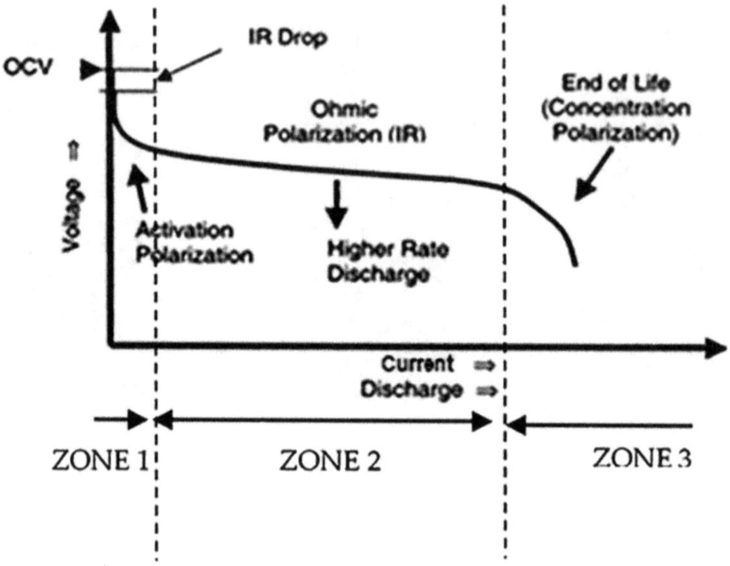

Figure 7. Equivalent circuit of a supercapacitor cell

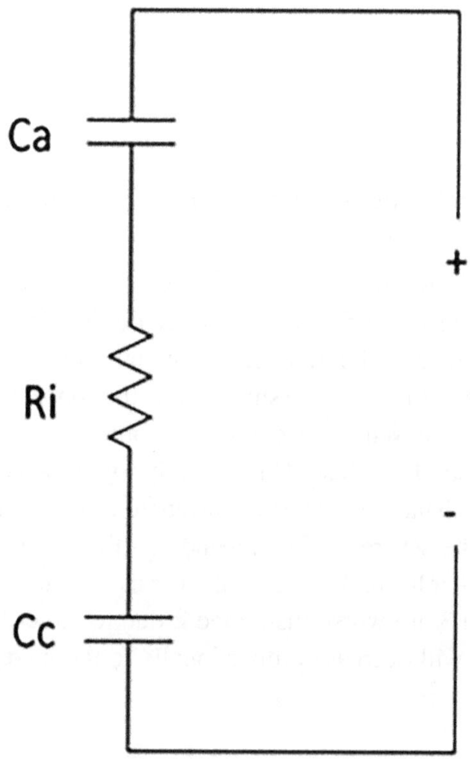

The above discussion proves that proper voltage and current over discharging periods would benefit to both cycle-life and performance of the batteries.

Characteristics of Super-Capacitors

Super-capacitors can store electrical energy but have much faster response than batteries. Their life cycle can easily exceed 1 million times. Their equivalent circuit is shown in Figure 7. The equations are used to describe the electrical performance of the supercapacitors is given by.

$$\frac{1}{C}=\frac{1}{C_a}+\frac{1}{C_c} \tag{5}$$

In general

Figure 8. Super capacitor module

Figure 9. Charge/discharge performance

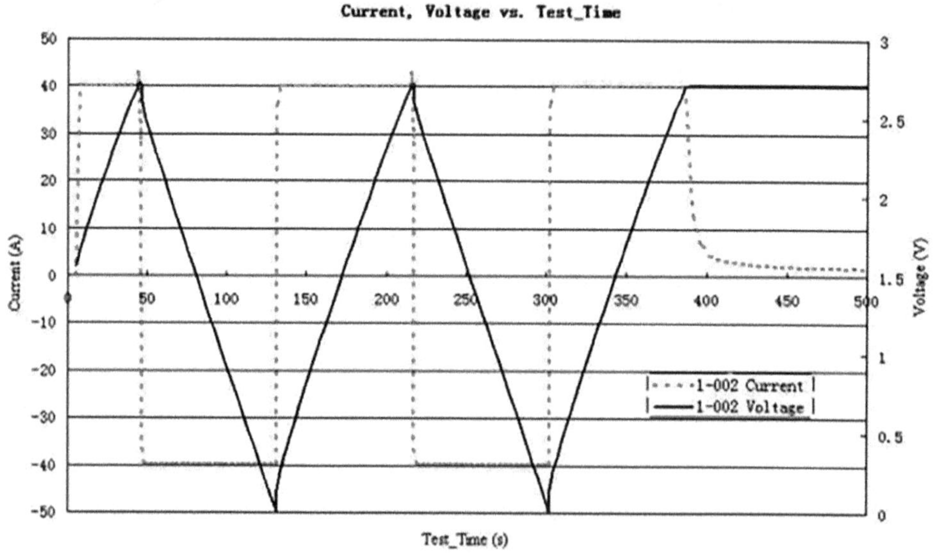

$$C_a = C_c \quad (6)$$

$$E_{storage} = \frac{1}{2}CV^2 \quad (7)$$

where R_i, C_a, and C_c represents the internal resistance, anode and cathode capacitances of the cell, respectively. $E_{storage}$ $E_{release}$, C, V are the stored energy, equivalent capacitance and open-circuit voltage of the supercapacitor. It is easily understood that the stored energy is proportional to the capacitance and the voltage squared while an increase in the voltage will be more significant than the increase in the capacitance.

Experimental Test of HEES System

This section presents the investigations of the initial experimentation of the HEES. The key aim is to develop a set of power system to charge/discharge automatically. In the system, the hybrid power sources consist of super capacitors and batteries. The load is represented by an AC resistive load (bulb board). A super capacitor module with high power density was connected in parallel to the lead acid battery group as energy devices. Therefore, the small-scale HEES system has both desired

Table 3. Specification of super capacitor (single unit)

Item	value	Note
Rated voltage	2.7V	
Surge voltage	2.85V	
Internal Resistance	0.0016 Ω	
Capacitance	1200F	
Energy stored (at V_R) E	4.4kJ (1.2Wh)	$E=1/2CV_R^2$
Nominal current (25°C)	235A	Discharge to 1/2 V_R within 5S
Maximum current (25°C)	>550A	Discharge to 1/2 V_R within 1S
Power density	5.1Wh/kg	
Maximum leakage current	8mA	
Operational temperature	-40 ~60 °C	
Life	90000h(at V_R, 25°C) 1000h (at V_R, 70°C)	
Cycle life	500,000	
Size	Ø50×110	
Volume	216ml	
Mass	240g	

power and energy advantages as proposed. The purpose of this test is used to obtain the characteristics of the HEES system.

The test bench of the HEES includes: (1) a HCC super capacitor module with 40F/60V; (2) 3 units of lead acid batteries (12V/12Ah for each); (3) inverter/ charger (HBC-3000PVA) ; (4) load bank (up to 600W) ; (5) data detection devices; (6) Controller (Simens S7-200 PLC) and monitoring software 'King view'.

In the test, a super capacitor module was selected to be one of main components in the energy storage system. This module was constructed by 27 super capacitors units (2.7V/1200F for each) connected in serial.

Figure 8 Illustrates the layout of this module and Figure 9 Charge/discharge performance

super capacitor unit. This test used 40A DC current to charge/discharge the super capacitor constantly. The voltage increased or decreased in a linear manner. Table 3 shows the specifications of the super capacitor unit. From the Table 3, it can be seen that HCC super capacitor has small internal resistance which incur current leakage over operation. Meanwhile, high power density of the super capacitor allows high-level current release transiently and the maximum transient current reaches as high as 550A within 1s for single unit. Long cycle life is another sound characteristic of a super capacitor where it is 500,000 in this case.

Figure 10. Load Bank

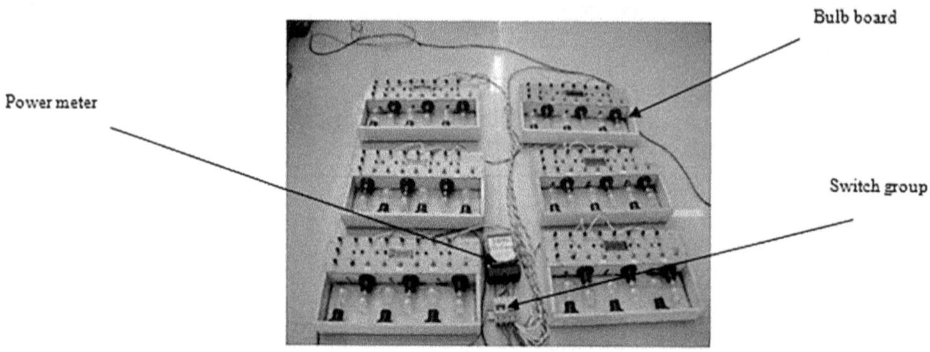

1. Load bank

The load bank in Figure 10 consisted of 6 groups of bulb board with 100W for each. Therefore, the load power can vary between 0-600W according to the demands. The alternation of power demands was accomplished by operating switch on or off. The power meter connected with load bank recorded the real-time power signal for further analysis.

2. Inverter/charger

As a central control device, HBC-P3000VA shown in the Figures 11 and 12 played a crucial role regarding to charging/discharging hybrid DC power sources (super capacitors and batteries). It accepted the AC current from the grid and transformed it into DC to charge DC devices and in turn to convert DC current into AC current back to the grid later on. Therefore, the inverter acted as inverter or charger with the capacity of bi-directional power delivery.

3. Signal detection devices

The test with DC hybrid sources aimed to detect their current, voltage alternation along with the charge/discharge operation. Therefore, the current and voltage sensors were employed in the signal measuring circuits. Figure 12 illustrates these connections.

4. Signal conversion control and display

The signals detected by sensors came out with analogue mode which needed to be transformed into digital signals before being analysed for further process. Figures

Figure 11. Inverter charger

Figure 12. Common connection with HBC-P3000VA

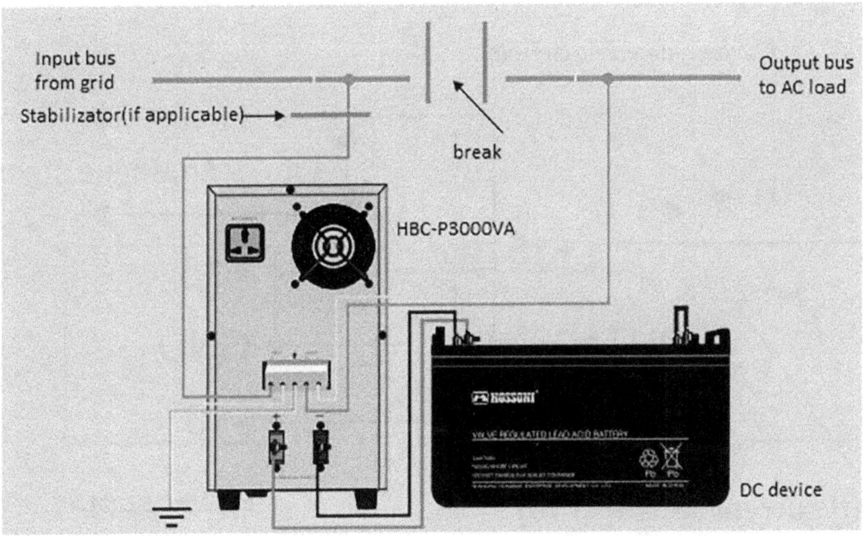

Table 4. S7-200 PLC

Item	Specification
CPU module	CUP-224
I/O module	EM231/EM232 (8 analogue in/4 analogue out)
Communication	RS485 (built-in)
Power module	220V In / 24V Out
Software	HMI-Software WinCC

13 and 14 illustrate current and voltage transducers in this study. In this test, Siemens S7-200 PLC (Programmable Logic Controller) extension module EM231 was selected for accepting analogue signals and transferring them for further calculation by CPU unit in the PLC. Meanwhile, these digital signals were displayed on a desktop by software Kingview which communicated with PLC. The experiments employed a Siemens S7-200 PLC as the controller to perform the charge/discharge tasks for HEES automatically. It accepted the signals from the data detection devices via I/O module and calculated the control output sending to the execution device relay 1 and relay 2. Table 3 lists all of the constituents of the PLC employed.

5. Experimental circuit design

Experimental bench was illustrated in Figure 15. An equivalent circuit is demonstrated in Figure 16. A super capacitor module was connected to 3 units of lead acid batteries (as energy devices and fed in the inverter/charger via DC input

Figure 13. Current measuring circuit

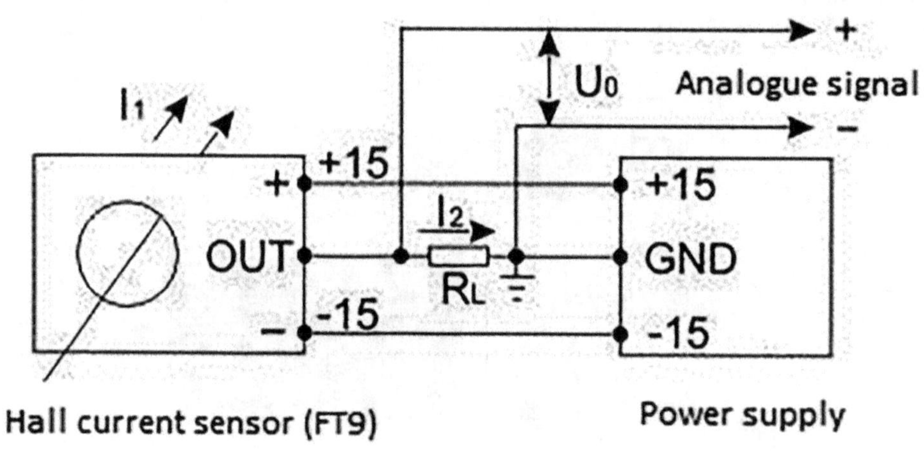

Figure 14. DC voltage measuring circuit

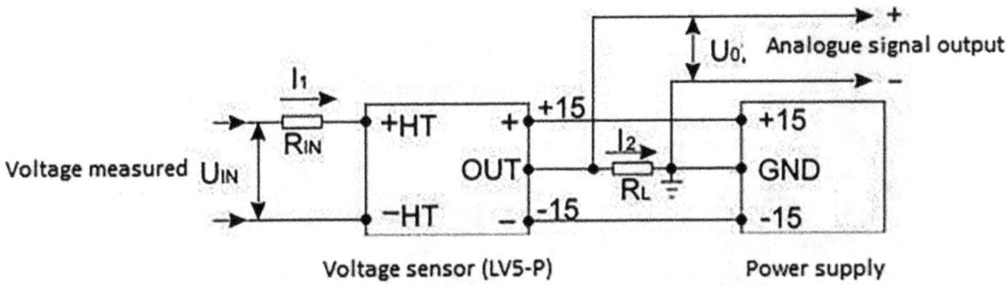

port. The bulb boards, as load bank linked with the inverter/charger via AC output port while the AC power from the grid took the role of the main electrical source to supply the HBC-P3000. HEES parameters were detected by the sensors placed in the DC circuits and sent to the PLC for calculation. A PC interacted with the hardware circuit via serial communication RS485 and interfaced with the software WINCC and the utility 'Kingview' for monitoring and control purposes. The programme implemented in a PLC was used to assess the data collected by input ports which linked with the sensors and DC meters and then drive actuator elements (relay 1 and relay 2) via output ports for charging and discharging the hybrid power system. Simultaneously, the data PLC received was sent to supervisory software 'Kingview' on a desktop to display.

6. Test plan and procedures

The experimentation bench facilitates to charge and discharge hybrid power sources automatically.

In the test, charge/discharge scenarios are programmed in the inverter controller. Based on the charge voltage level, the hybrid DC power components (supercapacitors and batteries) are charged with three stage methods. Over the discharge period, the voltage level is also the logical controlling condition. The discharge ends as the DC components decrease to the threshold as set.

Tests were carried out in two steps:

1. Charge batteries and super capacitors: relay1 on and relay 2 on
2. Discharge batteries and super capacitors: relay 1 off and relay 2 on

The inverter charges the DC sources (batteries and super capacitors) along with supplies electricity to the load bank. After batteries and super capacitors are fully charged, the inverter started to convert DC power originating from batteries and

Figure 15. Layout of the primary test bench

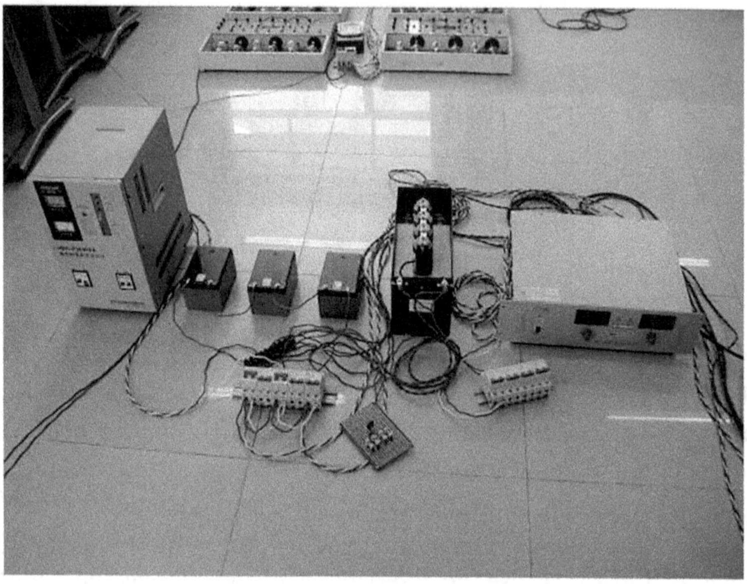

Figure 16. Circuit of the primary test

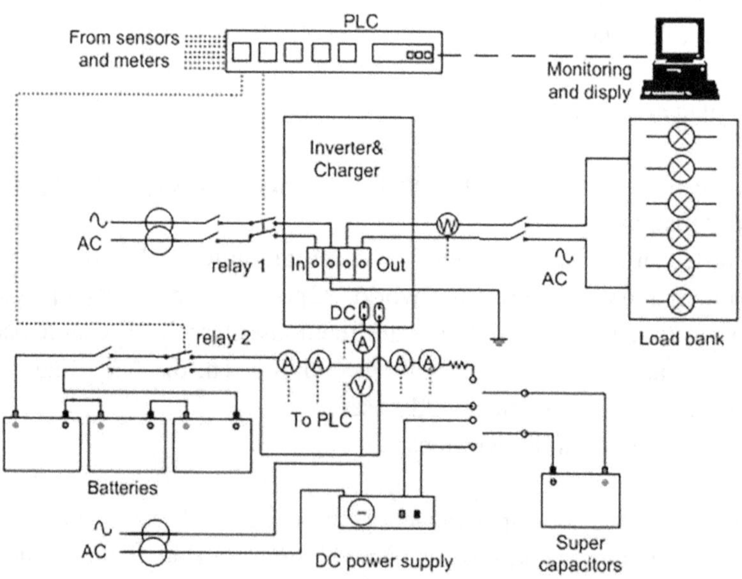

super capacitors into AC power to supply the load bank. During the operation, the load demand was 100watt.

7. Test results

Series of diagrams in Figures 17, 18 and 19 show voltage and current waveforms as a result of charging and discharging the DC sources. Figure 17 presents the experimental result of voltage curve of hybrid power sources during both the charging and discharging processes. From these figures, the charge process can be divided into two stages.

Three-stage charging is a widely used method for lead-acid batteries. It is the combination of constant-current and constant-voltage and float stages. The Batteries were charged with constant current at the first stage while with constant voltage at the second stage. Float charge provided small current to charge batteries. When the charging starts, an immediate high current is supplied to batteries and its voltage increased gradually. This is the first stage, constant-current charging. Afterwards, the batteries were charged with constant voltage for a period of time. In this stage, the current decayed till a pre-set value, charging process transferred from the second stage to the third stage the purpose of float charging is to charge the batteries with a low rate that is approximately equivalent to the internal losses of the battery over long-period operation.

In the tests, the batteries were charged by the first two stages of the three-stage charging method. Therefore, the voltage variation over the charging process can be divided into two parts.

Figure 17. Voltage waveforms of the DC sources

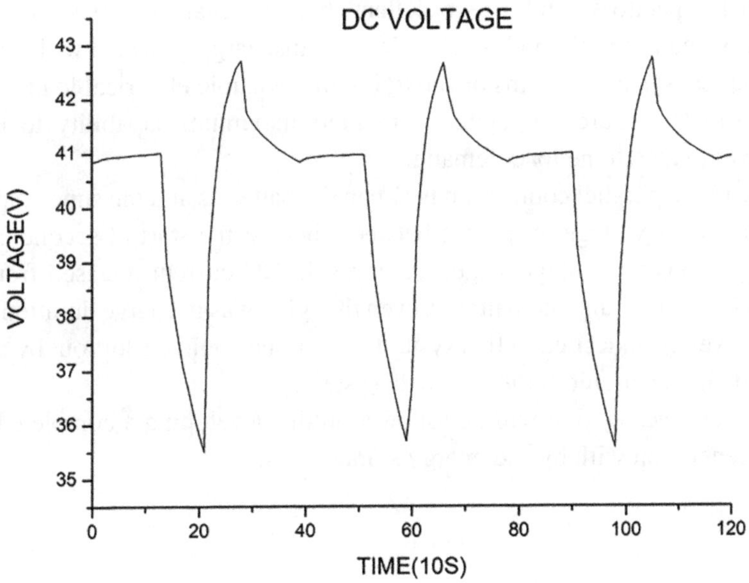

On the other side, voltage dropped swiftly in discharge course in the test because the hybrid DC sources offered the maximum power to the load. However, the total energy supplied by hybrid power sources was relatively limited. As a result, the voltage plunged when hybrid power sources discharged with relatively high power.

Figure 18 presents the current curves of the hybrid power sources being charged. From Figure 18, it can be seen that the superior of the super capacitors in comparison of the batteries'. The charging current for the super capacitor increased from 0 to 6A within 5s. In contrast, the charging current for the batteries only increased 1.8A.

On the other hand, there was undesired current rushing to batteries at the last stage of the charge duration. It is mainly due to the higher voltage of super capacitor over batteries at transient time. This undesired current can be seen in Figure 18.

Figure 19 shows the current curves of hybrid power sources being discharged. From the figure, it can be seen that during the period of discharge, the super capacitor module showed active characteristics and fast response capacity. Super capacitors can provide a higher current to the load instantly than batteries at the start of the discharging course. However, the current plunged along with the voltage plummet while the batteries' current increased dramatically.

Summary

The tests were conducted to gain in an understanding of the small-scale HEES system and were used later as a basis for developing the complete BMT-HEES system. Some points could be achieved according to the outcomes from it, including

- Super capacitors exhibit us excellent dynamic characteristics and they are responded promptly both in charging and discharging, which will benefit the complete system in terms of satisfying the dynamic electrical loads.
- Hybrid DC sources operated with their maximum capability to generate power to match the load demand.
- Due to the parallel connection between the batteries and the super capacitors, there was a voltage difference between them at the start of second charging stage (constant-voltage stage). As a result, DC current released from super capacitors rushing into batteries even though it was in charge duration, which, however, is undesired in this system. This issue could be sort out by adding a diode in the circuit in the complete system.
- Further investigation will be followed in the developing a complete bio-fuel tri generation with hybrid energy storages system.

Figure 18. Current waveforms of charging the DC sources

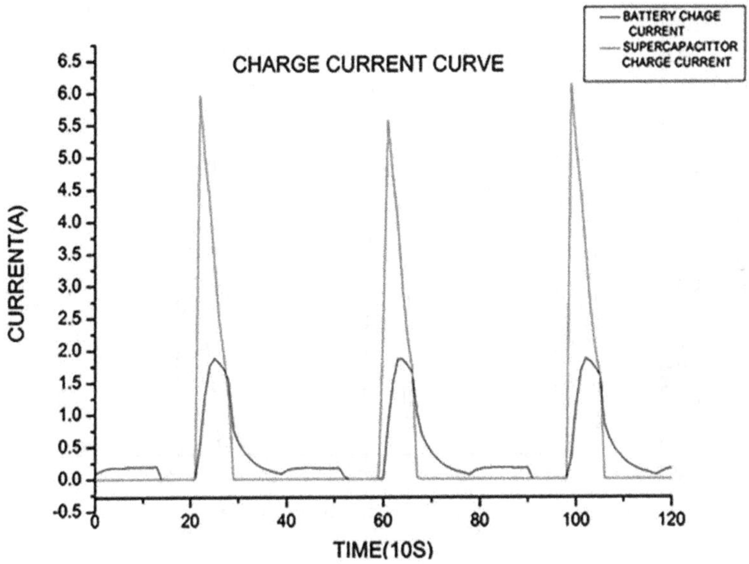

OTHER IMPORTANT ENERGY STORAGE SYSTEMS

Electricity is an important source for development of industrial society (Sandén, 2014). Electrical energy storage has a long period history with regard to peak shaving and emergency power supplying. Their capacities can range from large scale, medium to micro small sizes (Díaz-González et al., 2012). An increasing requirement exists for a huge-scale energy storage systems which is capable to maintain transmission of electrical grids and permit the steadfast

operation of renewable energy sources sporadically (Jansen et al., 2016). Power generation by combusting oil and coal produces a lot of carbon dioxide (Kermani & Morshed, 2003). So we have to change the way of generation by using renewables. On the way to renewable energy there is one block that is storage and so we have to understand what is going on in our world.

The Photovoltaic Energy

One of most promising technologies is photovoltaic. This technology is developed rapidly as much as 10 times in recent year (Dewald & Truffer, 2012; Surek, 2005; Zhao & Zhu, 2014). In the early nineties, United States was a leading country in PV installation while the whole world had generation about 100MW (Bull, 2001; Goetzberger et al., 2003). PV applications in Japan grow swiftly under the support

Figure 19.

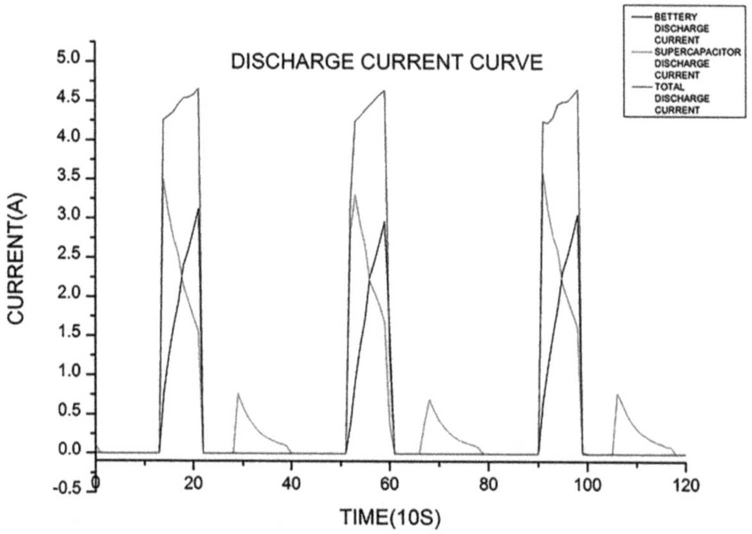

Figure 20. Global cumulative solar photovoltaic installed capacity

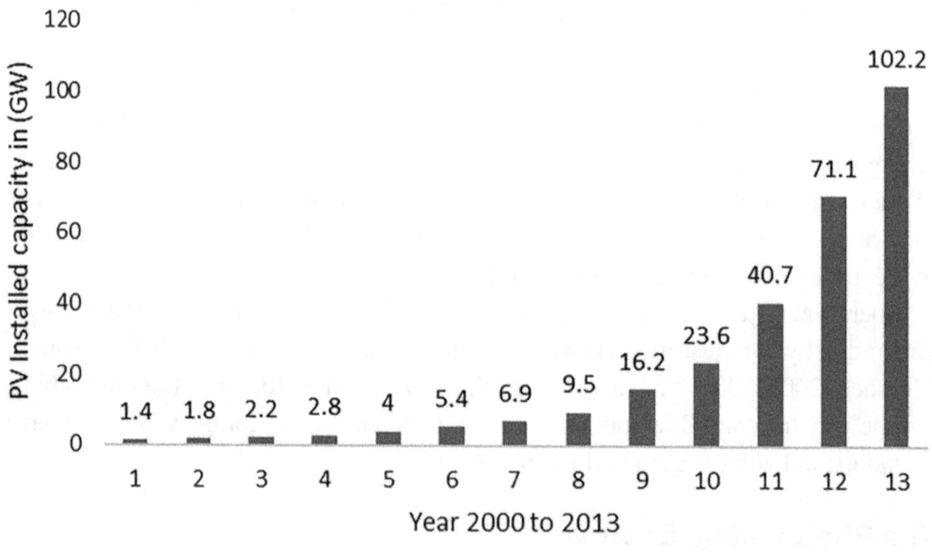

from government programme. This trend climbs continually until the beginning of the 21st century (Silva Fonseca et al., 2012). Germany started to increase PV development thereafter. Nowadays, there are over 100GWh of electricity coming from photovoltaic (Haas et al., 2013; Watt & Fechner, 2009). The global cumulative installed capacity from solar is shown in Figure 20. As shown in Figure 20, PV power generation grows over 10 times over the first 10 years. It thereafter raises more steep. Annual growth of 40% is estimated over the next decade. Global demand is anticipated to be satisfied by photovoltaic in the future.

Today, 0.5% of the electricity is from photovoltaic, which is a very small portion of power generation. This portion was less, only 0.003% in 1998. By 2028, this number will increase to 50% which means half of our energy will come from photovoltaic (Dincer, 2011).

Development of Photovoltaic by Countries

The interesting thing is that not only that happened in Germany, but Italy was the country with most installations in 2011 (Bergamasco & Asinari, 2011). India has the largest photovoltaic cells in the world, China has most of the production capacity (De et al., 2011; Grau et al., 2012) and the US has highest efficiency means there is global priority for photovoltaic (Hanna & Nozik, 2006), but there is a problem of night, we have to properly understand that then we can solve it. As in early morning the demand is low, as the people wake up and go to office and start their computers and other appliances till afternoon, the demand peaks and then it goes slowly down till evening and again increases while going to home. Now to close that gap the conventional power is always produced such an amount that is exactly matched to it that is great but we need to store it in a way coal, gas and so on. Now the sun which is always available during the day so we have abundant of energy, If we don't need so much energy at the day time then we have to store it during a day utilize it for night time, Now day by day more solar energy is coming near future, what will go on? There will be infinite energy at the day time even more energy if we can use though we have to waste that, and wasting that energy is not good idea and we should store that energy and use it at night time that is the basic idea. we need to check and build our energy resources, we need to check that, what do we really need, we need low price for kWh storage capacity for easiness otherwise it is not economic, and if it is not economic it won't work, the next thing is we need high efficiency because we produce all this electricity with quiet expensive thing like photovoltaic and wind turbines, and we want small footprint of all the changes to renewable is about environment friendly stuff, storage should also be environmentally friendly. So first case of battery, great thing is, if we used it every day but the low price is not here. Battery is about 200 $/KWh and they are already in high scale produce like every car has battery and there will not be a big change in near future (Dorrell et al., 2010). Small footprint is not here the reason we need metals and we have problem of environment during mining and disposal (Yender, 1998), but one thing is good, battery has 95% efficiency which a really a great (Olsen & Pasquale, 2007).

Pumped Hydro Energy Storage

Pumped hydro energy storage (PHES) is well established and business-wise innovation for utility-scale power storage which has been developed since mid-1890s. approximately 1000-1500MW ranges of pumped hydro energy storage system is being used all around the world with the efficiency in between 70 to 80% (Rehman et al., 2015). The requirement for aquatic consumption, exchangeable energy and environmental friendly goals promote PHES the most attractive technology in the recent years. It becomes increasingly popular in large-scale energy storage applications (Vieira & Ramos, 2008). A pumped hydro storages is an energy storage system in which water is reused between an higher reservoir to a lower reservoir (Lu et al., 2003). During the peak load operation, the combination of hydro-thermal is used to minimize the rate of fuel by means of activating hydro generator; during the light load period, water is pumped back into an upper reservoir to store extra energy (Conejo et al., 1990; Raul et al., 1984; Wood et al., 2012). Literature reveals that 7GW capability of PHES will be further added to the European grid in the upcoming years (Deane et al., 2010). Similar applications are planned in the other countries, such as the US and Japan.

Superconducting Magnetic Energy Storage (SMES)

The superconducting magnetic energy storage is an emerging technology (Lefebvre & Tezel, 2017), as the mutual storage of energy in electromagnetic arrangement. They have much higher capabilities by comparing to the capacitors in similar size (Dubal et al., 2013). The origin of SMES is the capacity of vitality in the magnetic field of a DC current streaming in a superconductor. The losses of energy are nearly zero in light of the fact that no resistance offered by superconductors during the flow of electrons. Although discharging time is short, extremely low heat losses is the significant advantage of a SMES. Nowadays, SMESs are less competitive to chemical batteries due to their low energy densities (Sørensen, 2017). Since it is conceivable to infuse and remove current rapidly all through superconducting coil-loops, SMES has been created for the usage in high power control devices to keep up the state of superconducting. Meanwhile, the appliance must be cooled to the temperature at which superconductivity is achieved. For low-temperature superconductors, fluid helium is essential. If a high-temperature superconductor is utilized, fluid nitrogen could be utilized (Hall & Bain, 2008). A SMES can be used as a tool for quality adjustment with high effectiveness (energy losses are related to the cooling framework, however, a total efficiency of 98% can be reached). Moreover, it can be cycled infinitely. The commercial micro devices are available from 1MW to 10MW. A SMES of 50MW consisting of more than 30 units is available in the US. In the

Europe, relevant research are under developing, aiming to improve superconducting materials for the SMES devices with less than 10MW (Hall & Bain, 2008).

Gas Energy

Germany demonstrates promising development regarding multi-energy systems, such as integrated systems with electricity-gas-hydrogen, or systems including electricity-hydrogen-chemical products (Dönitz, 1988; Weiland, 2006). Even natural gas is a commonly used energy source and infrastructure is also available, these multi-energy systems suffer from low efficiency due to multi-stage conversions from electricity to hydrogen and back to electricity. In this case, only 25% of the energy is useful while 75% of the energy is wasted (Gahleitner, 2013; Pöschl, et al., 2010). The potential solution to increase efficiency is to integrate PHES with multi-energy systems. It has high efficiency while energy conversion within the cycle of a PHES could be achieved with almost 99% (Zhao & Davison, 2009). However, a PHES requires specific geological conditions which limits its applications.

Other Alternatives Sources

Electrical energy has top vitality as it can be converted into other types of energy with small efforts. In addition, it can be utilized to regulate different types of energies (Odum, 1975). The variation in the use of electrical energy are altering along with the changes of the load types, user practices and their behaviour (Wang et al., 2011). The power generation from non-renewable energy resources and nuclear plants still occupy the largest portion. Moreover, the control of most inexhaustible sources of power (Renewable Sources) are not well-regulated and to a great extent subject to the natural environmental components even for the non-renewable energy source and atomic power plants, the generation measure, can't be changed rapidly enough to react to quick changes in the load demand. In this way, generation of the power supply and demand (utilization) are commonly not well-balanced with each other. The low-cost, accessible energy storage systems are desirable to enhance the energy efficiency of the electrical power grid for load-levelling, frequency regulation and penetration of large-scale of renewable energy resources, such as wind, solar (Abraham et. al., 1990; Amine et. al., 2010). The electrical energy storage (EES) frameworks in this manner can increment control unwavering quality and proficiency, remunerate the supply-request and satisfy the highest demand of power. Additionally, EES can be integrated with other assets, for example, streams and lakes, nuclear, solar and wind which are used to create power (Whittingham, 2012; Xiaoliang et, al., 2014).

CONCLUSION

This chapter has presented the development of hybrid energy storage system with focus on batteries and supercapacitor. HEES is these days the most popular components as a part of the electric vehicles, energy savings, and other sustainable power sources based applications. It is hard to use a single EES component which satisfies all necessities of electrical vitality storing and recovery processes. A HEES system consisting of at least two heterogeneous EES components, battery and supercapacitor has been experimentally tested. The design, structure, key components in the integrated system and characteristics of the energy storage systems are presented. This system can be applied to existing units of power conditioning applications, in order to enhance their abilities. Both current and voltage compensation are achieved by utilizing a single self-contained inverter topology with the use of HESS into the structure. Additionally, the present stress due to intermittent changes in demands are addressed by energy storage due to the involvement of a supercapacitor module, which also reduces the stress the battery units and enhances its life expectancy. Integrating with the grid can also be realised by an efficient hybrid energy storage system in the near future.

REFERENCES

Abraham, K., Pasquariello, D., & Willstaedt, E. (1990). N-butylferrocene for overcharge protection of secondary lithium batteries. Journal of the Electrochemical Society, 137(6).

Albers, J., Meissner, E., & Shirazi, S. (2011). Lead-acid batteries in micro-hybrid vehicles. *Journal of Power Sources*, *196*(8), 3993–4002. doi:10.1016/j.jpowsour.2010.11.094

Amine, K., Zhang, L., & Zhang, Z. (2010). *Develop and evaluate materials and additives that enhance thermal and overcharge abuse*. Department of Energy Advanced Battery Research Review.

Barnhart, C. J., & Benson, S. M. (2013). On the importance of reducing the energetic and material demands of electrical energy storage. *Energy & Environmental Science*, *6*(4), 1083–1092. doi:10.1039/c3ee24040a

Bergamasco, L., & Asinari, P. (2011). Scalable methodology for the photovoltaic solar energy potential assessment based on available roof surface area: Application to Piedmont Region (Italy). *Solar Energy*, *85*(5), 1041–1055. doi:10.1016/j.solener.2011.02.022

Bhatnagar, N., & Venkatesh, B. (2012). *Energy storage and power systems.* Paper presented at the Electrical & Computer Engineering (CCECE), 2012 25th IEEE Canadian Conference on. 10.1109/CCECE.2012.6334823

Borden, E., & Schill, W.-P. (2013). *Policy Efforts for the Development of Storage Technologies in the US and Germany*. Academic Press.

Bull, S. R. (2001). Renewable energy today and tomorrow. *Proceedings of the IEEE, 89*(8), 1216–1226. doi:10.1109/5.940290

Conejo, A. J., Caramanis, M. C., & Bloom, J. A. (1990). An efficient algorithm for optimal reservoir utilization in probabilistic production costing. *IEEE Transactions on Power Systems, 5*(2), 439–447. doi:10.1109/59.54550

De La Tour, A., Glachant, M., & Ménière, Y. (2011). Innovation and international technology transfer: The case of the Chinese photovoltaic industry. *Energy Policy, 39*(2), 761–770. doi:10.1016/j.enpol.2010.10.050

Deane, J. P., & Gallachóir, Ó. (2010). Techno-economic review of existing and new pumped hydro energy storage plant. *Renewable & Sustainable Energy Reviews, 14*(4), 1293–1302. doi:10.1016/j.rser.2009.11.015

Dewald, U., & Truffer, B. (2012). The local sources of market formation: Explaining regional growth differentials in German photovoltaic markets. *European Planning Studies, 20*(3), 397–420. doi:10.1080/09654313.2012.651803

Díaz-González, F., Sumper, A., Gomis-Bellmunt, O., & Villafáfila-Robles, R. (2012). A review of energy storage technologies for wind power applications. *Renewable & Sustainable Energy Reviews, 16*(4), 2154–2171. doi:10.1016/j.rser.2012.01.029

Dincer, F. (2011). The analysis on photovoltaic electricity generation status, potential and policies of the leading countries in solar energy. *Renewable & Sustainable Energy Reviews, 15*(1), 713–720. doi:10.1016/j.rser.2010.09.026

Ding, L.-X., Zheng, F.-L., Wang, J.-W., Li, G.-R., Wang, Z.-L., & Tong, Y.-X. (2012). Super-large dendrites composed of trigonal PbO 2 nanoplates with enhanced performances for electrochemical devices. *Chemical Communications, 48*(9), 1275–1277. doi:10.1039/C2CC15271A PMID:22179048

Dönitz, W., Dietrich, G., Erdle, E., & Streicher, R. (1988). Electrochemical high temperature technology for hydrogen production or direct electricity generation. *International Journal of Hydrogen Energy, 13*(5), 283–287. doi:10.1016/0360-3199(88)90052-3

Dorrell, D. G., Knight, A. M., Popescu, M., Evans, L., & Staton, D. A. (2010). *Comparison of different motor design drives for hybrid electric vehicles.* Paper presented at the Energy Conversion Congress and Exposition (ECCE). 10.1109/ECCE.2010.5618318

Dubal, D., Holze, R., & Kulal, P. (2013). Enhanced supercapacitive performances of hierarchical porous nanostructure assembled from ultrathin MnO2 nanoflakes. *Journal of Materials Science*, *48*(2), 714–719. doi:10.100710853-012-6783-6

Elliott, D. (2013). *A review of sustainable energy supply options*. Philadelphia: IOP Publishing. doi:10.1088/978-0-750-31040-6

Eyer, J., & Corey, G. (2010). Energy storage for the electricity grid: Benefits and market potential assessment guide. *Sandia National Laboratories*, *20*(10), 5.

Fletcher, S. (2011). *Bottled lightning: superbatteries, electric cars, and the new lithium economy*. Hill and Wang.

Gahleitner, G. (2013). Hydrogen from renewable electricity: An international review of power-to-gas pilot plants for stationary applications. *International Journal of Hydrogen Energy, 38*(5), 2039-2061.

Goetzberger, A., Hebling, C., & Schock, H.-W. (2003). Photovoltaic materials, history, status and outlook. *Materials Science and Engineering R Reports*, *40*(1), 1–46. doi:10.1016/S0927-796X(02)00092-X

Grau, T., Huo, M., & Neuhoff, K. (2012). Survey of photovoltaic industry and policy in Germany and China. *Energy Policy*, *51*, 20–37. doi:10.1016/j.enpol.2012.03.082

Haas, R., Lettner, G., Auer, H., & Duic, N. (2013). The looming revolution: How photovoltaics will change electricity markets in Europe fundamentally. *Energy*, *57*, 38–43. doi:10.1016/j.energy.2013.04.034

Hall, P. J., & Bain, E. J. (2008). Energy-storage technologies and electricity generation. *Energy Policy*, *36*(12), 4352–4355. doi:10.1016/j.enpol.2008.09.037

Hanna, M., & Nozik, A. (2006). Solar conversion efficiency of photovoltaic and photoelectrolysis cells with carrier multiplication absorbers. *Journal of Applied Physics*, *100*(7), 074510. doi:10.1063/1.2356795

Jansen, A. N., Vaughey, J. T., Chen, Z., Zhang, L., & Brushett, F. R. (2016). *Organic non-aqueous cation-based redox flow batteries*. Google Patents.

Jossen, A., Garche, J., & Sauer, D. U. (2004). Operation conditions of batteries in PV applications. *Solar Energy*, *76*(6), 759–769. doi:10.1016/j.solener.2003.12.013

Kermani, M., & Morshed, A. (2003). Carbon dioxide corrosion in oil and gas production—a compendium. *Corrosion*, *59*(8), 659–683.

Lefebvre, D., & Tezel, F. H. (2017). A review of energy storage technologies with a focus on adsorption thermal energy storage processes for heating applications. *Renewable & Sustainable Energy Reviews*, *67*, 116–125. doi:10.1016/j.rser.2016.08.019

Lu, N., Chow, J. H., & Desrochers, A. A. (2003). *Pumped-storage hydro-turbine bidding strategies in a competitive electricity market.* Paper presented at the Power Engineering Society General Meeting. 10.1109/PES.2003.1270415

Masaud, T. M., Lee, K., & Sen, P. (2010). *An overview of energy storage technologies in electric power systems: What is the future?* Paper presented at the North American Power Symposium (NAPS). 10.1109/NAPS.2010.5619595

Odum, H. T. (1975). Energy quality and carrying capacity of the earth. *Tropical Ecology*, *16*(1), 14.

Olabi, A. G. (2017). *Renewable energy and energy storage systems*. Academic Press.

Olsen, I. I., & Pasquale, N. B. (2007). *Battery and inverter configuration with increased efficiency*. Google Patents.

Park, M., Sun, H., Lee, H., Lee, J., & Cho, J. (2012). Lithium-air batteries: Survey on the current status and perspectives towards automotive applications from a battery industry standpoint. *Advanced Energy Materials*, *2*(7), 780–800. doi:10.1002/aenm.201200020

Pech, D., Brousse, T., Bélanger, D., & Guay, D. (2009). EQCM study of electrodeposited PbO 2: Investigation of the gel formation and discharge mechanisms. *Electrochimica Acta*, *54*(28), 7382–7388. doi:10.1016/j.electacta.2009.07.070

Pedram, M., Chang, N., Kim, Y., & Wang, Y. (2010). Hybrid electrical energy storage systems. *Proceedings of the 16th ACM/IEEE international symposium on Low power electronics and design.*

Pollet, B. G., Staffell, I., & Shang, J. L. (2012). Current status of hybrid, battery and fuel cell electric vehicles: From electrochemistry to market prospects. *Electrochimica Acta*, *84*, 235–249. doi:10.1016/j.electacta.2012.03.172

Pöschl, M., Ward, S., & Owende, P. (2010). Evaluation of energy efficiency of various biogas production and utilization pathways. *Applied Energy*, *87*(11), 3305–3321. doi:10.1016/j.apenergy.2010.05.011

Raul, N. S., & Necsulescu, C. (1984). Economics of Energy Storage Devices in Interconnected Systems-A New Approach. *IEEE Transactions on Power Apparatus and Systems*, *PAS-103*(6), 1217–1223. doi:10.1109/TPAS.1984.318452

Rehman, S., Al-Hadhrami, L. M., & Alam, M. M. (2015). Pumped hydro energy storage system: A technological review. *Renewable & Sustainable Energy Reviews*, *44*, 586–598. doi:10.1016/j.rser.2014.12.040

Sandén, B. A. (2014). *Systems Perspectives on Renewable Power 2014*. Academic Press.

Shapira, R., Nessim, G. D., Zimrin, T., & Aurbach, D. (2013). Towards promising electrochemical technology for load leveling applications: Extending cycle life of lead acid batteries by the use of carbon nano-tubes (CNTs). *Energy & Environmental Science*, *6*(2), 587–594. doi:10.1039/C2EE22970F

Silva Fonseca, J. G., Oozeki, T., Takashima, T., Koshimizu, G., Uchida, Y., & Ogimoto, K. (2012). Use of support vector regression and numerically predicted cloudiness to forecast power output of a photovoltaic power plant in Kitakyushu, Japan. *Progress in Photovoltaics: Research and Applications*, *20*(7), 874–882. doi:10.1002/pip.1152

Sørensen, B. E. (2017). *Renewable Energy: Physics, engineering, environmental impacts, economics and planning*. Academic Press.

Spiers, D. J., & Rasinkoski, A. D. (1995). Predicting the service lifetime of lead/acid batteries in photovoltaic systems. *Journal of Power Sources*, *53*(2), 245–253. doi:10.1016/0378-7753(94)01989-9

Spyker, R., & Nelms, R. (1997). *Evaluation of double layer capacitor technologies for high power and high energy storage applications*. Paper presented at the Industrial Electronics, Control and Instrumentation, 1997. IECON 97. 23rd International Conference on. 10.1109/IECON.1997.668434

Surek, T. (2005). Crystal growth and materials research in photovoltaics: Progress and challenges. *Journal of Crystal Growth*, *275*(1), 292–304. doi:10.1016/j.jcrysgro.2004.10.093

Vieira, F., & Ramos, H. (2008). Hybrid solution and pump-storage optimization in water supply system efficiency: A case study. *Energy Policy*, *36*(11), 4142–4148. doi:10.1016/j.enpol.2008.07.040

Wang, L., Han, Y., Feng, X., Zhou, J., Qi, P., & Wang, B. (2016). Metal–organic frameworks for energy storage: Batteries and supercapacitors. *Coordination Chemistry Reviews, 307*, 361–381. doi:10.1016/j.ccr.2015.09.002

Wang, Y., Kim, Y., Xie, Q., Chang, N., & Pedram, M. (2011). Charge migration efficiency optimization in hybrid electrical energy storage (HEES) systems. *Proceedings of the 17th IEEE/ACM international symposium on Low-power electronics and design*. 10.1109/ISLPED.2011.5993620

Watt, G., & Fechner, H. (2009). Photovoltaic market and industry trends–latest results from the IEA PVPS programme. *Elektrotechnik und Informationstechnik, 126*(9), 328-330.

Weiland, P. (2006). Biomass digestion in agriculture: A successful pathway for the energy production and waste treatment in Germany. *Engineering in Life Sciences, 6*(3), 302–309. doi:10.1002/elsc.200620128

Whittingham, M. S. (2012). History, evolution, and future status of energy storage. *Proceedings of the IEEE, 100*, 1518-1534. 10.1109/JPROC.2012.2190170

Winter, M., & Brodd, R. J. (2004). *What are batteries, fuel cells, and supercapacitors?* ACS Publications.

Wood, A. J., & Wollenberg, B. F. (2012). *Power generation, operation, and control*. John Wiley & Sons.

Xiaoliang, H., Tosiyoki, H., & Yoichi, H. (2014). *System design and converter control for super capacitor and battery hybrid energy system of compact electric vehicles*. Paper presented at the Power Electronics and Applications (EPE'14-ECCE Europe), 2014 16th European Conference on. 10.1109/EPE.2014.6911060

Yender, G. L. (1998). Battery recycling technology and collection processes. *Electronics and the Environment, 1998. ISEE-1998. Proceedings of the 1998 IEEE International Symposium on*.

Zhao, G., & Davison, M. (2009). Optimal control of hydroelectric facility incorporating pump storage. *Renewable Energy, 34*(4), 1064–1077. doi:10.1016/j.renene.2008.07.005

Zhao, Y., & Zhu, K. (2014). CH3NH3Cl-assisted one-step solution growth of CH3NH3PbI3: Structure, charge-carrier dynamics, and photovoltaic properties of perovskite solar cells. *The Journal of Physical Chemistry C, 118*(18), 9412–9418. doi:10.1021/jp502696w

Chapter 16
Energy Informatics Using the Distributed Ledger Technology and Advanced Data Analytics

Umit Cali
University of North Carolina at Charlotte, USA

Claudio Lima
Blockchain Engineering Council, USA

EXECUTIVE SUMMARY

The main drivers of the third industrial revolution era were the internet technologies and rise of renewable and distributed energy technologies. Transition to green and decentralized energy resources and digital transformation of the existing industrial infrastructure had been the biggest achievements of the third industrial revolution. The main drivers of the fourth era will be artificial intelligence (AI), quantum computing, advanced biotechnology, internet of things, additive manufacturing, and most importantly, distributed ledger technology (DLT). Energy forecasting such as wind and solar power forecasting models are the most common energy AI-based informatics applications in the energy sector. In addition, use of DLT is expected to be an industrial standard in various industrial sectors including energy business in the coming decade. This chapter emphasizes description of energy forecasting using AI and energy DLT and future developments and solutions to overcome challenges that are associated with standardization of the energy DLT applications.

DOI: 10.4018/978-1-5225-8559-6.ch016

BACKGROUND INFORMATION

The industrial revolution is now in the fourth phase. The first industrial revolution is started with the invention of the steam engines and matured with the extended infrastructure supported by derivative or follow-up technologies such as building the factories and railways. Electrification, invention of combustion engines and evolving the mass production were the major indicators of the second industrial revolution. The main drivers of the third industrial revolution era were the internet technologies, advance molecular biology, smart internet of every things, rise of renewable and distributed energy technologies. Transition to green and decentralized energy resources and digital transformation of the existing industrial infrastructure had been the biggest achievements of the third industrial revolution. Rifkin (2011;2013) described the third industrial revolution as “a new convergence of communication and energy” to design and operate a powerful new infrastructure. While the industry is evolving towards the third relation wave, the fourth industrial revolution stage is started. Schwab (2016) showed “Third used electronics and information technology to automate the mass production” and “now a fourth industrial revolution is building on the Third, the digital revolution that has been occurring since the middle of the last century. It is characterized by a fusion of technologies that is blurring the lines between the physical, digital and biological spheres.”. According to this definition, the industry is evolving to the cyber-physical-system (CPS) level, where the information and communication technologies (ICT) are converging to physical components such as mechanical and energy systems or in future biological systems. The main drivers of the Fourth era will be artificial intelligence, robotics, quantum computing, advanced biotechnology, internet of things (IoTs), additive manufacturing, fully autonomous vehicles and most importantly distributed ledger technology (DLT). There had been a very strong correlation between the fundamental changes in the field and use of energy, the monetary flow (economics aspects) and the social aspects, if the historical influences of the previous industrial evolutions are considered. Therefore, it is inevitable that the new Third and Fourth industrial revolution infrastructure will enable fundamental paradigm shifts in near future. Sharing economy and democratization of information and power (or any good or service in general) infrastructure will be central hot spots of the new scientific discussions. In this study, we attempted to discuss the use of two main (AI and DLT) Third and Fourth industrial era technologies in the field of energy.

SETTING THE STAGE

Currently, a major part of the world's energy need is satisfied by non-renewable fossil fuel resources such as coal, oil, and natural gas which are well known to release a large number of greenhouse gasses. However, use of clean and carbon-free energy resources such as wind and solar have become more dominant in the global energy mix over the last two decades (Cali, 2018). The forms of renewable energy, such as wind and solar power, are among the fastest growing forms in terms of use in countries like China, the United States, India, and Germany. Wind and solar power are different from non-renewable fossil fuel resources, since the "fuel" cannot be stored until such time that its power is needed in the power system. Besides, most of the decentralized energy resources are in the form of intermittent renewable energy resources (RES) which make it challenging to predict their power outputs (Cali, 2010; Eltawil et al., 2010). Concentrating solar power (CSP) and large-scale hydropower plants are exceptional in terms of intermittency and controllability, as the heat produced by the CSP plants and the water in the hydro-electric dams can be stored for several hours and be used for power generation when it is needed. Moreover, utilization of energy storage units in parallel to higher penetration of RES are becoming more popular and constitute an economically viable option to integrate the higher number of intermittent RES which can also be used to store electrical power and be operated to provide various services such as time-of-use bill management, demand charge reduction, energy arbitrage, frequency regulation, and peak shaving.

However, the grid and market integration of the high number of intermittent RES is still a great challenge to be solved. Advanced data analytics methods such as Auto-regressive Integrated Moving Average (ARIMA), Auto-regressive Integrated Moving Average with Explanatory Variable (ARIMAX), artificial neural networks (ANN), support vector machines (SVM), deep learning algorithms and extreme gradient boosting (XGBoost) algorithms give various opportunities to generate renewable energy forecasting systems which helps to integrate RES into power systems successfully.

In addition, consumer-owned decentralized generation of electrical power is another emerging trend in the power industry. Decentralization of the energy generation assets creates new market players such as prosumers. Prosumers can be both power producers and consumers. Distributed ledger and Blockchain technologies enable instant monetary and contractual transactions to occur over peer to peer (P2P) private and public networks by satisfying cyber security concerns. Both changes have the potential to empower a paradigm shift in the field of energy in the coming years.

CASE-1 DESCRIPTION

Renewable Energy Forecasting

Renewable energy forecasting methods have been integral parts of modern power energy systems and markets over the past two decades. Forecasting of wind speed, wind power, global irradiance, and solar power have been the most popular applications in energy forecasting. Most of the energy forecasting methods use advanced statistical or data analytics methods such as ARIMA, ARIMAX, ANN, SVM, deep learning algorithms and XGBoost algorithms. Various institutions and companies developed hybrid models which use a combination of various methods to increase the accuracy of forecasting. Wind and solar power forecasting models are two of the most used energy forecasting algorithms in the industry. Especially the transmission system operators (TSOs) and distribution systems operators (DSOs) utilize solar and wind power forecasting models to make more precise day-ahead and intraday power delivery operations in their daily business. DSOs also use such forecasting algorithms to reduce the probability of the local power congestions on critical spot. In addition to these, solar and wind power forecasting models are used in the operation of wind and solar farms where the outcome of their next day or hour power generation is estimated. The forecasted time series can be used for several purposes by the RES operators such as making accurate energy trading decisions on the power market or scheduling planned maintenance activities for time where less power generation is forecasted.

Classification of Renewable Energy Forecasting According to Forecast Horizon

Wind and solar power forecasting systems operate in various time-scales according to their applications and can be categorized into five groups:

- Ultra-short-term or minutes-ahead forecasting: From a few minutes to 1 hour ahead.
- Short-term forecasting: From 1 hour to several hours ahead.
- Day ahead forecasting: 24 hours ahead.
- Week(s) ahead forecasting: From 1 day to 1 week
- Seasonal or long-term forecasting: From 1 week to one year.

TSOs prefer the day-ahead prediction values to schedule next days' schedules in their operational regions to minimize the unbalances between the power generation and demand sides. They may also run intraday or hour ahead predictions to adjust set values on the power market. Power trading is allowed in some countries like Germany where wind farm operators may select their power purchase operations using either fix feed-in-tariffs (FIT) or direct marketing option. The operators selecting the FIT mode would not need wind power forecasting system for trading, but they can use the wind power forecasting systems for optimizing maintenance tasks. On the other hand, an operator who prefers to sell wind power output needs a very accurate wind power forecasting model to maximize its power trading revenues. Ultra-short-term energy forecasting applications are becoming more popular in recent years. Such kind of forecasting models can be categorized as remote sensing, time-series and data assimilation models can deliver the core data required for ultra-short-term or minute-scale renewable energy forecasting models. As in Wurth et al. 2019, minute-scale energy forecasting can be used for wind turbine, wind farm control, power grid balancing, energy and ancillary services market operations. Hour ahead and day ahead energy forecasting had been integral part of TSO and DSO operations since two decades. The forecast accuracy decreases as the forecast horizon extends. (Cali, 2010; Ghofrani et al., 2018). Weeks ahead and seasonal energy forecasting algorithms have lower accuracy due to the stochastic nature of wind and solar but can still be used for various energy sector related tasks and operations such as estimating the annual energy yield of the renewable energy projects, TSO and DSO system planning purposes. (Lin et al., 2015)

Wind Power Forecasting

The volatile and highly weather dependent nature of wind power generation causes inevitably some serious challenges in terms of grid and market integration. Wind power forecasting systems provide feasible solutions to mitigate negative impacts of fluctuating wind power resources. Wind power forecasting (WPF) has been overwhelmingly investigated over the past two decades (McCarthy, 1998; Troen et al., 1990; Cali, 2010; Lange, 2002; Carpenter et al., 2016; Kariniotakis et al., 2006; Nielsen et al., 1999; Folken et al., 2001; Giebel et. al., 2002). The Figure 1 demonstrates the overall structure of wind power forecasting systems. The input parameters can be categorized as dynamic and static. Static input parameters are essential, especially for physical WPF models. Power curves of the wind turbines installed capacity of wind farm, terrain information and wind farm layout (coordinates of the wind turbines) can be considered as the most important static input parameters. Dynamic input parameters, on the other hand, include a variable input dataset such as historical or actual numerical weather prediction (NWP), measured wind power, meteorological measurements from meteorological towers or stations, offshore

Figure 1. Overall structure of a wind power forecasting system

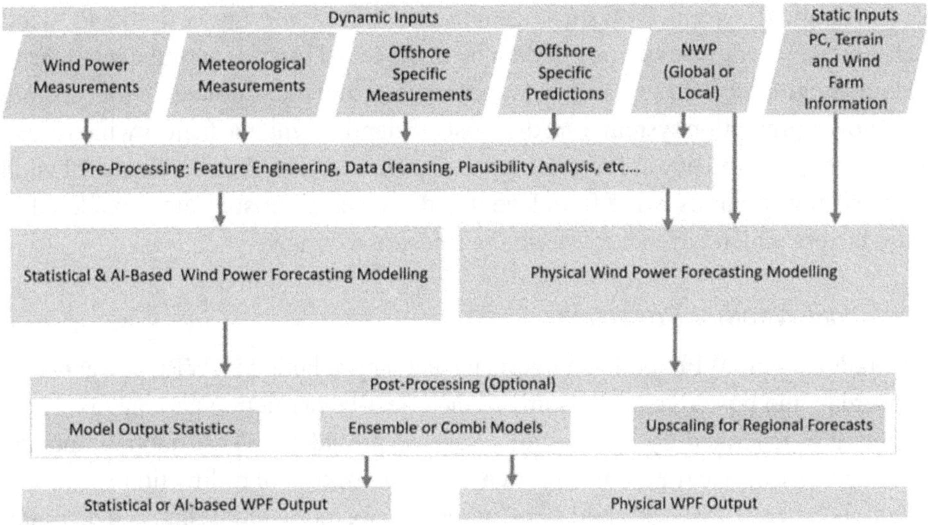

specific forecasting and measurement timeseries if the model is designed for the offshore environment. The second layer of the system is the pre-processing layer. Depending on the WPF model, some preliminary data cleansing, plausibility for missing or faulty data, and feature engineering procedures should be carried out in this stage. Feature engineering is responsible to transfer the raw or semi-raw data into a feature vector to increase the efficiency of the prediction system. This method can in particular boost the performance of artificial intelligence (AI) based models. Correlation analysis or principal component analysis (PCA) are commonly used to determine best fitting input parameters for forecasting models. PCA is among the most successful pre-processing techniques which uses an orthogonal transformation to convert large set of input variables into more meaningful and stronger data patterns. This enables one to reduce computational time and reach higher accuracy levels in forecasting. The third layer is the wind power forecasting layer (core) where various type of WPF models such as physical or statistical models can be deployed to generate the wind power forecasting output timeseries. Optionally, a post-processing layer can be used depending on the implementation of the WPF model. Ensemble models are used to merge various type of WPF timeseries to generate hybrid or combined models. Model output statistics (MOS) methods have the potential to reduce forecasting uncertainty. The regional WPF can be developed using special upscaling algorithms. The Figure 1 summarizes the overall structure of physical, statistical, AI-based and hybrid wind power forecasting approaches.

Factors Influencing Wind Power Forecasting Systems

Meteorological parameters such as numerical weather prediction and wind measurements, wind power output data from the SCADA system of the wind farm, and oceanographic data sets such as wave forecasts and measurements (for offshore wind power prediction systems) are the most influential dynamic factors which affect the accuracy of wind power forecasting systems. In addition, the coordinates of wind turbines, power curves, wind farm layout and site characteristics are considered as static factors which have an effect on the forecast accuracy.

Numerical Weather Prediction

Depending on the WPF model, numerical weather prediction (NWP), power curve, measured wind power value are some of the effective prediction parameters. NWP models are generated by meteorology service providers. These models include predicted metrological parameters such as wind speed, wind direction at various heights (such as 10, 50, 100 and 130 m), temperature, momentum flux, and pressure. Depending on the meteorology service provider and their modeling structure, NWP timeseries can be generated once or twice a day with two to several days forecast horizons.

Wind Power Measurement Dataset

Commercial wind turbines are advanced electromechanical systems which are operated by supervisory control and data acquisition (SCADA) system where critical operational data are also stored. Other type of critical information recorded by the SCADA system is the active wind power output of each individual wind turbine or the entire wind farm. Measured wind power output is an essential input parameter especially for the statistical and AI based wind power forecasting models.

Power Curve

A power curve represents the power generation characteristics versus various wind speed values. Each wind turbine has its own power curve depending on the manufacturer and the technical configuration of wind turbine. The power curves are determined by using field measurements where an anemometer is utilized on a measurement mast close to the wind turbine investigated. In general, the commercial wind turbines start generating power after receiving more than 3 or 4 m/sec wind speeds which is called the cut-in wind speed. At lower speeds, there is not sufficient torque on the wind turbine blades to beat the frictional forces and to make them rotate. The rated power indicates the installed capacity of the wind turbine where the effective wind power can be generated if the wind turbine will be operated in

Figure 2. The power curve of a sample 2 MW wind turbine

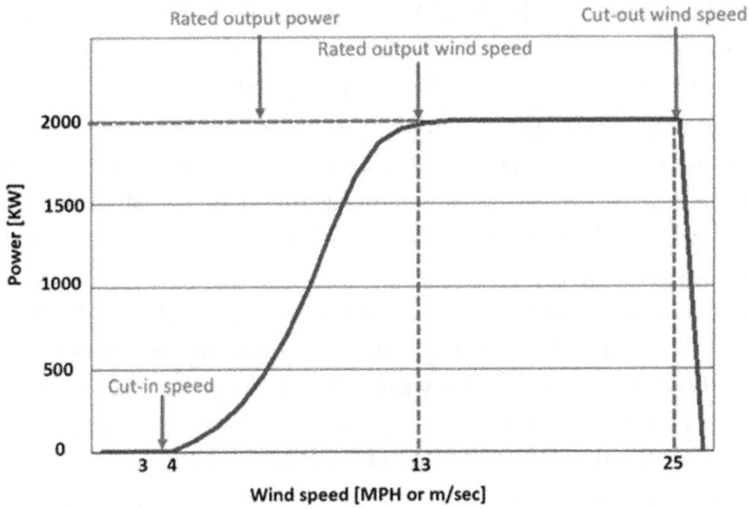

rated output wind speeds typically 12 to 18 m/sec (see Figure 2). After reaching the rated output wind speed point, the wind turbine is able to generate power with the maximum capacity until 25 m/sec the cut-out wind speed) where the entire system shall be switched off gradually to prevent damages to the mechanical components of wind turbine such as blades.

Wind Measurements

Wind information measured by anemometers can be used as an additional input parameter for statistical and AI-based wind power forecasting models. This information is generally used for the forecast horizons ranging from a couple of minutes to a couple of hours. Wind speed and direction are the most important wind measurement parameters. The wind measurement equipment can be mounted on special wind measurement masts which are in the vicinity of the wind farm. Wind turbines have the meteorological measurement instrumentation equipment on their nacelle. The reliability of nacelle-mounted wind measurements is lower due to distortion of observed wind speed values. Measured meteorology parameters are useful only for short-term WPF models which have maximum 8 hours forecast horizons.

Offshore Measurements and Operational Wave Forecasting Models

Even though most of the global renewable energy is currently harvested from onshore wind farms and solar power resources, offshore wind resources are gaining wide popularity among the other renewable energy resources. Offshore specific observations and measurements are the most influential input parameters for offshore wind power forecasting (OWPF) models. The Waverider buoy is one of the commonly used instruments to measure the sea or ocean waves which can capture the following physical parameters such as spectral peak wave period, Significant wave height and length. Optimally high-frequency and microwave (X-ray) based radars and satellite remote sensing systems can also be utilized to observe and analyze sea wave and tidal currents. Use of operational wave forecasting models from various providers like ECMWF as an additional input dataset dramatically increases the accuracy of the predictions dramatically

The observed or forecasted wave parameters which correlate best with the most important WPF input parameter (i.e., wind speed) can be chosen as an additional input to the OWPF models. Wind – wave interaction can be understood by means of the transfer of momentum, heat and other tidal currents which occurs at the atmosphere-sea/ocean interface. The wind at offshore and onshore regions features different characteristics. The main reason of difference in wind behavior can be explained by investigating additional physical interactions such as air flow and thermal stratification between sea surface and lower atmospheric boundaries. Thermal stratification is associated with the water depth, incoming heat and temperature degree of water column. The stability of thermal stratification regimes can be stable, unstable, or neutral depending on physical conditions. (Cali, 2010; Lange, 2002; Carpenter et al., 2016). An effective air-sea interaction model for computing offshore wind profiles was presented in (Kariniotakis et al., 2006). The proposed model named as inertially coupled wind profiles based on inertial coupling of the wave-field to a wave boundary layer with constant wind shear stress.

Classification of Wind Power Forecasting Models According to Modelling Approaches

There are four main types of renewable energy generation forecasting models: physical, statistical, AI based, and ensemble or hybrid models. The majority of forecasting customers use a hybrid approach where various forecasting models are operated in parallel to decrease the uncertainty of the prediction.

Physical Wind Power Forecasting

Classical physical WPF models are used to transform available wind speed profiles by optimizing the wind speed prediction on hub height level of wind turbines. Most of the classical physical models consider additional factors such as wake effects within a project area, layout of the wind farm under focus, and power curves of wind turbines. It is also essential to model the thermal stability of the atmosphere in addition to involving the terrain information of the project area such as coastal orography and surface roughness parameters. Physical models require no or less historical measured dataset to generate forecasting timeseries. This type of WPF models require complex physical modeling where all the necessary physical parameters must be considered very precisely. Therefore, poor physical modeling leads to poor WPF results. Besides, the resolution of used NWP dataset and terrain characteristics of wind farm have significant effects on the accuracy of physical WPF models. In general, physical WPF models require more computational power compared to the counterparts like statistical WPF models. In addition, poor availability of measured meteorological data to confirm NWP outputs reduces the forecast accuracy for wind farm locations (Tambke, 2006; Tambke et al., 2003).

(McCarthy, 1998; Troen et al., 1990) developed the early versions of the physical WPF models using NWP and local meteorological measurement datasets by EPRI in US and Risø National Laboratory in Denmark. Many other physical models were designed in various institutions such Zephyr, Previento (Germany) and WPPT (Denmark) after the first successful versions by (Nielsen et al., 1999; Folken et al., 2001; Giebel et al., 2002) respectively. GH (Garrad Hassan) Forecaster (UK), eWind (USA) and PowerSight (USA) are other commercial hybrid WPF tools which are commonly used by the power industry.

The adoption of the existing onshore WPF models can be accomplished by deploying considerable modification efforts. The atmospheric and environmental conditions of onshore and wind farms differ from their offshore counterparts. Therefore, offshore wind power forecasting models require more detailed modeling of offshore wind speed profiles considering variable roughness of the project location, thermal stratification, influence of the land-sea-discontinuity, land mass and wake effects. Thermal effects and sea surface roughness for offshore wind conditions has been explored by various authors (Lange, 2003; Watson et al., 2003; Pinson et al., 2012).

The Figure 3 shows the overview structure of physical onshore and offshore specific physical WPF models. In addition to various functionalities such as wind-wave interaction modelling (only for offshore WPF models), generation of local wind speed profiles normalized on the hub height levels of the wind turbines and modelling of wake effects, static and dynamic system variables are used as input dataset to perform physical WPF models. MOS and upscaling functions optional. Upscaling function might be necessary if several wind farm clusters are available and regional offshore WPF timeseries are needed. WPF timeseries might be deterministic or probabilistic depending on the need of WPF end user needs.

Figure 3. Structure of physical wind power forecasting models

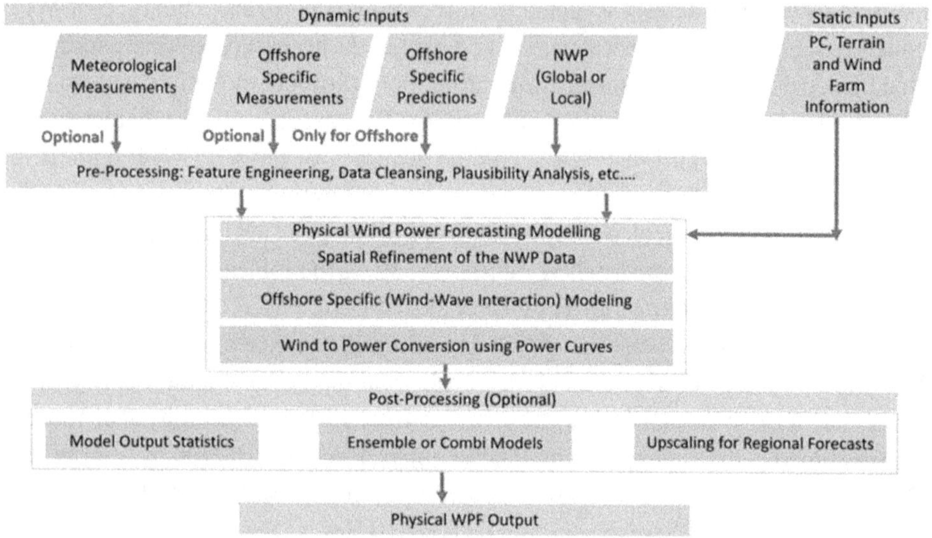

Statistical Wind Power Forecasting

Statistical WPF models need a sufficient amount of historical measured data such as wind power generation data and historical NWP data to find the relationship between measured wind power and NWP data to generate accurate forecasting outcomes. In general, statistical WPF models convert the NWP forecasts to wind power prediction timeseries using statistical methods such as autoregressive (AR), ARMA, and Kalman Filter. Brown et al. 1984 and Hill et al. 2012 developed the first statistical WPF based on AR approach where a transformation function based on the Gaussian distribution of wind speeds, measured power curve and a wind speed to power transform function based on the power law to predict the wind power output. (Tantareanu 1992), Kavasseri et al. (2009) and Torres et al. (2005) used ARMA and ARIMA models to predict and simulate wind speed timeseries. Kalman filters are another alternative approach to predict wind speed values (Wei, 2010). Louka et al. (2008) demonstrated that use of Kalman Filtering is not only improves the forecast accuracy of wind speed predictions but also the wind power forecasting results. Usually, at least one year of historical data would be sufficient to catch the seasonality effect, but, in some cases, three months of data may deliver acceptable results. Statistical WPF models are cheaper and easier to develop in comparison to other WPF models. Figure 4 demonstrates the overview of the statistical wind power forecasting models.

Figure 4. Structure of statistical wind power forecasting models

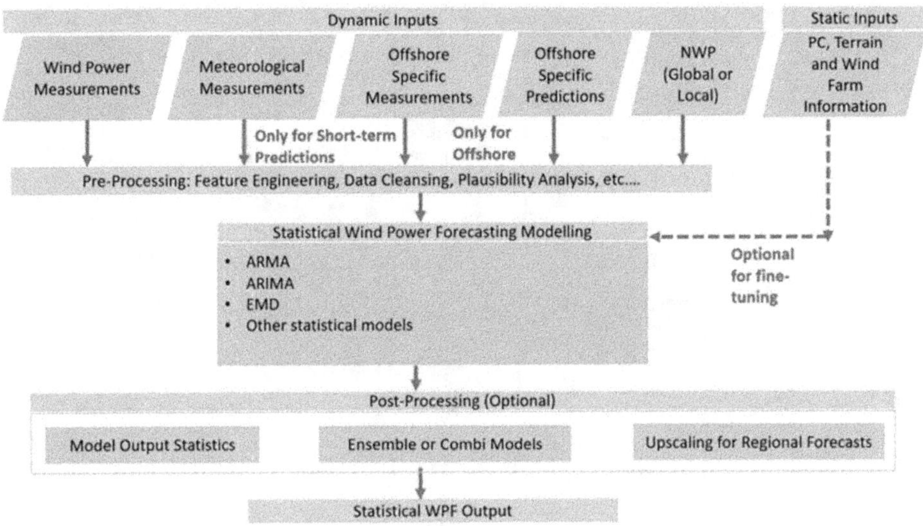

AI-Based Wind Power Forecasting

Similar to the statistical WPF models, AI-based models also derive a relation between the dynamic and static input parameters such as NWP and measured wind power timeseries information from the past to forecast wind power output for the future. However, AI-based WPF models utilize AI and machine learning (ML) algorithms instead of clearly defined statistical equations to solve non-linear and more complex engineering problems. The main advantage of the AI-based WPF models is their capability to predict the wind power output without knowing the physical conditions of the wind farm. Thus, detailed physical modelling is not required. Having said that, use of limited physical information static information helps to increase the accuracy of AI-based WPF models if they are integrated into the algorithms. A variety of AI and ML algorithms used to predict wind power output are listed as:

- Artificial neural networks (ANNs) (Cali, 2010; Wu, 2010; Sfetsos, 2002)
- Support vector machines (SVMs) (Zeng et al., 2011),
- Neuro-fuzzy networks (NFN) (Xia et al., 2010).

Several researchers have developed various AI-based WPF models over the last two decades for onshore wind farms (Wu, 2010; Sfetsos, 2002; Zeng et al., 2011; Xia et al., 2010; Giebel et al., 2011), whereas to date few WPF models have been derived for offshore environments. It is expected to observe an increasing trend in terms of research activities especially in offshore wind forecasting. The Figure 5 shows the overview of AI-based WPF models.

Figure 5. Overview of AI-based wind power forecasting models

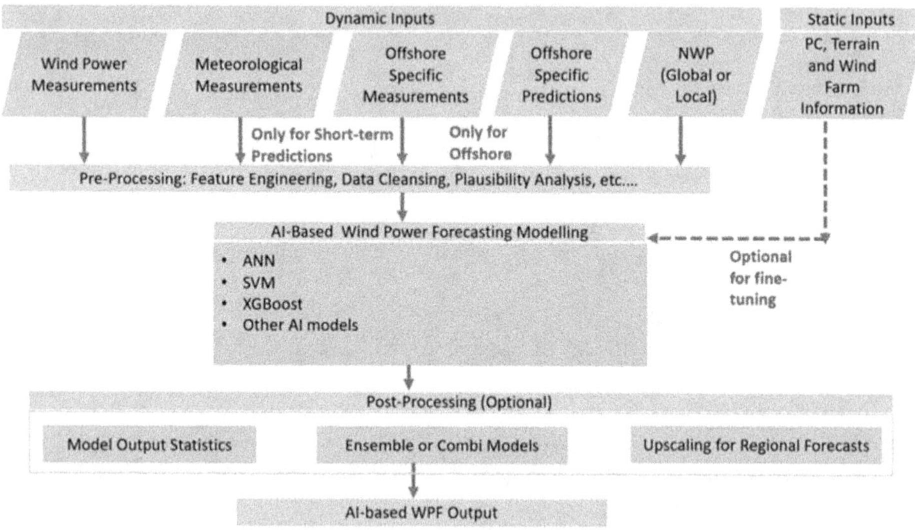

FNN based approaches were demonstrated for an offshore wind farm with an installed capacity of 5 MW in Tunø Knob, Denmark. The FNN based model was developed by ARMINES and delivered similar or even better results than the onshore WPF models in terms of performance and accuracy (Pinson et al., 2004). Pinson et al. (2012) developed a model (Adaptive Markov-switching autoregressive) for offshore wind power forecasting challenges where a new parameterization of the model coefficients is presented. The developed models yielded better results in comparison to autoregressive-model based forecasts.

Ensemble and Hybrid Wind Power Forecasting

Hybrid models usually combine multiple physical, statistical models and / or NWP datasets from various origins. This approach aims to exploit the best available outcome from multiple models to reach the highest available accuracy. Hybrid models increase redundancy but may require higher computational time to generate the combined prediction output. The ANEMOS project is one of the successful implementations of hybrid onshore and offshore WPF approaches (Tambke, 2006) and (ANEMOS, n.d.).

An ensemble based probabilistic offshore wind forecasting approach was introduced by Pinson et al. (2009) for another Danish offshore wind farm Horns Rev with an installed capacity of 160 MW. The proposed model employed local polynomial regression and orthogonal fitting methods to increase forecast accuracy. Another novel offshore WPF approach was developed using additional oceanographic parameters such as wave measurements and forecasts by Cali, (2010). This model proposed a hybrid model approach using various statistical forecasting methods and ensemble NWP dataset which increased the forecast accuracy. Wake adjusted physical power model and offshore specific WPF models were demonstrated in (Kurt, 2017).

Solar Power Forecasting

Beside wind power, solar power is another emerging energy harvesting technology type that constantly increases its share in the global energy mix constantly. Like wind energy resources, solar energy resources have a non-controllable and fluctuating power generation pattern. Power transmission and distribution companies have to deal with the intermittent power input from large amounts of solar energy resources. Therefore, the use of accurate solar power energy systems to integrate the large amount of solar energy resources to existing power grids is a critical task for utilities. In other words, utilities use the solar and wind power predictions to manage their power grid operations and keep their power system stable in a better way. Determining the ramping events, optimization of unit commitment and power transmission scheduling are mission critical daily tasks for power system operators where renewable energy forecasting models are integrated into their system. The performance of solar power is presented using statistical metrics such as normalized root means square error (nRMSE), mean absolute percentage error (MAPE) or mean absolute error (MEA).

Solar power and irradiance forecasting models basically use parallel underlying principles as the wind power and speed forecasting models do. Forecast horizons, type of input parameters such as solar power measurement timeseries, meteorological forecast datasets and use of similar statistical and AI based forecasting algorithms are some similarities between these two energy forecasting domains. However, due to the different nature of physical principles, there are some significant differences between solar and wind power related forecasting models. Forecasting the global horizontal irradiance (GHI) is the most important parameter which has higher grade of influence on the solar power generation predictions.

Solar power is the conversion of energy from sunlight using solar thermal, concentrated solar power (CSP) or photovoltaic (PV) units into electrical power. The energy conversion principle of CSP and solar thermal systems are based on thermodynamic rules and in most of the cases thermal or heat energy is harvested. CSP systems then convert the generated heat to electrical power using heat engines. On the other hand, the PV systems are used to generate electrical power using the photovoltaic effect. PV systems generate direct current (DC) power which varies depending on the sunlight's intensity or solar irradiance. The harvested DC power is then converted to AC power using power electronic devices.

Influential Factors for Solar Forecasting Systems

NWP Data

Meteorological data is the most essential dataset for energy forecasting systems. There are a number of types of meteorological data such as measured/observed and weather prediction data. Numerical weather prediction (NWP) is the most common input dataset for SPF systems. NWP models are operated by public or private weather services to predict the weather condition of their coverage area or globally by using physical laws and prognostic equations such as Navier–Stokes equations (Dutton, 1976). Global models predict the atmospheric status of the entire Earth. Global NWP models are generated by limited number of providers such as GFS (Global Forecast System) by NOAA (US National Oceanic and Atmospheric Administration) and the IFS (Integrated Forecast System) ECMWF (European Centre for Medium- Range Weather Forecasts). Data assimilation models are used to integrate the observed meteorological information from the meteorological stations and other ad hoc measurement systems to increase the accuracy of the NWP models. Mesoscale models are then operated after the global models to predict the atmospheric status of specific locations in the Earth. Most important NWP data are solar radiation density, solar irradiation, cloud coverage, temperature, humidity, pressure and wind speed information for SPF models. Irradiance predictions timeseries can be generated based on NWP. Having said that, many of NWP models can offer surface solar irradiance which can directly be used as input parameter to a SPF system. The spatial resolution of the NWP can vary depending on the local model and usually ranges between 27 km x 27 km down to 1 km x 1 km (Heinemann et al, 2006).

Solar Irradiance

Solar irradiance is a measure which defines the power per unit area received for the Sun in terms of electromagnetic radiation. Productivity of solar power systems are mainly depending on this parameter. Solar irradiation can be categorized in four groups: direct, diffuse, total and global irradiance respectively. Direct solar irradiation or direct normal irradiance (DNI) is the measure which represents the amount of solar beam radiation received by a surface (such as PV panel) perpendicularly as a straight line from the sun. Diffuse Horizontal Irradiance (DHI) is a measure of solar radiation received at the horizontal earth surface by following an indirect path from the sun passing through some atmospheric obstacles such as air molecules, cloud and aerosol particles. Total Solar Irradiance (TSI) is a measure of solar power which indicates the totality of the solar energy as a function of wavelength per unit area incident that arrives to the Earth's upper atmospheric layer. Global Horizontal

Irradiance (GHI) is the total solar radiation incident received from above by a surface horizontal to the ground. In other words, it is the sum of DHI and DNI by considering the cosine of the zenith angle.

Power Output Data

Solar power measurements can be provided by data acquisition and metering devices of small-scale solar PV installations or from SCADA systems of larger solar PV farms.

Sky Imager Data

Sky imagers are used to detect cloud movement images. Use of sky imagers for short-term solar power forecasting tasks can significantly increase the accuracy of the predictions. Furthermore, sky imagers are very efficient at capturing sudden changes in solar irradiance and ramping effects in solar power output. According to Chow et al. (2011), the most effective time horizons for sky imagers data usage are between 5 to 30 minutes (nowcasting).

Satellite Data

Satellites can provide information coming from the governmental or privately-owned satellites which are in the geostationary orbit. In particular the dynamic cloud movement information and surface irradiance are two main parameters which can be delivered by satellites (Hammer et al., 2003). Kühnert et al. (2013) applied operational satellite-based irradiance forecasting systems to SPF models successfully.

Classification of Solar Power Forecasting Models According to Modelling Approaches

Physical Methods

The physical SPF models are designed to convert GHI or solar irradiance and other meteorological variables in to predicted power output. The physical solar energy related forecasting models are based on NWP, total sky imager measurements, cloud observations delivered by satellites, and / or other relevant atmospheric measurements. Predicted solar irradiation is the most important variable for the physical SPF models. This chapter is dedicated to discuss only renewable energy forecasting approaches and prediction of meteorological parameters is not covered

Figure 6. Overall structure of physical solar power forecasting models

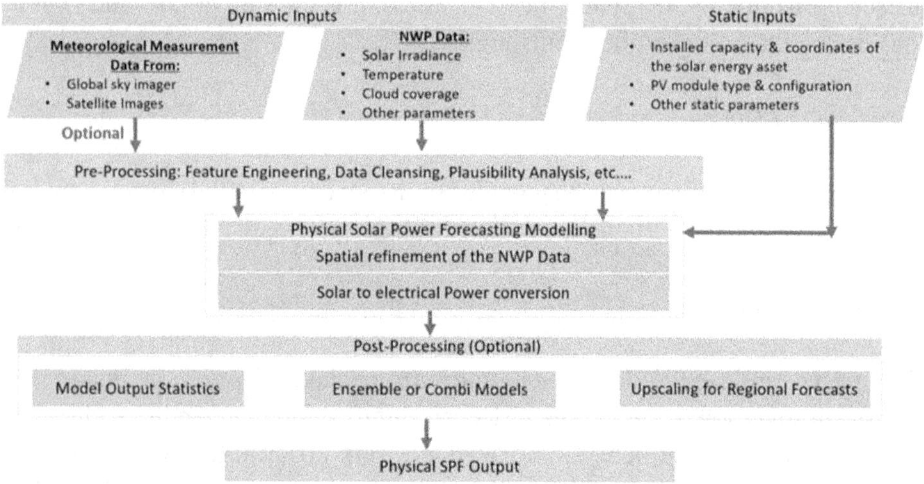

in detail. Static (location of PV farm, PV configuration including tilt angles and installed capacities) and dynamic input (NWP and meteorological observation sources such satellite images and global sky imager) parameters and data are inserted to the physical SPF modeling environment where the spatial refinement and most importantly solar energy conversion related equations deployed to generate the predicted solar power output of the target solar plant. Model output statistics (MOS) algorithms can also be used to execute regional solar predictions using special upscaling methodologies or to generate commination models. SPF might have various output representation types depending on the applications with respect to forecast horizon and spatiotemporal resolution. Depending on the methodology deterministic or probabilistic SPF timeseries can also be generated depending on the needs of the end users. Multiple researchers developed physical SPF using various methodologies (Remund et al., 2008; Larson et al., 2016; Lorenz et al., 2007; 2008; 2009; 2014). The main advantage of the physical SPF models is that, no historical input parameters are needed. On the other hand, developing physical models is a time-consuming process and lack of spatiotemporal information may lead generation very poor PV power prediction timeseries (Dolara, 2015). Depending on the use case, special upscaling functions can be deployed to generate the regional SPF timeseries forecasting (Lorenz et al., 2001). Lorenz et al. (2008) developed ensemble SPF models by spatially dispersed grid-connected PV systems. Figure 6 summarizes the overall structure of SPF models.

Statistical Methods

Solar energy related prediction models are based on historical data such as solar irradiance and power measurements to predict the future behavior of the PV power output. Statistical models have the ability to correct the systematic error values using statistical approaches. Auto-Regressive (AR) (Bacher et al., 2009), Auto regressive moving average (ARMA) (Chu et al., 2015), auto regressive integrated moving average (ARIMA) (Pedro and Coimbra, 2012) coupled autoregressive and dynamical systems (CARDS) (Boland, 2012) are commonly used statistical methods in solar forecasting. Bouzerdoum et al. (2013) integrated seasonal variability can be also integrated to the statistical models using Seasonal ARIMA (SARIMA). Solar energy forecasting prediction timeseries can be represented using deterministic and/or probabilistic approaches. Bessa et al. (2015) developed a probabilistic SPF models using Vector AR (VAR).

Artificial-Intelligence Solar Power and Irradiance Foresting Models

Like statistical SPF models, AI-based models also strongly relaying on historical dynamic variables. The ANNs are the most common AI-based SPF models use various machine learning configurations such as feedforward, recurrent wavelet and radial basis function neural networks (Yang, 2013; Yona, 2007; Oudjana, 2012).

The ANN consists of the input, hidden and output layers and links (neurons) between them. The input variables are inserted to the input layer. The hidden layer(s) processes the given input variables and applies various functions and forwards them to the output layer where the solar power predictions are generated. The neurons are responsible to link various layers and adjust the dynamically update weights which helps the entire system learn from the historical data. The neuron cell has two main functions: combination and activations functions respectively. The performance of SPF models is strongly correlated with the design of the ANN architecture. The available dataset is split into training and testing datasets. In some cases, validation dataset shall also be included to the model development procedure. Training dataset is used to let the learning algorithm learn from the historical (input) variables where the relationship between input parameters (mainly NWP dataset) and the target value (solar power measurements) is built by adjusting the synaptic weights of the neurons. The deviation between target input and output is calculated iteratively. The process targets to minimize the error between input and output. It is also possible to deploy more advanced NN models such as adaptive neuro fuzzy interference system

Figure 7. Overview of statistical and AI-based solar power forecasting models.

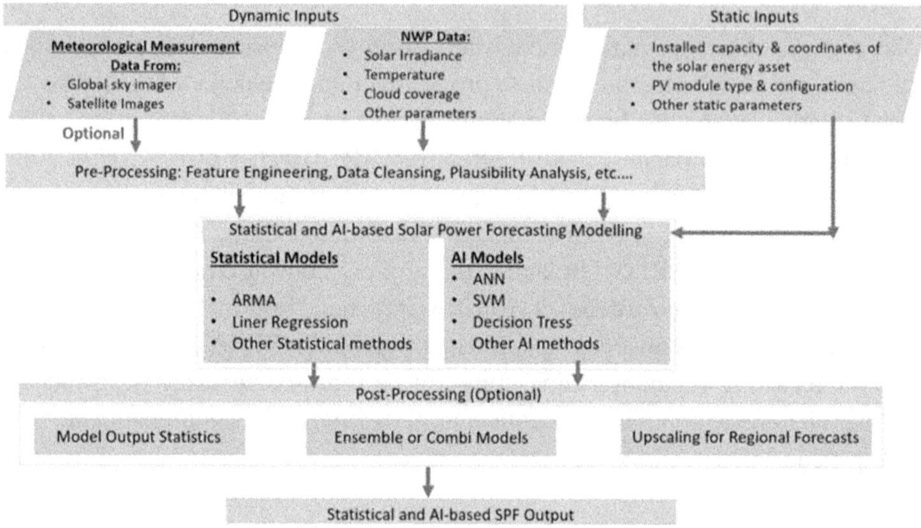

(ANFIS) which is based on Takagi-Sugeno fuzzy interference system principle. ANFIS based SPF models combines the ANN and fuzzy logic approaches where a set of IF-THEN procedures are integrated to the system to approximate the non-linear functions (Jang, 1991). Other AI approaches including support vector machines (SVM) (Shi et al., 2011), random forest and decision trees (Mohammed, 2015) has been used for developing AI-based SPF models. The Figure 7 shows the overview of the AI-based SPF models.

Ensemble and Hybrid Solar Power Forecasting

Deploying hybrid SPF models is another popular approach. Hybrid models can reduce the prediction errors and increase the redundancy of the system. Those approaches can be applied in pre or post-processing stages of the SPF systems. Applying feature engineering and optimization techniques such as principle component analysis (PCA) (Fonceca et al., 2014), genetic algorithm (GA), wavelet transform (WT), support vector machines (SVM), particle swarm optimization (PSO) has potential to increase the forecast accuracy and quality as pre-processing procedure. Hybrid approaches had also been developed using the following combinations such as GA + WT + PSO (Eseye et al., 2018) and GA +SVM (Wang, 2017). Bouzerdoum et al. (2013) used SARIMA and SVM and Zeng (2013) combined SVM and least-square (LS).

Green Energy Forecasting and Energy Storage

Weather dependent and variable RES, such as wind and solar power, are causing various challenges for grid operators. Essentially if sun is not shining or wind is not blowing expected or forecasted, or if the RES generation is higher than the forecasted values, the power grid operators face with some challenges to balance the energy supply and balance. Hence, higher amounts of RES penetration may influence the market and grid integration operations negatively. However combined use of energy forecasting systems and energy storage units have potential to minimize or mitigate this effect. As investment and operational costs of the energy storage technologies come down, combining wind, solar and energy storage offers new way to integrate renewables at competitive cost and higher efficiencies. Hodge et al. (2018) investigated the impact of combined use of energy forecasting and energy storage in terms of technical and as well as economic perfectives. According to their findings, interaction of energy forecasting and storage options provides significant economic savings and environmental benefits by reducing the curtailment times of RES and reducing amount of expensive fossil fuel based conventional balancing power plants.

CASE 2 DESCRIPTION

Distributed Ledger Technology in Green Energy Applications

Historically, centralized and fossil fuel-based energy generation resources has been utilized to satisfy the entire electricity demand. After the global oil crises in the mid-1970s, policy makers initiated to deploy push-policies to increase the share of domestic and clean energy resources by supporting renewable and decentralized energy resources, such as wind and solar energy. In parallel to decentralization and decarbonization processes of the power systems, liberalization of the energy markets had been completed in many countries. Due to the help of new support mechanisms developed by governments clean and decentralized energy resources have been promoted in many segments of the electrical power industry. Meanwhile, manufacturers of renewable energy system components are constantly working on highly efficient and lower cost clean and decentralized energy technologies. Furthermore, decentralized energy generation reduces the overall energy losses and decreases the carbon-dioxide emissions due to less physical distances between energy supply and demand sites. Besides new market players such as prosumers were born. Prosumers are the individuals or entities who produces and consumes a product or good at the same time. People have moved being only energy consumers to energy producers especially with the help of the roof top mounted behind the meter (BTM)

Figure 8. Decentralization of power systems and distributed ledger technology

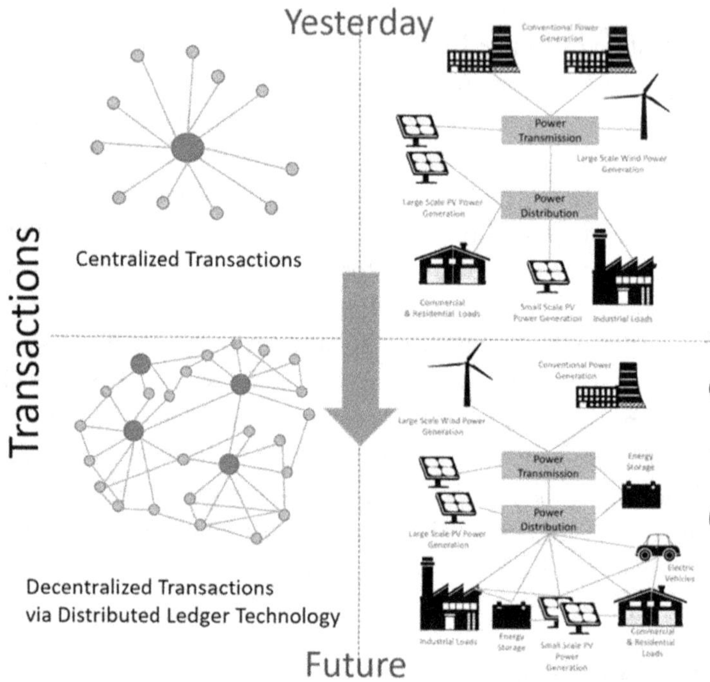

PV panels. Prosumer terminology is not limited to PV generation or individual persons but also it may cover the any type of distributed energy generation, energy storage services and commercial companies as well. At present, power markets are using merit-order methods to determine the electricity prices on a national or large-regional level which does not consider the small-scale local energy balance conditions. Current wholesale markets are not able to capture in [near] real-time to the intermittent power generation characteristics of RES (Bahrami et al., 2017). Future energy markets are expected to mimic the prosumer oriented local energy generation patterns and local market by deploying innovative local energy market mechanisms. Recent improvements in digitalization of energy systems using emerging information and communication technologies (ICT), advance data analytics algorithms, real-time controlling, and optimization methods helped the energy industry to design intelligent and smart energy systems. Real-time or close to real-time metering and track-recording the energy supply-demand nodes and monetary transactions are two key components of future decentralized energy markets.

Distributed Ledger Technologies (DLT), in particular Blockchain, are emerging technologies which will transform the future business and social consumer behavior in several industrial segments fundamentally. DLT and decentralization of the energy sector are two disruptive components of the future of decentralized power markets. DLT is a perfect match for future power systems and markets which has the capability to empower the decentralization, democratization, and transparency of the entire sector. (See Figure 8). DLT allows instant transfer of monetary records and smart contracts in a reliable way over various medium such as private, public and hybrid networks. In this chapter, overview of possible energy DLT use cases will be discussed briefly.

Distributed Ledger Technology

Distributed ledgers are distributed databases (ledgers) spread across several independent nodes (devices) that exists in multiple locations on a virtual network where the duplicates of each corresponding records (ledgers) are stored, shared and synchronized on each participating node.

In contrast to traditional systems which required a central authority and many cases multiple third parties, DLT does not require a central authority and designed to eliminate the unnecessary third parties in the system. It is possible to categorize

Figure 9. Overview of the DLT components
(Lima, 2018a; 2018b)

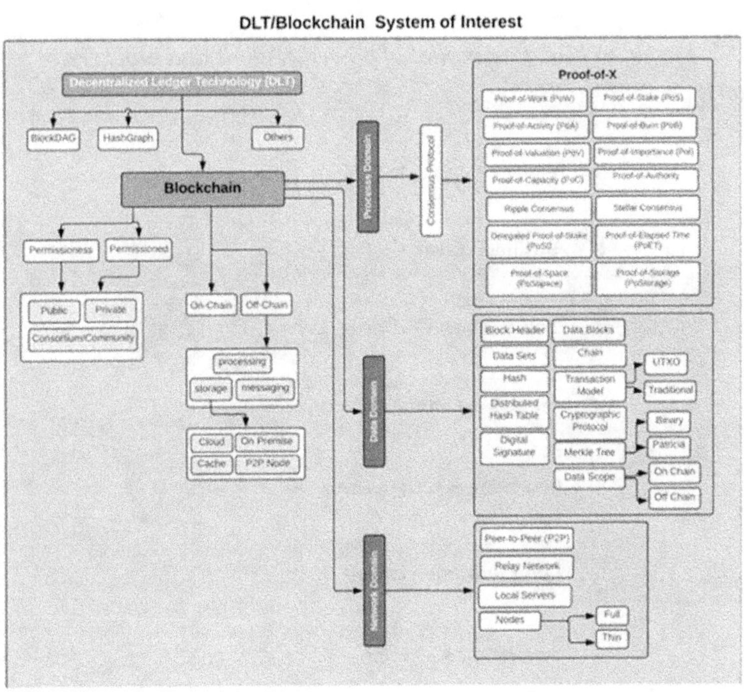

the DLT technologies under four main groups: Directed acyclic graph (DAG) or BlockDAG, Blockchain and hybrid or other types of DLT technologies. The blockchain has process, data and network domains. The consensus protocols are used to achieve and agreement between the distributed peers or nodes in the blockchain network and build the trust for the entire system. Proof-of-X (PoX) consensus mechanisms such as Proof-of-Work (PoW), Proof-of-Stake (PoS) and Proof-of-Authority (PoA) are the most known PoX protocols. The Data Domain contains the details about the block header, data blocks, distributed hash tables, Merkle tree, cryptographic protocols and other data related components of the system. The Network Domain enables the peer-to-peer communication between distrusted nodes via communication and networking protocols. BC networks can be permissionless and permissioned depending on the use case. All additional extensions such as applying artificial intelligence or enabling third party data storage and messaging services can be developed on the Off-chain domain. (See Figure 9)

Multi-Layer Approach: Power Systems and the DLT Interactions

A 5-layer Smart Grid Architecture Model (SGAM) was developed by the EU Mandate M/490's Reference Architecture working group (Smart, 2012). SGAM has been used as a reference model for several European Union R&D projects and researchers to accomplish a common understanding and standardization of smart

Figure 10. 7-Layer architecture model of power systems and blockchain technology interactions

grid modelling efforts. Various researchers used similar 4-layer architecture model without considering the use of blockchain technology in their studies. (Porsinger et al., 2016; 2017; Trefke et al., 2013; Daenekas et al., 2014).

There has been no other existing publication identified where the employment of blockchain technology is discussed on such multi-later energy model structures. In this section a 7-layer architecture energy model where the impacts and interactions of distirbuted ledger technology on possible energy related use-cases in a multi-domain complex structure are investigated. (see Figure 10)

Energy Policy and Regulatory Layer

The energy policy framework layer covers all regulatory and legislative issues in the field of energy. Policy makers creates legislative documents to satisfy the country's or state's energy safety and security strategy. All players on the market are responsible to follow the regulatory framework determined by the policy makers. Various countries such as Estonia and Luxemburg have already started to develop roadmaps for the integration of the blockchain technology related activities in their countries. In future, the BCT related energy regulatory activities is expected to attract other governments as well.

Business Layer

Investors in the field of energy, trading companies, and transmission and distribution system operators are traditional stakeholders of the existing power markets. Liberalization processes of the power market led diversified the power market players. In future, it is expected to observe new smaller and decentralized market players such as prosumers and other innovation service provider companies. Information and communication technologies such as artificial intelligence and DLT are two strong drivers of next generation business segments. Economic decisions are given by considering the energy related legislative documents and the economic metrics, which are used make the feasible investment decisions. The most popular energy economic indicators are return on investment (RoI), discounted payback periods (DPP), net present values (NPV), and levelized cost of electricity (LCoE). The cost and benefits analysis of blockchain technology can be considered one of the new fields for the energy economists and investors. The cost of required energy, communication and computation hardware and software are some of the cost elements of the future sharing economy where DLT will be utilized dominantly.

Power Markets and Pricing Layer

Electricity and the power market are a complex system which enables power flow, monetary and communication / data transactions between the buyer and seller of electrical power. Liberalization of the power markets have transformed power market rules during the last two decades. Market rules of the competitive power markets allow new independent power producers, non-utility power producers and new local power markets which allows to trade electrical power with two-directional manner. Future Peer-to-peer (P2P) power market place rules and structures are still under development. Blouin & Serrano, (2001) and Block et al. (2008) introduced various P2P market mechanism structures. Block et al. (2008), Ilic et al. (2012) and Liu et al. (2017) described the rules of P2P decentralized power markets based on DLT platform in their studies.

Control and Optimization Layer

Supervisory Control and Data Acquisition (SCADA) systems have been used as standard tool for the power industry for more than two decades. SCADA, and similar advanced software tools are utilized to control the operation of smart energy systems to include the decision-making and optimization algorithms deployed in this layer. It is expected that the data analytics such as the use of artificial intelligence (AI), and DLT will be integrated to the existing advanced decision-making processes in the coming years.

Information and Data Layer

The information and data layer is built on the communication layer and accommodates data processing, analysis and cyber security related tasks. Smart contracting and consensus mechanisms functionalities of DLT infrastructure is operational also under this layer.

Communication Layer

The Communication Layer is responsible for hosting various communication protocols which are used for interaction between different layers. The reliability, security, robustness, scalability, power consumption levels and economic viability of the communication technologies and effectivity of the used protocols are critical factors which should be considered for designing efficient smart power systems. The following communication protocols are used in energy systems:

- IEEE 802.3 - Ethernet
- IEEE 1901 - Power Line Communication (PLC)
- IEEE 802.15.3: High Rate Wireless Personal Area Network (WPAN)
- IEEE 802.11 – WIFI
- IEEE 802.15.4 – Zigbee
- IEEE 802.16x -Worldwide Interoperability for Microwave Access (WIMAX)
- IMT 2000 – Third generation wireless telecommunication technology 3G
- IMT 2000 – Fourth generation of broadband cellular network technology 4G

DLT uses the same layer for communication between the nodes.

Power Systems Layer

Physical components including power generation, transportation and distribution units on the generation side, consumers or prosumers on the demand side are operational under this layer.

Figure 11. Layers of Energy DLT
(Lima, 2018a; 2018b)

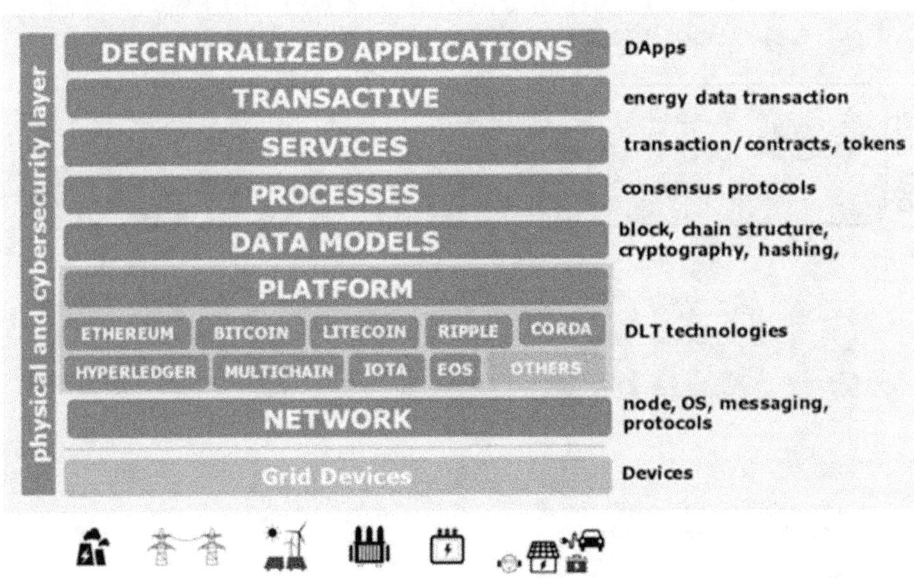

Blockchain / DLT in Energy Layered Model

Figure 11 shows an example of the IEEE DLT/Blockchain standards contribution and layers of Energy DLT (under development), identifying the key stakeholders, concerns, architectural viewpoints and system of interest, including the main elements and sub-systems to define and specify DLT/Blockchain technologies, mapped onto the IEEE 42010 framework (Lima, 2018a).

There are several potential Blockchain grid applications that will be deployed in the coming years, mostly focused on using Blockchain/DLT as an asset, transaction and event distributed registry. There is also a strong need to create standards in the Blockchain energy vertical (Lima, 2018b). With this proposition, the IEEE Standards Association (SA) created, in September 2018, the newest IEEE P2418.5 Blockchain/DLT in Energy Working Group Standards (Lima, 2018c), as the first global standards to address reference architecture, interoperability, and use cases.

Open Blockchain Energy (OBE) Framework

A new Open Blockchain Energy (OBE) Architecture Framework is proposed, and under consideration by the IEEE Standards, to create the first concepts on how to

Figure 12. Open Blockchain Energy (OBE) framework
(Lima, 2018a; 2018b)

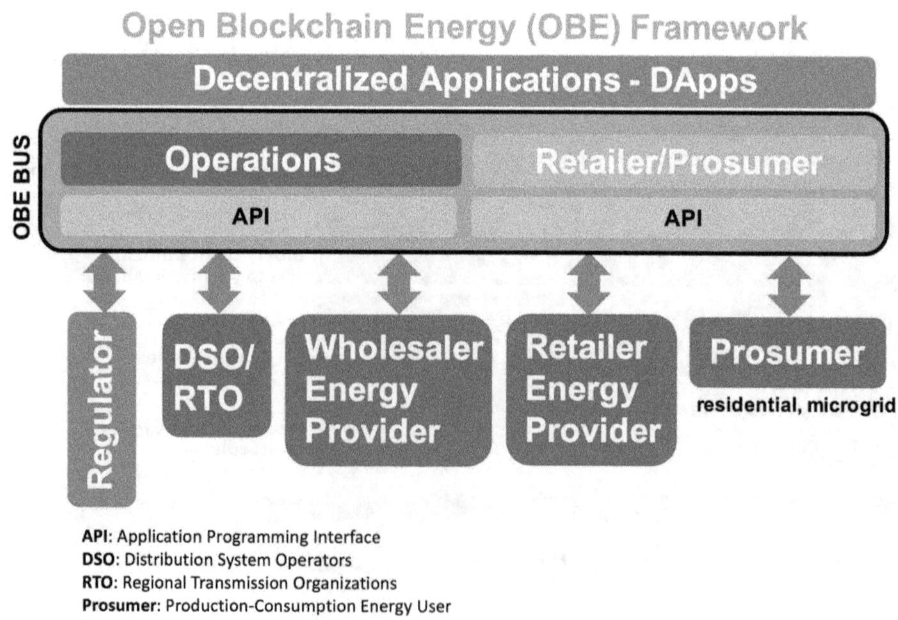

segment the Blockchain processes, functionalities, applications, and use cases in the energy grid. The Distributed Ledger OBE BUS contains two segments. One related to grid mission critical operation and the other on the prosumer or customer-facing side. Both will have DLT/ Blockchain segment specific open Application Programming Interface (APIs) to support multiple Blockchain grid application segmentation. The Operations segment can support high performance, high security and mission-critical industrial grid Blockchain operations, where Distribution System Operators (DSO) and Regional Transmission Organizations (RTO) can connect, as well as wholesaler energy providers, such as Independent Power Producer (IPPs). On the consumer-facing side of the grid, lots of Blockchain applications can be developed. Retailer energy providers, residential and microgrid producers and consumers of energy, called prosumers can be connected to OBE open APIs. For each grid segment, a set of distinguished Blockchain Decentralized Applications (DApps) can be developed. This framework can be further evolved and detailed to accommodate more specific grid domains and applications. An **Open Blockchain Energy (OBE) Reference Model** is needed to help drive new grid services, improve and optimize the existing ones and eventually help new regulation in the Energy sector. Fig. 12 shows the summary of the OBE framework. This is a working in progress.

Energy Blockchain Use Cases

There are several types of grid applications that can use Blockchain technology today. However, most use cases for Blockchain in Energy are still being defined, as the business models and value proposition evolves and are justified, more will be

Figure 13. Energy DLT use cases

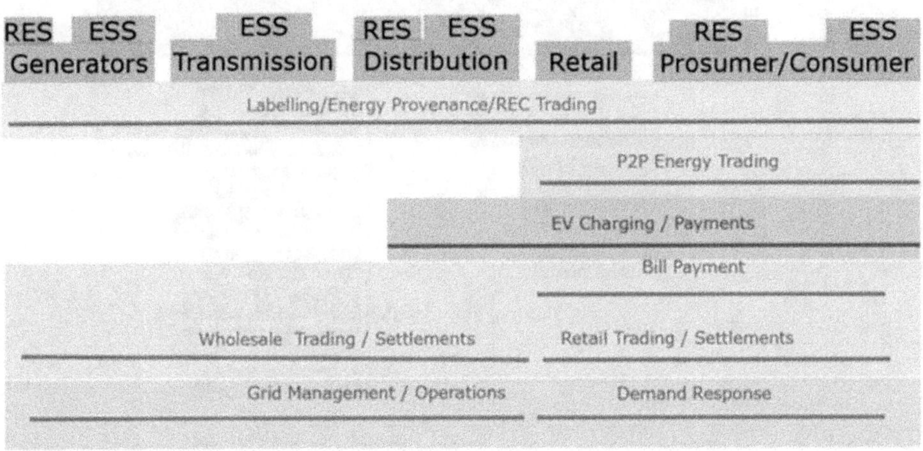

added soon. According to industrial and academic literature reviews, the following blockchain use cases are identified as the most popular in the energy sector: Wu & Tran, (2018) & World Energy Council. (2018).

- Labelling, Energy Provenance and Certification
- P2P Energy Trading
- EV Charging and Payments
- Wholesale Trading and Settlements
- Demand Response
- Energy Financing and ICO Crowdfunding
- Others

The following figure summarizes the entire energy system chain covering the main domains including generators, transmission, distribution, retail and prosumer/ consumer segments. Participation of RES and energy storage system (ESS) can be in different levels. Large-scale offshore or onshore wind farms can be connected to transmission line, lower capacity solar and wind resources are eligible to be connected to the distribution line. ESS can also be connected to distribution and transmission level depending on the need. Finally, especially BTM solar power and energy storage capacities can be connected to the low voltage levels on the prosumer/ consumer edge of the chain. The most popular energy blockchain use cases are segmented

Figure 14. OBE Application Segmentation
(Lima, 2018a; 2018b)

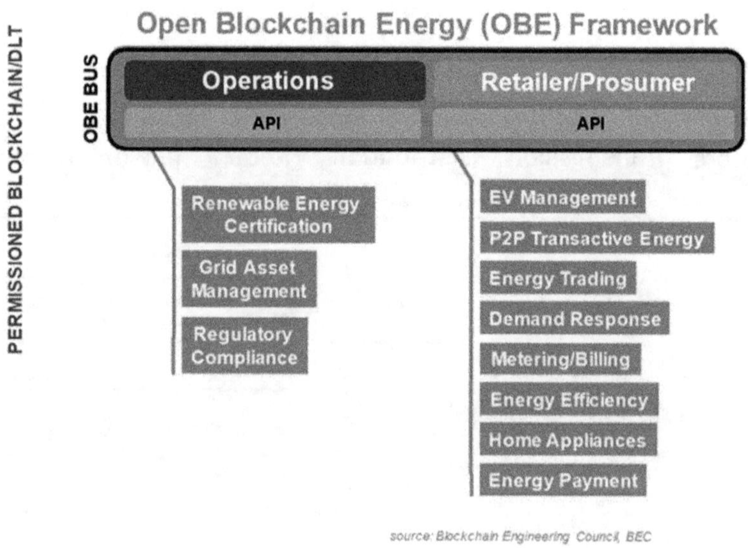

under the corresponding power market actors or domain. Most popular use cases are segmented under the main energy system chain in Figure.13.

Transactive P2P Retailer Electricity and Energy Trading Platforms in residential, enterprise-campus microgrid and municipal applications are almost 90% of the entire Blockchain grid applications available today. Most of these applications are still in their early stages, with lots of proof-of-concepts (POC) going on, and start-up companies creating these products and services. The remaining 10% of the Blockchain grid applications are mostly focused on the mission-critical grid operational side as it will be discussed next. (Lima, 2018a; 2018b)

The OBE Application Segmentation is shown in Figure 14. From the Retailer/ Prosumer side applications such, as Electric Vehicle (EV) charging station management, Peer-to-Peer Transactive Energy, Energy Trading, Demand Response, Smart Metering and Billing, Energy Efficiency applications, Blockchain-enabled home appliances and Blockchain tokenized energy payment solutions can be considered. From the grid operation side, more robust, reliable and secure applications are sought, such as renewable energy certification (REC) at the point of provenance in Megawatt Solar and Wind farms, grid asset management and end-to-end regulatory compliance. These are just a few examples of these grid applications using Blockchain/ DLT technologies.

CURRENT CHALLENGES FACING THE ORGANIZATION OR SOCIETY

The use of artificial intelligence (AI) and distributed ledger technologies (DLT) are among the most influential technologies which are expected to transform the future energy systems and businesses. In comparison to AI, DLT can be considered as an immature technology which needs additional time to be tested in various implementations and use cases in future. AI applications are becoming more dominant in our daily life in the field of personal assistant systems (like Apple's Siri), google translation and Amazon's Alexa applications. Energy industry is utilizing the AI based technologies since more than two decades. Currently especially wind and solar forecasting applications are integral parts of the energy market players' daily operators. In future, the energy industry will utilize the big data which are generated by various digital assets such as SCADA systems and phaser measurement units (PMUs) to identify the anomalies in the power systems or power plants. More advanced technologies will be deployed to form next generation predictive analytics applications which will reduce the costs and increase the overall efficiency of the entire system. New power market rules and infrastructure will be designed where the shared and central economy will be converged to each other. In addition, use of

DLT such as blockchain applications will be integrated in various segments of daily life (social impacts) as well as the energy industry. The socio-economic impacts of the foreseen paradigm shift will be expected to keep the scientists, industrial players and policy makers busy. The current DLT applications and infrastructure is still under development. There are still various challenges associated with the new technologies for instance, which use cases will be economically viable to deploy DLT, the validation of the desacralized transections on well known blockchain technologies such as Bitcoin and ETH are becoming more expensive due to higher hardware and energy consumption costs. Some issues in terms of interoperability of various DLT infrastructure between each other and lack of standardization can be considered as other challenges.

SOLUTIONS AND RECOMMENDATIONS

IEEE initiated a new working group (IEEE P2418.5 Blockchain/DLT in Energy) recently in Q4 2019 which will be responsible to design the standardization framework in this field. Lack of interoperability and other related issues will be handled by developing new protocols and off-chain applications. Off-chain applications will enable other type of third-party applications including the AI related product and services which will exponentially increase the capability and acceptability of the DLT applications in future. Third generation DLT applications will be more intelligent and be able to capture the socio-technical domain as well.

REFERENCES

ANEMOS. (n.d.). *ANEMOS Project Web Page*. Retrieved from http://anemos.cma.fr/

Bacher, P., Madsen, H., & Nielsen, H. (2009). Online short-term solar power forecasting. *Solar Energy*, *83*(10), 1772–1783. doi:10.1016/j.solener.2009.05.016

Bahrami, S., & Amini, M. H. (2017). *A decentralized framework for real-time energy trading in distribution networks with load and generation uncertainty*. arXiv preprint arXiv:1705.02575

Bessa, R., Trindade, A., Silva, C., & Miranda, V. (2015). Probabilistic Solar Power Forecasting in Smart Grids using Distributed Information. *Electrical Power Energy Systems*, *72*, 16–23. doi:10.1016/j.ijepes.2015.02.006

Block, C., Neumann, D., & Weinhardt, C. (2008). A Market Mechanism for Energy Allocation in Micro-chp Grids. *Proceedings of the 41st annual Hawaii international conference on system sciences*, 172–172. 10.1109/HICSS.2008.27

Blouin, M. R., & Serrano, R. (2001). A Decentralized Market with Common Values Uncertainty: Non-steady States. *The Review of Economic Studies*, *68*(2), 323–346. doi:10.1111/1467-937X.00171

Boland, J., Korolkiewicz, M., Agrawal, M., & Huang, J. (2012). Forecasting Solar Radiation on Short Time Scales using a Coupled Autoregressive and Dynamical System (CARDS) model. *Australian Solar Energy Society*, *87*, 136–149.

Bouzerdoum, M., Mellit, A., & Massi, P. A. (2013). A hybrid model (SARIMA-SVM) for Short-term Power Forecasting of a Small-scale Grid-connected Photovoltaic Plant. *Solar Energy*, *98*, 226–235. doi:10.1016/j.solener.2013.10.002

Brown, B. G., Katz, R. W., & Murphy, A. H. (1984). Time Series Models to Simulate and Forecast Wind Speed and Wind Power. *Journal of Climate and Applied Meteorology*, *23*(8), 1184–1195. doi:10.1175/1520-0450(1984)023<1184:TSMTSA>2.0.CO;2

Cali, Ü. (2010). *Grid and Market Integration of Large-Scale Wind Farms Using Advanced Wind Power Forecasting: Technical and Energy Economic Aspects*. Kassel University Press GmbH.

Carpenter, J. R., Merckelbach, L., Callies, U., Clark, S., Gaslikova, L., & Baschek, B. (2016). Potential Impacts of Offshore Wind Farms on North Sea Stratification. *PLoS One*, *11*(8). doi:10.1371/journal.pone.0160830

Chow, C. W., Urquhart, B., & Lave, M. (2011). Intra-hour Forecasting with a Total Sky Imager at the UC3 San Diego Solar Energy Testbed. *Solar Energy*, *85*, 2881–2893.

Chu, Y., Urguhart, B., Gohari, S., Pedro, H., Kleissl, J., & Coimbra, C. (2015). Short-term Reforecasting of Power Output from a 48 MWe Solar PV Plant. *Solar Energy*, *112*, 68–77. doi:10.1016/j.solener.2014.11.017

Daenekas, C., Neureiter, C., Rohjans, S., Uslar, M., & Engel, D. (2014). Towards a model-driven-architecture process for smart grid projects. *Digital Enterprise Design & Management, ser. Advances in Intelligent Systems and Computing*, *261*, 47–58. doi:10.1007/978-3-319-04313-5_5

Dolara, A., Leva, S., & Manzolini, G. (2015b). Comparison of Different Physical Models for PV Power Output Prediction. *Solar Energy*, *119*, 83–99. doi:10.1016/j.solener.2015.06.017

Dutton, J. A. (1976). *The Ceaseless wind: an Introduction to the Theory of Atmospheric Motion*. New York: McGraw-Hill.

Eltawil, M. A., & Zhao, Z. (2010). Grid-connected Photovoltaic Power Systems. Technical and Potential Problems-A Review. *Renewable & Sustainable Energy Reviews*, *14*(1), 112–129. doi:10.1016/j.rser.2009.07.015

Eseye, A., Zhang, J., & Zheng, D. (2018). Short-term Photovoltaic Solar Power Forecasting using a Hybrid Wavelet-PSO-SVM Model Based on SCADA and Meteorological Information. *Elsevier, Renewable. Energy Journal*, *118*, 357–367.

Focken, U., Lange, M., & Waldl, H. P. (2001). Previento – A Wind Power Prediction System with an Innovative Upscaling Algorithm. *Proceedings of the European Wind Energy Conference*, 826-829.

Ghofrani, M., & Alolayan, M. (2018). Time Series and Renewable Energy Forecasting. *IntechOpen*, *10*, 5772.

Giebel, G., Brownsword, R., Kariniotakis, G., Denhard, M., & Draxl, C. (2011). The State-Of-The-Art in Short-Term Prediction of Wind Power: A Literature Overview (2nd ed.). ANEMOS.plus.

Giebel, G., Landberg, L., Nielsen, T. S., & Madsen, H. (2002). The Zephyr Project– The Next Generation Prediction System. In *Proceedings of the 2001 European Wind Energy Conference, EWEC* (*Vol. 1*, pp. 777-780). Academic Press.

Hammer, A., Heinemann, D., Hoyer, C., Kuhlemann, R., Lorenz, E., Müller, R., & Beyer, H. G. (2003). Solar Energy Assessment using Remote Sensing Technologies. *Elsevier. Remote Sensing of Environment*, *86*(3), 423–432. doi:10.1016/S0034-4257(03)00083-X

Heinemann, D., Lorenz, E., & Girodo, M. (2006). Forecasting of Solar Radiation. Solar Energy Resource Management for Electricity Generation from Local Level to Global Scale, 223–233.

Hill, D. C., McMillan, D., Bell, K. R. W., & Infield, D. (2012). Application of Auto-regressive Models to U.K. Wind Speed Data for Power System Impact Studies. *IEEE Transactions on Sustainable Energy*, *3*(1), 134–141. doi:10.1109/TSTE.2011.2163324

Hodge, B., Martinez-Anido, C., Wang, Q., Chartan, E., Florita, A., & Kiviluoma, J. (2018). The Combined Value of Wind and Solar Power Forecasting Improvements and Electricity Storage. *Applied Energy*, *214*, 1–15. doi:10.1016/j.apenergy.2017.12.120

Ilic, D., Da Silva, P. G., Karnouskos, S., & Griesemer, M. (2012). An energy market for trading electricity in smart grid neighbourhoods. *6th IEEE international conference on digital ecosystems technologies (DEST),* 1–6.

Jang, R. J. (1991). Fuzzy Modeling Using Generalized Neural Networks and Kalman Filter Algorithm. *Proceedings of the 9th National Conference on Artificial Intelligence, 14,* 762–767.

Kariniotakis, G., Halliday, J., Brownsword, R., Marti, I., Palomares, A. M., Cruz, I., & Lange, M. (2006, February). Next Generation Short-Term Forecasting of Wind Power–Overview of the ANEMOS Project. In *European Wind Energy Conference, EWEC 2006*. Academic Press.

Kavasseri, R., & Seetharaman, K. (2009). Day-ahead Wind Speed Forecasting using f-ARIMA Models. *Renewable Energy, 34*(5), 1388–1393. doi:10.1016/j.renene.2008.09.006

Kühnert, J., Lorenz, E., & Heinemann, D. (2013) Satellite-based Irradiance and Power Forecasting for the German Energy Market in Solar Energy Forecasting and Resource Assessment. Solar Energy Forecasting and Resource Assessment, 267-295.

Kurt, M. (2017). *Development of an Offshore Specific Wind Power Forecasting System* (PhD thesis). Kassel, Germany: Kassel University Press.

Lange, B. (2003). *Modelling the marine boundary layer for offshore wind power utilisation* (Doctoral dissertation). Universität Oldenburg.

Lange, B. (2003). Importance of Thermal Effects and Sea Surface Roughness for Offshore Wind Resource Assessment. *European Wind Energy Conference EWEC.*

Larson, D., Nonnenmacher, L., & Coimbra, C. F. M. (2016). Day-ahead Forecasting of Solar Power Output from Photovoltaic Plants in the American Southwest. *Renewable Energy, 91*, 11–20. doi:10.1016/j.renene.2016.01.039

Lima, C. (2018a). DLT/Blockchain Architecture and Reference Frameworks. *2018 IEEE NIST Global Blockchain Summit.*

Lima, C. (2018b). Developing Open and Interoperable Distributed Ledger Technology (DLT)/ Blockchain Standards. *IEEE Special Publication, 51*, 106–111.

Lima, C. (2018c). *IEEE P2418.5 Blockchain in Energy WG Standards.*

Lin, Y., Kruger, U., Zhang, J., Wang, Q., Lamont, L., & El Chaar, L. (2015). Seasonal Analysis and Prediction of Wind Energy Using Random Forests and ARX Model Structures. *IEEE Transactions on Control Systems Technology*, *23*(5), 1109. doi:10.1109/TCST.2015.2389031

Liu, N., Yu, X., Wang, C., Li, C., Ma, L., & Lei, J. (2017). Energy-Sharing Model with Price-Based Demand Response for Microgrids of Peer-to-Peer Prosumers. *IEEE Transactions on Power Systems*, *32*(5), 3569–3583. doi:10.1109/TPWRS.2017.2649558

Lorenz, E., Heinemann, D., & Kurz, C. (2012). Local and regional photovoltaic power prediction for large scale grid integration: Assessment of a new algorithm for snow detection. *Progress in Photovoltaics: Research and Applications*, *20*(6), 760–769. doi:10.1002/pip.1224

Lorenz, E., Heinemann, D., Wickramarathne, H., Beyer, H. G., & Bofinger, S. (2007). Forecast of ensemble power production by grid-connected PV systems. *20th European PV Conference*.

Lorenz, E., Hurka, J., Karampela, G., Heinemann, D., Beyer, H. G., & Schneider, M. (2008). Qualified forecast of ensemble power production by spatially dispersed grid-connected PV systems. *23rd European Photovoltaic Solar Energy Conference*.

Lorenz, E., Kühnert, J., Wolff, B., Hammer, A., Kramer, O., & Heinemann, D. (2014). PV Power Predictions on Different Spatial and Temporal Scales Integrating PV Measurements, Satellite Data and Numerical Weather Predictions. *29th EUPVSEC*, 22–26.

Lorenz, E., Scheidsteger, T., Hurka, J., Heinemann, D., & Kurz, C. (2011). Regional PV Power Prediction for Improved Grid Integration. *Progress in Photovoltaics: Research and Applications*, *19*(7), 757–771. doi:10.1002/pip.1033

Louka, P., Galanis, G., Siebert, N., Kariniotakis, G., Katsafados, P., Pytharoulis, I., & Kallos, G. (2008). Improvements in Wind Speed Forecasts for Wind Power Prediction Purposes using Kalman filtering. *Journal of Wind Engineering and Industrial Aerodynamics*, *96*(12), 2348–2362. doi:10.1016/j.jweia.2008.03.013

McCarthy, E. F. (1998). *Wind Speed Forecasting in the Central California Wind Resource Area. Paper presented in the*. Burlingame, CA: EPRI-DOE-NREL Wind Energy Forecasting Meeting.

Mohammed, A.A., Yaqub, W., & Aung, Z. (2015). Probabilistic forecasting of solar power: an ensemble learning approach. *Intelligent Decision Technologies / Smart Innovational Systems Technologies, 39,* 449–458.

Nielsen, T. S., Madsen, H., & Tofting, J. (1999). Experiences with statistical methods for wind power prediction. *1999 European Wind Energy Conference and Exhibition*.

Oudjana, S. H., Hellal, A., & Mahamed, I. H. (2012). Short term photovoltaic power generation forecasting using neural network. *Environment and Electrical Engineering (EEEIC), 2012 11th International Conference*, 706–711.

Pedro, H. T., & Coimbra, C. F. (2012). Assessment of forecasting techniques for solar power production with no exogenous inputs. *Solar Energy*, *86*(7), 2017–2028. doi:10.1016/j.solener.2012.04.004

Pinson, P., & Madsen, H. (2009). Ensemble-based Probabilistic Forecasting at Horns Rev, *Wind Energy special issue Offshore*. *Wind Energy (Chichester, England)*, *12*(2), 137–155. doi:10.1002/we.309

Pinson, P., & Madsen, H. (2012). Adaptive modelling and forecasting of offshore wind power fluctuations with Markov-switching autoregressive models. *Journal of Forecasting*, *31*(4), 281–313. doi:10.1002/for.1194

Pinson, P., Ranchin, T., & Kariniotakis, G. (2004, March). Short-term wind power prediction for offshore wind farms Evaluation of Fuzzy-Neural network based models. *Global Windpower Conference*.

Porsinger, T., Janik, P., Leonowicz, Z., & Gono, R. (2016). Component modelling for microgrids. In *Proceedings of the 2016 IEEE 16th Internationa Conference on Environment and Electrical Engineering (EEEIC)*. IEEE. 10.1109/EEEIC.2016.7555869

Porsinger, T., Janik, P., Leonowicz, Z., & Gono, R. (2017). Modelling and Optimizing in Microgrids. *Energies*, *10*(4), 523. doi:10.3390/en10040523

Remund, J., Schilter, C., Dierer, S., Stettler, S., & Toggweiler, P. (2008). Operational forecast of PV production. In *23rd European Photovoltaic Solar Energy Conference* (pp. 3138-3140). Academic Press.

Rifkin, J. (2011). *The Third Industrial Revolution: How Lateral Power Is Transforming Energy, the Economy, and the World*. Palgrave MacMillan.

Rifkin, J. (2013). The Third Industrial Revolution: How Lateral Power is Transforming Energy, The Economy, and The World. Basingstoke, UK: Palgrave Macmillan.

Schwab, K. (2016). *The Fourth Industrial Revolution*. World Economic Forum.

Sfetsos, A. (2002). A Novel Approach for the Forecasting of Mean Hourly Wind Speed Time Series. *Renewable Energy*, *27*(2), 163–174. doi:10.1016/S0960-1481(01)00193-8

Shi, J., Lee, W. J., Liu, Y., Yang, Y., & Wang, P. (2011). Forecasting power output of photovoltaic system based on weather classification and support vector machine. *IEEE Industry Applications Society Annual Meeting (IAS).*

Smart. (2012). *Smart Grid Coordination Group, Smart Grid Reference Architecture.* CEN-CENELEC-ETSI, Tech. Rep.

Tambke, J. (2006). Short-term Forecasting of Offshore Wind Farms Production – Developments of the Anemos Project. *Proc. of the European Wind Energy Conference 2006*, 27.

Tambke, J., Lange, M., Focken, U., & Heinemann, D. (2003). Previento meets Horns Rev - Short-term Wind Power Prediction - Adaptation to Offshore Sites in CD. *Proceedings of the 2003 European Wind Energy Association Conference, EWEC'03.*

Tantareanu, C. (1992). *Wind Prediction in Short Term: A first Step for a Better Wind Turbine Control.* Nordvestjysk Folkecenter for Vedvarende Energi.

Torres, J. L., Garcia, A., De Blas, M., & De Francisco, A. (2005). Forecast of hourly average wind speed with ARMA models in Navarre (Spain). *Solar Energy*, *79*(1), 65–77. doi:10.1016/j.solener.2004.09.013

Trefke, J., Rohjans, S., Uslar, M., Lehnhoff, S., Nordstrom, L., & Saleem, A. (2013). *Smart Grid Architecture Model Use Case Management in a large European Smart Grid Project. In 4th IEEE European Innovative Smart Grid Technologies.* ISGT.

Troen, I., & Landberg, L. (1990). Short-Term Prediction of Local Wind Conditions. *Proceedings of the European Community Wind Energy Conference*, 76-78.

Wang, J., Ran, R., Song, Z., & Sun, J. (2017). Short-Term Photovoltaic Power Generation Forecasting Based on Environmental Factors and GA-SVM. *Journal of Electrical Engineering & Technology*, *12*(1), 64–71. doi:10.5370/JEET.2017.12.1.064

Watson, S. J., & Montavon, C. (2003). CFD modelling of the wind climatology at a potential offshore farm site. *Proc. Europ. Wind Energy Conf. EWEC.*

Wei, Z., & Weimin, W. (2010, March). Wind speed forecasting via ensemble Kalman Filter. In *2010 2nd International Conference on Advanced Computer Control* (Vol. 2, pp. 73-77). IEEE.

World Energy Council. (2018). World Energy Insights Brief 2018, Is Blockchain in Energy Driving an Evolution or a Revolution? London: Author.

Wu, J., & Tran, N. K. (2018). Application of Blockchain Technology in Sustainable Energy Systems: An Overview. *Journal of Sustainability, MDPI, 10*(9), 3067–3089. doi:10.3390u10093067

Wu, Y. K., Lee, C. Y., Tsai, S. H., & Yu, S. N. (2010). Actual Experience on the Short-Term Wind Power Forecasting at Penghu-From an Island Perspective. *Proceedings of the 2010 International Conference on Power System Technology*, 1-8.

Würth, I., Valldecabres, L., Simon, E., Möhrlen, C., Uzunoğlu, B., Gilbert, C., ... Kaifel, A. (2019). Minute-Scale Forecasting of Wind Power—Results from the Collaborative Workshop of IEA Wind Task 32 and 36. *Energies, 12*(4), 712. doi:10.3390/en12040712

Xia, J. R., Zhao, P., & Dai, Y. P. (2010) Neuro-Fuzzy Networks for Short-Term Wind Power Forecasting. *Proceedings of the International Conference on Power System Technology*, 1-5.

Yang, Y., & Dong, L. (2013, August). Short-term PV generation system direct power prediction model on wavelet neural network and weather type clustering. In *2013 5th International Conference on Intelligent Human-Machine Systems and Cybernetics* (Vol. 1, pp. 207-211). IEEE. 10.1109/IHMSC.2013.56

Yona, A., Senjyu, T., Saber, A. Y., Funabashi, T., Sekine, H., & Kim, C. H. (2007) Application of neural network to one-day-ahead 24 hours generating power forecasting for photo- voltaic system. *IEEE Intelligent Systems Applications to Power Systems, 2007. ISAP 2007. International Conference*, 1–6.

Zeng, J., & Qiao, W. (2013). Short-term Solar Power Prediction using a Support Vector Machine. *Renewable Energy, 52*, 118–127. doi:10.1016/j.renene.2012.10.009

Zeng, J. W., & Qiao, W. (2011). Support Vector Machine-Based Short-Term Wind Power Forecasting. *Proceedings of the IEEE/PES Power Systems Conference and Exposition*, 1-8. 10.1109/PSCE.2011.5772573

ADDITIONAL READING

Bessa, R., Möhrlen, C., Fundel, V., Siefert, M., Browell, J., Haglund El Gaidi, S., & Kariniotakis, G. (2017). Towards improved understanding of the applicability of uncertainty forecasts in the electric power industry. *Energies, 10*(9), 1402. doi:10.3390/en10091402

Bossavy, A., Girard, R., & Kariniotakis, G. (2013). Forecasting ramps of wind power production with numerical weather prediction ensembles. *Wind Energy (Chichester, England), 16*(1), 51–63. doi:10.1002/we.526

Foley, A. M., Leahy, P. G., Marvuglia, A., & McKeogh, E. J. (2012). Current methods and advances in forecasting of wind power generation. *Renewable Energy, 37*(1), 1–8. doi:10.1016/j.renene.2011.05.033

Peng, Z., Yu, D., Huang, D., Heiser, J., Yoo, S., & Kalb, P. (2015). 3D cloud detection and tracking system for solar forecast using multiple sky imagers. *Solar Energy, 118*, 496–519. doi:10.1016/j.solener.2015.05.037

Pinson, P., Nielsen, H. A., Madsen, H., & Kariniotakis, G. (2009). Skill forecasting from ensemble predictions of wind power. *Applied Energy, 86*(7-8), 1326–1334. doi:10.1016/j.apenergy.2008.10.009

Rahman, S., & Bhatnagar, R. (1988). An expert system-based algorithm for short term load forecast. *IEEE Transactions on Power Systems, 3*(2), 392–399. doi:10.1109/59.192889

Yang, H., Kurtz, B., Nguyen, D., Urquhart, B., Chow, C. W., Ghonima, M., & Kleissl, J. (2014). Solar irradiance forecasting using a ground-based sky imager developed at UC San Diego. *Solar Energy, 103*, 502–524. doi:10.1016/j.solener.2014.02.044

KEY TERMS AND DEFINITIONS

Artificial Intelligence: Artificial intelligence (AI) is man-made intelligence used and demonstrated by machines and devices in contrast to biological intelligence.

Blockchain: Blockchain is a decentralized, immutable, secure data repository or digital ledger where the data is chronologically recorded. The initial block named as Genesis. It is a chain of immutable data blocks what has anonymous individuals as nodes who can transact securely using cryptology. Blockchain technology is subset of distributed ledger technology.

Distributed Ledger Technology: Distributed ledger technology (DLT), in particular blockchain, is emerging technologies which is based on a consensus mechanism where the digital records or ledgers are stored in decentralized repository in immutable manner. DLT is supported by advanced encrypted algorithms to satisfy the cyber-security requirements and build the trust. DLT is designed to eliminate central authority to validate and store the ledgers. This system also eliminates the unnecessary third parties and reduce the business costs.

Numerical Weather Prediction: Numerical weather prediction (NWP) data consists of predictions of meteorological variables such as wind speed, wind direction, temperature. Pressure and solar radiation parameters for the next a couple of days. NWP data is the most essential dataset for energy forecasting systems.

Prosumers: Prosumers are the individuals or entities who produces and consumes a product or good at the same time. People have moved being only energy consumers to energy producers especially with the help of the roof top mounted behind the meter (BTM) PV panels. Prosumer terminology is not limited to PV generation or individual persons but also it may cover the any type of distributed energy generation, energy storage services and commercial companies as well.

Renewable Energy Forecasting: Renewable energy forecasting is sub-category of energy forecasting which focuses on the forecasting of renewable energy resources' output in various forecasting horizons between a couple of seconds to multiple years.

Smart Contract: Smart contracts are the digital equivalent of a legal contract between two parties or nodes in digital world. Smart contracts are self-executing contracts with the term of agreement written in to a code within a distributed ledger technology network

APPENDIX 1: ACRONYMS

AI: Artificial Intelligence
ANFIS: Adaptive Neuro Fuzzy Interference System
ANN: Artificial Neural Networks
ARIMA: Auto-regressive Integrated Moving Average
ARIMAX: Auto-regressive Integrated Moving Average with Explanatory Variable
BTM: Behind the Meter
CPS: Cyber-Physical-Systems
CSP: Concentrated Solar Power
DAG: Directed Acyclic Graph
DC: Direct Current
DL: Deep Learning
DLT: Distributed Ledger Technology
DSO: Distribution System Operator
DHI: Diffuse Horizontal Irradiance
DPP: Discounted Payback Period
ESS: Energy Storage System
EV: Electric Vehicle
FIT: Feed-in-Tariff
FNN: Fuzzy Neural Networks
GA: Genetic Algorithms
GHI: Global horizontal Irradiance
ICT: Information and Communication Technologies
ICO: Initial Coin Offering
IPP: Independent Power Producer
IRR: Internal Rate of Return
IoT: Internet of Things
kWh: Kilowatt Hour
LCOE: Levelized Cost of Electricity
MAPE: Mean Absolute Percentage Error
MEA: Mean Absolute Error
MOS: Model Output Statistics
NPV: Net Present Value
NWP: Numerical Weather Prediction
OBE: Open Blockchain Energy
P2P: Peer-to-Peer
PCA: Principal Component Analysis
PMU: Phaser Measurement Unit

PSO: Particle Swarm Optimization
PV: Photovoltaics
PoA: Proof-of-Authority
PoS: Proof-of-Stake
PoW: Proof-of-Work
REC: Renewable Energy Certificate
RES: Renewable Energy Sources
RMSE: Root Mean Square Error
RTO: Regional Transmission Organization
RoI: Return on Investment
SCADA: Supervisory Control and Data Acquisition
SPF: Solar Power Forecasting
SVM: Support Vector Machines
TSI: Total Solar Irradiance
TSO: Transmission System Operator
WPF: Wind Power Forecasting
XGBoost: Extreme Gradient Boosting

APPENDIX 2: ANALYSIS AND DISCUSSION QUESTIONS

1. What is the influence of the new components of industrial revolution such as digitalization on the modern power systems?

   We are witnessing the fourth phase of the industrial revolution. The first industrial revolution was started with the invention of the steam engines. The first industrial revolution continued with the additional infrastructure supported by next generation technologies such as advanced mass production methodologies (modern factory building and operations), mass transportation (railway infrastructure and locomotives) and electrification. This phase named as second industrial revolution. The main pilars of the third industrial revolution era were the use of semiconductor devices, internet technologies, advance molecular biology, smart internet of every things, rise of renewable and distributed energy technologies. Transition to green and decentralized energy resources and digital transformation of the existing industrial infrastructure had been the biggest achievements of the third industrial revolution. While the industry is evolving towards the third relation wave, the fourth industrial revolution stage is started. In other words, decarbonization, decentralization and digitalization components yielded modern smart energy systems. The industry is evolving to the cyber-physical-system (CPS) level, where the information and communication technologies (ICT) are meeting with physical components such as mechanical and energy systems or in future biological systems. The main drivers of the Fourth era will be artificial intelligence, robotics, quantum computing, advanced biotechnology, internet of things (IoTs), additive manufacturing, fully autonomous vehicles and most importantly distributed ledger technology (DLT). Next generation power systems will be more autonomous and democratized with the help of new derivative technologies.

2. Why do the modern power systems need energy forecasting tools?

Currently, a major part of the world's energy need is satisfied by non-renewable fossil fuel resources such as coal, oil, and natural gas which are well known to release a large number of greenhouse gasses. However, use of clean and carbon-free energy resources such as wind and solar have become more dominant in the global energy mix over the last two decades. Wind and solar power are different from non-renewable fossil fuel resources, since the harvested energy have to be consumed in the real-time market or the surplus energy to be stored until such time that its power is needed in the power system. Current energy storage systems still need to be further developed in order to reach economic viability limits in terms of power grid operations. Besides, most of the decentralized energy resources are in the form of intermittent renewable energy resources (RES) which make it challenging to predict their power outputs. Energy prediction models are essential for the grid operators to satisfy the supply – demand balance. Therefore, renewable energy forecasting methods have been integral parts of modern power energy systems and markets over the past two decades. Forecasting of wind speed, wind power, global irradiance, and solar power have been the most popular applications in energy forecasting. Wind and solar power forecasting models are two of the most used energy forecasting algorithms in the industry. Especially the transmission system operators (TSOs) and distribution systems operators (DSOs) utilize solar and wind power forecasting models to make more precise day-ahead and intraday power delivery operations in their daily business. DSOs also use such forecasting algorithms to reduce the probability of the local power congestions on critical spot. In addition to these, solar and wind power forecasting models are used in the operation of wind and solar farms where the outcome of their next day or hour power generation is estimated. The forecasted time series can be used for several purposes by the RES operators such as making accurate energy trading decisions on the power market or scheduling planned maintenance activities for time where less power generation is forecasted.

3. How can distributed ledger technology disrupt the power systems and markets?

Distributed Ledger Technologies (DLT), in particular Blockchain, are emerging technologies which will transform the future business and social consumer behavior in several industrial segments fundamentally. DLT and decentralization of the energy sector are two disruptive components of the future of decentralized power markets. DLT is a perfect match for future power systems and markets which has the capability to empower the decentralization, democratization, and transparency of the entire sector. DLT allows instant transfer of monetary records and smart contracts in a reliable way over various medium such as private, public and hybrid networks. Energy sector investors, trading companies, and transmission and distribution system operators are traditional stakeholders of the existing power markets. Liberalization processes of the power market led diversified the power market players. In future, it is expected to observe new smaller and decentralized market players such as prosumers and other innovation service provider companies. Information and communication technologies such as artificial intelligence and DLT are two strong drivers of next generation business segments. Economic decisions are given by considering the energy related legislative documents and the economic metrics, which are used make the feasible investment decisions. The most popular energy economic indicators are return on investment (RoI), discounted payback periods (DPP), net present values (NPV), and levelized cost of electricity (LCoE). The cost and benefits analysis of blockchain technology can be considered one of the new fields for the energy economists and investors. The cost of required energy, communication and computation hardware and software are some of the cost elements of the future sharing economy where DLT will be utilized dominantly. It is expected to observe new markets and market players in the future power market which will utilize the DLT as core technology for their business models. Those business models will be categorized as next generation sharing energy economy models.

Chapter 17
Green Building Technologies

Jeremy Gibberd
The Council for Scientific and Industrial Research (CSIR), South Africa

EXECUTIVE SUMMARY

Buildings are responsible for 40% of global energy use and produce over a third of global greenhouse gas emissions. These impacts are being acknowledged and addressed in specialist building design techniques and technologies that aim to reduce the environmental impacts of buildings. These techniques and technologies can be referred to collectively as green building technologies. This chapter describes green building technologies and shows why they are vital in addressing climate change and reducing the negative environmental impacts associated with built environments. A structured approach is presented which can be applied to identify and integrate green building technologies into new and existing buildings. By combining global implications with technical detail, the chapter provides a valuable guide to green building technologies and their role in supporting a transition to a more sustainable future.

INTRODUCTION

Green building technologies describe technologies and techniques used in built environments to minimize environmental impacts, such as climate change while ensuring that buildings are able to accommodate the functions they have been designed for, and are comfortable and productive to live and work in.

Given the onset of climate change, green building technologies must also now ensure that built environments can continue to support their required functions and maintain comfortable conditions under projected future climatic conditions

DOI: 10.4018/978-1-5225-8559-6.ch017

Therefore, in order to understand green technologies in buildings, it important to understand the relationship between built environments and the natural environment. In particular, it is important to ascertain the role that technology plays in this relationship, as this has the potential to increase impacts and environmental damage or to avoid damage and create beneficial impacts.

This understanding can be used to identify and develop, 'green technologies' which can be applied in built environments to reduce environmental impacts. It is also important in understanding how these technologies can be adopted and integrated within a larger built environment scheme and processes.

As the application and integration of these technologies in built environments can be complex, it is also valuable to define structured processes which can be used to integrate technologies effectively into the planning, design, construction and management of built environments.

This chapter on green building technologies therefore is structured in the following parts:

- **Climate Change:** This describes climate change and the role green building technologies play in both climate change mitigation and adaptation.
- **Occupant Comfort and Productivity:** This describes the nature of occupant comfort and productivity and how green building technologies can be used to enhance this in buildings.
- **Green Building Technologies:** This describes technologies and techniques in buildings which can be applied to achieve occupant comfort and productivity in buildings while minimizing environmental impacts. It focusses on energy efficient technologies and passive design techniques related to occupant comfort and productivity.
- **Integrating green building technologies**: This section describes a structured approach that can be used to support the integration of green technologies in buildings. It focusses methodologies that support the selection and application of technologies that is responsive not only to global environmental concerns but also to local environmental, social and economic issues.

CLIMATE CHANGE

Climate change has been identified as one of the most significant global issues (Hamin and Gurran, 2009). Climate change describes changes to the climate associated with human activity (Intergovernmental Panel on Climate Change, 2015). These changes are also referred to as global warming and are caused by the accumulation of greenhouse gasses in the upper atmosphere. Gases such as carbon dioxide, methane, nitrous oxide and chlorofluorocarbons are known as greenhouse gases because they trap heat from the sun and reduce the extent to which this heat from the sun is reradiated into space from the earth. Increasing the quantities of these gases results in a stronger 'greenhouse effect' as more heat is retained, leading to higher global temperatures.

Increases in carbon dioxide levels in the atmosphere, therefore, has a direct impact on global warming and climate change. Rapid increases in carbon dioxide are attributed to increases in the burning of fossil fuels and a loss in vegetation that sequestrates carbon dioxide. Both of these activities are linked to built environments. Built environments have a physical footprint and new urban areas and cities result in losses in farmland and natural vegetation, and therefore sequestration capacity. Built environments also consume energy in construction and operation from power stations that burn fossil fuels.

Human activities have now resulted in levels of carbon dioxide in the atmosphere increasing from 280 parts per million to 400 parts per million. The International Panel on Climate Change (IPCC) warns that this trend is unsustainable and that further increases will very severe consequences, as follows:

Continued emission of greenhouse gases will cause further warming and long-lasting changes in all components of the climate system, increasing the likelihood of severe, pervasive and irreversible impacts for people and ecosystems. Limiting climate change would require substantial and sustained reductions in greenhouse gas emissions which, together with adaptation, can limit climate change risks (Pachauri et al, 2014).

Climate Change Projections

In order to understand the implications of climate change, the IPCC has developed scenarios based on anthropogenic greenhouse gas emissions related to population size, economic activity, lifestyle, energy use, land use patterns, technology and climate policy. These are scenarios are described in terms of Representative Concentration Pathways (RCPs) which define contributions to atmospheric radiative forcing in the year 2100 relative to pre-industrial values as a result of additional greenhouse gases. RCP 8.5, therefore, indicates that additional greenhouses will result in an increase +8.5 W/m^2 to the earth's radiation budget. Implementing mitigation measures which reduce greenhouse gas emissions, therefore, result in lower RCPs. The RCPs used for climate change projections are listed below:

- **RCP2.6:** This represents a stringent mitigation scenario
- **RCP4.5:** This represents an intermediate scenario

- **RCP6.0:** This represents an intermediate scenario
- **RCP8.5:** This represents very limited mitigation and very high GHG emissions

Based on these scenarios the following global changes in global mean surface temperatures are projected by the IPCC (Pachauri et al, 2014).

- **2016 - 2035:** Global mean surface temperatures will increase 0.3°C to 0.7°C over the period 2016–2035 relative to the period 1986–2005. This impact is projected to be similar for all four RCPs.
- **2035 – 2050:** Over this period global mean surface temperature increases are likely to exceed 1.5°C for RCP4.5 and exceed 2°C for RCP6.0 and RCP8.5 relative to the period 1850–1900.
- **2050 – 2100:** Over this period global mean surface temperature are likely to be 0.3°C to 1.7°C higher for RCP2.6, 1.1°C to 2.6°C higher for RCP4.5, 1.4°C to 3.1°C higher for RCP6.0 and 2.6°C to 4.8°C higher for RCP8.5.

These projections indicate that very significant increases in temperature are projected from 2035 and that that increases of 5°C are possible by 2100. Increases in global surface temperatures will have many impacts, which differ depending on location. These impacts will include:

- Higher average temperatures
- Droughts and an increase in heat waves
- Higher sea levels as ice caps melt
- Changes in agricultural and food production as growing seasons change
- Irregular rainfall patterns with increases in some areas, and reductions in others
- Increases in storms, hurricanes and flooding

MITIGATION AND ADAPTATION

The onset of climate change means that green building technologies are now required not only minimize built environment contributions to climate change (mitigation) but are also required to help buildings adapt to projected climate changes (adaptation).

Mitigation aims to reduce greenhouse gas emissions in order to slow down, and ultimately, stop climate change. Adaptation aims to ensure that the unavoidable changes that will occur as a result of climate change are accommodated and their negative impacts minimised (Hamin and Gurran, 2009; Intergovernmental Panel on Climate Change, 2007).

Achieving these dual objectives in the selection and development of building technologies is not always possible. For instance, building technologies developed to respond to increased wind speeds, storms and floods projected under climate change projections may require additional materials and a heavier structure (increased adaptation). This more robust construction, however, may also be made from more sophisticated heavier materials that have larger carbon emissions associated with their manufacture compared to conventional construction thereby resulting in increased contributions to climate change (reduced mitigation).

However, integrated design processes can be used to develop green building technologies that achieve synergies and address both mitigation and adaptation (Hamin and Gurran, 2009).

An example can illustrate this. The development of renewable energy micro-grid for smaller human settlements can be used to reduce carbon emissions and achieve climate change mitigation objectives. At the same time, this micro-grid reduces the burden on the main power grid and can help avoid widespread power outages if there are severe storms. The decentralized system therefore also contributes to climate change adaptation.

Green building technologies therefore need to minimize built environment contributions to global warming (climate change mitigation) as well as ensuring that they are resilient to future climate changes that are projected (climate change adaptation). At the same time they must ensure that the appropriate levels of occupant comfort and productivity in buildings are achieved.

OCCUPANT COMFORT AND PRODUCTIVITY

As green building technologies aim to minimize environmental impacts while ensuring that buildings accommodate required functions and are productive to live and work in it is important to understand how functions and productive working and living environments are defined.

Buildings are complex and accommodate a wide range of primary functions. They can also accommodate secondary functions. An example of this is a service station or gas station that sells fuels for cars. While this is the primary function, service stations can also provide a valuable economic and social role by providing an accessible retail outlet that can be walked to by local community members.

The range of building types and functions accommodated by buildings is large so cannot be discussed in this chapter. However, it is important to note that green technologies must not only support the primary role of buildings but also, where possible, support valuable secondary roles. In addition, to ensuring that they support their primary and secondary functions, the building must ensure that internal environmental conditions within the building are conducive to health and productivity.

Minimum conditions for comfortable, health and productivity are usually defined in legislation such as building regulations and occupational health and safety acts. These differ from country to country but generally cover aspects such as lighting, views, space provision, ventilation and emergency egress requirements. Green building technologies must ensure that these minimum requirements are fully met.

Comfort, health and productivity are also defined in standards and guides that aim to capture, in more nuanced terms, optimum conditions for human health, comfort and productivity. Examples of these conditions are outlined below.

Thermal Comfort

Research has shown that there is a direct link between temperature, humidity and occupant comfort and performance. Mendell and Heath (2005) show how students report increasing discomfort and deteriorating performance on tasks as temperatures and humidity increase. Controlling temperature has also been found by Wagocki and Wyon (2007) to improve learning efficiency in students.

The ANSI/ASHRAE Standard 55-2004 Thermal Environmental Conditions for Human Occupancy can be used to define thermal comfort in buildings (American Society of Heating, Refrigerating and Air-Conditioning Engineers, 2004). It is particularly applicable for buildings that include green technologies as it has an emphasis on low energy and passively cooled and heated buildings.

The standard defines thermal comfort in terms of an adaptive comfort model. This means that temperatures deemed to be comfortable for indoor environments will vary with outdoor temperatures. This relationship is shown in figure 1. Thus, temperatures for indoor comfort for 9 out of 10 individuals will be within a band of 24 and 30°C, when monthly mean outdoor air temperatures are 33°C. Similarly, temperatures for indoor comfort for 9 out of 10 individuals will be within a band of 17 and 22°C, when monthly mean outdoor air temperatures are 5°C.

The adaptive comfort concept can be used to achieve very significant energy savings as temperatures within buildings can move up and down with external temperatures and do not have to be restricted within a narrow band and has significant implications for the type of green building technologies that can be developed and applied in buildings.

Figure 1. Adaptive comfort model
(American Society of Heating, Refrigerating and Air-Conditioning Engineers, 2004)

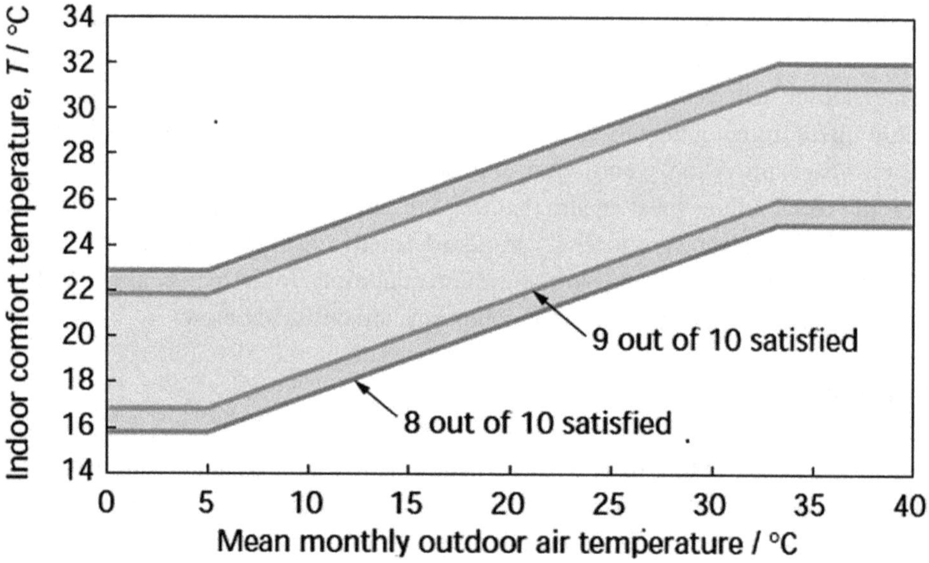

Indoor Air Quality and Movement

There is strong evidence to show that productivity and health of occupants relate to ventilation (Barrett et al, 2015). Wargocki et al (1999), show that increasing ventilation rates from 5 to 10l/s/student improve the performance of students doing schoolwork by 15%.

Similarly, significant improvements in productivity of 5 to 10% were found to be achieved through improving ventilation in offices (Wargocki et al, 2005)

Fresh clean air has been shown to have important beneficial health and productivity impacts and at least 10-15 litres of fresh air per person per second should be provided in internal environments. Care should be taken that opening windows, air intakes and outlets where fresh air is sourced for interiors are located to avoid pollutants and re-circulation of air.

Slowly moving air (under 0.5 meters per second) is also normally consider conducive to comfort. However, discomfort can be experienced as wind speeds increase and air movement in pedestrian areas should not exceed 5m/s while wind speeds of 20m/s are regarded as dangerous.

Indoor Pollutant Source Control

Health can also be adversely affected by exposure to hazardous chemicals from technology used in the building and from off-gassing from materials and finishes within the building. For instance, activities associated with contaminants should be separated from occupied spaces or separate exhaust ventilation systems provided. Examples include activities such as large-scale photocopying and printing as well as the handling and storage of chemicals.

In addition, volatile organic compounds (VOCs) are now associated with many modern materials. These compounds are off-gassed from materials contaminating internal air quality and resulting in negative health impacts. Modern materials that can contain these compounds include paints, adhesives, composite boards and veneer and plywood products. These types of products should, therefore, be screened to ensure they do not have harmful effects on human health.

Daylighting

Work by Tanner (2009) shows that good lighting significantly impacts reading, vocabulary and science test costs. A comprehensive study by Loisos (1996) shows quantifiable improvements in student performance with good daylighting. His study indicates that students with good daylight achieve 20% better in maths tests and 16% better in reading tests than students in classrooms with poor daylighting

Views

Views and contact with external environments have been shown to be important for occupant comfort and wellbeing. Views are achieved by ensuring that occupied spaces have windows that open on to external environments and views are not obstructed both within and immediately outside the buildings. It means that occupied spaces must also be physically close (a maximum of 7m) to external windows so that views are experienced.

Glare

Poor artificial lighting and window design can lead to glare which can strain eyes and be uncomfortable. Glare can be avoided through careful lighting design and through controlling the size, location and shading of external fenestration.

Acoustics

Disturbance from external noise and internal acoustics of spaces have been found to be a significant influence on occupant comfort and productivity. Crandell and Smaldino (2000) show that the acoustics of classrooms is a significant critical factor in the academic and psychosocial achievement of children.

External Amenity

External amenity refers to the character of local environments around a building. This includes the quality of landscaping and the availability of comfortable external spaces that can be used as a break from activities undertaken inside buildings. Wells and Evans (2003) and White (2004) show that exposure to nature has significant benefits for the development of children. Comfortable external spaces, particularly landscaped with vegetation, therefore, can make a valuable contribution to occupant wellbeing and comfort.

Local Control

Occupant wellbeing is also influenced by the extent to which users have control of their environments. A study of office environments by Wilson and Hedge (1987) showed symptoms of ill-health reduce and productivity increases, with perceived individual control over local environments. Therefore green building technologies should provide local control over temperature, lighting and ventilation as this is likely to contribute to occupant satisfaction.

Surveys and Sensors

The quality of indoor environments can be monitored through surveys and sensors. Surveys such as occupant comfort surveys of occupants can be used to determine the extent to which occupants experience comfortable and productive conditions. Field measurements using instruments, such as lux meters and thermocouples, can also be used to ascertain conditions within internal environments at particular points of the day or week. Permanent sensors, such as temperature, lighting and carbon dioxide sensors can also be installed to provide real-time data. It is important that monitoring and evaluation processes linked to accurate, comprehensive, and up-to-date data on key aspects of occupant comfort and productivity are carried out and used to refine the selection, development and operation of green building technologies.

GREEN BUILDING TECHNOLOGIES

In many modern buildings, the conventional approach to achieving the functions of the building and ensuring that it is comfortable and healthy has been to install powered mechanical plant and systems. These systems, such as lifts, lighting, water heating cooking, office and equipment, pumps and air conditioning systems, use significant amounts of energy it is estimated that, as a result, buildings consume about 40% of all energy globally (USGBC, 2008). Despite these negative impacts there has been considerable resistance to adopting greener technologies (Chan et al, 2018; Wadu Mesthrige & Kwong, 2018).

Mechanical plant and equipment found in older buildings are inefficient and poorly controlled and estimates indicate that savings of up to 30% of energy consumption can be achieved at little, or no, cost through better management and the application of more efficient green technologies.

Energy targets in new buildings are becoming more stringent and an increasing number of buildings are aiming to be carbon neutral (Gething & Puckett, 2013). This means that the building generates sufficient energy from on-site renewable energy systems, such as photovoltaic systems, to meet its own needs. These buildings are very energy efficient and have large renewable energy and battery storage systems that are capable of generating and storing sufficient energy to enable the building to run continuously without mains power (Gething & Puckett, 2013).

Incorporating more energy efficient greener technologies in buildings has a wide range of benefits, including:

- Reduced carbon emissions and therefore global warming impacts
- Reduced impact of mains power outages
- Reduced negative health impacts associated with pollution from coal-fired power stations
- Reduced operational costs
- Improved internal environments as a result of better daylighting, natural ventilation and local control.

As a result, there is an increasing choice of green technologies and techniques that can be applied in buildings to reduce energy consumption and environmental impacts. These are described next.

The Building Envelope

Green building technologies can be classified as technologies that relate to the building envelope, that is the walls, the roof, the floors and openings such as windows and doors, and those that related to environmental control such as heating, cooling and lighting systems.

Building envelope green building technologies are combined in sophisticated designs to ensure that comfortable and productive environments are achieved with very little or no energy. These technologies and techniques are described next.

Daylighting

In addition to passive environmental control strategies, a range of other green building technologies can draw on ambient conditions to create comfortable, productive internal environments in buildings. Technologies that draw on natural daylighting are particularly important in this respect as daylight not only can be used to providing sufficient lighting for required activities in buildings to take place, it has also been shown to be important for human health and wellbeing.

Good daylighting in buildings reduce the requirement for artificial lighting and therefore can reduce energy consumption and associated carbon emissions. Daylight strategies should consider the following factors.

- Access to daylighting: Landscaping and building location to ensure good access to daylight
- Depth of the building: The depth of the building should be limited to ensure that internal spaces which cannot be day lit are limited in area. A general rule of thumb is that daylight quality will be reasonable within the space 2h from a window, where h is the height of the head of the window from floor level.
- Type of glazing: Selection of glazing to allow good daylight penetration.
- Light shelves: The use of daylight shelves to enable daylight penetration deeper into the building.
- Internal colour: The choice of colour and finishes to improve the internal reflectance of spaces.

Windows

A review of passive environmental control and daylighting strategies indicate the important role that windows play in providing comfortable internal conditions. Increasing the technology used in windows is becoming more sophisticated. Advanced green building technologies such as opening mechanisms linked to sensors can be programmed to open and close windows depending on internal and external conditions to ensure optimum internal conditions are achieved. Characteristics of windows that can be used to support energy efficiency are outlined below:

- Where windows are being used as part of a cross ventilation strategy, the size and location of opening sections should be designed to ensure that breeze paths through the building are direct and are guided through areas, equipment and people that need ventilation and cooling.
- Where possible, glazing should be avoided on east and west facades to avoid unwanted heat gains.
- Glazing should also be placed where it provides views and higher up in walls to support good daylighting.
- Naturally ventilated buildings should have an equivalent opening area (of windows or doors) of at least 5 – 10% of the floor area.
- Light coloured chamfered reveals help reduce the contrast between windows and surrounding walls reducing glare and improving daylighting.
- Windows with opening sections at both high and low-level benefit from being able to use the stack effect to create air movement and can be used to vent hot air out of the top of rooms and draw in cooler air in at low levels.
- Window opening controls should be designed to give occupants control over their local environment. This can be done by having regularly spaced windows and providing at least one opening section per 5 m of façade. Window controls should also cater for people with disabilities.

Glass

More sophisticated glazing can be used to increase the thermal and light transmittance performance of windows. The following characteristics of glass can be used to promote energy efficient design.

- **Visible Light Transmittance:** Glass with high light transmittance supports good daylight within buildings, reducing the requirement for artificial lighting.
- **Solar Heat Gain Coefficient:** This is the proportion of total solar radiation that is transferred through the glass at normal incidence. Glass with lower coefficients reduces the amount of heat gain from the sun.
- **U-Values:** Increasing U-values, for instance by using double glazing, reduces heat losses and gains.

Shading Devices

Windows, however, can also be the weak point in low energy buildings as they usually have lower thermal performance relative to walls and roofs. It is therefore important to ensure that these are correctly sized, located and shaded. Shading devices help avoid unwanted heat gains from the sun through windows. These are designed by modelling sub patterns through the day and year and designing shading to cut off sun to the window during periods when there is a danger of unwanted heat gain (for instance during summer) and allowing this into the building during winter when this heat gain is beneficial and warms the building (for instance as part of a solar direct gain strategy)

The following issues should be considered in the design and application of solar shading:

- In general, horizontal shading elements are appropriate on northern facades and vertical moveable louvres are suitable on east and west facades.
- Spaces with computer screens may require solar access to be avoided altogether because of visual and glare problems

Doors

In a similar way to windows (see above) external doors can be used to ventilate and cool buildings. Their use on very hot or very cold conditions, however, can introduce large volumes of cold or hot air affecting comfort and requirements for heating and cooling. Air movement through external doors can be reduced in the following ways:

- Revolving doors and lobbies can be used to reduce the amount of air moving in and out of the building.
- Well fitted insulated doors with perimeter seals can be used.
- Automatic closers can be used to minimise the duration which doors are open

Building Envelope and Roof Colour

The colour of the building envelop and roof can be used as part of a passive environmental control strategy (Gibberd, 2009). Darker colours absorb heat so can be used where additional heat or warming is required. Lighter colours absorb less heat and are therefore suitable when heat gains need to be avoided. A range of green building technologies has been developed, such as ‘cool roofs’ based on this principle and have been shown to be effective in significantly increasing comfort and reducing energy requirements in buildings (Cotana et al, 2014).

Table 1 shows the absorbencies of different colours. In warm climates, colours with a value of 0.45 or lower are recommended for roofs as this helps limit heat gain through the roof.

Table 1. Absorbencies of different colours

Colour	Value
Slate (dark grey)	0.9
Red, green	0.75
Yellow, buff	0.6
Zinc aluminium—dull	0.55
Galvanised steel—dull	0.55
Light grey	0.45
Off-white	0.35
Light cream	0.3

Insulation

Insulation in building envelopes is a particularly important green building technology as it reduces heat flows through the building envelope. This enables other elements such as windows to be used to control temperatures more closely and effectively to achieve comfortable internal conditions. Integrating this technology in buildings should be informed by the following:

- Most heat losses and gains are through the roof of buildings; this is, therefore, the place where insulation will have the most effect.
- In order to maximise the thermal flywheel effect in buildings, insulation should be located on the outside of high thermal mass envelopes and building structures.
- The thermal resistance of building envelopes can be increased easily and at low cost through design. For instance, the incorporation of an air gap within a wall build-up or planting creepers on an external façade use the insulative properties of air to increase the thermal resistance of the building envelope.
- Care should be taken to ensure that insulation is as continuous as possible within a building envelope and gaps and thermal bridges should be avoided.

Environmental Control

Environmental control green building technologies draw on ambient energy sources such as sunlight and wind to create comfortable internal conditions in buildings with very little or no energy. Where energy is used, such as in active heating and cooling, the technologies applied are highly efficient and are controlled to minimize energy consumption. These technologies and techniques are described next.

Figure 2. A report from Climate Consultant showing climate for Johannesburg in relation to passive environmental control strategies
(UCLA, 2018)

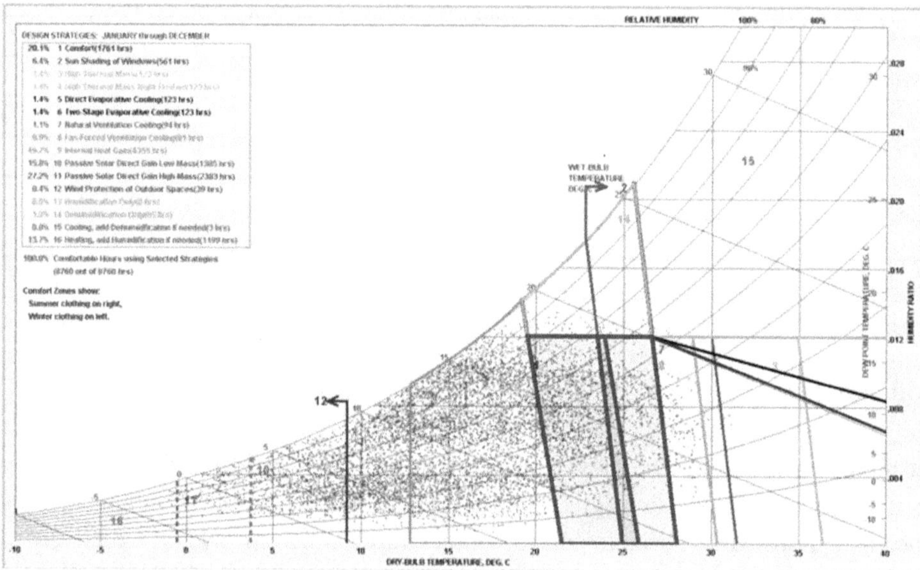

Passive Environmental Control

Passive environmental control design is a specialized field and relies on a detailed understanding of climate and green building technologies that can be applied to draw on ambient sources of energy, such as the sun, to heat, cool and ventilate buildings.

Understanding local climate and identifying appropriate strategies is supported by software and climate analysis tools such as Climate Consultant (UCLA, 2018). Figure 2, from Climate Consultant, shows the climate for Johannesburg, South Africa on a diagram. Analysis of the diagram can then be used to ascertain appropriate passive environmental control techniques.

Figure 2 shows that for much of the year, climatic conditions (the green dots) are within the comfort envelope (the blue zones). This means that no additional heating or cooling is required. It also shows that temperatures during warmer months can be addressed through thermal mass, night time flushing and fan-forced ventilation.

The diagram also indicates that cold conditions are experienced for significant periods of the year and proposes that these are addressed through internal heat gain and direct and indirect solar gain strategies.

Most importantly, it confirms that mechanical cooling and heating are not be required. This type of analysis is valuable as it ensures that there is a good understanding of local climate conditions and that green building technology is carefully selected, developed and applied in a way that is responsive to local conditions. In this way, buildings can avoid the need for mechanical heating and cooling thereby reducing operating costs and environmental impacts. A selection of passive environmental control strategies is described next which demonstrate the role that green building technologies can play in minimizing environmental impacts and energy consumption while providing comfortable internal environments.

Direct Solar Gain

Sunlight is a free heat source that can be used to avoid the requirement for heating in buildings during cold conditions. This works by allowing sunlight to enter buildings to warm high thermal mass areas, such as exposed masonry walls or tiled floors. The thermal mass then stores this heat during the day and releases it gradually, this can keeps buildings warm over a considerable period.

Direct solar gain should not be used where people are working on computers as glare may be a problem. Direct solar gain is particularly appropriate for non-working spaces such as circulation and pause areas near working spaces where heat can be stored and shared with adjacent areas. The following factors should be considered in harnessing solar gain.

- Location and orientation: Location and orientation of the building to ensure good solar access at the right times of the year
- Building envelope: Openings, glazing (and possible blinds and curtains) in the building should be designed to direct solar access to the right area and retain heat gathered.
- Material and finishes: The location, colour and type of finishes should be selected to provide good thermal storage

Indirect Solar Gain

Indirect solar gain systems are more complex, but more also more controllable than direct solar gain systems. These systems use the sun to warm high thermal mass materials such as rock or water under glass within a purpose-designed container. Heat from this container is then stored and circulated to the building using air or water as a medium. These systems can be complex to design and key considerations include:

- Location: Location of the indirect system to ensure that is near the building and has good solar access.

- Sizing: The collection area, thermal storage capacity and heat circulating system needs to be sized correctly.

Cross Ventilation

Cross ventilation is an energy efficient way of cooling buildings and providing ventilation in areas where there are moderate breezes and limited, or no, noise and air pollution. Air flow through the building removes heat and brings in fresh air. The following factors should be considered developing buildings with cross ventilation:

- Landscaping and building layout: Care should be taken to expose facades with opening windows to breezes and to avoid these being in the 'wind shadow' of other buildings and obstructions.
- Depth of the buildings: The depth of the building should not be more than 12-15m.
- Internal spatial layout: Air movement should be directed around people and the 'breeze path' between windows on opposite walls be made a direct as possible to ensure that air movement is effective.

Stack Effect Systems

A stack effect system uses tall vertical spaces and the physical tendency of warm air to rise, to ventilate and warm or cool buildings. Rising air within a vertical space, such as an atrium or solar chimney, draws air at a lower level into buildings, ventilating it. This air can be drawn from a cool source, cooling the building, or from a warmer source, warming the building. The following factors should be considered in developing stack effect systems.

- The stack effect chimney: The taller this space is the more powerful the system will be. A minimum of 9m is usually required for effective systems.
- Location: the vertical space should be located adjacent to spaces to be cooled or heated
- Heat sources: Stack effect systems can be assisted from heat sources such as the sun, people and equipment. The design and location of the stack effect system should harness these heat sources.
- Controls: To control air movement and the extent of heating and cooling.

Night Time Flushing

Night time cooling relies on the diurnal range to cool buildings. The diurnal range refers to the difference in maximum and minimum temperatures over the period of a day (24hours). Where this is high, for instance in deserts which experience very high daytime temperatures and low night-time temperatures, night-time flushing can be very effective. It works by opening the building during the night an allowing cooler air to flush warm air out and cool the thermal mass of a building. The cooler thermal mass then keeps conditions in the building comfortably cool during the day. The following factors should be considered in designing for night time cooling.

- **Openings:** The design and location of openings should enable good airflow at night through the building. Airflow should be directed around thermal mass in order to remove heat at night.
- **Security:** Care should be taken to avoid compromising security.
- **Thermal Mass:** The location of thermal mass within the building where it can act as heat-sink during the day and be cooled by night time ventilation.

Mechanical Systems

Where the climate, or the building type, mean that passive environmental control strategies cannot be used as the sole means of heating and cooling buildings, mechanical systems may be used.

However, it is important that these are designed carefully and are as energy efficient as possible as the energy costs in a typical air-conditioned building are usually at least double the energy costs and associated CO2 emissions of a building with passive environmental control (Carbon Trust, 2006). In addition, capital and maintenance costs are also usually much higher (Carbon Trust, 2006).

Figure 3 below provides a simple way of ascertaining whether mechanical heating, cooling and ventilation actually is required and if it is, what type should be used.

Where mechanical systems are used, a number of green building technologies and techniques can be applied to reduce energy consumption, and therefore carbon emissions, associated with these systems. These are described next.

Zoning

Zoning a building into different areas depending on ventilation and heating and cooling requirements enable mechanical systems to be used more efficiently.

For instance, areas with high ventilation requirements and heat gains such as kitchens can be zoned and dealt with separately to other areas such as passageways,

Figure 3. Factors influencing the type of environmental control system
(Adapted from Carbon Trust, 2006)

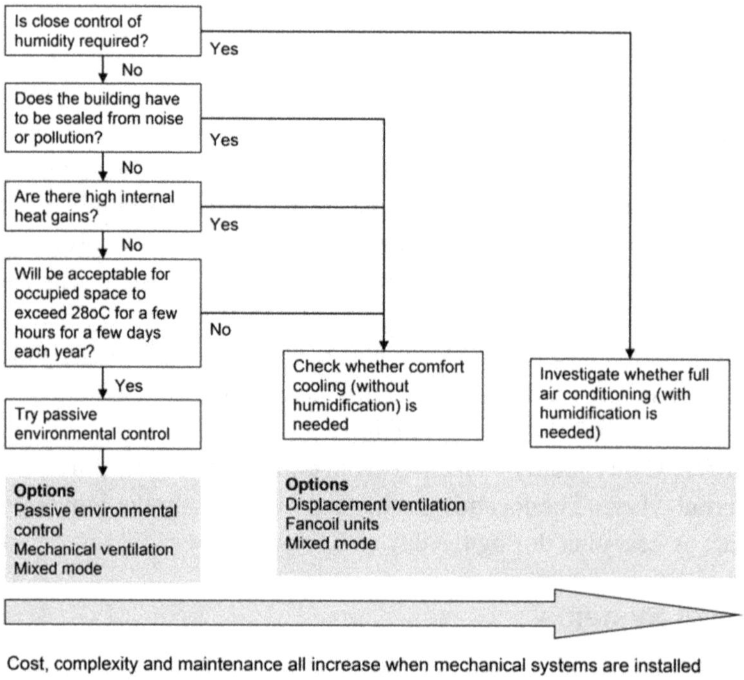

or storage where ventilation requirements are lower. Zoning also allows heating, cooling and ventilation to different areas of the building to be reduced to match requirements more closely. This approach allows parts of the systems to be turned down or off to react to local conditions leading to significant savings.

This approach also enables systems to be more responsive to local conditions. For instance, heat gains on an East façade may be high in the morning and therefore this area may require additional cooling. This will change by the late afternoon, when the sun on a West façade may result in the need for additional cooling in this area.

Pre-Heating and Pre-Cooling

Heating and cooling loads in mechanical systems can be reduced by preheating or precooling fresh air entering the system. For example, a heat exchanger can be used to extract heat from exhaust air from a kitchen or gym and use this to warm incoming fresh air that is being used to heat and ventilate other areas of the building.

Natural or Economy Cycle

In most climates, there are periods where heating and cooling is not required. During these periods significant energy savings can be achieved by avoiding and cooling and heating and instead, just using systems to ventilate the building. This is sometimes referred to a 'natural' or 'economy' cycle and can be used to achieve substantial energy savings.

Mechanical Ventilation

By law, areas such as toilets are required to be ventilated. This can be achieved through opening windows or mechanical ventilation. Where possible, mechanical ventilation should be avoided. Where this is not possible, such as in internal kitchens, underground parking, toilets and server and print rooms the following measures can be adopted to reduce energy consumption and environmental impacts:

- Spaces required ventilation should be located as near as possible to an external wall in order to minimise the distance air has to be extracted.
- Ducting design and fan specification should be carried out to minimise energy consumption
- Controls such as movement sensors, CO2 monitors and timers should be used to ensure that spaces are not over-ventilated when not in use.

INTEGRATING GREEN TECHNOLOGIES INTO BUILDINGS

Environmental impacts should be addressed as soon as possible in the design process of buildings. Addressing this effectively can help ensure that appropriate green technologies are selected, developed and applied in all aspects of the building including the choice of site, the size of the building and detailed design of the building envelope, systems and interior.

Integrating green building technologies can be supported through a structured approach to the development of new buildings and the upgrading and refurbishment of existing buildings. This structured approach is described in terms of a number of stages below. These require particular actions to be taken and designate responsible parties.

A very important consideration is ensuring that environmental impact and energy is considered, and addressed, at the onset of the project. Figure 3, shows that the potential for addressing environmental impacts and energy efficiency and therefore achieving energy savings is very high at the beginning of the project. This, however,

Figure 4. Energy savings and the design process

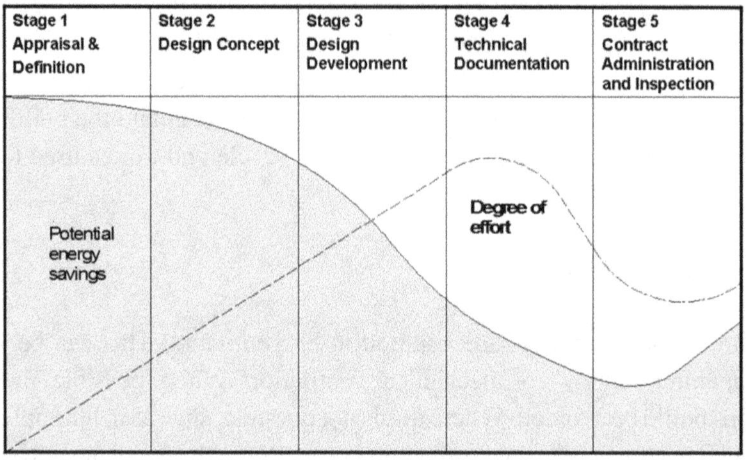

drops rapidly the later this is addressed and the dashed line, shows that significant additional effort, and often costs, are required if this is only addressed at a later stage in the development of projects. The diagram confirms that if early strategic decisions are wrong, the potential energy savings will be reduced and significantly more effort will be required to achieve energy savings.

A Structured Approach

A structured approach can be used to help ensure that green building technologies are effectively integrated into existing and new buildings and that low energy and environmental impact performance is achieved. This approach is structured around the following key principles.

- **Knowing your Building Type and Functions:** The built environment professional team that is appointed and building users understand how energy is used in the building. This includes studies of existing consumption in the existing building, or for new buildings, studies of energy efficient, low impact buildings with a similar size and function. These studies should be used to provide useful targets and approaches that can be drawn on.
- **Explicit Targets:** Early in the design process, challenging energy efficiency targets based on studies above, should be set for the building and agreed on by all stakeholders including the full professional team, user and the client. These targets should exceed good practice benchmarks and standards and should be linked to evidence-based climate, environmental and other commitments.

- **Integrated design:** Concept approaches and design development should be carried out in an Iterative and lateral way to ensure that a wide range of innovative options is also considered. Optimum approaches should then developed to enable high performance, integrated solutions to be developed.
- **Specialists:** Where appropriate, specialist input is sought and used to inform the design. Examples include the use of passive environmental control and modelling expertise to develop low energy, passive strategies. Façade

Table 2. Stage one: appraisal and definition of the project

Activity	By
Select a built environment professional team with low environmental impact and passive design experience and skills. Ideally, this team should have worked together and are conversant with integrated design processes.	Client
Provide a brief to the built environment professional team which outlines key environmental and energy targets and request that this is developed further to ensure that targets are both detailed and challenging.	Client
Undertake background studies on environmental impacts and energy consumption patterns of the existing building or if a new building is being proposed, of buildings which are a similar size and accommodate similar functions. Establish key factors that affect environmental impacts and energy consumption within the existing building or proposed design including work patterns, environmental conditions and transportation patterns of both people and goods. Evaluate strategic options for reducing environmental impacts and energy consumption including site locations near public transport or residential areas, home working building management techniques such as hot-desking and the sharing of facilities with other local buildings.	Built environment professional team
Develop detailed environmental and energy targets for the building and outline the implications of pursuing these to the client. Implications could include urban site location, reduce building size, more flexible thermal conditions and a requirement for specialist consultants and modelling.	Built environment professional team
Undertake feasibility studies and analysis to identify sites and or buildings that will achieve environmental and energy targets	Built environment professional team

Table 3. Stage two: Design concept

Activity	By
Analyse environmental aspects of the site in order to understand how designs and green building technologies can work with these to reduce impacts and energy consumption.	Built environment professional team
Develop concept designs and identify green building technologies that may achieve proposed targets.	Built environment professional team
Check, through modelling and calculations, that proposed approach will achieve targets. Report on progress.	Built environment professional team
Check that targets are being achieved.	Client/independent adviser

Table 4. Stage three: Design development

Activity	By
Develop detailed designs that aim to achieve energy targets	Built environment professional team
Check, through modelling and calculations, that proposed approach will achieve targets. Report on progress	Built environment professional team
Check that targets are being achieved.	Client/independent adviser

Table 5. Stage four: Technical documentation

Activity	By
Develop detailed designs and green building technologies that will achieve targets. Ensure that tender and contractual documentation requires the contractor and relevant suppliers to contribute, as required, in order to achieve targets.	Built environment professional team
Check, through modelling, calculations and inspections that targets are being achieved. Report on progress.	Built environment professional team
Confirm that targets are being achieved.	Client/independent adviser

Table 6. Stage five: Contract administration and inspection

Activity	By
Ensure that the completed intervention/installation/building achieves targets. Put in place systems that enable close control and monitoring of environmental impacts and energy consumption in the building. Issue manuals and technical information that detail the targets and explain how green building technologies should be operated to achieve targets. Report on progress.	Built environment professional team
Provide facilities management training using the training manuals and technical information developed by the team (see above) to ensure that there are strong systems to support low environmental impact and energy efficient operation of the building.	Built environment professional team / Facilities management
Develop induction training for new occupants of the building to ensure that they understand the building's systems and green technologies and will use these to achieve targets.	Facilities management / Human resources
Carry out a Post Occupancy Evaluation to confirm that building, systems, green building technologies, management and occupants are working together to achieve required targets. If necessary, take action to address problems and ensure integrated and efficient performance.	Client/independent adviser

engineers and glass specialists may also be appointed to optimize building envelope designs which are energy efficient and optimize comfort. Similarly, specialist urban designers and landscape architects may be appointed to develop site layouts and built form that conserves and enhance natural environmental features and support increased energy efficiency.

- **Responsive Design:** The approach ensures that the design of green technologies in the building responds to, and works with, features of a site and local climate rather than against this. For example, technologies may respond to topography and existing vegetation to achieve optimum access to natural ventilation and light. This requires a detailed analysis of the site and can be supported by modelling and simulation.
- **Modelling and an Iterative Design Process:** Having set explicit and challenging targets, the built environment professional team must make sure that these are achieved, or exceeded. This can be supported through calculations and modelling which demonstrate that selected technologies and techniques will achieve the required targets. Modelling options also is a useful way to identify the best technologies and in an iteratively way to develop these.

CONCLUSION AND APPLICATIONS

Green building technologies describe technologies and techniques used in built environments to minimize environmental impacts while ensuring that buildings are able to accommodate the functions they have been designed for and are comfortable and productive to live and work in.

As buildings are responsible for more than 40 per cent of global energy use and produce over a third of global greenhouse gas emissions it is particularly important to ensure that they become more energy efficient and reduce their contribution to global warming.

Green building technologies not only have to have a mitigation function (to reduce carbon emissions), they also have an adaptation function (to help building accommodate project climate changes) as the earth is already experiencing global warming.

These requirements can make the selection, development and application of green building technologies in buildings complex. This chapter addresses this complexity by providing a structured approach to the integration of green building technologies in buildings. This structured approach is described in the following key areas.

Firstly, the relationship between built environments and the natural environments are defined. In particular, linkages between energy use in buildings, carbon dioxide emissions and global warming are set out.

Secondly, the potential role of green building technologies in reducing environmental impacts and reducing energy consumption is described. In particular, the importance of ensuring that the buildings accommodate the required functions and are conducive to health and productivity while reducing impacts, is established.

Thirdly, a structured stage-by-stage process is described which can be used to identify, refine, and integrate green building technologies into buildings to ensure that low environmental impact, energy efficient building are achieved.

In this way, the chapter provides useful insight that can be used by people wishing to understand how environmental impacts can be reduced in buildings through the selection, design and integration of green building technologies.

REFERENCES

American Society of Heating, Refrigerating and Air-Conditioning Engineers. (2004). *Thermal environmental conditions for human occupancy*. Author.

Barrett, P. S., Zhang, Y., Davies, F., & Barrett, L. C. (2015). *Clever classrooms: Summary report of the HEAD project*. University of Salford.

Carbon Trust. (2006). *Energy Saving Factsheets*. Retrieved from www.carbontrust.co.uk

Chan, A. P. C., Darko, A., Olanipekun, A. O., & Ameyaw, E. E. (2018). Critical barriers to green building technologies adoption in developing countries: The case of Ghana. *Journal of Cleaner Production*, *172*, 1067–1079. doi:10.1016/j.jclepro.2017.10.235

Cotana, F., Rossi, F., Filipponi, M., Coccia, V., Pisello, A. L., Bonamente, E., ... Cavalaglio, G. (2014). Albedo control as an effective strategy to tackle Global Warming: A case study. *Applied Energy*, *130*, 641–647. doi:10.1016/j.apenergy.2014.02.065

Crandell, C. C., & Smaldino, J. J. (2000). Classroom acoustics for children with normal hearing and with hearing impairment. *Language, Speech, and Hearing Services in Schools*, *31*(4), 362–370. doi:10.1044/0161-1461.3104.362 PMID:27764475

Gething, B., & Puckett, K. (2013). *Design for climate change*. Royal Institute of British Architecture.

Hamin, E. M., & Gurran, N. (2009). Urban form and climate change: Balancing adaptation and mitigation in the US and Australia. *Habitat International*, *33*(3), 238–245. doi:10.1016/j.habitatint.2008.10.005

Intergovernmental Panel on Climate Change. (2007). *Synthesis report, fourth assessment report*. International Panel on Climate Change and Cambridge University Press.

Intergovernmental Panel on Climate Change. (2015). *Climate change 2014: Mitigation of climate change*. Cambridge University Press.

Loisos, G. (1999). *Day Lighting in Schools. An Investigation in the Relationship between Daylighting and Human Performance*. Pacific Gas and Electric Company.

Mendell, M., & Heath, G. (2005). Do indoor pollutants and thermal conditions in schools influence student performance? A critical review of the literature. *Indoor Air, 15*(1), 27–52. doi:10.1111/j.1600-0668.2004.00320.x PMID:15660567

Pachauri, RK., Allen, MR., Barros, VR., Broome, J., Cramer, W., Christ, R., & Dubash, NK. (2014). *Climate change 2014: synthesis report.* Contribution of Working Groups I, II and III to the fifth assessment report of the Intergovernmental Panel on Climate Change.

Tanner, C. (2009). Effects of school design on student outcomes. *Journal of Educational Administration, 47*(3), 381–399. doi:10.1108/09578230910955809

UCLA. (2018). *Climate Consultant.* Retrieved from http://www.energy-design-tools.aud.ucla.edu/climate-consultant/request-climate-consultant.php

Wadu Mesthrige, J., & Kwong, H. Y. (2018). Criteria and barriers for the application of green building features in Hong Kong. *Smart and Sustainable Built Environment, 7*(3/4), 251–276. doi:10.1108/SASBE-02-2018-0004

Wargocki, P., Wyon, D. P., Baik, Y. K., Clausen, G., & Fanger, P. O. (1999). Perceived air quality, sick building syndrome(SBS) and productivity in two different pollution loads. *Indoor Air, 9*(3), 165–179. doi:10.1111/j.1600-0668.1999.t01-1-00003.x PMID:10439554

Wargocki, P., Wyon, D. P., Matysiak, B., & Irgens, S. (2005). The effects of classroom air temperature and outdoor air supply on performance of school work by children. Proceedings of Indoor Air, 1(1), 368-372.

Wells, N., & Evans, G. (2003). Nearby Nature: A Buffer of Life Stress among Rural Children. *Environment and Behavior, 35*(3), 311–330. doi:10.1177/0013916503035003001

White, R. (2004). *Young children's relationship with nature: Its importance to children's development & the earth's future.* White Hutchinson Leisure & Learning Group.

ADDITIONAL READING

Edenhofer, O., Pichs-Madruga, R., Sokona, Y., Seyboth, K., Matschoss, P., Kadner, S., & von Stechow, C. (2011). *IPCC Special Report on Renewable Energy Sources and Climate Change Mitigation*. Cambridge, UK: Cambridge University Press. doi:10.1017/CBO9781139151153

Hallegatte, S. (2009). Strategies to adapt to an uncertain climate change. *Global Environmental Change*, *19*(2), 240–247. doi:10.1016/j.gloenvcha.2008.12.003

Holmes, M. J., & Hacker, J. N. (2007). Climate change, thermal comfort and energy: Meeting the design challenges of the 21st century. *Energy and Building*, *39*(7), 802–814. doi:10.1016/j.enbuild.2007.02.009

Karimpour, M., Belusko, M., Xing, K., Boland, J., & Bruno, F. (2015). Impact of climate change on the design of energy efficient residential building envelopes. *Energy and Building*, *87*, 142–154. doi:10.1016/j.enbuild.2014.10.064

Lombardi, D. R., Leach, J. M., & Rogers, C. D. F. (2012). *Designing Resilient Cities*. IHS BRE Press.

Roaf, S., Crichton, D., & Nicol, F. (2009). *Adapting buildings and cities for climate change: a 21st-century survival guide*. Routledge.

The Met Office. (2018) *Climate Science*. Retrieved from https://www.metoffice.gov.uk/climate-guide

USGBC. (2011). *Green Building and Climate Change Resilience, Understanding Impacts and Preparing for Changing Conditions*. Retrieved from https://www.usgbc.org/Docs/Archive/General/Docs18496.pdf

Wang, W., Zhang, S., Su, Y., & Deng, X. (2018). Key Factors to Green Building Technologies Adoption in Developing Countries: The Perspective of Chinese Designers. *Sustainability*, *10*(11), 4135. doi:10.3390u10114135

Wilson, S., & Hedge, A. (1987). *The Office Environment Survey: A Study of Building Sickness*. London: Building Use Studies.

Yin, S., & Li, B. (2018). Transferring green building technologies from academic research institutes to building enterprises in the development of urban green building: A stochastic differential game approach. *Sustainable Cities and Society*, *39*, 631–638. doi:10.1016/j.scs.2018.03.025

Zhang, Y., & Feng, T. (2018). Research on Green Building Index System Based on Low Carbon Ecological Design in the Detailed Planning. *Chemical Engineering Transactions*, *66*, 559–564.

Zhao, X., Zuo, J., Wu, G., & Huang, C. (2019). A bibliometric review of green building research 2000–2016. *Architectural Science Review*, *62*(1), 74–88. doi:10.1080/00038628.2018.1485548

KEY TERMS AND DEFINITIONS

Adaptation: Adaptation refers to adjustments to respond to climate change.

Air Conditioning: A mechanical system installed in a building to control the temperature and humidity of the air by heating or cooling.

Climate: Climate describes the average and variations of weather in a region over long periods of time.

Climate Change: Describes a change in the state of the climate that persists for an extended period. This is identified by changes to the mean, for instance, mean temperatures over a long period.

Emissions: Are substances released into the air and are measured by their concentrations, or parts per million, in the atmosphere.

Envelope: The external elements of the building such as the walls, windows, and roofs.

Glazing: Windows, glazed doors or other transparent and translucent elements including their frames (such as glass bricks, glazed doors, etc.) located in the building fabric.

Global Warming: Global warming is the warming trend that the earth has experienced over the past century.

Greenhouse Gas: The main greenhouse gases are water vapor (H2O), carbon dioxide (CO2), methane (CH4), nitrous oxide (N2O), hydrofluorocarbons (HFCs), perfluorocarbons (PFCs), and sulfur hexafluoride (SF6).

HVAC: Heating, ventilation, and air conditioning.

Mitigation: Mitigation involves taking actions to reduce greenhouse gas emissions and to enhance sinks, such as carbon sequestration, in order to reduce the extent of global warming.

Mixed-Mode Ventilation: A hybrid solution where natural ventilation systems are assisted by mechanical systems to achieve improved ventilation and comfort criteria. Complementary and zone-mixed strategies are most commonly adopted.

Natural Ventilation: Ventilation provided by thermal, wind, or diffusion effects through doors windows or other intentional openings in the building.

Occupants: People who occupy the building on a normal working day.

R-Value: Used in the construction industry to denote the measurement of the thermal resistance of a material.

RCPs: Representative concentration pathways are defined according to their contribution to atmospheric radiative forcing in the year 2100 relative to pre-industrial values. An RCP 8.5, therefore, represents the addition to the earth's radiation budget as a result of an increase in GHGs of +8.5 W/m^2.

Renewable Energy: Is energy from sources that will renew themselves within our lifetime. Renewable energy sources include wind, sun, water, biomass (vegetation), and geothermal heat.

Shading Coefficient: A measure of the solar gain performance of windows. It is the ratio of the solar energy transmitted and convected by the window to the solar energy transmitted and convected by clear 3 mm glass.

Solar Heat Gain Coefficient (SHGC): A measure of the amount of solar radiation (heat) passing through the entire window, including the frame. SHGC is expressed as a number between 0 and 1.0. The lower the SHGC the better.

Thermal Mass: A term to describe the ability of building materials to store heat.

Thermal Resistance: The resistance to heat transfer across a material. Thermal resistance is measured as an R-value. The higher the R-value the better the ability of the material to resist heat flow.

Ventilation: The process of supplying air to or removing air from a space for the purpose of controlling air contaminant levels, humidity, or temperature within the space.

Ventilation Opening: An opening in the external wall, floor, or roof of a building designed to allow air movement into or out of the building by naturally-driven ventilation through a permanent opening, an openable part of a window, a door, or other device which can be held open.

Watt (W): A unit for power (P) or the rate at which work is performed.

Compilation of References

ABB. (2009). *Towards a Smarter Grid—ABB's vision for the power system of the future*. Retrieved from http://www02.abb.com/db/db0003/db002698.nsf/0/e30fc9d5f79d4ae8c12579e2002a4209/%24file/Toward_a_smarter_grid_Julþ09.pdf

Abdmouleh, Z., Alammari, R., & Gastli, A. (2015). Review of policies encouraging renewable energy integration and best practices. *Renewable & Sustainable Energy Reviews*, *45*, 249–262. doi:10.1016/j.rser.2015.01.035

Abdullah, M. A., Chew, B. C., & Hamid, S. R. (2016). Benchmarking key success factors for the future green airline industry. *Procedia: Social and Behavioral Sciences*, *224*, 246–253. doi:10.1016/j.sbspro.2016.05.456

Abraham, K., Pasquariello, D., & Willstaedt, E. (1990). N-butylferrocene for overcharge protection of secondary lithium batteries. Journal of the Electrochemical Society, 137(6).

Abramskiehn, D., Hallmeyer, K., Trabacchi, C., Escalante, D., Netto, M., Cabrera, M., & Vasa, A. (2017). *Supporting National Development Banks to Drive Investment in Nationally Determined Contributions of Brazil, Mexico and Chile*. Washington, DC: Inter-American Development Bank.

Abubakar Mas'ud, A., Vernyuy Wirba, A., Muhammad-Sukki, F., Albarracin, R., Abu-Bakar, S. H., Munir, A. B., & Aini Bani, N. (2016). A review on the recent progress made on solar photovoltaic in selected countries of sub-Saharan Africa. *Renewable & Sustainable Energy Reviews*, *62*, 441–452.

Adbmouleh, Z., Rashid A.M., & Gastli, A. (2015). Review of policies encouraging renewable energy integration & best practices. *Renewable & Sustainable Energy Reviews*, *45*, 249–262.

Adnan, O., Mustafa, E., Yilser, D., Ozgur, C., & Feridun, H. (2017). Evaluation of sulfonated polysulfone/zirconium hydrogen phosphate composite membranes for direct methanol fuel cells. *Electrochimica Acta*, *256*, 196–210. doi:10.1016/j.electacta.2017.10.002

Aga, G. A., & Peria, M. S. M. (2014). *International remittances and financial inclusion in Sub-Saharan Africa*. Policy Research Working Paper 6991. Washington, DC: World Bank Group (Development Research Group).

Agrawal, A., Wollenberg, L., & Persha, L. (2014). Governing Mitigation in Agriculture-Forest Landscapes. *Global Environmental Change*, *29*, 270–326. doi:10.1016/j.gloenvcha.2014.10.001

Ahlborg, H., & Hammar, L. (2014). Drivers and barriers to rural electrification in Tanzania and Mozambique – Grid-extension, off-grid, and renewable energy technologies. *Renewable Energy*, *62*, 117–124.

Akhtar, F., Lodhi, S. A., & Khan, S. S. (2015). Permaculture approach: Linking ecological sustainability to business strategies. *Management of Environmental Quality*, *26*(6), 795–809. doi:10.1108/MEQ-01-2015-0001

Akhtar, F., Lodhi, S. A., Khan, S. S., & Sarwar, F. (2016). Incorporating permaculture and strategic management for sustainable ecological resource management. *Journal of Environmental Management*, *179*, 31–37. doi:10.1016/j.jenvman.2016.04.051 PMID:27155728

Al-Badi, A. H., Malik, A., & Gastli, A. (2009). Assessment of renewable energy resources potential in Oman and identification of barrier to their significant utilization. *Renewable & Sustainable Energy Reviews*, *13*, 2734–2739.

Albers, J., Meissner, E., & Shirazi, S. (2011). Lead-acid batteries in micro-hybrid vehicles. *Journal of Power Sources*, *196*(8), 3993–4002. doi:10.1016/j.jpowsour.2010.11.094

Allayev, K.R., Khokhlov, V.A., Sytdykov, R.A., & Titova, J.O. (2013). *Elektromekhanicheskiye i gidromekhanicheskiye protsessy v gidroenergeticheskikh ustanovkakh* [Electromechanical and hydro-mechanical processes in hydropower plants]. Tashkent: Fan va texnologiya.

Amabile, J. (2011). Boulder's Energy Future [Editorial Advisory Board]. *The Daily Camera.* Retrieved from LexisNexis Academic.

Amankwah-Amoah, J. (2015). Solar energy in Sub-Saharan Africa: The challenges and opportunities of technological leapfrogging. *Thunderbird International Business Review*, *57*(1), 15–31.

Amatayakul, W., & Berndes, G. (2012). Determining factor for the development of CDM biomass power projects. *Energy for Sustainable Development*, *16*(2), 197–203.

American Society of Heating, Refrigerating and Air-Conditioning Engineers. (2004). *Thermal environmental conditions for human occupancy*. Author.

Americas Electric Cooperatives. (2018). *We Are America's Electric Cooperatives*. Retrieved from https://www.electric.coop/our-mission/we-are-americas-electric-cooperatives/

Amine, K., Zhang, L., & Zhang, Z. (2010). *Develop and evaluate materials and additives that enhance thermal and overcharge abuse*. Department of Energy Advanced Battery Research Review.

Amorós, J. E., & Bosma, N. (2014). *Global entrepreneurship monitor 2013 global report: Fifteen years of assessing entrepreneurship across the globe*. London: GEM.

Andersen, S. M. (2017). *Experiences with carbon taxation - Denmark, Sweden and Europe*. Retrieved from http://pure.au.dk/portal/files/111427005/CapeTown1_ANDERSEN.pdf

Anderson, J. L., Lowry, M. V., & Thomte, J. C. (2001). Hydraulic and water quality modeling of distribution systems: What are the trends in the U.S. and Canada? In *Proceedings of the AWWA Annual Conference*. Denver, CO: American Water Works Association (AWWA).

Anderson, T. L., & Leal, D. R. (2001). *Free Market Environmentalism*. New York, NY: Palgrave. doi:10.1057/9780312299736

Andersson, T., Carlsen, J., & Getz, D. (2002). Family business goals in the tourism and hospitality sector: Case studies and cross-case analysis from Australia, Canada, and Sweden. *Family Business Review*, *15*(2), 89–106. doi:10.1111/j.1741-6248.2002.00089.x

ANEMOS. (n.d.). *ANEMOS Project Web Page*. Retrieved from http: //anemos.cma.fr/

APP (Africa Progress Panel). (2014). *Finance and banking in Africa: Extracts from the Africa progress report 2014*. Geneva: Africa Progress Panel.

Appadurai, A. (2000). Grassroots globalization and the research imagination. *Public Culture*, *12*(1), 1–19. doi:10.1215/08992363-12-1-1

Aquilaa, G., Pamplona, E., Queiroz, A. R., Junior, P., & Fonseca, M. (2017). An overview of incentive policies for the expansion of renewable energy generation in electricity power systems and the Brazilian experience. *Renewable & Sustainable Energy Reviews*, *70*, 1090–1098. doi:10.1016/j.rser.2016.12.013

Arikiez, M., Grasso, F., & Zito, M. (2016). Heuristic algorithm for coordinating smart houses in MicroGrid. *2015 IEEE International Conference on Smart Grid Communications, SmartGridComm 2015*, 49–54. 10.1109/SmartGridComm.2015.7436275

Arnold, E. R. (2013). Xcel Energy Can't Answer the Hard Questions [Opinion]. *The Daily Camera*. Retrieved from LexisNexis Academic.

Aronoff, C., & Ward, J. (2011). *Family business values: How to assure a legacy of continuity and success*. New York, NY: Palgrave MacMillan. doi:10.1007/978-1-137-51208-6

Asian Development Bank. (2014). *Technical Assistance for the South Asia Economic Integration Partnership – Power Trading in Bangladesh and Nepal (TA 8658-REG)*. Manilla: Author.

Asif, M., & Barua, D. (2011). Salient features of the Grameen Shakti renewable energy program. *Renewable & Sustainable Energy Reviews*, *15*(9), 5063–5067.

Azuri Technologies. (2015). *Pay as you go solar systems*. Retrieved from http://www.azuri-technologies.com/info-hub/azuri-quad

Bacher, P., Madsen, H., & Nielsen, H. (2009). Online short-term solar power forecasting. *Solar Energy*, *83*(10), 1772–1783. doi:10.1016/j.solener.2009.05.016

Bahrami, S., & Amini, M. H. (2017). *A decentralized framework for real-time energy trading in distribution networks with load and generation uncertainty*. arXiv preprint arXiv:1705.02575

Baizeng, F., Min, S. K., Jung, H. K., Min, Y. S., Yan, J. W., Haijiang, W., ... Jong, S. Y. (2011). High Pt loading on functionalized multiwall carbon nanotubes as a highly efficient cathode electrocatalyst for proton exchange membrane fuel cells. *Journal of Materials Chemistry*, *21*(22), 8066–8073. doi:10.1039/c1jm10847f

Bank of Canada. (2018). *Annual Exchange Rates*. Retrieved from: https://www.bankofcanada.ca/rates/exchange/annual-average-exchange-rates/

Bansal, P. (2005). Evolving sustainably: A longitudinal study of corporate sustainable development. *Strategic Management Journal*, *26*(3), 197–218. doi:10.1002mj.441

Barbose, G. (2017). *U.S. Renewables Portfolio Standards: 2017 Annual Status Report*. Retrieved from https://emp.lbl.gov/publications/us-renewables-portfolio-standards-0

Barbose, G., Wiser, R., Heeter, J., Mai, T., Bird, L., Bolinger, M., ... Millstein, D. (2016). A retrospective analysis of benefits and impacts of U.S. renewable portfolio standards. *Energy Policy*, *96*, 645–660. doi:10.1016/j.enpol.2016.06.035

Barnhart, C. J., & Benson, S. M. (2013). On the importance of reducing the energetic and material demands of electrical energy storage. *Energy & Environmental Science*, *6*(4), 1083–1092. doi:10.1039/c3ee24040a

Barrett, P. S., Zhang, Y., Davies, F., & Barrett, L. C. (2015). *Clever classrooms: Summary report of the HEAD project*. University of Salford.

Barry, J. (2005). Ecological Modernization. In J. Dryzek (Ed.), *Debating the Earth, 3030-321*. York, UK: Oxford University Press.

Baumeister, S., & Onkila, T. (2017). An eco-label for the airline industry? *Journal of Cleaner Production*, *142*, 1368–1376. doi:10.1016/j.jclepro.2016.11.170

Baumol, W. J., & Oates, W. E. (1998). *The theory of environmental policy*. Cambridge, UK: Cambridge University Press.

Baumol, W. J., Panzar, J. C., & Willig, T. D. (1982). *Contestable markets and the theory of industry structure*. New York, NY: Harcourt Brace Jovanovich.

Bayod-Rújula, A. A. (2009). Future development of the electricity systems with distributed generation. *Energy*, *34*(3), 377–383. doi:10.1016/j.energy.2008.12.008

Bayrakc, A. G., & Kocar, G. (2012). Utilization of renewable energies in Turkey's agriculture. *Renewable & Sustainable Energy Reviews*, *16*(1), 618–633. doi:10.1016/j.rser.2011.08.027

Bazilian, M., Nussbaumer, P., Rogner, H., Brew-Hammond, A., Foster, V., Pachauri, S., ... Kammen, D. (2012). Energy access scenarios to 2030 for the power sector in sub-Saharan Africa. *Utilities Policy*, *20*(1), 1–16.

Bebbington, J. (2001). Sustainable Development: A Review of the International Development Business and Accounting Literature. *Accounting Forum*, *25*(2), 128–157. doi:10.1111/1467-6303.00059

Beck, F., & Martinot, E. (2004). Renewable Energy Policies and Barriers. Encyclopedia of Energy, 5, 365-383.

Beck, U. (1995). *Ecological politics in an age of risk*. Malden, MA: Blackwell Publishers Inc.

Belash, I. G. (2010). Causes of the accident of the hydraulic unit No. 2 at the Sayano-Shushenskaya HPP. *Hydrotechnical construction*, *3*, 25–31.

Bergamasco, L., & Asinari, P. (2011). Scalable methodology for the photovoltaic solar energy potential assessment based on available roof surface area: Application to Piedmont Region (Italy). *Solar Energy*, *85*(5), 1041–1055. doi:10.1016/j.solener.2011.02.022

Berlemann, M., & Jahn, V. (2015). Regional importance of Mittelstand firms and innovation performance. *Regional Studies*, *50*(11), 1819–1833. doi:10.1080/00343404.2015.1058923

Berrone, P., Cruz, C., & Gomez-Mejia, L. R. (2012). Socioemotional wealth in family firms theoretical dimensions, assessment approaches, and agenda for future research. *Family Business Review*, *25*(3), 258–279. doi:10.1177/0894486511435355

Bessa, R., Moreira, C., Silva, B., & Matos, M. (2014). Handling renewable energy variablity and uncertainty in power system operations. *Wiley Interdisciplinary Reviews*, *3*(2), 156–178. doi:10.1002/wene.76

Bessa, R., Trindade, A., Silva, C., & Miranda, V. (2015). Probabilistic Solar Power Forecasting in Smart Grids using Distributed Information. *Electrical Power Energy Systems*, *72*, 16–23. doi:10.1016/j.ijepes.2015.02.006

Bhardwaj, B. R. (2016). Role of green policy on sustainable supply chain management: A model for implementing corporate social responsibility (CSR). *Benchmarking: An International Journal*, *23*(2), 456–468. doi:10.1108/BIJ-08-2013-0077

Bhatnagar, N., & Venkatesh, B. (2012). *Energy storage and power systems.* Paper presented at the Electrical & Computer Engineering (CCECE), 2012 25th IEEE Canadian Conference on. 10.1109/CCECE.2012.6334823

Bhoye, M., Purohit, S. N., Trivedi, I. N., Pandya, M. H., Jangir, P., & Jangir, N. (2016). Energy Management of Renewable Energy Sources in a Microgrid using Cuckoo Search Algorithm. *IEEE Students' Conference on Electrical, Electronics and Computer Science*.

Bian, B., Kuzlu, M., Pipattanasomporn, M., & Rahman, S. (2014). *Analysis of communication schemes for advanced metering infrastructure (AMI)*. National Harbor, MD: IEEE PES General Meeting Conference and Exposition. doi:10.1109/PESGM.2014.6939562

Biasca, D. (2013). Don't Sign Xcel's Misleading Petition [Opinion]. *The Daily Camera*. Retrieved from LexisNexis Academic.

Bird, L., Cochran, J., & Wang, X. (2014). *Wind and Solar Energy Curtailment: Experience and Practices in the United States*. National Renewable Energy Laboratory. Retrieved from https://www.nrel.gov/docs/fy14osti/60983.pdf

Bird, L., Bolinger, M., Gagliano, T., Wiser, R., Brown, M., & Parsons, B. (2005). Policies and market factors driving wind power development in the United States. *Energy Policy*, *33*(11), 1397–1407. doi:10.1016/j.enpol.2003.12.018

Birol, F. (2010). *World Energy Outlook*. International Energy Agency. Retrieved from http://www.oecd.org/berlin/46389140.pdf

Block, C., Neumann, D., & Weinhardt, C. (2008). A Market Mechanism for Energy Allocation in Micro-chp Grids. *Proceedings of the 41st annual Hawaii international conference on system sciences*, 172–172. 10.1109/HICSS.2008.27

Block, J. H., & Wagner, M. (2014). The effect of family ownership on different dimensions of corporate social responsibility: Evidence from large US firms. *Business Strategy and the Environment*, *23*(7), 475–492. doi:10.1002/bse.1798

Bloomberg New Energy Finance. (2018). *Clean energy investment trends, 2Q 2018*. BloombergNEF.

Bloomberg. (2018). *Electric Utilities: Public Service Company of Colorado*. Retrieved from https://www.bloomberg.com/research/stocks/private/snapshot.asp?privcapId=298400

Bloomberg. (2018, October 25). *Africa's biggest wind farm nears transmission as line is tested*. Retrieved from https://www.bloomberg.com/news/articles/2018-08-31/africa-s-biggest-wind-farm-nears-transmission-as-line-is-tested

Blouin, M. R., & Serrano, R. (2001). A Decentralized Market with Common Values Uncertainty: Non-steady States. *The Review of Economic Studies*, *68*(2), 323–346. doi:10.1111/1467-937X.00171

Bohringer, C., Rivers, N. J., Rutherford, T., & Wigle, R. (2012). Green Jobs and Renewable Electricity Policies: Employment Impacts of Ontario's Feed - in - Tariff. *The B.E. Journal of Economic Analysis & Policy*, *12*(1), 1–40. doi:10.1515/1935-1682.3217

Boland, J., Korolkiewicz, M., Agrawal, M., & Huang, J. (2012). Forecasting Solar Radiation on Short Time Scales using a Coupled Autoregressive and Dynamical System (CARDS) model. *Australian Solar Energy Society*, *87*, 136–149.

Boloukat, M., & Foroud, A. (2016). Stochastic-based resource expansion planning for a grid-connected microgrid using interval linear programming. *Energy*, *113*, 776–787. doi:10.1016/j.energy.2016.07.099

Borden, E., & Schill, W.-P. (2013). *Policy Efforts for the Development of Storage Technologies in the US and Germany*. Academic Press.

Bouzerdoum, M., Mellit, A., & Massi, P. A. (2013). A hybrid model (SARIMA-SVM) for Short-term Power Forecasting of a Small-scale Grid-connected Photovoltaic Plant. *Solar Energy*, *98*, 226–235. doi:10.1016/j.solener.2013.10.002

Bracco, S., Delfino, F., Pampararo, F., Robba, M., & Rossi, M. (2015). A Dynamic Optimization-based Architecture for Polygeneration Microgrids with Tri-Generation, Renewables, Storage Systems and Electrical Vehicles. *Energy Conversion and Management*, *96*, 511–520. doi:10.1016/j.enconman.2015.03.013

Brasseur, G. P., Gupta, M., Anderson, B. E., Balasubramanian, S., Barrett, S., Duda, D., ... Halthore, R. N. (2016). Impact of aviation on climate: FAA's aviation climate change research initiative (ACCRI), phase ii. *Bulletin of the American Meteorological Society*, *97*(4), 561–583. doi:10.1175/BAMS-D-13-00089.1

Brehm, S., & Brehm, J. W. (1981). *Psychological reactance: a theory of freedom and control.* New York: Academic Press.

Brian, C. (2001). An introduction to fuel cells and hydrogen technology. Vancouver, Canada: Academic Press.

Brice, W. D., & Richardson, J. (2009). Culture in family business: A two-country empirical investigation. *European Business Review*, *21*(3), 246–262. doi:10.1108/09555340910956630

Brid, L., Cochran, J., & Wang, X. (2014). *Wind and solar curtailment: Experience and practices in the United States*. National Renewable Energy Laboratory, Technical Report NREL/TP-6A20-60983. Accessed on March 10, 2018, from: http://www.nrel.gov/docs/fy14osti/60983.pdf

British Petroleum. (2018). *Statistical Review of World Energy*. Retrieved June 10, 2018, from BP Global: https://www.bp.com/en/global/corporate/energy-economics/statistical-review-of-world-energy.html

Brooks, K. P., Snowden-Swan, L. J., Jones, S. B., Butcher, M. G., Lee, G. S., Anderson, D. M., . . . Burton, F. (2016). Low-Carbon Aviation Fuel through the Alcohol to Jet Pathway. In Biofuels for Aviation (pp. 109-150). Academic Press. doi:10.1016/B978-0-12-804568-8.00006-8

Brown, B. G., Katz, R. W., & Murphy, A. H. (1984). Time Series Models to Simulate and Forecast Wind Speed and Wind Power. *Journal of Climate and Applied Meteorology*, *23*(8), 1184–1195. doi:10.1175/1520-0450(1984)023<1184:TSMTSA>2.0.CO;2

Brown, D., & McGranahan, G. (2016). The urban informal economy, local inclusion and achieving a global green transformation. *Habitat International*, *53*, 97–105.

Brudermann, T., Reinsberger, K., Orthofer, A., Kislinger, M., & Posch, A. (2013). Photovoltaics in agriculture: A case study on decision making of farmers. *Energy Policy*, *61*, 96–103. doi:10.1016/j.enpol.2013.06.081

Brundtland, G. H. (1987). *Our Common Future. The World Commission on Environment and Development*. Oxford, UK: Oxford University Press.

Brundtland, G. H. (1987). *Our common future: Report of the world commission on environment and development (UN Publication No. A/42/427)*. United Nations.

Buchan, D. (2012). *The Energiewende - Germany's Gamble. The Oxford Institute of Energy Studies.* Retrieved from https://www.oxfordenergy.org/wpcms/wp-content/uploads/2012/06/SP-261.pdf

Buchner, B. K., Oliver, P., Wang, X., Carswell, C., Meattle, C., & Mazza, F. (2017). *Global Landscape of Climate Finance 2017*. Venice: Climate Policy Initiative.

Bugaje, I. (2006). Renewable energy for sustainable development in Africa: A review. *Renewable & Sustainable Energy Reviews*, *10*(6), 603–612. doi:10.1016/j.rser.2004.11.002

Bull, S. R. (2001). Renewable energy today and tomorrow. *Proceedings of the IEEE*, *89*(8), 1216–1226. doi:10.1109/5.940290

Bundesverband Erneuerbare Energie. (n.d.). *BEE forecast: slightly lower EEG surcharge in 2018*. Retrieved October 26, 2018, from: https://www.bee-ev.de/home/presse/mitteilungen/detailansicht/bee-prognose-leicht-sinkende-eeg-umlage-2018/

Burke, K. (1985). Dramatism and logology. In D. L. Sills (Ed.), International encyclopedia of the social sciences (pp. 445–452). London, UK: Macmillan Publishers. doi:10.1080/01463378509369584

Burness, A. (2017). Boulder says municipalization ruling adds $23M in city costs. *The Daily Camera*. Retrieved from http://www.dailycamera.com/news/boulder/ci_31325514/boulder-says-municipalization-ruling-adds-23m-city-costs

Burris, P. (2013). We Should Re-vote on the Muni [Opinion]. *The Daily Camera*. Retrieved from LexisNexis Academic.

Busgen, U., & Durrschmidt, W. (2009). The expansion of electricity generation from renewable energies in Germany - A review based on Renewable Energy Sources Act Progress Report 2007 and the new German feed - in legislation. *Energy Policy*, *37*(7), 2536–2545.

Business and Financial Times. (2018). *Lake Turkana wind farm produce first megawatts*. Business and Financial Times.

Butterfield, A. (2010). Franchise Agreement, Pay the Same Get More [Opinion]. *The Daily Camera*. Retrieved from LexisNexis Academic.

Butterfield, A. (2011). Boulder's Energy Future [Editorial Advisory Board]. *The Daily Camera*. Retrieved from LexisNexis Academic.

Cabrera-Suarez, K., Saa-Perez, P., & Garcia-Almeida, D. (2001). The succession process from a resource- and knowledge-based view of the family firm. *Family Business Review*, *14*(1), 37–48. doi:10.1111/j.1741-6248.2001.00037.x

Cali, Ü. (2010). *Grid and Market Integration of Large-Scale Wind Farms Using Advanced Wind Power Forecasting: Technical and Energy Economic Aspects*. Kassel University Press GmbH.

Carbon Trust. (2006). *Energy Saving Factsheets*. Retrieved from www.carbontrust.co.uk

Carlsen, J., Getz, D., & Ali-Knight, J. (2001). The environmental attitudes and practices of family businesses in the rural tourism and hospitality sectors. *Journal of Sustainable Tourism*, *9*(4), 281–297. doi:10.1080/09669580108667403

Carpenter, J. R., Merckelbach, L., Callies, U., Clark, S., Gaslikova, L., & Baschek, B. (2016). Potential Impacts of Offshore Wind Farms on North Sea Stratification. *PLoS One*, *11*(8). doi:10.1371/journal.pone.0160830

Casado-Asensio, J., & Steurer, R. (2014). Integrated strategies on sustainable development, climate change mitigation and adaptation in Western Europe: Communication rather than coordination. *Journal of Public Policy*, *34*(03), 437–473.

Castells, M. (1983). *The City and the Grassroots: A cross-cultural theory of urban social movements*. Los Angeles, CA: University of California Press.

Centrum. (2016). *Centum to build 12MW plants for Two Rivers Mall*. Retrieved from http://www.businessdailyafrica.com/Corporate-News/Centum-to-build-12MW-plants-for-Two-Rivers-Mall/539550-3047216-5a3rylz/index.html

CEPA. (2014). Policy risk in renewable energy investments in developing countries. London: UK Department of Energy and Climate Change (DECC).

Chalkova, E., Wang, C., Komarneni, S., Lee, J., Fedkin, M., & Lvov, S. (2009). Composite proton conductive membranes for elevated temperature and reduced relative humidity PEMFC. *Electrochemical society. Transactions*, *25*, 1141–1150.

Chan, A. P. C., Darko, A., Olanipekun, A. O., & Ameyaw, E. E. (2018). Critical barriers to green building technologies adoption in developing countries: The case of Ghana. *Journal of Cleaner Production*, *172*, 1067–1079. doi:10.1016/j.jclepro.2017.10.235

Chapman, M. (2016). Sustaining Reductions in Aircraft Emissions for Canada's Major Airlines. In *Managing in a VUCA World* (pp. 175–193). Cham: Springer. doi:10.1007/978-3-319-16889-0_12

Chen, W., Dollar, D., & Tang, H. (2015). *Why is China investing in Africa? Evidence from the firm level*. Washington, DC: The Brookings Institution.

Chen, W.-M., Kim, H., & Yamaguchi, H. (2014). Renewable energy in eastern Asia: Renewable energy policy review and comparative SWOT analysis for promoting renewable energy in Japan, South Korea, and Taiwan. *Energy Policy*, *74*, 319–329. doi:10.1016/j.enpol.2014.08.019

Chen, Z., Holmberg, B., Li, W., Wang, X., Deng, W., Munoz, R., & Yan, Y. (2006). Nafion/zeolite nanocomposite membrane by in situ crystallization for a direct methanol fuel cell. *Chemistry of Materials*, *18*(24), 5669–5675. doi:10.1021/cm060841q

Chichilnisky, G. (1996). An axiomatic approach to sustainable development. *Social Choice and Welfare*, *13*(2), 231–257. doi:10.1007/BF00183353

Chirambo, D. (2016b). Integrating microfinance, climate finance and climate change adaptation: A Sub-Saharan Africa perspective. In Climate Change Adaptation, Resilience and Hazards (pp. 195-207). Heidelberg, Germany: Springer.

Chirambo, D. (2016a). Moving past the rhetoric: Policy considerations that can make Sino-African relations to improve Africa's climate change resilience and the attainment of the sustainable development goals. *Advances in Climate Change Research*, *7*(4), 253–263.

Chow, C. W., Urquhart, B., & Lave, M. (2011). Intra-hour Forecasting with a Total Sky Imager at the UC3 San Diego Solar Energy Testbed. *Solar Energy*, *85*, 2881–2893.

Chrisman, J. J., Chua, J. H., Pearson, A. W., & Barnett, T. (2012). Family involvement, family influence, and family_centered non_economic goals in small firms. *Entrepreneurship Theory and Practice*, *36*(2), 267–293. doi:10.1111/j.1540-6520.2010.00407.x

Chrisman, J. J., & Patel, P. C. (2012). Variations in R&D investments of family and nonfamily firms: Behavioral agency and myopic loss aversion perspectives. *Academy of Management Journal*, *55*(4), 976–997. doi:10.5465/amj.2011.0211

Christ-Janer, K. (2013). Municipal Energy is Not the Way to Create Real Change [Letter to the editor]. *The Daily Camera*. Retrieved from http://www.dailycamera.com/letters/ci_24418805/karey-christ-janer-municipal-energy-is- not-way

Christy Webber Landscapes. (n.d.). *Millennium park*. Retrieved from http://www.christywebber.com/projects/millennium-park/

Chua, J. H., Chrisman, J. J., & Sharma, P. (1999). Defining the family business by behavior. *Entrepreneurship Theory and Practice*, *23*(4), 19–40. doi:10.1177/104225879902300402

Chu, Y., Urguhart, B., Gohari, S., Pedro, H., Kleissl, J., & Coimbra, C. (2015). Short-term Reforecasting of Power Output from a 48 MWe Solar PV Plant. *Solar Energy*, *112*, 68–77. doi:10.1016/j.solener.2014.11.017

City of Boulder. (2018). *Boulder's Climate Commitment*. Retrieved from https://bouldercolorado.gov/climate

City of Chicago. (2017, January). *Chicago Sustainable Development Policy*. Retrieved from https://www.cityofchicago.org/city/en/depts/dcd/supp_info/sustainable_development/chicago-sustainable-development-policy-update.html

Climate Bonds Initiative. (2018, April 5). *From billions to trillions – Your 2-minute takeaway on our Annual Conference 2018*. Retrieved from Climatebonds.net: https://www.climatebonds.net/2018/04/billions-trillions-%E2%80%93-your-2-minute-takeaway-our-annual-conference-2018

Cobbinah, P. B., Erdiaw-Kwasie, M. O., & Amoateng, P. (2015). Africa's urbanisation: Implications for sustainable development. *Cities (London, England)*, *47*, 62–72.

Çoker, B., Cathoglu, H., & Birgin, O. (2010). Conceptions of students about renewable energy sources: A need to teach based on contextual approaches. *Procedia: Social and Behavioral Sciences*, *2*(2), 1488–1492. doi:10.1016/j.sbspro.2010.03.223

Coleman, E. (2015). Common property endowments in the trust game: Experimental evidence from Bulgaria. *Journal of Theoretical Politics*, *28*(1), 27–43. doi:10.1177/0951629814568400

Colorado Public Utilities Commission. (2004). *Decision No. C05-0049 – Docket No. 04A-214E: In the Matter of the Application of PSCo for Approval of its 2003 Least-Cost Resource Plan; Docket No. 04A-215E: In the Matter of the Application of PSCo for an Order Approving a Regulatory Plan to Support the Company's 2003 Least-Cost Resource Plan; Docket No. 04A-216E*. Author.

Conejo, A. J., Caramanis, M. C., & Bloom, J. A. (1990). An efficient algorithm for optimal reservoir utilization in probabilistic production costing. *IEEE Transactions on Power Systems*, *5*(2), 439–447. doi:10.1109/59.54550

Cookson, T., Lang, N., & Thornton, E. (2008). Adjustable speed drives applied to large AC induction motor and pump systems. In *Proceedings of the Twenty-fourth International Pump Users Symposium*, (pp. 75-80). Houston, TX: GA Valves. Retrieved January 20, 2016, from http://www.gavalves.co.uk

Cory, K., Couture, T., & Kreycik, C. (2009). *Feed-in Tariff Policy: Design, Implementation, and RPS Policy Interactions*. Retrieved from https://www.nrel.gov/docs/fy09osti/45549.pdf

Cotana, F., Rossi, F., Filipponi, M., Coccia, V., Pisello, A. L., Bonamente, E., ... Cavalaglio, G. (2014). Albedo control as an effective strategy to tackle Global Warming: A case study. *Applied Energy*, *130*, 641–647. doi:10.1016/j.apenergy.2014.02.065

Couture, T., & Cory, K. (2013). *State Clean Energy Policies Analysis (SCEPA) Project: An Analysis of Renewable Energy Feed-in Tariffs in the United States*. National Renewable Energy Laboratory (NREL). Accessed October 27, 2018 from: https://www.nrel.gov/docs/fy09osti/45551.pdf

Craig, J., & Dibrell, C. (2006). The natural environment, innovation, and firm performance: A comparative study. *Family Business Review*, *19*(4), 275–288. doi:10.1111/j.1741-6248.2006.00075.x

Crandell, C. C., & Smaldino, J. J. (2000). Classroom acoustics for children with normal hearing and with hearing impairment. *Language, Speech, and Hearing Services in Schools*, *31*(4), 362–370. doi:10.1044/0161-1461.3104.362 PMID:27764475

Czarniawska, B. (1999). *Writing management: Organization theory as a literary genre*. Oxford, UK: Oxford University Press. doi:10.1093/acprof:oso/9780198296140.001.0001

Daenekas, C., Neureiter, C., Rohjans, S., Uslar, M., & Engel, D. (2014). Towards a model-driven-architecture process for smart grid projects. *Digital Enterprise Design & Management, ser. Advances in Intelligent Systems and Computing*, *261*, 47–58. doi:10.1007/978-3-319-04313-5_5

Danish Energy Agency. (2017). *Denmark's Energy and Climate Outlook*. Retrieved from https://ens.dk/sites/ens.dk/files/Analyser/denmarks_energy_and_climate_outlook_2017.pdf

Daping, H., Chao, Z., Cheng, X., Niancai, C., Huaiguang, L., Shichun, M., & Mu, P. (2011). Polyaniline-Functionalized Carbon Nanotube Supported Platinum Catalysts. *Langmuir, 27*(9), 5582–5588. doi:10.1021/la2003589 PMID:21476530

Datsyuk, V., Kalyva, M., Papagelis, K., Parthenios, J., Tasis, D., Siokou, A., ... Galiotis, C. (2008). Chemical oxidation of multiwalled carbon nanotubes. *Carbon, 46*(6), 833–840. doi:10.1016/j.carbon.2008.02.012

Davidescu, S. (2017). The Europeanization of renewable energy policy in Romania. In I. Solorio & H. Jorgens (Eds.), *A guide to EU renewable energy policy: Comparing Europeanization and domestic policy change in EU member states* (pp. 204–223). Cheltenham, UK: Edward Elgar Publishing. doi:10.4337/9781783471560.00021

Davis, M., & Costyk, D. (2005). *IEEE Guide for Design, Operation, and Integration of Distributed Resource Island Systems with Electric Power Systems*. IEEE.

Day, T., Röser, F., & Kurdziel, M. (2016). *Conditionality of Intended Nationally Determined Contributions (INDCs)*. Berlin: International Partnership on Mitigation and MRV and New Climate Institute.

De La Tour, A., Glachant, M., & Ménière, Y. (2011). Innovation and international technology transfer: The case of the Chinese photovoltaic industry. *Energy Policy, 39*(2), 761–770. doi:10.1016/j.enpol.2010.10.050

de Oliveira, J. A. P. (2009). The implementation of climate change related policies at the subnational level: An analysis of three countries. *Habitat International, 33*(3), 253–259.

Deane, J. P., & Gallachóir, Ó. (2010). Techno-economic review of existing and new pumped hydro energy storage plant. *Renewable & Sustainable Energy Reviews, 14*(4), 1293–1302. doi:10.1016/j.rser.2009.11.015

Department of Energy. (n.d.). *Business Energy Investment Tax Credit*. Retrieved October 26, 2018, from https://www.energy.gov/savings/business-energy-investment-tax-credit-itc

Department of Regulatory Agencies. (2014). *Mission and History. About Us.* Retrieved from http://cdn.colorado.gov/cs/Satellite/DORA- PUC/CBON/DORA/1251627010991

Devenyi, R., & Mladenova, I. (2012). *International Markets for Renewable Energy Certificates (RECs). Sustainable Real Estate Roundtable: Member Briefing.* Retrieved from http://sustainround.com/library/sites/default/files/SRER_Member%20Briefing_International%20Markets%20for%20Renewable%20Energy%20Certificates_2012-07-16.pdf

Dewald, U., & Truffer, B. (2012). The local sources of market formation: Explaining regional growth differentials in German photovoltaic markets. *European Planning Studies, 20*(3), 397–420. doi:10.1080/09654313.2012.651803

Di Francia, G. (2013). The impact of recycling policies on the photovoltaic Levelized Cost of the Electricity. In *Proceedings of the 2013 International Conference on Renewable Energy Research and Applications (ICRERA)*. Piscataway, NJ: IEEE 10.1109/ICRERA.2013.6749894

Díaz-González, F., Sumper, A., Gomis-Bellmunt, O., & Villafáfila-Robles, R. (2012). A review of energy storage technologies for wind power applications. *Renewable & Sustainable Energy Reviews*, *16*(4), 2154–2171. doi:10.1016/j.rser.2012.01.029

Dincer, F. (2011). The analysis on photovoltaic electricity generation status, potential and policies of the leading countries in solar energy. *Renewable & Sustainable Energy Reviews*, *15*(1), 713–720. doi:10.1016/j.rser.2010.09.026

Dincer, I. (2000). Renewable energy and sustainable development: A crucial review. *Renewable & Sustainable Energy Reviews*, *4*(2), 157–175. doi:10.1016/S1364-0321(99)00011-8

Ding, L.-X., Zheng, F.-L., Wang, J.-W., Li, G.-R., Wang, Z.-L., & Tong, Y.-X. (2012). Super-large dendrites composed of trigonal PbO 2 nanoplates with enhanced performances for electrochemical devices. *Chemical Communications*, *48*(9), 1275–1277. doi:10.1039/C2CC15271A PMID:22179048

Dolara, A., Leva, S., & Manzolini, G. (2015b). Comparison of Different Physical Models for PV Power Output Prediction. *Solar Energy*, *119*, 83–99. doi:10.1016/j.solener.2015.06.017

Dönitz, W., Dietrich, G., Erdle, E., & Streicher, R. (1988). Electrochemical high temperature technology for hydrogen production or direct electricity generation. *International Journal of Hydrogen Energy*, *13*(5), 283–287. doi:10.1016/0360-3199(88)90052-3

Dorrell, D. G., Knight, A. M., Popescu, M., Evans, L., & Staton, D. A. (2010). *Comparison of different motor design drives for hybrid electric vehicles.* Paper presented at the Energy Conversion Congress and Exposition (ECCE). 10.1109/ECCE.2010.5618318

Dou, C., Zhang, Z., Yue, D., & Zheng, Y. (2017). MAS-Based Hierarchical Distributed Coordinate Control Strategy of Virtual Power Source Voltage in Low-Voltage Microgrid. *IEEE Access: Practical Innovations, Open Solutions*, *5*, 11381–11390. doi:10.1109/ACCESS.2017.2717493

Dovers, S. R., & Handmer, J. W. (1992). Uncertainty, sustainability and change. *Global Environmental Change*, *2*(4), 262–276. doi:10.1016/0959-3780(92)90044-8

Downey, L. (2015). *Inequality, Democracy and the Environment*. New York, NY: NYU Press. doi:10.18574/nyu/9781479850723.001.0001

Drexhage, J., & Murphy, D. (2010). *Sustainable Development: from Brundtland to Rio 2012*. United Nations, Background Paper. Retrieved June 2015 from http://www.un.org/wcm/webdav/site/ climatechange/shared/gsp/docs/GSP1-6_Background%20on%20Sustainable%20Devt.pdf

Dubal, D., Holze, R., & Kulal, P. (2013). Enhanced supercapacitive performances of hierarchical porous nanostructure assembled from ultrathin MnO2 nanoflakes. *Journal of Materials Science*, *48*(2), 714–719. doi:10.100710853-012-6783-6

Duran, P., Kammerlander, N., Van Essen, M., & Zellweger, T. (2016). Doing more with less: Innovation input and output in family firms. *Academy of Management Journal*, *59*(4), 1224–1264. doi:10.5465/amj.2014.0424

Dutra, R. M., & Szklo, A. S. (2008). Incentive policies for promoting wind power production in Brazil: Scenarios for the Alternative Energy Sources Incentive Program (PROINFA) under the New Brazilian electric power sector regulation. *Renewable Energy*, *33*(1), 65–76. doi:10.1016/j.renene.2007.01.013

Dutton, J. A. (1976). *The Ceaseless wind: an Introduction to the Theory of Atmospheric Motion*. New York: McGraw-Hill.

EASA (European Airline Safety Association). (2016). European Aviation. *Environment Reporter, 2016*. Available online https://ec.europa.eu/transport/sites/transport/files/european-aviation-environmental-report-2016-72dpi.pdf

EC - European Commission. (2013). *Report from the Commission to the European Parliament, the Council, the European Economic and Social Committee and the Committee of the Regions* (SWD-175 final). Accessed October 24, 2018, from: http://ec.europa.eu/transparency/regdoc/rep/1/2013/EN/1-2013-175-EN-F1-1.Pdf

EC - European Commission. (2016). *Proposal for a Directive of the European Parliament and of the Council on the promotion of the use of energy from renewable sources*. Accessed October 20, 2018 from: http://eur-lex.europa.eu/legal-content/EN/ALL/?uri=CELEX%3A52016PC0767R%2801%29

EC - European Commission. (2018a). *Europe Leads the Global Clean Energy Transition: Commission Welcomes Ambitious Agreement on Further Renewable Energy Development in the EU* (EC-Statement). Accessed on February 8, 2019 from: http://europa.eu/rapid/press-release_STATEMENT-18-4155_en.htm

EC - European Commission. (2018b). *EU Energy in Figures: Energy Statistical Pocketbook*. Accessed October 25, 2018 from: https://ec.europa.eu/energy/en/data/energy-statistical-pocketbook

Ecotec Research Consulting Ltd. (2001). *Aphrodite Mourelatou European Environment Agency, "Renewable energies: success stories*. Copenhagen: European Environment Agency.

Edenhofer, O., Pichs-Madruga, R., Sokona, Y., Seyboth, K., Matschoss, P., Kadner, S., … von Stechow, C. (Eds.). (2011). IPCC special report on renewable energy sources and climate change mitigation. Cambridge, UK: Cambridge University Press.

Edison Electric Utilities. (2018). *Our Members*. Retrieved from http://www.eei.org/about/members/Pages/default.aspx

Egelston, A. E. (2013). *Sustainable development: A history*. New York, NY: Springer. doi:10.1007/978-94-007-4878-1

Eisenhardt, K. M. (1989). Building theories from case study research. *Academy of Management Review*, *14*(4), 532–550. doi:10.5465/amr.1989.4308385

Eisenhardt, K. M., & Graebner, M. E. (2007). Theory building from cases: Opportunities and challenges. *Academy of Management Journal*, *50*(1), 25–32. doi:10.5465/amj.2007.24160888

Elamathi, S., Nithyakalyani, G., Sangeetha, D., & Ravichandran, S. (2008). Preparation and evaluation of ionomeric membranes based on sulfonated-poly (styrene_isobutylene_styrene) membranes for proton exchange membrane fuel cells (PEMFC). *Ionics*, *14*(5), 377–385. doi:10.100711581-007-0163-2

Elliott, D. (2013). *A review of sustainable energy supply options*. Philadelphia: IOP Publishing. doi:10.1088/978-0-750-31040-6

Eltawil, M. A., & Zhao, Z. (2010). Grid-connected Photovoltaic Power Systems. Technical and Potential Problems-A Review. *Renewable & Sustainable Energy Reviews*, *14*(1), 112–129. doi:10.1016/j.rser.2009.07.015

Engelken, M., Römer, B., Drescher, M., Welpe, I. M., & Picot, A. (2016). Comparing drivers, barriers, and opportunities of business models for renewable energies: A review. *Renewable & Sustainable Energy Reviews*, *60*, 795–809.

Environmental Defense Fund. (2012). Strategic Plan: Leading Transformational Change: Strategic Plan 2010-2014. *Environmental Defense Fund*. Retrieved from http://www.edf.org/sites/default/files/10585_strategic_plan_2010-2014_0.pdf

Environmental Protection Agency. (2012). How Does Electricity Affect the Environment? *Environmental Protection Agency*. Retrieved from http://www.epa.gov/cleanenergy/energy-and-you/affect/index.html

Environmental Protection Agency. (2018). Sources of Green House Gas emissions. *Environmental Protection Agency*. Retrieved from https://www.epa.gov/ghgemissions/sources-greenhouse-gas-emissions

EP - European Parliament. (2009). *Directive 2009/28/EC of the European Parliament and of the Council of 23 April 2009 on the promotion of the use of energy from renewable sources and amending and subsequently repealing Directives 2001/77/EC and 2003/30/EC. EUR-Lex*. Accessed on October 25, 2018 from: https://eur-lex.europa.eu/legal-content/EN/ALL/?uri=CELEX:32009L0028

Epstein, P. R., Buonocore, J. J., Eckerle, K., Hendryx, M., Stout, B. M. III, Heiberg, R., ... Glustron, L. (2011). Mining Coal, Mounting Costs: The Life Cycle Consequences of Coal. *Annals of the New York Academy of Sciences*, *1219*(1), 73–98. doi:10.1111/j.1749-6632.2010.05890.x PMID:21332493

Eseye, A., Zhang, J., & Zheng, D. (2018). Short-term Photovoltaic Solar Power Forecasting using a Hybrid Wavelet-PSO-SVM Model Based on SCADA and Meteorological Information. *Elsevier, Renewable. Energy Journal*, *118*, 357–367.

European Commission. (2001). *Green Paper: Promoting a European Framework for Corporate Social Responsibility. 28 Official Journal C 340*. Brussels: Treaty on European Union.

European Commission. (2002). Communication - Corporate Social Responsibility: A Business Contribution to Sustainable Development. COM(2002) 347 final, Brussels.

Eurostat, European Commission. (2014). Smarter, Greener, More Inclusive? - Indicators to support the Europe 2020 strategy. In I. Savova (Ed.), *Climate change and energy* (pp. 73-91). Accessed on October 26, 2018 at: https://ec.europa.eu/eurostat/documents/3217494/5777461/KS-02-13-238-EN.PDF/1a6fa7e5-85b7-40aa-987e-6a6d049ad723

Eurostat, European Commission. (2018). *Share of renewable energy in gross final energy consumption by sector.* Accessed October 25, 2018 from: https://ec.europa.eu/eurostat/web/products-datasets/-/sdg_07_40

Evans, A., Strezov, V., & Evans, T. J. (2012). Assessment of utility energy storage options for increased renewable energy penetration. *Renewable & Sustainable Energy Reviews*, *16*(6), 4141–4147. doi:10.1016/j.rser.2012.03.048

Eves, D. (2010). Let the People Vote on Xcel Franchise [Opinion]. *The Daily Camera.* Retrieved from LexisNexis Academic.

EY. (2015). *The YieldCo structure - Unlocking the value in power generation assets.* EY.

Eyer, J., & Corey, G. (2010). Energy storage for the electricity grid: Benefits and market potential assessment guide. *Sandia National Laboratories*, *20*(10), 5.

Fagiani, R., Barquin, J., & Hakvoort, R. (2013). Risk-based assessment of the cost-efficiency and the effectivity of renewable energy support schemes: Certificate markets versus feed-in-tariffs. *Energy Policy*, *55*, 648–661. doi:10.1016/j.enpol.2012.12.066

Fanco, B. (2005). *PEM Fuel Cells: Theory and Practices, Academic press sustainable world series*. Elsevier Academic Press.

Faria, H. Jr, Trigoso, F. B. M., & Cavalcanti, J. A. M. (2017). Review of distributed generation with photovoltaic grid connected systems in Brazil: Challenges and prospects. *Renewable & Sustainable Energy Reviews*, *75*, 469–475. doi:10.1016/j.rser.2016.10.076

Federal Foreign Office. (n.d.). *The German Energiewende.* Retrieved from https://www.auswaertiges-amt.de/blob/610620/5d9bfec0ab35695b9db548d10c94e57d/the-german-energiewende-data.pdf

Federal Ministry of Economic Affairs and Energy. (n.d.). *Information portal Renewable Energies.* Retrieved October 26, 2018, from https://www.erneuerbare-energien.de/EE/Navigation/DE/Recht-Politik/Das_EEG/das_eeg.html

FEMA. (n.d.). Retrieved from https://www.fema.gov/

Ferguson, R. S., & Lovell, S. T. (2013, August). *Engaging ecological literacy from the ground up: The influence and impact of the permaculture movement.* Paper presented at 98th Ecological Society of America Annual Meeting: Sustainable pathways: Learning from the past and shaping the future, Minneapolis, MN.

Ferguson, R. S., & Lovell, S. T. (2013, April). Permaculture for agroecology: Design, movement, practice, and worldview. A review. *Agronomy for Sustainable Development, 34*(2), 251–274. doi:10.100713593-013-0181-6

Ferguson, R. S., & Lovell, S. T. (2015, December). Grassroots engagement with transition to sustainability diversity and modes of participation in the international permaculture movement. *Ecology and Society, 20*(4), 39. doi:10.5751/ES-08048-200439

Fink, S., Mudd, C., Porter, K., & Morgernstern, B. (2009). *Wind Energy Curtailment Case Studies, May 2008 – May 2009.* Retrieved from https://www.nrel.gov/docs/fy10osti/46716.pdf

Fisher, D. R., & Freudenburg, W. R. (2001). Ecological modernization and its critics: Assessing the past and looking toward the future. *Society & Natural Resources, 14*(8), 701–709. doi:10.1080/08941920152524891

Fletcher, S. (2011). *Bottled lightning: superbatteries, electric cars, and the new lithium economy.* Hill and Wang.

FOCAC (The Forum on China-Africa Co-operation). (2016). *The Forum on China-Africa Cooperation Johannesburg Action Plan (2016-2018).* Retrieved from http://www.focac.org/eng/ltda/dwjbzjjhys_1/t1327961.htm

Focken, U., Lange, M., & Waldl, H. P. (2001). Previento – A Wind Power Prediction System with an Innovative Upscaling Algorithm. *Proceedings of the European Wind Energy Conference,* 826-829.

Foxon, T., & Pearson, P. (2008). Overcoming barriers to innovation and diffusion of cleaner technologies: Some features of a sustainable innovation policy regime. *Journal of Cleaner Production, 16*(1), 148–161.

Frangoul, A. (2018, April 05). *Google says it's biggest corporate buyer of renewable energy on the planet.* Retrieved from https://www.cnbc.com/2018/04/05/google-says-its-the-biggest-corporate-buyer-of-renewable-energy-on-the-planet.html

Frederickson, G. H., Smith, K. B., Larimer, C. W., & Licari, M. J. (2012). *The Public Administration Theory Primer* (2nd ed.). Boulder, CO: Westview Press.

FS-UNEP & BNEF. (2018). *Global trends in renewable energy investment 2018.* Author.

Fu, R., Feldman, D., Margolis, R., Woodhouse, M., & Ardani, K. (2017). U.S. Solar Photovoltaic System Cost Benchmark: Q1 2017. *National Renewable Energy Laboratory.* Retrieved from: https://www.nrel.gov/docs/fy17osti/68925.pdf

Gaddy, B., Sivaram, V., & O'Sullivan, F. (2016). *Venture capital and cleantech: The wrong model for clean energy innovation.* Author.

Gahleitner, G. (2013). Hydrogen from renewable electricity: An international review of power-to-gas pilot plants for stationary applications. *International Journal of Hydrogen Energy, 38*(5), 2039-2061.

Gambhir, A., Sant, G., & Deshmukh, R. (2011). India's Solar Mission: Procurement and Auctions. *Economic and Political Weekly*, *46*(28), 22–26.

Gamula, G., Hui, L., & Peng, W. (2013). Development of renewable energy technologies in Malawi. *International Journal of Renewable Energy Technology*, *2*(2), 44–52.

Garaud, R., Jain, S., & Tuertscher, P. (2008). Incomplete by design and designing for incompleteness. *Organization Studies*, *29*(3), 351–371. doi:10.1177/0170840607088018

Garcia-Hernandez, A., Wilcox, M., & Moore, T. (2010) Hydraulic modelling and simulation of pumping systems. *Proceedings of the Twenty-sixth International Pump Users Symposium*, 81-88.

Gething, B., & Puckett, K. (2013). *Design for climate change*. Royal Institute of British Architecture.

Getz, D., & Carlsen, J. (2000). Characteristics and goals of family and owner-operated businesses in the rural tourism and hospitality sectors. *Tourism Management*, *21*(6), 547–560. doi:10.1016/S0261-5177(00)00004-2

Getz, D., & Petersen, T. (2005). Growth and profit-oriented entrepreneurship among family business owners in the tourism and hospitality industry. *International Journal of Hospitality Management*, *24*(2), 219–242. doi:10.1016/j.ijhm.2004.06.007

Gharavi, H., & Ghafurian, R. (2011). Smart grid: The electric energy system of the future. *Proceedings of the IEEE*, *99*(6), 917–921. doi:10.1109/JPROC.2011.2124210

Ghofrani, M., & Alolayan, M. (2018). Time Series and Renewable Energy Forecasting. *IntechOpen*, *10*, 5772.

Ghosh, P., Dhole, C. K., Ganguly, S., Banerjee, D., & Kargupta, K. (2018). Phosphosilicate gel-sulfonated poly(ether ether ketone) nanocomposite membrane for polymer electrolyte membrane fuel cell. *Materials Today: Proceedings*, *5*, 2186–2192.

Giebel, G., Brownsword, R., Kariniotakis, G., Denhard, M., & Draxl, C. (2011). The State-Of-The-Art in Short-Term Prediction of Wind Power: A Literature Overview (2nd ed.). ANEMOS.plus.

Giebel, G., Landberg, L., Nielsen, T. S., & Madsen, H. (2002). The Zephyr Project–The Next Generation Prediction System. In *Proceedings of the 2001 European Wind Energy Conference, EWEC* (*Vol. 1*, pp. 777-780). Academic Press.

Gladwin, T. N., Kennelly, J. J., & Krause, T. S. (1995). Shifting paradigms for sustainable development: Implications for management theory and research. *Academy of Management Review*, *20*(4), 874–907. doi:10.5465/amr.1995.9512280024

Glanzberg, J. (n.d.). *An open letter and plea to the permaculture community.* Retrieved from http://patternmind.org/an-open-letter-and-plea-to-the-permaculture-community/

Glemarec, Y. (2012). Financing off-grid sustainable energy access for the poor. *Energy Policy*, *47*(S1), 87–93.

Global C. C. S. Institute. (n.d.). *PROINFA*. Retrieved from https://hub.globalccsinstitute.com/publications/analysis-regulatory-framework-wind-power-generation-brazil-summary-report/1-proinfa

Goetzberger, A., Hebling, C., & Schock, H.-W. (2003). Photovoltaic materials, history, status and outlook. *Materials Science and Engineering R Reports*, *40*(1), 1–46. doi:10.1016/S0927-796X(02)00092-X

GoM (Government of Malawi). (2015). *Republic of Malawi Intended Nationally Determined Contribution.* Environmental Affairs Department, Lilongwe. Retrieved from http://www4.unfccc.int/submissions/INDC/Published%20Documents/Malawi/1/MALAWI%20INDC%20SUBMITTED%20TO%20UNFCCC%20REV%20pdf.pdf

Gomez-Mejıa, L. R., Haynes, K. T., Nunez-Nickel, M., Jacobson, K. J., & Moyano-Fuentes, J. (2007). Socioemotional wealth and business risks in family-controlled firms: Evidence from Spanish olive oil mills. *Administrative Science Quarterly*, *52*(1), 106–137. doi:10.2189/asqu.52.1.106

Gottlieb, R. (2005). *Forcing the Spring: The Transformation of the American Environmental Movement. NW.* Washington, DC: Island Press.

Gould, K. A., Pellow, D. N., & Schnaiberg, A. (2004). Interrogating the Treadmill of Production: Everything You Wanted to Know About the Treadmill but were Afraid to Ask. *Organization & Environment*, *17*(3), 296–316. doi:10.1177/1086026604268747

Government of India. (n.d.). *National Action Plan for Climate Change*. Retrieved from http://www.moef.nic.in/downloads/home/Pg01-52.pdf

Government of the Republic of Kiribati. (2009). *Kiribati National Energy Policy*. Retrieved from http://www.mfed.gov.ki/sites/default/files/Kiribati%20National%20Energy%20Policy.pdf

Government of the Republic of Kiribati. (2015). *Intended Nationally Determined Contribution*. Retrieved from http://www4.unfccc.int/Submissions/INDC/Published

Grau, T., Huo, M., & Neuhoff, K. (2012). Survey of photovoltaic industry and policy in Germany and China. *Energy Policy*, *51*, 20–37. doi:10.1016/j.enpol.2012.03.082

Greenlee, B. (2008). People Power [Opinion]. *The Daily Camera.* Retrieved from LexisNexis Academic.

Greenlee, B. (2010). Power to the People [Opinion]. *The Daily Camera.* Retrieved from LexisNexis Academic.

GRI. (2018). *About GRI*. Available online: https://www.globalreporting.org/Information/about-gri/Pages/default.aspx

Gronhovd, D. (2018). Xcel's Electric Resource Plan Conflicts with Boulder's Climate Goals [Opinion]. *The Daily Camera.* Retrieved from LexisNexis Academic.

Growing Home. (n.d.). Retrieved from http://growinghomeinc.org/about-us/

GTM Research. (2018). *Trends in Solar Technology and System Prices*. Presented at GTM 2018 Solar Summit Breakfast Briefing.

Gujba, H., Thorne, S., Mulugetta, Y., Rai, K., & Sokona, Y. (2012). Financing low carbon energy access in Africa. *Energy Policy*, *47*(S1), 71–78.

Gumbo, T. (2014). Scaling up sustainable renewable energy generation from municipal solid waste in the African continent: Lessons from eThekwini, South Africa. *Journal of Sustainable Development*, *12*(1), 46–62.

Gupta, S., Beninger, S., & Ganesh, J. (2015). A hybrid approach to innovation by social enterprises: Lessons from Africa. *Social Enterprise Journal*, *11*(1), 89–112.

Gutru, R., & Santoshkumar, D. B. (2018). Amino acid functionalized graphene oxide based nanocomposite membrane electrolytes for direct methanol fuel cells. *Journal of Membrane Science*, *551*, 1–11. doi:10.1016/j.memsci.2018.01.026

Haas, R., Lettner, G., Auer, H., & Duic, N. (2013). The looming revolution: How photovoltaics will change electricity markets in Europe fundamentally. *Energy*, *57*, 38–43. doi:10.1016/j.energy.2013.04.034

Hagmann, C., Semeijn, J., & Vellenga, D. B. (2015). Exploring the green image of airlines: Passenger perceptions and airline choice. *Journal of Air Transport Management*, *43*, 37–45. doi:10.1016/j.jairtraman.2015.01.003

Hahn, E. (2014). *A permaculture primer.* Retrieved from https://midwestpermaculture.com/wordpress/wp-content/uploads/2015/12/Midwest-Permaculture-Permaculture-Primer.pdf

Hallegatte, S., Bangalore, M., Bonzanigo, L., Fay, M., Narloch, U., & Rozenberg, J. (2014). *Climate change and poverty: An analytical framework.* Policy Research Working Paper 7126. Washington, DC: World Bank Group.

Hallegatte, S., Bangalore, M., Bonzanigo, L., Fay, M., Kane, T., Narloch, U., ... Vogt-Schilb, A. (2016). *Shock waves: Managing the impacts of climate change on poverty. Climate Change and Development Series.* Washington, DC: World Bank.

Hall, J. K., Daneke, G. A., & Lenox, M. J. (2010). Sustainable development and entrepreneurship: Past contributions and future directions. *Journal of Business Venturing*, *25*(5), 439–448. doi:10.1016/j.jbusvent.2010.01.002

Hall, J., & Vredenburg, H. (2003). The Challenges of Innovating for Sustainable Development. *MIT Sloan Management Review*, *45*(1), 61–68.

Hall, P. J., & Bain, E. J. (2008). Energy-storage technologies and electricity generation. *Energy Policy*, *36*(12), 4352–4355. doi:10.1016/j.enpol.2008.09.037

Hamilton, B. A., Pullins, S., Miller, J., Renz, B., & Hanley, M. (2010). Smart grid principal characteristic enables new products, services, and markets. *US DOE/NETL*. Retrieved from http://www.smartgridinformation. info/pdf/1267_doc_1.pdf

Hamilton, C., Gemenne, F., & Bonneuil, C. (2015). *The anthropocene and the global environmental crisis: rethinking modernity in a new epoch*. London: Routledge. doi:10.4324/9781315743424

Hamin, E. M., & Gurran, N. (2009). Urban form and climate change: Balancing adaptation and mitigation in the US and Australia. *Habitat International*, *33*(3), 238–245. doi:10.1016/j.habitatint.2008.10.005

Hammer, A., Heinemann, D., Hoyer, C., Kuhlemann, R., Lorenz, E., Müller, R., & Beyer, H. G. (2003). Solar Energy Assessment using Remote Sensing Technologies. *Elsevier. Remote Sensing of Environment*, *86*(3), 423–432. doi:10.1016/S0034-4257(03)00083-X

Han, H., Yu, J., & Kim, W. (2019). Environmental corporate social responsibility and the strategy to boost the airline's image and customer loyalty intentions. *Journal of Travel & Tourism Marketing*, 1–13.

Hanna, M., & Nozik, A. (2006). Solar conversion efficiency of photovoltaic and photoelectrolysis cells with carrier multiplication absorbers. *Journal of Applied Physics*, *100*(7), 074510. doi:10.1063/1.2356795

Hansen, S. (2012). *Cultivating the Grassroots: A Winning Approach for Environment and Climate Funders*. National Committee for Responsive Philanthropy.

Hartman, B, Tayer, J. & Toor, W. (2017). Time to Change Course on Muni [Opinion]. *The Daily Camera*. Retrieved from LexisNexis Academic.

Hartman, B. (2013). For Real Climate Change Impact, work with Xcel [Guest column]. *The Daily Camera*. Retrieved from http://www.dailycamera.com/guest- opinions/ci_24340299/guest-column-real-climate-change-impact-work-xcel

Hart, S. L., & Milstein, M. B. (1999). Global Sustainability and the Creative Destruction of Industries. *MIT Sloan Management Review*, *41*(1), 23–33.

Harvey, G., Williams, K., & Probert, J. (2013). Greening the airline pilot: HRM and the green performance of airlines in the UK. *International Journal of Human Resource Management*, *24*(1), 152–166. doi:10.1080/09585192.2012.669783

Ha, S., Hale, T., & Ogden, P. (2016). Climate finance in and between developing countries: An emerging opportunity to build on. *Global Policy*, *7*(1), 102–108.

Hathaway, M. D. (2015). Agroecology and permaculture: Addressing key ecological problems by rethinking and redesigning agricultural systems. *Journal of Environmental Studies and Sciences*, *6*(2), 239–250. doi:10.100713412-015-0254-8

Heck, R. K., & Stafford, K. (2001). The vital institution of family business: Economic benefits hidden in plain sight. In G. K. McCann & N. Upton (Eds.), Destroying myths and creating value in family business (pp. 9–17). Deland, FL: Stetson University.

Heck, R. K., & Trent, E. S. (1999). The prevalence of family business from a household sample. *Family Business Review*, *12*(3), 209–219. doi:10.1111/j.1741-6248.1999.00209.x

Heinemann, D., Lorenz, E., & Girodo, M. (2006). Forecasting of Solar Radiation. Solar Energy Resource Management for Electricity Generation from Local Level to Global Scale, 223–233.

Hemenway, T. (2009). *Gaia's garden*. White River Junction, VT: Chelsea Green Publishing.

Hemmati, R. (2018). Optimal cogeneration and scheduling of hybrid hydro-Thermal-wind-solar system incorporating energy storage systems. *Journal of Renewable and Sustainable Energy*, *10*(1), 014102. doi:10.1063/1.5017124

Hengge, K., Heinzl, C., Perchthaler, M., Varley, D., Lochner, T., & Scheu, C. (2017). Unraveling micro- and nanoscale degradation processes during operation of high-temperature polymer-electrolyte-membrane fuel cells. *Journal of Power Sources*, *364*, 437–448. doi:10.1016/j.jpowsour.2017.08.042

Henrich, C., & Waissbein, O. (2017). *Belarus: Derisking renewable energy investment - Selecting public instruments to promote wind energy investment in Belarus*. Minsk: GEF.

Hernadi, P. (1987). Literary interpretation and the rhetoric of the human sciences. In J. S. Nelson, A. Megill, & D. N. McCloskey (Eds.), The rhetoric of the human sciences (pp. 263–275). Madison, WI: University of Wisconsin Press.

Hesse, H. C., Martins, R., Musilek, P., Naumann, M., Truong, C. N., & Jossen, A. (2017). Economic optimization of component sizing for residential battery storage systems. *Energies*, *10*(7), 835. doi:10.3390/en10070835

He, Y., Xu, Y., Pang, Y., Tian, H., & Wu, R. (2016). A regulatory policy to promote renewable energy consumption in China: Review and future evolutionary path. *Renewable Energy*, *89*, 695–705. doi:10.1016/j.renene.2015.12.047

Hill, D. C., McMillan, D., Bell, K. R. W., & Infield, D. (2012). Application of Auto-regressive Models to U.K. Wind Speed Data for Power System Impact Studies. *IEEE Transactions on Sustainable Energy*, *3*(1), 134–141. doi:10.1109/TSTE.2011.2163324

Hinnen, G., Hille, S. L., & Wittmer, A. (2017). Willingness to Pay for Green Products in Air Travel: Ready for Take-Off? *Business Strategy and the Environment*, *26*(2), 197–208. doi:10.1002/bse.1909

Hiteva, R. (2009, August). *Untangling the puzzle of energy policy in Bulgaria.* Paper presented at Energy Vulnerability and Urban Transitions, University of Manchester, UK. Accessed on November 3, 2014 at: http://urban-energy.org/2013/08/29/guest-contribution-untangling-the-puzzle-of-energy-policy-in-bulgaria

Hiteva, R., & Maltby, T. (2017). Hitting the target but missing the point: Failing and succeeding in the Bulgarian energy sector. In I. Solorio & H. Jorgens (Eds.), *A guide to EU renewable energy policy: Comparing Europeanization and domestic policy change in EU member states* (pp. 224–246). Cheltenham, UK: Edward Elgar Publishing. doi:10.4337/9781783471560.00022

Hodge, B., Martinez-Anido, C., Wang, Q., Chartan, E., Florita, A., & Kiviluoma, J. (2018). The Combined Value of Wind and Solar Power Forecasting Improvements and Electricity Storage. *Applied Energy*, *214*, 1–15. doi:10.1016/j.apenergy.2017.12.120

Hogarth, J. R. (2012). Promoting diffusion of solar lanterns through microfinance and carbon finance: A case study of FINCA-Uganda's solar loan programme. *Energy for Sustainable Development*, *16*(4), 430–438.

Horio, B. M., Kumar, V., Levin, D. J., & Sung, P. E. (2016). Modeling Carbon Tax Policy Impacts on US Commercial Airlines using Agent-Based Modeling and Crowdsourced Data. In *AIAA Modeling and Simulation Technologies Conference* (p. 4302). AIAA.

Hove, H. (2004). Critiquing sustainable development: A meaningful way of mediating the development impasse? *Undercurrent*, *1*(1), 48–54.

Howarth, R. W., Santoro, R., & Ingraffea, A. (2011). Methane and the greenhouse-gas footprint of natural gas from shale formations. *Climatic Change*, *106*(4), 679–690. doi:10.100710584-011-0061-5

Howlett, M. (2014). Why are policy innovations rare and so often negative? Blame avoidance and problem denial in climate change policy-making. *Global Environmental Change*, *29*, 395–403.

Howorth, C., Rose, M., Hamilton, E., & Westhead, P. (2010). Family firm diversity and development: An introduction. *International Small Business Journal*, *28*(5), 437–451. doi:10.1177/0266242610373685

Huenteler, J., Schmidt, T. S., & Kanie, N. (2012). Japan's post Fukushima challenge – implications from the German experience on renewable energy policy. *Energy Policy*, *45*, 6–11. doi:10.1016/j.enpol.2012.02.041

Hussain, H., Javaid, N., Iqbal, S., Hasan, Q., Aurangzeb, K., & Alhussein, M. (2018). An Efficient Demand Side Management System with a New Optimized Home Energy Management Controller in Smart Grid. *Energies*, *11*(1), 190. doi:10.3390/en11010190

Hvelplund, F., Østergaard, A., & Meyer, N. (2017). Incentives and barriers for wind power expansion and system integration in Denmark. *Energy Policy*, *107*, 573–584. doi:10.1016/j.enpol.2017.05.009

Hydro Ottawa. (2018). *Business Rates*. Retrieved from: https://static.hydroottawa.com/documents/business/rates_brochures/rates_business_e.pdf

IEA (International Energy Agency). (2016). *Boosting the Power Sector in Sub-Saharan Africa: China's involvement*. Paris: IEA.

IEA. (2017). *Early-stage venture capital for energy innovation: Financing models, trends and implications for policy*. IEA.

IEA. (2018). *World energy investment 2018*. IEA.

IESO, Independent Electricity System Operator. (2018). *Data Directory*. Retrieved from: http://ieso.ca/en/Power-Data/Data-Directory

IIED. (2016). *Informality and inclusive green growth: Evidence from 'The biggest private sector' event 2016*. London: IIED.

Ilic, D., Da Silva, P. G., Karnouskos, S., & Griesemer, M. (2012). An energy market for trading electricity in smart grid neighbourhoods. *6th IEEE international conference on digital ecosystems technologies (DEST),* 1–6.

Ilyin, D., Shestopalova, T., Vaskov, A., & Ko, A. (2019). Strategies for Protection Systems of Wind Turbines with Doubly Fed Induction Generator: Control Strategies and Techniques for Fault Ride Through of Doubly Fed Induction Wind Generator. In Renewable Energy and Power Supply Challenges for Rural Regions (pp. 267-288). Hershey, PA: IGI Global.

Imran, R. J., Arockiados, T., Rajalakshmi, N., & Ramaprabhu, S. (2010). Nanostructured Pt dispersed on graphene multiwalled carbon nanotube hybrid nanomaterials as electrocatalyst for PEMFC. *Journal of the Electrochemical Society, 157*(6), B874–B879. doi:10.1149/1.3374353

Intergovernmental Panel on Climate Change (IPCC). (2018, October 6). *Global warming of 1.5° C* (UN Publication No. IPCC SR1.5). Retrieved from http://report.ipcc.ch/sr15/pdf/sr15_spm_final.pdf

Intergovernmental Panel on Climate Change. (2007). *Synthesis report, fourth assessment report*. International Panel on Climate Change and Cambridge University Press.

Intergovernmental Panel on Climate Change. (2015). *Climate change 2014: Mitigation of climate change*. Cambridge University Press.

International Energy Agency, International Renewable Energy Agency, United Nations Statistics Division, World Bank Group, & World Health Organisation. (2018). *Tracking SDG7: The Energy Progress Report*. Retrieved from http://www.irena.org/-/media/Files/IRENA/Agency/Publication/2018/May/SDG7_Tracking_report_executive_summary_2018.pdf

International Energy Agency. (2017a). *Energy Access Outlook 2017 – From Poverty to Prosperity. World Energy Outlook Special Report*. Retrieved from https://www.iea.org/publications/freepublications/publication/WEO2017SpecialReport_EnergyAccessOutlook.pdf

International Energy Agency. (2017b). *CO2 emissions from fuel combustion*. Retrieved from http://www.indiaenvironmentportal.org.in/files/file/CO2 Emissions from Fuel Combustion 2018 Highlights.pdf

International Energy Agency. (2019). *IEA/IRENA Global renewable energy policies and measures database*. Retrieved from https://www.iea.org/policiesandmeasures/renewableenergy/?country=Canada

International Energy Agency. (n.d.a). *Programme of Incentives for Alternative Electricity Sources – PROINFA*. Retrieved from https://www.iea.org/policiesandmeasures/pams/brazil/name-21963-en.php?s=&s=

International Energy Agency. (n.d.b) *Renewable Electricity Quota and Assessment Method.* Retrieved October 26, 2018, from https://www.iea.org/policiesandmeasures/pams/china/name-170554-en.php

International Renewable Energy Agency. (2016a). *The Power to Change: Solar and Wind Cost Reduction Potential to 2025.* Abu Dhabi: The International Renewable Energy Agency.

International Renewable Energy Agency. (2016b). *Letting in the Light: How Solar Photovoltaics Will Revolutionise the Electricity System.* Abu Dhabi: The International Renewable Energy Agency.

International Renewable Energy Agency. (2017a). *Renewable Energy Statistics 2017.* Abu Dhabi: The International Renewable Energy Agency.

International Renewable Energy Agency. (2017b). *Kiribati Integrated Energy Roadmap: 2017-2025.* Abu Dhabi: The International Renewable Energy Agency.

International Renewable Energy Agency. (2017c). *Trends in Photovoltaic Applications 2017.* Abu Dhabi: The International Renewable Energy Agency.

International Renewable Energy Agency. (2018). *Renewable Power Generation Costs in 2017.* Abu Dhabi: The International Renewable Energy Agency.

International Renewable Energy Agency. (n.d.). Retrieved September 24, 2018, from International Renewable Energy Agency: http://resourceirena.irena.org/gateway/dashboard/index.html

Iovine, A., Damm, G., De Santis, E., & Di Benedetto, M. D. (2017). Management Controller for a DC MicroGrid integrating Renewables and Storages. *IFAC-PapersOnLine*, *50*(1), 90–95. doi:10.1016/j.ifacol.2017.08.016

IPCC. (2015). *Climate change 2014 synthesis report: fifth assessment report.* Geneva: IPCC.

IRENA (International Renewable Energy Agency). (2015). *Roadmap for a renewable energy future.* Abu Dhabi: IRENA.

IRENA Abu Dhabi. (2015). *Renewable Energy Prospects: United States of America, REmap 2030 analysis.* Retrieved from www.irena.org/remap

IRENA. (2016). *Unlocking renewable energy investment: The role of risk mitigation and structured finance.* International Renewable Energy Agency.

IRENA. (2018). *Corporate Sourcing of Renewables: Market and Industry Trends.* IRENA.

ISA. (2014). *Information sharing environment-information interoperability framework (I2F)* (Version 0.5). Retrieved from http://www. ise.gov/sites/default/files/FINAL%20-%20ISE_I2F_v0%205.pdf

Iychettira, K., Hakvoort, R. A., & Linares, P. (2017). Towards a comprehensive policy for electricity from renewable energy: An approach for policy design. *Energy Policy*, *106*, 169–182. doi:10.1016/j.enpol.2017.03.051

James, H. S. Jr. (1999). What can the family contribute to business? Examining contractual relationships. *Family Business Review*, *12*(1), 61–71. doi:10.1111/j.1741-6248.1999.00061.x

Jang, R. J. (1991). Fuzzy Modeling Using Generalized Neural Networks and Kalman Filter Algorithm. *Proceedings of the 9th National Conference on Artificial Intelligence*, *14*, 762–767.

Jansen, A. N., Vaughey, J. T., Chen, Z., Zhang, L., & Brushett, F. R. (2016). *Organic non-aqueous cation-based redox flow batteries*. Google Patents.

Janssens-Maenhout, G., Crippa, M., Guizzardi, D., Muntean, M., Schaaf, E., … Schure, K.M. (2017). *Fossil CO2 and GHG emissions of all world countries*. Retrieved from http://edgar.jrc.ec.europa.eu/booklet2017/CO2_and_GHG_emissions_of_all_world_countries_booklet_online.pdf

Javadi, F. S., Rismanchi, B., Sarraf, M., Afshar, O., Saidur, R., Ping, H., & Rahim, A. (2013). Global policy of rural electrification. *Renewable & Sustainable Energy Reviews*, *19*(C), 402–416.

Javaid, N., Naseem, M., Rasheed, M. B., Mahmood, D., Khan, S. A., Alrajeh, N., & Iqbal, Z. (2017). A new heuristically optimized Home Energy Management controller for smart grid. *Sustainable Cities and Society*, *34*, 211–227. doi:10.1016/j.scs.2017.06.009

Jayachandran, M., & Ravi, G. (2017). Design and Optimization of Hybrid Micro-Grid System. *Energy Procedia*, *117*, 95–103. doi:10.1016/j.egypro.2017.05.111

Jeppesen, T., & Nielsen, L. (2003). Tradable Green Certificates in selected European countries - overview and assessment. *Energy Policy*, *31*(1), 3–14. doi:10.1016/S0301-4215(02)00112-X

Jevons, W. S. (1875). *The Coal Question: An Inquiry Concerning the Progress of the Nation, and the Probable Exhaustion of Our Coal Mines*. London, UK: Macmillan & Co.

Jiajun, W., Geping, Y., Yuyan, S., Zhenbo, W., & Yunzhi, G. (2008). Investigation of further improvement of platinum catalyst durability with highly graphitized carbon nanotubes support. *The Journal of Physical Chemistry C*, *112*(15), 5784–5789. doi:10.1021/jp800186p

Jick, T. D. (1979). Mixing qualitative and quantitative methods: Triangulation in action. *Administrative Science Quarterly*, *24*(4), 602–611. doi:10.2307/2392366

Jing, L., Guoxiao, X., Xingying, L., Jie, X., Zhao, L., & Weiwei, C. (2018). Effect of nano-size of functionalized silica on overall performance of swelling-filling modified Nafion membrane for direct methanol fuel cell application. *Applied Energy*, *213*, 408–414. doi:10.1016/j.apenergy.2018.01.052

Johnson, O., Muhoza, C., Osano, P., Senyagwa, J., & Kartha, S. (2017). *Catalysing investment in sustainable energy infrastructure in Africa: overcoming financial and non-financial constraints*. Stockholm Environment Institute Working Paper No. 2017-03. Nairobi: Stockholm Environment Institute.

Joint MDBs Report. (2018). *Joint Report on Multilateral Development Banks Climate Finance*. Author.

Jones, L. (2017). *Renewable energy integration: Practical management of variability, uncertainty, and flexibility in power grids* (2nd ed.). Academic Press, Elsevier Inc.

Jossen, A., Garche, J., & Sauer, D. U. (2004). Operation conditions of batteries in PV applications. *Solar Energy*, *76*(6), 759–769. doi:10.1016/j.solener.2003.12.013

Jurgen, G., & Ludwig, J. (2015). Applications of Fuel Cell Technology: Status and Perspectives. *The Electrochemical Society Interface*, 39–43.

Kabeer, N. (2018). *China Ends Wind Feed-In Tariffs and Opts for Auctions*. Retrieved October 26, 2018, from https://mercomindia.com/china-ends-wind-feed-in-tariffs/

Kaijage, E., Nyagawa, S., Best, S., Cosmas, R., Temba, S., Mtwanga, B., & Mahanga, N. (2017). *Money is Power: Tracking finance flows for decentralised energy access in Tanzania*. IIED Working Paper. London: The International Institute for Environment and Development (IIED).

Kamalov, T.S. (2014). *Chastotno-reguliruyemyy elektroprivod nasosnykh stantsiy sistem mashinnogo orosheniya* [Frequency-regulated electric drive of pumping stations of machine irrigation systems]. Tashkent: Fan.

Kaminker, C., & Stewart, F. (2012). *The role of institutional investors in financing clean energy*. Paris: OECD Publishing.

Kammerlander, N., Dessì, C., Bird, M., Floris, M., & Murru, A. (2015). The Impact of Shared Stories on Family Firm Innovation A Multicase Study. *Family Business Review*, *28*(4), 332–354. doi:10.1177/0894486515607777

Kandpal, T., & Broman, L. (2014). Renewable energy education: A global status review. *Renewable & Sustainable Energy Reviews*, *34*, 300–324. doi:10.1016/j.rser.2014.02.039

Karakezi, S., & Kithyoma, W. (2003). *Renewable energy in Africa: prospects and limits*. Paper presented at Republic of Senegal and United Nations Workshop for African Energy Experts on Operationalizing the NEPAD Energy Initiative, Dakar, Senegal.

Kariniotakis, G., Halliday, J., Brownsword, R., Marti, I., Palomares, A. M., Cruz, I., & Lange, M. (2006, February). Next Generation Short-Term Forecasting of Wind Power–Overview of the ANEMOS Project. In *European Wind Energy Conference, EWEC 2006*. Academic Press.

Katiraei, F., Iravani, M. R., & Lehn, P. W. (2005). Micro-grid autonomous operation during and subsequent to islanding process. *IEEE Transactions on Power Delivery*, *20*(1), 248–257. doi:10.1109/TPWRD.2004.835051

Kavasseri, R., & Seetharaman, K. (2009). Day-ahead Wind Speed Forecasting using f-ARIMA Models. *Renewable Energy*, *34*(5), 1388–1393. doi:10.1016/j.renene.2008.09.006

Keiren. (2013, October 17). The many benefits of hügelkultur. *Inspiration Green and Permaculture magazine*. Retrieved from https://www.permaculture.co.uk/articles/many-benefits-hugelkultur

Kelley, K. (2016). City Approves 'Carbon Tax' in Effort to Reduce Gas Emissions. *The New York Times*. Retrieved from https://www.nytimes.com/2006/11/18/us/18carbon.html

Kelly, R., Sirr, L., & Ratcliffe, R. (2004). Futures thinking to achieve sustainable development at local level in Ireland. *Foresight*, *6*(2), 80–90. doi:10.1108/14636680410537547

Kermani, M., & Morshed, A. (2003). Carbon dioxide corrosion in oil and gas production—a compendium. *Corrosion*, *59*(8), 659–683.

Khalilpour, R., & Vassallo, A. (2016). Planning and operation scheduling of PV-battery systems: A novel methodology. *Renewable & Sustainable Energy Reviews*, *53*, 194–208. doi:10.1016/j.rser.2015.08.015

Kharchenko, V., Panchenko, V., Tikhonov, P. V., & Vasant, P. (2018) Cogenerative PV thermal modules of different design for autonomous heat and electricity supply. In Handbook of Research on Renewable Energy and Electric Resources for Sustainable Rural Development (pp. 86-119). Hershey, PA: IGI Global. doi:10.4018/978-1-5225-3867-7.ch004

Khashimov, A.A. & Abidov, K.G. (2012) *Energoeffektivnyye sposoby samozapuska elektroprivodov nasosnykh stantsiy* [Energy-efficient ways to self-start electric drives of pumping stations]. Tashkent: Fan va texnologiya.

Khokhlov, A. V., Khokhlov, V. A., & Titova, J. O. (2018) Energy Saving and Safe Operating Modes of the Large Irrigative Pumping Stations. In Handbook of Research on Renewable Energy and Electric Resources for Sustainable Rural Development (pp. 176-203). Hershey, PA: IGI Global. doi:10.4018/978-1-5225-3867-7.ch008

Khokhlov, A.V., Khokhlov, V.A. & Titova, J.O. (2014) *Rezhimy raboty nasosnykh stantsiy Dzhizakskogo kaskada* [Operating modes of Djizak pumping stations stage. Monograph.]. Tashkent: Fan va texnologiya.

Khokhlov, V.A., Khokhlov, A.V. & Titova, J.O. (2011) *Regulirovaniye rezhimov raboty struynykh nasosov* [Regulation of operating modes of jet pumps. Monograph.]. Tashkent: Fan va texnologiya.

Khokhlov, A. V., Khokhlov, V. A., & Titova, J. O. (2015). *Rezhimy raboty nasosnykh stantsiy Karshinskogo kaskada* [Operating modes of Karshi pumping stations stage]. Tashkent: Navruz.

Kinzig, A., Ehrlich, P., Alston, L., Arrow, K., Barrett, S., Buchman, T., ... Saari, D. (2013). Social Norms and Global Environmental Challenges: The Complex Interaction of Behaviors, Values, and Policy. *Bioscience*, *63*(3), 164–175. doi:10.1525/bio.2013.63.3.5 PMID:25143635

Kirchoff, J. F., Tate, W. L., & Mollenkopf, D. A. (2016). The impact of strategic organizational orientations on green supply chain management and firm performance. *International Journal of Physical Distribution & Logistics Management*, *46*(3), 269–292. doi:10.1108/IJPDLM-03-2015-0055

Kiser, L., & Ostrom, E. (1982). The Three Worlds of Action: A Metatheoretical Synthesis of Institutional Approaches. In E. Ostrom (Ed.), *Strategies of Political Inquiry* (pp. 179–222). Beverly Hills, CA: Sage.

Kittner, N., Lill, F., Kammen, D. M. (n.d.). Energy storage deployment and innovation for the clean energy transition. *Nature Energy, 2.* Doi:10.1038/nenergy.2017.125

Kluger, J. (2018, October 8). Why we keep ignoring even the most dire climate change warnings. *Time.* Retrieved from http://time.com/5418690/why-ignore-climate-change-warnings-un-report/?utm_medium=social&xid=time_socialflow_facebook&utm_source=facebook.com&utm_campaign=time

Kondarev, G. (2013). Bulgaria: energy efficiency and renewable sources. Position of the Bulgarian Coalition for Sustainable use of the EU Funds and CEE Bankwatch network: Funding for clean energy through the Structural and Cohesion Funds for the programming period 2014-2020. *CEE: Bankwatch network.* Accessed September 9, 2014 from: http://bankwatch.org/sites/default/files/shadow-BG-EE-RES.pdf

König, A., Kammerlander, N., & Enders, A. (2013). The family innovator's dilemma: How family influence affects the adoption of discontinuous technologies by incumbent firms. *Academy of Management Review, 38*(3), 418–441. doi:10.5465/amr.2011.0162

Konrad, T. (2016, May 13). *The YieldCo boom and bust: The consequences of greed and a return to normalcy.* Retrieved from greentechmedia.com: https://www.greentechmedia.com/articles/read/the-yieldco-boom-and-bust-the-consequences-of-greed#gs.0D19Jx0

Kotlar, J., & De Massis, A. (2013). Goal setting in family firms: Goal diversity, social interactions, and collective commitment to family_centered goals. *Entrepreneurship Theory and Practice, 37*(6), 1263–1288. doi:10.1111/etap.12065

Kühnert, J., Lorenz, E., & Heinemann, D. (2013) Satellite-based Irradiance and Power Forecasting for the German Energy Market in Solar Energy Forecasting and Resource Assessment. Solar Energy Forecasting and Resource Assessment, 267-295.

Kurt, M. (2017). *Development of an Offshore Specific Wind Power Forecasting System* (PhD thesis). Kassel, Germany: Kassel University Press.

Kuzlu, M., & Pipattanasomporn, M. (2013). Assessment of communication technologies and network requirements for different smart grid applications. *Proceedings of the IEEE Innovative Smart Grid Technologies (ISGT) PES Conference, 1*(6), 24–7. 10.1109/ISGT.2013.6497873

Lake Turkana Wind Power. (2018, October 25). Retrieved from https://ltwp.co.ke/: https://ltwp.co.ke/

Lange, B. (2003). *Modelling the marine boundary layer for offshore wind power utilisation* (Doctoral dissertation). Universität Oldenburg.

Lange, B. (2003). Importance of Thermal Effects and Sea Surface Roughness for Offshore Wind Resource Assessment. *European Wind Energy Conference EWEC.*

Langer, K., Decker, T., & Menrad, K. (2017). Public participation in wind energy projects located in Germany: Which form of participation is the key to acceptance? *Renewable Energy, 112*, 63–73. doi:10.1016/j.renene.2017.05.021

Larsen, T. (2012). *Water hammer in pumped sewer mains*. Aalborg, Denmark: University of Aalborg, Department of Civil Engineering.

Larson, D., Nonnenmacher, L., & Coimbra, C. F. M. (2016). Day-ahead Forecasting of Solar Power Output from Photovoltaic Plants in the American Southwest. *Renewable Energy*, *91*, 11–20. doi:10.1016/j.renene.2016.01.039

Lasseter, R. H., & Paigi, P. (2004). Microgrid: A Conceptual Solution. *PESC Record - IEEE Annual Power Electronics Specialists Conference*, *6*, 4285–4290. 10.1109/PESC.2004.1354758

Lasseter, R., Akhil, A., Marnay, C., Stevens, J., Dagle, J., Guttromson, R., … Eto, J. (2002). White Paper on Integration of Distributed Energy Resources - The MicroGrid Concept. *Consortium for Electric Reliability Technology Solutions (CERTS) April*, 1–27.

Lau, L. C., Lee, K. T., & Mohamed, A. R. (2012). Global warming mitigation and renewable energy policy development from the Kyoto Protocol to the Copenhagen Accord – a comment. *Renewable & Sustainable Energy Reviews*, *16*(7), 5280–5284.

Le Breton-Miller, I., & Miller, D. (2006). Why do some family businesses out_compete? Governance, long-term orientations, and sustainable capability. *Entrepreneurship Theory and Practice*, *30*(6), 731–746. doi:10.1111/j.1540-6520.2006.00147.x

Le Breton-Miller, I., & Miller, D. (2008). To grow or to harvest? Governance, strategy and performance in family and lone founder firms. *Journal of Strategy and Management*, *1*(1), 41–56. doi:10.1108/17554250810909419

Lee, D. S., Fahey, D. W., Forster, P. M., Newton, P. J., Wit, R. C., Lim, L. L., ... Sausen, R. (2009). Aviation and global climate change in the 21st century. *Atmospheric Environment*, *43*(22-23), 3520–3537. doi:10.1016/j.atmosenv.2009.04.024

Lee, D. S., Pitari, G., Grewe, V., Gierens, K., Penner, J. E., Petzold, A., ... Iachetti, D. (2010). Transport impacts on atmosphere and climate: Aviation. *Atmospheric Environment*, *44*(37), 4678–4734. doi:10.1016/j.atmosenv.2009.06.005

Lee, K. C., Tsai, W. H., Yang, C. H., & Lin, Y. Z. (2017). An MCDM approach for selecting green aviation fleet program management strategies under multi-resource limitations. *Journal of Air Transport Management*, *68*, 76–85. doi:10.1016/j.jairtraman.2017.06.011

Lefebvre, D., & Tezel, F. H. (2017). A review of energy storage technologies with a focus on adsorption thermal energy storage processes for heating applications. *Renewable & Sustainable Energy Reviews*, *67*, 116–125. doi:10.1016/j.rser.2016.08.019

Lemaire, X., & Kerr, D. (2016). *Waste Management: Innovative solutions for municipalities. SAMSET Policy Brief*. London: UCL Energy Institute.

Leznov, B. S. (2006). *Energosberezhenie i reguliruemyy privod v nasosnykh i vozdushnykh ustanovkakh* [*Energy saving and adjustable drive for pump and blower sets*]. Moscow: Energoatomizdat.

Li, Ch., Ma, Sh., Liu, H., Wu, H., Gao, P., & Yang, F. (2016). *Research on scheme of pumping energy-saving based on computer control technology.* MATEC Web of Conferences. Retrieved March 21, 2017, from http://www.matec-conferences.org

Lima, C. (2018a). DLT/Blockchain Architecture and Reference Frameworks. *2018 IEEE NIST Global Blockchain Summit.*

Lima, C. (2018c). *IEEE P2418.5 Blockchain in Energy WG Standards.*

Lima, C. (2018b). Developing Open and Interoperable Distributed Ledger Technology (DLT)/ Blockchain Standards. *IEEE Special Publication, 51*, 106–111.

Linowes, L. (2017). *Denmark's Anti-Wind Problem: Wind News Update*. Retrieved October 27, 2018, from https://www.masterresource.org/uncategorized/wind-news-update-denmarks-anti-wind-problem/

Lin, Y., Kruger, U., Zhang, J., Wang, Q., Lamont, L., & El Chaar, L. (2015). Seasonal Analysis and Prediction of Wind Energy Using Random Forests and ARX Model Structures. *IEEE Transactions on Control Systems Technology, 23*(5), 1109. doi:10.1109/TCST.2015.2389031

Lipp, J. (2007). Lessons for effective renewable electricity policy from Denmark, Germany and the United Kingdom. *Energy Policy, 35*(11), 5481–5495. doi:10.1016/j.enpol.2007.05.015

Liu, N., Yu, X., Wang, C., Li, C., Ma, L., & Lei, J. (2017). Energy-Sharing Model with Price-Based Demand Response for Microgrids of Peer-to-Peer Prosumers. *IEEE Transactions on Power Systems, 32*(5), 3569–3583. doi:10.1109/TPWRS.2017.2649558

Liu, X., Zhou, D., Zhou, P., & Wang, Q. (2017). Dynamic carbon emission performance of Chinese airlines: A global Malmquist index analysis. *Journal of Air Transport Management, 65*, 99–109. doi:10.1016/j.jairtraman.2017.09.009

Loisos, G. (1999). *Day Lighting in Schools. An Investigation in the Relationship between Daylighting and Human Performance*. Pacific Gas and Electric Company.

Lorenz, E., Kühnert, J., Wolff, B., Hammer, A., Kramer, O., & Heinemann, D. (2014). PV Power Predictions on Different Spatial and Temporal Scales Integrating PV Measurements, Satellite Data and Numerical Weather Predictions. *29th EUPVSEC*, 22–26.

Lorenz, E., Heinemann, D., & Kurz, C. (2012). Local and regional photovoltaic power prediction for large scale grid integration: Assessment of a new algorithm for snow detection. *Progress in Photovoltaics: Research and Applications, 20*(6), 760–769. doi:10.1002/pip.1224

Lorenz, E., Heinemann, D., Wickramarathne, H., Beyer, H. G., & Bofinger, S. (2007). Forecast of ensemble power production by grid-connected PV systems. *20th European PV Conference.*

Lorenz, E., Hurka, J., Karampela, G., Heinemann, D., Beyer, H. G., & Schneider, M. (2008). Qualified forecast of ensemble power production by spatially dispersed grid-connected PV systems. *23rd European Photovoltaic Solar Energy Conference.*

Lorenz, E., Scheidsteger, T., Hurka, J., Heinemann, D., & Kurz, C. (2011). Regional PV Power Prediction for Improved Grid Integration. *Progress in Photovoltaics: Research and Applications*, *19*(7), 757–771. doi:10.1002/pip.1033

Louka, P., Galanis, G., Siebert, N., Kariniotakis, G., Katsafados, P., Pytharoulis, I., & Kallos, G. (2008). Improvements in Wind Speed Forecasts for Wind Power Prediction Purposes using Kalman filtering. *Journal of Wind Engineering and Industrial Aerodynamics*, *96*(12), 2348–2362. doi:10.1016/j.jweia.2008.03.013

Lu, N., Chow, J. H., & Desrochers, A. A. (2003). *Pumped-storage hydro-turbine bidding strategies in a competitive electricity market.* Paper presented at the Power Engineering Society General Meeting. 10.1109/PES.2003.1270415

Lumpkin, G. T., & Brigham, K. H. (2011). Long-term orientation and intertemporal choice in family firms. *Entrepreneurship Theory and Practice*, *35*(6), 1179–1197. doi:10.1111/j.1540-6520.2011.00495.x

Luna, J. M. (2014). *A mixed-methods study of sustainability education* (Doctoral dissertation).

Luna, J. M., Dávila, E. R., & Reynoso-Morris, A. (2018). Pedagogy of permaculture and food justice. *The Journal of Educational Foundations*, *31*(1 & 2), 57–85.

Luo, G., Dan, E., Zhang, X., & Guo, Y. (2018). Why the Wind Curtailment of Northwest China Remains High. *Sustainibility*, *10*(3), 570. doi:10.3390u10020570

Luthra, S., Kumar, S., Garg, D., & Haleem, A. (2015). Barriers to renewable/sustainable energy technologies adoption: Indian perspective. *Renewable & Sustainable Energy Reviews*, *41*, 762–776. doi:10.1016/j.rser.2014.08.077

Lynes, J. K., & Dredge, D. (2006). Going green: Motivations for environmental commitment in the airline industry. A case study of Scandinavian Airlines. *Journal of Sustainable Tourism*, *14*(2), 116–138. doi:10.1080/09669580608669048

Lynn, L. Jr, Heinrich, C. J., & Hill, C. J. (2000). Studying Governance and Public Management: Challenges and Prospects. *Journal of Public Administration: Research and Theory*, *10*(2), 233–262. doi:10.1093/oxfordjournals.jpart.a024269

MacDougall, H., Tomosk, S., & Wright, D. J. (2018). Geographic Maps of the Impact of Government Incentives on the Economic Viability of Solar Power. *Renewable Energy*, *122*, 497–506. doi:10.1016/j.renene.2017.12.108

Macintosh, A., & Wallace, L. (2009). International aviation emissions to 2025: Can emissions be stabilised without restricting demand? *Energy Policy*, *37*(1), 264–273. doi:10.1016/j.enpol.2008.08.029

Malesios, C., Skouloudis, A., Dey, P. K., Abdelaziz, F. B., Kantartzis, A., & Evangelinos, K. (2018). (in press). The impact of SME sustainability practices and performance on economic growth from a managerial perspective: Some modeling considerations and empirical analysis results. *Business Strategy and the Environment*. doi:10.1002/bse.2045

Mapping the Renewable Energy Sector to the Sustainable Development Goals: An Atlas. (2018). Retrieved from http://unsdsn.org/wp-content/uploads/2019/01/190108-mapping-renewables-report_WEB.pdf

Margolis, R. M., & Kammen, D. M. (1999). Underinvestment: The energy technology and R&D policy challenge. *Science*, *285*, 690–692. PMID:10426983

Mariaud, A., Acha, S., Ekins-Daukes, N., Shah, N., & Markides, C. N. (2017). Integrated optimisation of photovoltaic and battery storage systems for UK commercial buildings. *Applied Energy*, *199*, 466–478. doi:10.1016/j.apenergy.2017.04.067

Marincioni, F., Appiotti, F., Pusceddu, A., & Byrne, K. (2013). Enhancing resistance and resilience to disasters with microfinance: Parallels with ecological trophic systems. *International Journal of Disaster Risk Reduction*, *4*, 52–62.

Maruffi, BL., & Petri, R., & Malindretos, J. (2013). Corporate Social Responsibility and the Competitive Advantage of Multinational Corporations: What is the Right Balance? *The Journal of Global Business Issues*, *7*(2), 69–81.

Masaud, T. M., Lee, K., & Sen, P. (2010). *An overview of energy storage technologies in electric power systems: What is the future?* Paper presented at the North American Power Symposium (NAPS). 10.1109/NAPS.2010.5619595

Masters, G. (2004). *Renewable and efficient electric power systems*. Hoboken, NJ: John Wiley & Sons, Ltd. doi:10.1002/0471668826

Matsubara, H. (2017). *Renewable Energy Status in Japan and the World, Sustainable Zone 2016, April 2017*. Retrieved from https://www.japanfs.org/en/news/archives/news_id035824.html

Mawdsley, E. (2018). 'From billions to trillions': Financing the SDGs in a world 'beyond aid'. *Dialogues in Human Geography*, *8*(2), 191–195.

Mayer, R., Ryley, T., & Gillingwater, D. (2015). Eco-positioning of airlines: Perception versus actual performance. *Journal of Air Transport Management*, *44*, 82–89. doi:10.1016/j.jairtraman.2015.03.003

Mazzucato, M., & Semieniuk, G. (2018). Financing renewable energy: Who is financing what and why it matters. *Technological Forecasting and Social Change*, *127*, 8–22. doi:10.1016/j.techfore.2017.05.021

McCarthy, E. F. (1998). *Wind Speed Forecasting in the Central California Wind Resource Area. Paper presented in the*. Burlingame, CA: EPRI-DOE-NREL Wind Energy Forecasting Meeting.

McGinnis, M. (2011). An introduction to IAD and the language of the Ostrom Workshop: A simple guide to a complex framework. *Policy Studies Journal: the Journal of the Policy Studies Organization*, *39*(1), 169–183. doi:10.1111/j.1541-0072.2010.00401.x

McKibben, B. (2012). Global Warming's Terrifying New Math. *Rolling Stone*. Retrieved from http://www.rollingstone.com/politics/news/global-warmings-terrifying-new-math-20120719

McLaren, J., Anderson, K., Laws, N., Gagnon, P., DiOrio, N., & Li, X. (2018). Identifying critical factors in the cost-effectiveness of solar and battery storage in commercial buildings. *National Renewable Energy Laboratory.* Retrieved from: https://www.nrel.gov/docs/fy18osti/70813.pdf

Meadows, D. H., Meadows, D. L., Randers, J., & Behrens, W. W. III. (1972). *The Limits to Growth*. Hanover, NH: Signet.

MEE - Ministry of Economy and Energy. (2011, June). *Energy Strategy of the Republic of Bulgaria until 2020 for reliable, efficient, and clean energy.* Republic of Bulgaria. Accessed October 13, 2014: https://www.me.government.bg/files/useruploads/files/epsp/23_energy_strategy2020Eng_.pdf

MEE - Ministry of Economy and Energy. (2013, December). *Second National Report on Bulgaria's Progress in the Promotion and Use of Energy from Renewable Sources drawn up under Article 22(1) of Directive 2009/28/EC on the promotion of the use of energy from renewable sources and in accordance with the Template for Member States' progress reports under Directive 2009/28/EC.* Republic of Bulgaria. Accessed October 12, 2014 from: http://www.etipbioenergy.eu/images/Article_22_Bulgaria_report_EN.pdf

MEET - Ministry of Economy, Energy, and Tourism. (2011). *Re-submitted National Renewable Energy Action Plan (NREAP) Drawn up in Accordance with the Template for National Renewable Energy Action Plans as Set Out in Directive 2009/28/EC of the European Parliament and of the Council.* Accessed May 13, 2014 from: https://me.government.bg/library/index/download/lang/en/fileId/575

Mei, X. W., Fan, X., Hong, F. S., Qi, L., Kateryna, A., Eric, A. S., & Jian, X. (2011). Nanoscale graphite-supported Pt catalysts for oxygen reduction reactions in fuel cells. *Electrochimica Acta*, *56*(5), 2566–2573. doi:10.1016/j.electacta.2010.11.019

Meltzer, E. (2010). *Boulder Leaders Divided on Xcel Agreement. The Daily Camera*. Retrieved from LexisNexis Academic.

Menanteau, P., Finon, D., & Lamy, M. L. (2003). Prices quantities: Choosing policies for promoting the development of RE. *Energy Policy*, *31*(8), 799–812.

Mendell, M., & Heath, G. (2005). Do indoor pollutants and thermal conditions in schools influence student performance? A critical review of the literature. *Indoor Air*, *15*(1), 27–52. doi:10.1111/j.1600-0668.2004.00320.x PMID:15660567

Mendonca, M., Lacey, S., & Hvelplund, F. (2009). Stability, participation and transparency in renewable energy policy:Lessons from Denmark and the United States. *Policy and Society*, *27*(4), 379–398. doi:10.1016/j.polsoc.2009.01.007

Mercader-Moyano, P. (Ed.). (2017). *Sustainable development and renovation in architecture, urbanism, and engineering*. New York, NY: Springer. doi:10.1007/978-3-319-51442-0

Merei, G., Moshovel, J., Magnor, D., & Sauer, D. U. (2016). Optimization of self-consumption and techno-economic analysis of PV-battery systems in commercial applications. *Applied Energy*, *168*, 171–178. doi:10.1016/j.apenergy.2016.01.083

Migdadi, Y. K. A. A. (2016). Identifying the best practices in the green operations strategy of leading mobile phone producers. *International Journal of Business Excellence*, *9*(1), 92–112. doi:10.1504/IJBEX.2016.073377

Migdadi, Y. K. A. A. (2018). Identifying the best practices of airlines' green operations strategy: A cross-regional worldwide survey. *Environmental Quality Management*, *28*(1), 21–32. doi:10.1002/tqem.21575

Miller, J. (2009). Structuring the smart grid framework: application of complex systems engineering. *US DOE/NETL Modern Grid Team.* Retrieved from http://www.smartgrid.gov/sites/default/files/pdfs/structuring_smart_grid_framework_05-2009.pdf

Miller, J., Pullins, S., & Bossart, S. (2008). (NETL). *The modern grid.* Retrieved from http://wpui.wisc.edu/programs/Institute%20Lunches/Smart_Grid/Presentations/Miller.pdf

Miller, D., & Le Breton-Miller, I. (2005). Management insights from great and struggling family businesses. *Long Range Planning*, *38*(6), 517–530. doi:10.1016/j.lrp.2005.09.001

Min, C. K., Gyeong, S. H., & Rodney, S. R. (2009). Epoxide reduction with hydrazine on graphene: A first principles study. *The Journal of Chemical Physics*, *131*(6), 064704–064709. doi:10.1063/1.3197007 PMID:19691400

Minghui, W., Ge, L., Xuejun, C., Yuxiang, F., Huan, Z., Gaoxu, W., ... Yunqing, L. (2018). Self-crosslinked organic-inorganic nanocomposite membranes with good methanol barrier for direct methanol fuel cell applications. *Solid State Ionics*, *315*, 71–76. doi:10.1016/j.ssi.2017.12.001

Ming, Z., Ximei, L., Na, L., & Song, X. (2013). Overall review of renewable energy tariff policy in China: Evolution, implementation, problems and countermeasures. *Renewable & Sustainable Energy Reviews*, *25*, 260–271. doi:10.1016/j.rser.2013.04.026

Ministry of Power. (2012). *Scheme for Financial Restructuring of State Distribution Companies (DISCOMS).* Retrieved from http://powermin.nic.in/sites/default/files/uploads/Financial_restructuring_of_State_Distribution_Companies_discoms_Oct2012.pdf

Mirza, U. K., Ahmad, N., Harijan, K., & Majeed, T. (2009). Identifying and addressing barriers to renewable energy development in Pakistan. *Renewable & Sustainable Energy Reviews*, *13*, 927–931.

Missemer, A. (2012). William Stanley Jevons' *The Coal Question* (1865), beyond the rebound effect. *Ecological Economics*, *82*, 97–103. doi:10.1016/j.ecolecon.2012.07.010

Mohammed, A.A., Yaqub, W., & Aung, Z. (2015). Probabilistic forecasting of solar power: an ensemble learning approach. *Intelligent Decision Technologies / Smart Innovational Systems Technologies, 39,* 449–458.

Mol, A. P. J. (1997). Ecological Modernization: Industrial Transformations and Environmental Reform. In R. Michael & G. Woodgate (Eds.), The International Handbook of Environmental Sociology (pp. 138-149). Northampton, MA: Edward Elgar.

Mol, A. P. J. (2000). The Environmental Movement in an Era of Ecological Modernization. *Geoforum, 31*(1), 45–56. doi:10.1016/S0016-7185(99)00043-3

Mollison, B. (1988). *Permaculture: A designers' manual.* Tagari Publications.

Mollison, B., & Holmgren, D. (1978). *Permaculture one: A perennial agriculture for human settlements*. Melbourne, Australia: Transworld.

Moore, S. (2008). *Key features of meter data management systems*. Retrieved from https://www.itron.com/na/PublishedContent/Key%20MDM%20Features%20Whitepaper_FINAL.pdf

Muhammad-Sukki, F., Abu-Bakar, S. H., Munir, A. B., Yasin, S. H., Ramirez-Iniguez, R., McMeekin, S. G., ... Tahar, R. M. (2014). Feed-in tariff for solar photovoltaic: The rise of Japan. *Renewable Energy, 68*, 636–643. doi:10.1016/j.renene.2014.03.012

Müller, B. (2016). *Two unconventional options to enhance multilateral climate finance: Shares of proceeds and crowdfunding*. Oxford, UK: European Capacity Building Initiative.

Musoni. (2016). *Musoni microfinance system.* Retrieved from http://musonisystem.com/

Mutisya, E., & Yarime, M. (2014). *Microcredit for the development of the bottom of the pyramid segment: Impact of access to financial services on microcredit clients, institutions and urban sustainability*. Working Paper Series N° 199. Tunis: African Development Bank.

Nagoda, S. (2015). New discourses but same old development approaches? Climate change adaptation policies, chronic food insecurity and development interventions in northwestern Nepal. *Global Environmental Change, 35*, 570–579.

NASA's Jet Propulsion Lab. (2018). *Global climate change: Vital signs of the planet.* Retrieved from https://climate.nasa.gov/

Nattinee, K., Kodchakorn, V., Thirayu, C., Sairung, C., Katesara, P., Anuvat, S., & Karnthidaporn, W. (2018). Preparation of sulfonated zeolite ZSM-5/sulfonated polysulfone composite membranes as PEM for direct methanol fuel cell application. *Solid State Ionics, 319*, 278–284. doi:10.1016/j.ssi.2018.02.019

Natural Resources Canada. (2018). *Canadian Weather Year for Energy Calculation*. Retrieved from: http://climate.weather.gc.ca/prods_servs/engineering_e.html

New Zealand Government & European Union. (2016). *Pacific Energy Country Profiles*. New Zealand Ministry of Foreign Affairs and Trade.

Niang, I., Ruppel, O. C., Abdrabo, M. A., Essel, A., Lennard, C., Padgham, J., & Urquhart, P. (2014). Africa. In *Climate Change 2014: Impacts, adaptation, and vulnerability. Part B: Regional Aspects. Contribution of Working Group II to the Fifth Assessment Report of the Intergovernmental Panel on Climate Change*. Cambridge, UK: Cambridge University Press.

Nielsen, T. S., Madsen, H., & Tofting, J. (1999). Experiences with statistical methods for wind power prediction. *1999 European Wind Energy Conference and Exhibition.*

Nieto, M. J., Santamaria, L., & Fernandez, Z. (2015). Understanding the innovation behavior of family firms. *Journal of Small Business Management*, *53*(2), 382–399. doi:10.1111/jsbm.12075

Nikolaev, A., & Konidari, P. (2017). Development and assessment of renewable energy policy scenarios by 2030 for Bulgaria. *Renewable Energy*, *111*, 7922–802. doi:10.1016/j.renene.2017.05.007

Nordensvärd, J., & Urban, F. (2015). The stuttering energy transition in Germany: Wind energy policy and feed-in-tariff lock-in. *Energy Policy*, *82*, 156–165. doi:10.1016/j.enpol.2015.03.009

Nordhaus, T., & Shellenberger, M. (2004). *The Death of Environmentalism: Global Warming Politics in a Post-Environment World*. The Breakthrough Institute.

North, D. C. (1990). *Institutions, Institutional Change and Economic Performance*. Cambridge, UK: Cambridge University Press. doi:10.1017/CBO9780511808678

Novinite. (2014, October 31). *Bulgaria's National Electric Company Reports BGN 425M Loss in end-Sept*. Accessed October 20, 2018 from: https://www.novinite.com/articles/164465/Bulgaria's+National+Electric+Company+Reports+BGN+425M+Loss+in+end-Sept

Nowicki, R. (2016, April 12). *How to transition to a sustainable future at home*. Retrieved from https://www.youtube.com/watch?v=_ZIjgcdCjBc&feature=youtu.be

Nowicki, R. (n.d.). *The land office*. Retrieved from http://www.holdtonature.com/Hold_To_Nature/Home.html

NREL, National Renewable Energy Lab. (2018). *Distributed Generation Energy Technology Operations and Maintenance Costs*. Retrieved from: https://www.nrel.gov/analysis/tech-cost-om-dg.html

NSI - National Statistical Institute. (2017). Overall Energy Balance Sheets for 2011, 2012, 2013, 2014, 2015, 2016. *Bulletin on the state and development of the energy sector in the Republic of Bulgaria*. Accessed October 22, 2018 at: http://www.nsi.bg/sites/default/files/files/data/timeseries/Energy4.1_en.xls

NSI - National Statistical Institute. (2018, August). *Bulgaria's 2016 final energy consumption by type of fuel (Total=9.7 Mtoe) (EU Country data sheets for Bulgaria)*. Accessed August 18, 2017 from: https://ec.europa.eu/energy/en/data-analysis/energy-statistical-pocketbook)

Nygaard, I., Handsen, U. E., Mackenzie, G. A., & Pederson, M. b. (2017). Measures for diffusion of solar PV in selected Africa countries. *International Journal of Sustainable Energy*, *36*(7), 707–721.

O'Toole, L. J., & Meier, K. (1999). Modeling the Impact of Public Management: Implications of Structural Context. *Journal of Public Administration: Research and Theory*, *9*(4), 505–526. doi:10.1093/oxfordjournals.jpart.a024421

Odgaard, O. (2000). Workshop on Best Practices in Policies and Measures. *The Green Electricity Market in Denmark: Quotas, Certificates and International Trade*.

Odum, H. T. (1975). Energy quality and carrying capacity of the earth. *Tropical Ecology*, *16*(1), 14.

OECD. (2013). *The role of banks, equity markets and institutional investors in long-term financing for growth and development: Report for G20 leaders*. OECD Publishing.

OECD. (2015). *Mapping channels to mobilise institutional investment in sustainable energy*. Paris: OECD Publishing.

OECD. (2016). *Annual survey of large pension funds and public pension reserve funds: Report on pension funds' long-term investment*. Paris: OECD Publishing.

OECD. (2017a). *Breaking silos: Actions to develop infrastructure as an asset class and address the information gap - an Agenda for G20*. Paris: OCED Publishing.

OECD. (2017b). *Investing in climate, investing in growth*. Paris: OECD.

OECD. (2017c). *Investment governance and the integration of environmental, social and governance factors*. Paris: OECD Publishing.

OECD. (2017d). *Selected good practices for risk allocation and mitigation in infrastructure in APEC economies*. Paris: OECD Publishing.

OECD. (2018). *Making blended finance work for the Sustainable Development Goals*. Paris: OECD Publishing.

Oertwig, N., Galeitzke, M., Schmieg, H. G., Kohl, H., Jochem, R., Orth, R., & Knothe, T. (2017). Integration of Sustainability into the Corpory Strategy. In R. Stark, G. Seliger & J. Bonvoisin (Eds.), Sustainable Manufacturing. Challenge Solution and Implementations Perspective (pp. 175–200). SpringerOpen.

Office of the Auditor General of Ontario. (2011). *Annual Report: Chapter 3: Reports on value-for-money audits and reviews*. Retrieved from: http://www.auditor.on.ca/en/content/annualreports/arbyyear/ar2011.html

Office of the Auditor General of Ontario. (2015). *Annual Report: Chapter 3: Reports on value-for-money audits and reviews*. Retrieved from: http://www.auditor.on.ca/en/content/annualreports/arbyyear/ar2015.html

Ogasawara, J. (2008). *Overview of the Green Power Certification System*. Retrieved from https://eneken.ieej.or.jp/en/data/pdf/493.pdf

Ohunakin, O. S., Adaramola, M. S., Oyewola, O. M., & Fagbenle, R. O. (2014). Solar energy applications and development in Nigeria: Drivers and barriers. *Renewable & Sustainable Energy Reviews*, *32*, 294–301.

Olabi, A. G. (2017). *Renewable energy and energy storage systems*. Academic Press.

Olsen, I. I., & Pasquale, N. B. (2007). *Battery and inverter configuration with increased efficiency*. Google Patents.

Onishi, N. (2015). *China pledges $60 billion to aid Africa's development.* Retrieved from https://www.nytimes.com/2015/12/05/world/africa/china-pledges-60-billion-to-aid-africas-development.html

Ontario Energy Board. (2018). *What initiatives are available?* Retrieved from: https://www.oeb.ca/industry/tools-resources-and-links/information-renewable-generators/what-initiatives-are-available

Organisation for Economic Cooperation and Development & World Economic Forum. (2015). *A How-To Guide for Blended Finance.* Retrieved from http://www3.weforum.org/docs/WEF_Blended_Finance_How_To_Guide.pdf

Ostrom, E. (2009). *A polycentric approach for coping with climate change.* World Bank Policy Research Working Paper Series 5095. Washington, DC: World Bank.

Ostrom, E. (1990). *Governing the Commons: The evolution of institutions for collective action.* New York: Cambridge University Press. doi:10.1017/CBO9780511807763

Ostrom, E. (1999). Institutional Rational Choice: An Assessment of the Institutional Analysis and Development Framework. In P. Sabatier (Ed.), *Theories of the Policy Process* (pp. 21–64). Boulder, CO: Westview Press.

Ostrom, E. (2005). *Understanding Institutional Diversity.* Princeton University Press.

Ostrom, E. (2008). *Polycentric systems as one approach for solving collective-action problems.* doi:10.2139srn.1304697

Ostrom, E. (2010). Beyond markets and states: Polycentric governance of complex economic systems. *The American Economic Review, 100*(3), 1–33.

Ostrom, V., & Ostrom, E. (1971). Public Choice: A Different Approach to the Study of Public Administration. *Public Administration Review, 24*(2), 203–216. doi:10.2307/974676

Oudjana, S. H., Hellal, A., & Mahamed, I. H. (2012). Short term photovoltaic power generation forecasting using neural network. *Environment and Electrical Engineering (EEEIC), 2012 11th International Conference*, 706–711.

Oxford Community Energy Cooperative. (n.d.). *What is a renewable energy co-operative?* Retrieved from http://www.oxford-cec.ca/page-1741085

Ozturk, M., & Yuksel, Y. E. (2016). Energy structure of Turkey for sustainable development. *Renewable & Sustainable Energy Reviews, 53*, 1259–1272. doi:10.1016/j.rser.2015.09.087

Pachauri, RK., Allen, MR., Barros, VR., Broome, J., Cramer, W., Christ, R., & Dubash, NK. (2014). *Climate change 2014: synthesis report.* Contribution of Working Groups I, II and III to the fifth assessment report of the Intergovernmental Panel on Climate Change.

Padhy & Jena. (n.d.). Introduction to Smart Grid. *NPTEL online videos.* Retrieved from https://onlinecourses.nptel.ac.in/noc18_ee42/

Padmavathi, R., Sandhya, D. A., Saranya, N., Gnanasundaram, P., & Sangeetha, D. (2015). Synthesis and characterization of Pt supported on multiwalled carbon nanotubes for improved catalytic performance in fuel cell Applications. *Journal of Porous Materials*, *22*(3), 647–658. doi:10.100710934-015-9937-5

Padmavathi, R., & Sangeetha, D. (2013). Design of novel SPEEK-based proton exchange membranes by self-assembly method for fuel cells. *Ionics*, *19*(10), 1423–1436. doi:10.100711581-013-0867-4

Padmavathi, R., & Sangeetha, D. (2013). Synthesis and characterization of electrospun carbon nanofibersupported Pt catalyst for fuel cells. *Electrochimica Acta*, *112*, 1–13. doi:10.1016/j.electacta.2013.08.078

Palizban, O., & Kauhaniemi, K. (2015). Hierarchical control structure in microgrids with distributed generation: Island and grid-connected mode. *Renewable & Sustainable Energy Reviews*, *44*, 797–813. doi:10.1016/j.rser.2015.01.008

Park, J. (2016). Clean energy entrepreneurship in Sub-Saharan Africa. *Global Entrepreneurship: Past, Present & Future, 29*, 257-277.

Park, J., & Brooks, C. (2015). Local flood resiliency in an era of global climate change: Understanding the multi-sectoral policy dimensions. *Vermont Journal of Environmental Law*, *17*(2), 160–177.

Park, M., Sun, H., Lee, H., Lee, J., & Cho, J. (2012). Lithium-air batteries: Survey on the current status and perspectives towards automotive applications from a battery industry standpoint. *Advanced Energy Materials*, *2*(7), 780–800. doi:10.1002/aenm.201200020

Parks, R., & Oakerson, R. (2000). Regionalism, Localism, and Metropolitan Governance: Suggestions from the Research Program on Local Public Economies. *State & Local Government Review*, *32*(3), 169–179. doi:10.1177/0160323X0003200302

Parr, A. (2014). *The wrath of capital: neoliberalism and climate change*. New York: Columbia University Press.

Parra, D., Norman, S. A., Walker, G. S., & Gillott, M. (2017). Optimum community energy storage for renewable energy and demand load management. *Applied Energy*, *200*, 358–369. doi:10.1016/j.apenergy.2017.05.048

Pathak, M., & Bhaskar, U. (2018). *India defers largest global solar tender on bidders concerns*. Retrieved from https://www.livemint.com/Industry/uhzOpWbfficUpzEpLujveL/India-defers-largest-global-solar-tender-on-bidders-concerns.html

Patton, M. (1990). *Qualitative evaluation and research methods* (2nd ed.). Newbury Park, CA: Sage.

Pech, D., Brousse, T., Bélanger, D., & Guay, D. (2009). EQCM study of electrodeposited PbO 2: Investigation of the gel formation and discharge mechanisms. *Electrochimica Acta*, *54*(28), 7382–7388. doi:10.1016/j.electacta.2009.07.070

Pedram, M., Chang, N., Kim, Y., & Wang, Y. (2010). Hybrid electrical energy storage systems. *Proceedings of the 16th ACM/IEEE international symposium on Low power electronics and design.*

Pedro, H. T., & Coimbra, C. F. (2012). Assessment of forecasting techniques for solar power production with no exogenous inputs. *Solar Energy*, *86*(7), 2017–2028. doi:10.1016/j.solener.2012.04.004

Peiying, Z., Ming, S., Shuhua, X., & Dong, Z. (2011). Experimental study on the reducibility of graphene oxide by hydrazine hydrate. *Physica B, Condensed Matter*, *406*(3), 498–502. doi:10.1016/j.physb.2010.11.022

Perera, F. P. (2017). Multiple Threats to Child Health from Fossil Fuel Combustion: Impacts of Air Pollution and Climate Change. *Environmental Health Perspectives*, *125*(2), 141–148. doi:10.1289/EHP299 PMID:27323709

Permaculture Design Principles. (n.d.). Retrieved from https://permacultureprinciples.com/principles/

Permaculture Ethics. (n.d.). Retrieved from https://permacultureprinciples.com/ethics/

Petrova, V. (2018). *EC approves plan for mixed solar/wind tenders in Denmark.* Retrieved October 29, 2018, from https://renewablesnow.com/news/ec-approves-plan-for-mixed-solarwind-tenders-in-denmark-623981/

Pinson, P., & Madsen, H. (2009). Ensemble-based Probabilistic Forecasting at Horns Rev, *Wind Energy special issue Offshore. Wind Energy (Chichester, England)*, *12*(2), 137–155. doi:10.1002/we.309

Pinson, P., & Madsen, H. (2012). Adaptive modelling and forecasting of offshore wind power fluctuations with Markov-switching autoregressive models. *Journal of Forecasting*, *31*(4), 281–313. doi:10.1002/for.1194

Pinson, P., Ranchin, T., & Kariniotakis, G. (2004, March). Short-term wind power prediction for offshore wind farms Evaluation of Fuzzy-Neural network based models. *Global Windpower Conference.*

PMR (Partnership for Market Readiness). (2013). *Recent development of the Joint Crediting Mechanism (JCM)/Bilateral Offset Credit Mechanism (BOCM).* Retrieved from https://www.thepmr.org/system/files/documents/JCM.pdf

Polack, A. R. (2010). *Drivers and Barriers of Renewable Energy in the Electrification of Vanuatu.* Murdoch, Australia: Murdoch University, School of Engineering and Energy.

Pollet, B. G., Staffell, I., & Shang, J. L. (2012). Current status of hybrid, battery and fuel cell electric vehicles: From electrochemistry to market prospects. *Electrochimica Acta*, *84*, 235–249. doi:10.1016/j.electacta.2012.03.172

Pomerance, S. (2013). Working Group Process is More Show than Tell [Opinion]. *The Daily Camera.* Retrieved from LexisNexis Academic.

Porsinger, T., Janik, P., Leonowicz, Z., & Gono, R. (2016). Component modelling for microgrids. In *Proceedings of the 2016 IEEE 16th Internationa Conference on Environment and Electrical Engineering (EEEIC)*. IEEE. 10.1109/EEEIC.2016.7555869

Porsinger, T., Janik, P., Leonowicz, Z., & Gono, R. (2017). Modelling and Optimizing in Microgrids. *Energies*, *10*(4), 523. doi:10.3390/en10040523

Pöschl, M., Ward, S., & Owende, P. (2010). Evaluation of energy efficiency of various biogas production and utilization pathways. *Applied Energy*, *87*(11), 3305–3321. doi:10.1016/j.apenergy.2010.05.011

Poullikkas, A. (2013). A comparative assessment of net metering and feed in tariff schemes for residential PV systems. *Sustainable Energy Technologies and Assessments*, *3*, 1–8.

Power Finance Corporation (PFC). (2016). *The Performance of State Power Utilities for the years 2012-13 to 2014-15*. Retrieved from http://www.pfc.gov.in/Default/ViewFile/?id=1490186954263_Report%20on%20Performance%20of%20State%20Power%20Utilities%202012-13%20to%202014-15.pdf&path=Page

Press Information Bureau. (2017). *India gets Lowest Wind Tariff of Rs. 2.64 per kWh in second Wind Auction of 1000 MW*. Retrieved from http://pib.nic.in/newsite/PrintRelease.aspx?relid=171394

Proctor, C. (2014). Xcel Proposes New Solar Power Option for Colorado Customers. *Denver Business Journal*. Retrieved from http://www.bizjournals.com/denver/blog/earth_to_power/2014/04/xcel-proposes- newsolar-power-option-for-colorado.html?page=all

Public Power. (2018). *Stats and Facts: Three Types of Electric Utilities*. Retrieved from https://www.publicpower.org/public-power/stats-and-facts

Quitzow, L., Canzler, W., Grundmann, P., Leibenath, M., Moss, T., & Rave, T. (2016). The German Energiewende - What's Happening? Introducing the Special Issue. *Utilities Policy*, *41*, 1–9. doi:10.1016/j.jup.2016.03.002

Raizman, M. (2011). Boulder's Energy Future [Editorial Advisory Board]. *The Daily Camera*. Retrieved from LexisNexis Academic.

Ralon, P., Taylor, M., Ilas, A., Diaz-Bone, H., & Kairas, K. P. (2017). Electricity Storage and Renewables: Costs and Markets to 2030. International Renewable Energy Agency, Abu Dhabi. Retrieved from file:///C:/Users/Vasundhara/Downloads/IRENA_Electricity_Storage_Costs_2017.pdf

Ralph, T. R. (1997). Proton Exchange Membrane Fuel Cells: Progress in cost reduction of the key components. *Platinum Metals Review*, *41*(3), 102–113.

Ramli, A. M., Hiendro, A., & Ssennoga, T. (2015, June). Economic analysis of PV/diesel hybrid system with flywheel energy storage. *Renewable Energy*, *78*, 398–405. doi:10.1016/j.renene.2015.01.026

Ramli, M. A. M., & Twaha, S. (2015). Analysis of renewable energy feed-in tariffs in selected regions of the globe: Lessons for Saudi Arabia. *Renewable & Sustainable Energy Reviews*, *45*(C), 649–661.

RAS. (2017). *Sardegna in cifre 2017. Cagliari: Servizio della Statistica Regionale.* Retrieved from http://www.sardegnastatistiche.it/documenti/12_103_20170727151245.pdf

Raszmann, E., Baker, K., Shi, Y., & Christensen, D. (2017). Modeling stationary lithium-ion batteries for optimization and predictive control. *National Renewable Energy Laboratory*. Retrieved from: https://www.nrel.gov/docs/fy17osti/67809.pdf

Raul, N. S., & Necsulescu, C. (1984). Economics of Energy Storage Devices in Interconnected Systems-A New Approach. *IEEE Transactions on Power Apparatus and Systems*, *PAS-103*(6), 1217–1223. doi:10.1109/TPAS.1984.318452

Ravi, K., Mohamed, M., & Keith, S. (2014). Sulfonated polyether ether ketone-sulfonated graphene oxide composite membranes for polymer electrolyte fuel cells. *Royal Society of Chemistry Advances*, *4*, 617–621.

Reganold, J. P., Glover, J. D., Andrews, P. K., & Hinman, H. R. (2001). Sustainability of three apple production system. *Nature*, *410*(6831), 926–930. doi:10.1038/35073574 PMID:11309616

Regulatory Indicators for Sustainable Energy. (2018). *Kiribati Country Energy Profile*. Retrieved from http://rise.esmap.org/countries%20Documents/Kiribati/1/INDC_KIRIBATI.pdf

Rehman, S., Al-Hadhrami, L. M., & Alam, M. M. (2015). Pumped hydro energy storage system: A technological review. *Renewable & Sustainable Energy Reviews*, *44*, 586–598. doi:10.1016/j.rser.2014.12.040

Reilly, K. (2018, October 8). Here's what humanity must do immediately to prevent catastrophic climate change, according to the new U.N. report. *Time*. Retrieved from http://time.com/5418577/what-humanity-do-limit-climate-change/

Remund, J., Schilter, C., Dierer, S., Stettler, S., & Toggweiler, P. (2008). Operational forecast of PV production. In *23rd European Photovoltaic Solar Energy Conference* (pp. 3138-3140). Academic Press.

Rey, J. M., Marti, P., Velasco, M., Miret, J., & Castilla, M. (2017). Secondary Switched Control with no Communications for Islanded Microgrids. *IEEE Transactions on Industrial Electronics*, *64*(11), 8534–8545. doi:10.1109/TIE.2017.2703669

Rifkin, J. (2013). The Third Industrial Revolution: How Lateral Power is Transforming Energy, The Economy, and The World. Basingstoke, UK: Palgrave Macmillan.

Rifkin, J. (2011). *The Third Industrial Revolution: How Lateral Power Is Transforming Energy, the Economy, and the World.* Palgrave MacMillan.

Rimanoczy, I., & Pearson, T. (2010). Role of HR in the new world of sustainability. *Industrial and Commercial Training*, *42*(1), 11–17. doi:10.1108/00197851011013661

Rising Tide North America. (2011). Confronting the Root Causes of Climate Change. *A Primer from Rising Tide North America.*

Roggero, M., Bisaro, A., & Villamayor-Tomas, S. (2018). Institutions in the climate adaptation literature: A systematic literature review through the lens of the Institutional Analysis and Development framework. *Journal of Institutional Economics*, *14*(3), 423–448. doi:10.1017/S1744137417000376

Rondinelli, D. A., & London, T. (2003). How Corporations and Environmental Groups Cooperate: Assessing Cross-sector Alliances and Collaborations. *The Academy of Management Executive*, *17*(1), 61–76.

Roper, T. (2005). Small Island States – Setting an Example on Green Energy Use. *Review of European Community & International Environmental Law*, *14*(2), 108–116.

Röser, F., Day, T., & Kurdziel, M. (2016). *After Paris: What is next for Intended Nationally Determined Contributions (INDCs)?* Berlin: International Partnership on Mitigation and MRV and New Climate Institute.

Rowe, B. R., Haynes, G. W., & Stafford, K. (1999). The contribution of home-based business income to rural and urban economies. *Economic Development Quarterly*, *13*(1), 66–77. doi:10.1177/089124249901300109

Rubel, H., Pieper, C., Zenneck, J., & Sunak, Y. (2017). *How batteries and solar power are disrupting electricity markets*. Retrieved March 3, 2018, from https://www.bcg.com/publications/2017/energyenvironment-how-batteries-and-solar-power-are-disruptingelectricity-markets.aspx

RuddockB. (2013). *Plant guilds.* Retrieved from https://midwestpermaculture.com/eBook/Plant%20Guilds%20eBooklet%20-%20Midwest%20Permaculture.pdf

Ryan, S., Hebdon, C., & Dafoe, J. (2014). Energy research and the contributions of the social sciences: A contemporary examination. *Energy Research & Social Science*, *3*, 186–197. doi:10.1016/j.erss.2014.07.012

Sachs, J. D. (2015). *The age of sustainable development*. New York, NY: Columbia University Press. doi:10.7312ach17314

Saft. (2018). *Lithium-ion battery life*. Retrieved from www.saftbatteries.com

Sage, A. P. (1999). Sustainable Development: Issues in Information, Knowledge, and Systems Management. *Information, Knowledge, Systems Management*, *1*(3-4), 185–223.

Sahagun, L. (2011). Sierra Club Leader Departs Amid Discontent Over Group's Direction. *Los Angeles Times*. Retrieved from http://articles.latimes.com/2011/nov/19/local/la-me-sierra-club-20111119

Salman, K. (2017). Introduction to the Smart Grid – Concepts, Technologies and Evolution. The Institution of Engineering and Technology.

Salman, S. K., Jiang, F., & Rogers, W. J. S. (1994). Effects of wind power generators on the voltage control of utility's distribution networks. *Wind Engineering*, *18*(4), 181–187.

Sandén, B. A. (2014). *Systems Perspectives on Renewable Power 2014*. Academic Press.

Sands, W. (2004, September 29). Green Power Push Stirs up Sparks: Amendment 37 Mixes it Up Statewide. *The Durango Telegraph*. Retrieved from http://www.durangotelegraph.com/04-09-23/cover_story.html

Sanoh, A., Kocaman, A., Kocal, S., Sherpa, S., & Modi, V. (2014). The economics of clean energy resource development and grid interconnection in Africa. *Renewable Energy*, *62*(C), 598–609.

Sava, T., Rizea, C., & Flood, I. (2010). Benefits of the implementation and certification for environmental management system SMEs in Romania. *Quality*, *2*, 248–254.

Scarlat, N., Motola, V., Dallemand, J. F., Monforti-Ferrario, F., & Mofor, L. (2015). Evaluation of energy potential of Municipal Solid Waste from African urban areas. *Renewable & Sustainable Energy Reviews*, *50*, 1269–1286.

Schmid, G. (2012). The development of renewable energy power in India: Which policies have been effective? *Energy Policy*, *45*, 317–326. doi:10.1016/j.enpol.2012.02.039

Schnaiberg, A., Pellow, D. N., & Weinberg, A. (2000). The Treadmill of Production and the Environmental State. The Environmental State Under Pressure, 10, 15-32.

Schuman, S., & Lin, A. (2012). Chinese Renewable Energy Law and its impact on renewable power in China: Progress, Challenges and recommendation for improving implementation. *Energy Policy*, *51*, 89–109. doi:10.1016/j.enpol.2012.06.066

Schwab, K. (2016). *The Fourth Industrial Revolution*. World Economic Forum.

Schwerhoff, G., & Sy, M. (2017). Financing renewable energy in Africa – Key challenge of the sustainable development goals. *Renewable & Sustainable Energy Reviews*, *75*, 393–401.

Scrieciu, S. S., Barker, T., & Ackerman, F. (2015). Pushing the boundaries of climate economics: Critical issues to consider in climate policy analysis. *Ecological Economics*, *85*, 155–165.

Sehkyu, P., Yuyan, S., Haiying, W., Peter, C. R., Vilayanur, V. V., Silas, A. T., ... Yong, W. (2011). Design of graphene sheets-supported Pt catalyst layer in PEM fuel cells. *Electrochemistry Communications*, *13*(3), 258–261. doi:10.1016/j.elecom.2010.12.028

Sekine, Y., & Goldie-Scott, L. (2017). *2017 Global energy storage forecast*. Retrieved December 22, 2017, from Bloomberg database.

Sen, S., & Ganguly, S. (2017). Opportunities, barriers and issues with renewable energy development – A discussion. *Renewable & Sustainable Energy Reviews*, *69*, 1170–1181.

Seyfang, G., & Smith, A. (2007). Grassroots innovations for sustainable development: Towards a new research and policy agenda. *Environmental Politics*, *16*(4), 584–603. doi:10.1080/09644010701419121

Sfetsos, A. (2002). A Novel Approach for the Forecasting of Mean Hourly Wind Speed Time Series. *Renewable Energy*, *27*(2), 163–174. doi:10.1016/S0960-1481(01)00193-8

Shahidullah, A. K. M., & Haque, C. E. (2014). Environmental orientation of small enterprises: Can microcredit-assisted microenterprises be "green"? *Sustainability*, *6*(6), 3232–3251.

Shanker, M. C., & Astrachan, J. H. (1996). Myths and realities: Family businesses' contribution to the US economy—a framework for assessing family business statistics. *Family Business Review*, *9*(2), 107–123. doi:10.1111/j.1741-6248.1996.00107.x

Shapira, R., Nessim, G. D., Zimrin, T., & Aurbach, D. (2013). Towards promising electrochemical technology for load leveling applications: Extending cycle life of lead acid batteries by the use of carbon nano-tubes (CNTs). *Energy & Environmental Science*, *6*(2), 587–594. doi:10.1039/C2EE22970F

Sharma, P., Chrisman, J. J., & Chua, J. H. (2003). Predictors of satisfaction with the succession process in family firms. *Journal of Business Venturing*, *18*(5), 667–687. doi:10.1016/S0883-9026(03)00015-6

Sharma, P., & Manikutty, S. (2005). Strategic divestments in family firms: Role of family structure and community culture. *Entrepreneurship Theory and Practice*, *29*(3), 293–311. doi:10.1111/j.1540-6520.2005.00084.x

Shear, M. D. (2017, June 1). Trump will withdraw U.S. from Paris Climate Agreement. *The New York Times*. Retrieved from https://www.nytimes.com/2017/06/01/climate/trump-paris-climate-agreement.html

Shereef, R. M., & Khaparde, S. A. (2015). A unified REC market and composite RPO scheme for promotion of renewable energy in India. *International Journal of Sustainable Energy*, *36*(6), 606–618. doi:10.1080/14786451.2015.1075988

Shi, J., Lee, W. J., Liu, Y., Yang, Y., & Wang, P. (2011). Forecasting power output of photovoltaic system based on weather classification and support vector machine. *IEEE Industry Applications Society Annual Meeting (IAS)*.

Short, J. C., Payne, G. T., Brigham, K. H., Lumpkin, G. T., & Broberg, J. C. (2009). Family firms and entrepreneurial orientation in publicly traded firms: A comparative analysis of the S&P 500. *Family Business Review*, *22*(1), 9–24. doi:10.1177/0894486508327823

Shrimali, G., & Reicher, D. (2017). *Instruments to mitigate financial risk in indian renewable energy investments*. Stanford Precourt Institute for Energy.

Shrimali, G., Trivedi, S., Srinivasan, S., Goel, S., & Nelson, D. (2016). Cost-effective policies for reaching India's 2022 renewable targets. *Renewable Energy*, *93*, 255–268. doi:10.1016/j.renene.2016.02.062

Shrivastava, P. (1995). The role of corporations in achieving ecological sustainability. *Academy of Management Review*, *20*(4), 936–960. doi:10.5465/amr.1995.9512280026

Sieminksi, A. (2013). International energy outlook 2013. Washington, DC: Independent Statistics and Analysis, US Energy Information Administration.

Silva Fonseca, J. G., Oozeki, T., Takashima, T., Koshimizu, G., Uchida, Y., & Ogimoto, K. (2012). Use of support vector regression and numerically predicted cloudiness to forecast power output of a photovoltaic power plant in Kitakyushu, Japan. *Progress in Photovoltaics: Research and Applications*, *20*(7), 874–882. doi:10.1002/pip.1152

Singh, A., & Trivedi, A. (2016). Sustainable green supply chain management: Trends and current practices. *Competitiveness Review*, *26*(3), 265–288. doi:10.1108/CR-05-2015-0034

Sinha, A., Shahbaz, M., & Balsalobre, D. (2017). Exploring the relationship between energy usage segregation and environmental degradation in N-11 countries. *Journal of Cleaner Production*, *168*, 1217–1229. doi:10.1016/j.jclepro.2017.09.071

Sipos Aktorik. (n.d.). Retrieved January 20, 2016, from http://www.sipos.de

SITRADE (Société Ivoirienne de Traitement des Déchets). (2009). *Abidjan Municipal Solid Waste-To-Energy Project. CDM Project Design Document for the Abidjan Municipal Solid Waste-To-Energy Project (Version 5)*. SITRADE.

Smart. (2012). *Smart Grid Coordination Group, Smart Grid Reference Architecture*. CEN-CENELEC-ETSI, Tech. Rep.

Smith, D. J. (2016). The sustainable and green engine (SAGE) – Aircraft engine of the future? *The International Journal of Entrepreneurship and Innovation*, *17*(4), 256–262. doi:10.1177/1465750316672601

Soanes, M., Rai, N., Steele, P., Shakya, C., & Macgregor, J. (2017). *Delivering real change: Getting international climate finance to the local level*. IIED Working Paper. London: The International Institute for Environment and Development (IIED).

Solar Sisters. (2019). *About solar sisters*. Retrieved from https://solarsister.org/about-us/

Solar, S. (2018). *Scatec Solar closes financing for Mozambique's first large scale solar plant*. Retrieved from https://www.scatecsolar.com/Investor/Stock-exchange-notices/Scatec-Solar-closes-financing-for-Mozambique-s-first-large-scale-solar-plant

Solorio, I., & Jörgens, H. (2017). *A guide to EU renewable energy policy: Comparing Europeanization and domestic policy change in EU member states*. Cheltenham, UK: Edward Elgar Publishing. doi:10.4337/9781783471560

SOO (Seeds of Opportunity). (2016). *Beneficiary led climate change resilience building programme*. Retrieved from http://seedsofopportunity.org/activities-programmes/beneficiary-led-climate-change-resilience-building-programme-blccrbp/

Sørensen, B. E. (2017). *Renewable Energy: Physics, engineering, environmental impacts, economics and planning*. Academic Press.

Soubeiga, S., & Strauss, J. (2013). *Financial sector policy note: Financing small and medium-sized businesses in Burkina Faso*. Washington, DC: World Bank.

Soulopoulos, N. (2018). Cost Projections for Batteries, Vehicles and Total Cost of Operation. *Bloomberg New Energy Finance*. Retrieved from: https://www.iea.org/media/workshops/2018/aces/NikolasSOULOPOULOSBNEF.pdf

Spaargaren, G., & Mol, A. P. J. (1992). Sociology, environment, and modernity: Ecological modernization as a theory of social change. *Society & Natural Resources*, *5*(4), 323–344. doi:10.1080/08941929209380797

Spaargaren, G., & Van Vliet, B. (2000). Lifestyles, consumption and the environment: The ecological modernization of domestic consumption. *Environmental Pollution*, *9*(1), 50–76.

Speth, J. (2008). Environmental Failure: A Case for a New Green Politics. *Yale Environment360*. Retrieved from http://e360.yale.edu/feature/environmental_failure_a_case_for_a_new_green_politics/20 75/

Spiers, D. J., & Rasinkoski, A. D. (1995). Predicting the service lifetime of lead/acid batteries in photovoltaic systems. *Journal of Power Sources*, *53*(2), 245–253. doi:10.1016/0378-7753(94)01989-9

Springett, D. (2003). Business conceptions of sustainable development. A perspective from critical theory. *Business Strategy and the Environment*, *12*(2), 71–86. doi:10.1002/bse.353

Spyker, R., & Nelms, R. (1997). *Evaluation of double layer capacitor technologies for high power and high energy storage applications*. Paper presented at the Industrial Electronics, Control and Instrumentation, 1997. IECON 97. 23rd International Conference on. 10.1109/IECON.1997.668434

Stafford, E. R., Polonsky, M. J., & Hartman, C. L. (2000). Environmental NGO- Business Collaboration and Strategic Bridging: A Case Analysis of the Greenpeace-Foron Alliance. *Business Strategy and the Environment*, *9*(2), 122–135. doi:10.1002/(SICI)1099-0836(200003/04)9:2<122::AID-BSE232>3.0.CO;2-C

Starik, M., & Rands, G. P. (1995). Weaving an integrated web: Multilevel and multisystem perspective of ecologically sustainable organizations. *Academy of Management Review*, *20*(4), 908–935. doi:10.5465/amr.1995.9512280025

Stokes, L. (2013). The politics of renewable energy policies: The case of feed-in tariffs in Ontario, Canada. *Energy Policy*, *56*, 490–500. doi:10.1016/j.enpol.2013.01.009

Suberu, M. Y., Mustafa, M. W., Bashir, N., Muhamad, N. A., & Mokhtar, A. S. (2013b). Power sector renewable energy integration for expanding access to electricity in sub-Saharan Africa. *Renewable & Sustainable Energy Reviews*, *25*, 630–642.

Suberu, M., Mustafa, M., & Bashir, N. (2013a). Status of renewable energy consumption and developmental challenges in Sub-Sahara Africa. *Renewable & Sustainable Energy Reviews*, *27*, 453–463.

Suddaby, R. (2006). From the editors: what grounded theory is not. Academy of Management Journal, 49(4), 633–642.

Sun, S., & Liu, J. (2018). Research on DC Boost Type of Photovoltaic Power Generation System. *IOP Conference Series. Materials Science and Engineering*, *382*(3). doi:10.1088/1757-899X/382/3/032045

Surek, T. (2005). Crystal growth and materials research in photovoltaics: Progress and challenges. *Journal of Crystal Growth*, *275*(1), 292–304. doi:10.1016/j.jcrysgro.2004.10.093

Sustainable Development Goal. (n.d.). Retrieved from https://sustainabledevelopment.un.org/sdg7

Sustainable Development Goals. (n.d.) Retrieved from https://sustainabledevelopment.un.org/sdgs

Sustainable Energy for All (SE4ALL). (2015). *Scaling Up Finance for Sustainable Energy Investments – Report of the SE4ALL Advisory Board's Finance Committee – 2015*. Retrieved from https://www.seforall.org/sites/default/files/SE4All-Advisory-Board-Finance-Committee-Report.pdf

Sutton, P. W. (2007). *The environment: a sociological introduction*. London: Polity Press.

Szodruch, J., Grimme, W., Blumrich, F., & Schmid, R. (2011). Next generation single-aisle aircraft –requirements and technological solutions. *Journal of Air Transport Management*, *17*(1), 33–39. doi:10.1016/j.jairtraman.2010.10.007

Tambke, J. (2006). Short-term Forecasting of Offshore Wind Farms Production – Developments of the Anemos Project. *Proc. of the European Wind Energy Conference 2006*, 27.

Tambke, J., Lange, M., Focken, U., & Heinemann, D. (2003). Previento meets Horns Rev - Short-term Wind Power Prediction - Adaptation to Offshore Sites in CD. *Proceedings of the 2003 European Wind Energy Association Conference, EWEC'03*.

Tanaka, Y., Chapman, A., Sakurai, S., & Tezuka, T. (2017). Feed-in Tariff Pricing and Social Burden in Japan: Evaluating International Learning through a Policy Transfer Approach. *Social Sciences*, *6*(4), 27. doi:10.3390ocsci6040127

Tanner, C. (2009). Effects of school design on student outcomes. *Journal of Educational Administration*, *47*(3), 381–399. doi:10.1108/09578230910955809

Tan, R. H. G., & Chow, T. L. (2016). A comparative study of feed in tariff and net metering for UCSI University North Wing Campus with 100kW solar photovoltaic system. *Energy Procedia*, *100*, 86–91.

Tantareanu, C. (1992). *Wind Prediction in Short Term: A first Step for a Better Wind Turbine Control*. Nordvestjysk Folkecenter for Vedvarende Energi.

Technical Report of the Main Directorate of Pumping Stations Energy and Communications of the Ministry of Ministry of Agriculture and Water Resources of the Republic of Uzbekistan. (2006). Tashkent.

Teoh, L. E., & Khoo, H. L. (2016). Green air transport system: An overview of issues, strategies and challenges. *KSCE Journal of Civil Engineering*, *20*(3), 1040–1052. doi:10.100712205-016-1670-3

Thakur, T., Deshmukh, S. G., Kaushik, S., & Kulshrestha, M. (2005). Impact assessment of the Electricity Act 2003 on the Indian Power Sector. *Energy Policy*, *33*(9), 1187–1198. doi:10.1016/j.enpol.2003.11.016

The Denver Post. (2008). Xcel's new 'green' plan has cost savings in mind [Editorial]. *The Denver Post*. Retrieved from http://www.denverpost.com/editorials/ci_10180088

The International Institute for Sustainable Development. (2012). *Sustainable development timeline.* Retrieved from https://www.iisd.org/pdf/2012/sd_timeline_2012.pdf

The New Climate Economy. (2018). *Unlocking the inclusive growth story of the 21st Century.* Retrieved from: https://newclimateeconomy.report/2018/wp-content/uploads/sites/6/2018/09/NCE_2018_FULL-REPORT.pdf

The Plant. (n.d.). Retrieved from https://www.bubblydynamics.com/the-plant/

Theesfeld, I. (2004). Constraints on Collective Action in a Transitional Economy: The Case of Bulgaria's Irrigation Sector. *World Development*, *32*(2), 251–271. doi:10.1016/j.worlddev.2003.11.001

Tomar, A. (2011). *Introduction to Zig Bee Technology*. Retrieved from https://www.element14.com/community/servlet/JiveServlet/previewBody/37177-102-1-219424/Introduction%20to%20Zigbee%20Technology.pdf

Tomosk, S., Haysom, J. E., Hinzer, K., Schriemer, H., & Wright, D. J. (2017). Mapping the Geographic Distribution of the Economic Viability of Photovoltaic Load Displacement Projects in SW USA. *Renewable Energy*, *107*, 101–112. doi:10.1016/j.renene.2017.01.049

Tomosk, S., Haysom, J. E., & Wright, D. J. (2017). Quantifying Economic Risk in Photovoltaic Power Projects. *Renewable Energy*, *109*, 422–433. doi:10.1016/j.renene.2017.03.031

Tomoya, H., Kazuya, M., & Mitsuru, U. (2009). Sulfonated aromatic hydrocarbon polymers as proton exchange membranes for fuel cells. *Polymer*, *50*(23), 5341–5357. doi:10.1016/j.polymer.2009.09.001

Tompkins, E. L., & Amundsen, H. (2008). Perceptions of the effectiveness of the United Nations Framework Convention on Climate Change in advancing national action on climate change. *Environmental Science & Policy*, *11*(1), 1–13.

Torres, J. L., Garcia, A., De Blas, M., & De Francisco, A. (2005). Forecast of hourly average wind speed with ARMA models in Navarre (Spain). *Solar Energy*, *79*(1), 65–77. doi:10.1016/j.solener.2004.09.013

Trabacchi, C., & Mazza, F. (2015). *Emerging solutions to drive private investment in climate resilience*. Venice: Climate Policy Initiative.

Trefke, J., Rohjans, S., Uslar, M., Lehnhoff, S., Nordstrom, L., & Saleem, A. (2013). *Smart Grid Architecture Model Use Case Management in a large European Smart Grid Project. In 4th IEEE European Innovative Smart Grid Technologies*. ISGT.

Troen, I., & Landberg, L. (1990). Short-Term Prediction of Local Wind Conditions. *Proceedings of the European Community Wind Energy Conference*, 76-78.

U.S. EIA. (2013, May). *Feedin tariff: A policy tool encouraging deployment of renewable electricity technologies*. NREL technical report, State Clean Energy. Policies Analysis Project. Accessed October 26, 2018 from: https://www.nrel.gov/docs/fy09osti/45551.pdf

U.S. Geological Survey (USGS). (n.d.). Retrieved from https://www.usgs.gov/

U.S. Partnership for Renewable Energy Finance. (2017). *Tax credits, tax equity and alternatives to spur clean energy financing*. ACORE.ORG. Retreived from: https://acore.org/wp-content/uploads/2017/12/Tax-Credits-Tax-Equity-for-Clean-Energy-Financing.pdf

Ubi, E. N. (2014). Foreign aid and development in Sino-African relations. *Journal of Developing Societies*, *30*(3), 243–272.

UCLA. (2018). *Climate Consultant*. Retrieved from http://www.energy-design-tools.aud.ucla.edu/climate-consultant/request-climate-consultant.php

UN (United Nations). (2015). *Transforming our world: the 2030 Agenda for Sustainable Development*. New York: United Nations.

UN. (2018). *Suitable development agenda*. Available online: https://www.un.org/sustainabledevelopment/development-agenda/

United Nations Climate Change. (2018, June 20). *Extreme weather continues in 2018: A continuing call to climate action*. Retrieved from https://unfccc.int/news/extreme-weather-continues-in-2018-a-continuing-call-to-climate-action

United Nations Conference on Trade and Development. (n.d.). Retrieved October 24, 2018, from United Nations Conference on Trade and Development Classifications: http://unctadstat.unctad.org/EN/Classifications.html

United Nations Economic and Social Council (UN ESC). (2017, May). *Progress towards the Sustainable Development Goals*. Paper presented at the July 2017 session of the Economic and Social Council, New York, NY.

United Nations Framework Convention on Climate Change. (2008). *Report of the Conference of Parties on its thirteenth session*. Retrieved from https://unfccc.int/resource/docs/2007/cop13/eng/06a01.pdf

United Nations Intergovernmental Committee of Experts on Sustainable Development Financing. (2014). *Report of the Intergovernmental Committee of Experts on Sustainable Development Financing*. New York, NY: United Nations.

United Nations New York. (2018). *The Sustainable Development Goals Report.* Retrieved from https://unstats.un.org/sdgs/files/report/2018/TheSustainableDevelopmentGoalsReport2018-EN.pdf

Upton, N., Teal, E. J., & Felan, J. T. (2001). Strategic and business planning practices of fast growth family firms. *Journal of Small Business Management, 39*(1), 60–72. doi:10.1111/0447-2778.00006

USDA Forest Service. (2018, August 5). *Active fire mapping program.* Retrieved from https://fsapps.nwcg.gov/#

Usikalu, M. R. (2009). Health impact of climate change due to combustion of fossil fuel. *International Journal of Physical Sciences, 4*(13), 880–884.

Vaccaro, B. A., Popov, M., Villacci, D., & Terzija, V. (2011). An Integrated Framework for Smart Microgrids Modeling, Communication, and Verification. *Proceedings of the IEEE, 99*(1), 119–132. doi:10.1109/JPROC.2010.2081651

Val-Matic Valve & Manufacturing Corp. (n.d.). Retrieved January 20, 2016, from http://www//valmatic.com

van Huijstee, M., Pollock, L., Glasbergen, P., & Leroy, P. (2011). Challenge for NGOs Partnering with Corporations: WWF Netherlands and the Environmental Defense Fund. *Environmental Values, 20*(1), 43–74. doi:10.3197/096327111X12922350166030

Veteto, J. R., & Lockyer, J. (2008). Environmental anthropology engaging permaculture: Moving theory and practice toward sustainability. *Culture & Agriculture, 30*(1&2), 47–58. doi:10.1111/j.1556-486X.2008.00007.x

Viederman, S. (1993). *The economics and economy of sustainability: Five capitals and three pillars, talk delivered to Delaware Estuary Program.* New York: Noyes Foundation.

Vieira, F., & Ramos, H. (2008). Hybrid solution and pump-storage optimization in water supply system efficiency: A case study. *Energy Policy, 36*(11), 4142–4148. doi:10.1016/j.enpol.2008.07.040

Vijay, R. (2006, Spring). Fuel Cells. *The Electrochemical Society Interface.*

Villalonga, B., & Amit, R. (2006). How do family ownership, control and management affect firm value? *Journal of Financial Economics, 80*(2), 385–417. doi:10.1016/j.jfineco.2004.12.005

Villamayor-Tomas, S. (2017). Disturbance features, coordination and cooperation: An institutional economics analysis of adaptations in the Spanish irrigation sector. *Journal of Institutional Economics, 14*(3), 1–26.

Villamayor-Tomas, S., Grundmann, P., Epstein, G., Evans, T., & Kimmich, C. (2015). The Water-Energy-Food Security Nexus through the Lenses of the Value Chain and the Institutional Analysis and Development Frameworks. *Water Alternatives, 8*(1), 735–755.

Vishnevskii, K. P. (n.d.). *Transient processes of pumping stations. Calculations of hydraulic shock in pressure pipelines*. Retrieved January 29, 2016, from http://library.fsetan.ru/doc/perehodnyie-protsessyi-nasosnyih-stantsij-raschetyi-gidravlicheskogo-udara-v-napornyih-truboprovodah-posobie-k-vsn-33-2212-87-meliorativnyie-sistemyi-i-sooruzheniya-nasosnyie-stantsii-normyi-proektirovaniya

Vladimir, S. B. (2012). Fuel cells Problems and solutions (2nd ed.). Moscow, Russia: Electrochemical Society.

Vladimirov, M., Georgiev, A., & Kolarova, S. (2018). *Development of Small-Scale Renewable Energy Sources in Bulgaria: Legislative and Administrative Challenges*. Sofia, Bulgaria: Center for the Study of Democracy.

Wadu Mesthrige, J., & Kwong, H. Y. (2018). Criteria and barriers for the application of green building features in Hong Kong. *Smart and Sustainable Built Environment*, *7*(3/4), 251–276. doi:10.1108/SASBE-02-2018-0004

Walsh, B. (2014, March 31). Warming world threatens us all, warns U.N. report. *Time*. Retrieved from http://time.com/43118/climate-change-global-warming-united-nations/

Wang, J., Ran, R., Song, Z., & Sun, J. (2017). Short-Term Photovoltaic Power Generation Forecasting Based on Environmental Factors and GA-SVM. *Journal of Electrical Engineering & Technology*, *12*(1), 64–71. doi:10.5370/JEET.2017.12.1.064

Wang, L., Han, Y., Feng, X., Zhou, J., Qi, P., & Wang, B. (2016). Metal–organic frameworks for energy storage: Batteries and supercapacitors. *Coordination Chemistry Reviews*, *307*, 361–381. doi:10.1016/j.ccr.2015.09.002

Wang, Y., Kim, Y., Xie, Q., Chang, N., & Pedram, M. (2011). Charge migration efficiency optimization in hybrid electrical energy storage (HEES) systems. *Proceedings of the 17th IEEE/ACM international symposium on Low-power electronics and design*. 10.1109/ISLPED.2011.5993620

Wargocki, P., Wyon, D. P., Matysiak, B., & Irgens, S. (2005). The effects of classroom air temperature and outdoor air supply on performance of school work by children. Proceedings of Indoor Air, 1(1), 368-372.

Wargocki, P., Wyon, D. P., Baik, Y. K., Clausen, G., & Fanger, P. O. (1999). Perceived air quality, sick building syndrome(SBS) and productivity in two different pollution loads. *Indoor Air*, *9*(3), 165–179. doi:10.1111/j.1600-0668.1999.t01-1-00003.x PMID:10439554

Watson, S. J., & Montavon, C. (2003). CFD modelling of the wind climatology at a potential offshore farm site. *Proc. Europ. Wind Energy Conf. EWEC*.

Watt, G., & Fechner, H. (2009). Photovoltaic market and industry trends–latest results from the IEA PVPS programme. *Elektrotechnik und Informationstechnik, 126*(9), 328-330.

WB - World Bank. (2013). *Republic of Bulgaria: Power sector rapid assessment*. Washington, DC: World Bank. Retrieved from http://documents.worldbank.org/curated/en/391271468020330576/Republic-of-Bulgaria-power-sector-rapid-assessment

WB - World Bank. (2015). *Adapting to higher energy costs: Findings from qualitative studies in Europe and central Asia.* ACS12511. Washington DC: World Bank. Retrieved from http://documents.worldbank.org/curated/en/582021468245408944/pdf/ACS12511-WP-REPLACEMENT-Adapting-to-Energy-Costs-web.pdf

Wei, Z., & Weimin, W. (2010, March). Wind speed forecasting via ensemble Kalman Filter. In *2010 2nd International Conference on Advanced Computer Control* (Vol. 2, pp. 73-77). IEEE.

Weiland, P. (2006). Biomass digestion in agriculture: A successful pathway for the energy production and waste treatment in Germany. *Engineering in Life Sciences*, *6*(3), 302–309. doi:10.1002/elsc.200620128

Wells, N., & Evans, G. (2003). Nearby Nature: A Buffer of Life Stress among Rural Children. *Environment and Behavior*, *35*(3), 311–330. doi:10.1177/0013916503035003001

Welter, F., Baker, T., Audretsch, D. B., & Gartner, W. B. (2017). *Everyday Entrepreneurship—A Call for Entrepreneurship Research to Embrace Entrepreneurial Diversity*. Thousand Oaks, CA: SAGE Publications. doi:10.1111/etap.12258

White, R. (2004). *Young children's relationship with nature: Its importance to children's development & the earth's future.* White Hutchinson Leisure & Learning Group.

Whittingham, M. S. (2012). History, evolution, and future status of energy storage. *Proceedings of the IEEE, 100*, 1518-1534. 10.1109/JPROC.2012.2190170

Will, F., Tay, G., Becker, A., Carnelly, D., Eychenne, F., & Hornung, M. (2016). Green Airlines 2025: Environment and Sustainability in Commercial Aviation – A Scenario Study. In *16th AIAA Aviation Technology, Integration, and Operations Conference* (p. 3756). AIAA.

Williams, V., Noland, R. B., & Toumi, R. (2002). Reducing the climate change impacts of aviation by restricting cruise altitudes. *Transportation Research Part D, Transport and Environment*, *7*(6), 451–464. doi:10.1016/S1361-9209(02)00013-5

Wilson, A. (2013). Xcel's Energy Model Outdated [Opinion]. *The Daily Camera.* Retrieved from LexisNexis Academic.

Wilson, B. (2017, August 15). *Residential design for Midwest permaculture home.* Retrieved from https://midwestpermaculture.com/2017/08/residential-design-for-midwest-permaculture-home/

Winslow Food Forest. (n.d.). Retrieved from https://www.winslowfoodforest.com/about/

Winter, M., & Brodd, R. J. (2004). *What are batteries, fuel cells, and supercapacitors?* ACS Publications.

Wiser, R., Namovicz, C., & Smith, M. G. (2007). The Experience with Renewable Portfolio Standards in the United States. *The Electricity Journal*, *20*(4), 8–20. doi:10.1016/j.tej.2007.03.009

Wolf, S. (2018, October 8). *Report details brutal cost of climate inaction.* Retrieved from https://medium.com/center-for-biological-diversity/report-details-brutal-cost-of-climate-inaction-8a5a038bd6e0

Wood, A. J., & Wollenberg, B. F. (2012). *Power generation, operation, and control.* John Wiley & Sons.

Worku, M. Y., Hassan, M. A., & Abido, M. A. (2019). Real Time Energy Management and Control of Renewable Energy based Microgrid in Grid Connected and Island Modes. *Energies, 12*(2), 276. doi:10.3390/en12020276

Worland, J. (2018, October 8). Scientists just laid out paths to solve climate change: We aren't on track to do any of them. *Time.* Retrieved from http://time.com/5418134/ipcc-climate-change-report-2030-crisis/

World Bank. (2019a). *Scaling solar active engagements.* Retrieved from https://www.scalingsolar.org/

World Bank. (2019b). *Energy sector management assistance program annual report 2018.* Washington, DC: World Bank.

World Business Council for Sustainable Development. (2000). Corporate Social Responsibility: Making Good Business Sense. Geneva, Switzerland: Author.

World Commission on Environment and Development. (1987). *Our common future.* Oxford, UK: Oxford University Press.

World Energy Council. (2016). *World Energy Resources 2016 Summary.* Retrieved from https://www.worldenergy.org/wp-content/uploads/2016/10/World-Energy-Resources-Full-report-2016.10.03.pdf

World Energy Council. (2018). World Energy Insights Brief 2018, Is Blockchain in Energy Driving an Evolution or a Revolution? London: Author.

World Vision. (2018, August 1). *From the field: Facts, FAQs, and how to help.* Retrieved from https://www.worldvision.org/

Wouters, C., Fraga, E., & James, A. (2015). An energy integrated, multi-microgrid, MILP (mixed-integer linear programming) approach for residential distributed energy system – planning – A south Australian case-study. *Energy, 85,* 30–44. doi:10.1016/j.energy.2015.03.051

Wright, D. J., Badruddin, S., & Robertson-Gillis, C. (2018). Micro-Tracked CPV Can Be Cost Competitive With PV in Behind-The-Meter Applications With Demand Charges. *Frontiers in Energy Research, 6*(97). doi:10.3389/fenrg.2018.00097

Wu, J., & Tran, N. K. (2018). Application of Blockchain Technology in Sustainable Energy Systems: An Overview. *Journal of Sustainability, MDPI, 10*(9), 3067–3089. doi:10.3390u10093067

Würth, I., Valldecabres, L., Simon, E., Möhrlen, C., Uzunoğlu, B., Gilbert, C., ... Kaifel, A. (2019). Minute-Scale Forecasting of Wind Power—Results from the Collaborative Workshop of IEA Wind Task 32 and 36. *Energies*, *12*(4), 712. doi:10.3390/en12040712

Wu, Y. K., Lee, C. Y., Tsai, S. H., & Yu, S. N. (2010). Actual Experience on the Short-Term Wind Power Forecasting at Penghu-From an Island Perspective. *Proceedings of the 2010 International Conference on Power System Technology*, 1-8.

Xcel Energy. (2014). *2013 Owned and Purchased Energy: Public Service Company of Colorado Power Supply Mix*. Retrieved from http://www.xcelenergy.com/About_Us/Our_Company/Power_Generation/Power_Generation_Fuel_Mix_-_PSCo

Xcel Energy. (2015). Xcel Energy is the nation's No. 1 wind power provider. *Wind Power on our System*. Retrieved from http://www.xcelenergy.com/Environment/Renewable_Energy/Wind/Wind_Power_on_Our_System

Xcel Energy. (2017). Our Company. *About Us*. Retrieved from http://www.xcelenergy.com/About_Us/Our_Company/Service_Areas

Xia, J. R., Zhao, P., & Dai, Y. P. (2010) Neuro-Fuzzy Networks for Short-Term Wind Power Forecasting. *Proceedings of the International Conference on Power System Technology*, 1-5.

Xiaoliang, H., Tosiyoki, H., & Yoichi, H. (2014). *System design and converter control for super capacitor and battery hybrid energy system of compact electric vehicles*. Paper presented at the Power Electronics and Applications (EPE'14-ECCE Europe), 2014 16th European Conference on. 10.1109/EPE.2014.6911060

Xu, J., Khan, M., Mumtaz, M., Shahid, M., Habib, S., Hashmi, K., & Tang, H. (2018). A Hierarchical Control Methodology for Renewable DC Microgrids Supporting a Variable Communication Network Health. *Electronics (Basel)*, *7*(12), 418. doi:10.3390/electronics7120418

Yang, Y., & Dong, L. (2013, August). Short-term PV generation system direct power prediction model on wavelet neural network and weather type clustering. In *2013 5th International Conference on Intelligent Human-Machine Systems and Cybernetics* (Vol. 1, pp. 207-211). IEEE. 10.1109/IHMSC.2013.56

Yan, W., Cui, Z., & Gil, M. J. Á. (2016). Assessing the impact of environmental innovation in the airline industry: An empirical study of emerging market economies. *Environmental Innovation and Societal Transitions*, *21*, 80–94. doi:10.1016/j.eist.2016.04.001

Yaxiang, L., Lianqin, W., Kathrin, P., Mo, Q., Maria, M. T., John, V., & Qiong, C. (2017). Halloysite-derived nitrogen doped carbon electrocatalysts for anion exchange membrane fuel cells. *Journal of Power Sources*, *372*, 82–90. doi:10.1016/j.jpowsour.2017.10.037

Ye, L.-C., Rodrigues, J. F., & Lin, H. X. (2017). Analysis of feed-in tariff policies for solar photovoltaic in China 2011 -2016. *Applied Energy*, *203*, 496–505. doi:10.1016/j.apenergy.2017.06.037

Yender, G. L. (1998). Battery recycling technology and collection processes. *Electronics and the Environment, 1998. ISEE-1998. Proceedings of the 1998 IEEE International Symposium on.*

Ye, Q., Jiaqi, L., & Mengye, Z. (2017). *Wind Curtailment in China and Lessons from the United States*. Brookings: Tsinghua Centre for Public Policy. Retrieved from https://www.brookings.edu/wp-content/uploads/2018/03/wind-curtailment-in-china-and-lessons-from-the-united-states.pdf

Ying, O., Wen, C. T., Shin, C. J., Fu, S. C., Jie, W., Hai, L., ... Chunli, G. (2018). Novel composite polymer electrolyte membrane using solid superacidic sulfated zirconia - Functionalized carbon nanotubes modified chitosan. *Electrochimica Acta, 264*, 251–259. doi:10.1016/j.electacta.2018.01.131

Yin, R. K. (1984). *Case study research: Design and methods* (2nd ed.). Thousand Oaks, CA: Sage Publications.

Yin, R. K. (2009). *Case Study Research. Design and methods*. Thousand Oaks, CA: SAGE.

Yin, R. K. (2012). *Applications of case study research* (3rd ed.). Thousand Oaks, CA: SAGE.

Yona, A., Senjyu, T., Saber, A. Y., Funabashi, T., Sekine, H., & Kim, C. H. (2007) Application of neural network to one-day-ahead 24 hours generating power forecasting for photo- voltaic system. *IEEE Intelligent Systems Applications to Power Systems, 2007. ISAP 2007. International Conference*, 1–6.

Yu, C. C., & Jhao, R. C. (2011). Effects of surface chemical states of carbon nanotubes supported Pt nanoparticles on performance of proton exchange membrane fuel cells. *International Journal of Hydrogen Energy*, *36*(11), 6826–6831. doi:10.1016/j.ijhydene.2011.02.114

Yu, Y. (2014). Climate finance, Africa and China's role. *Africa East-Asian Affairs*, *1*, 36–57.

Zadek, S. (2006). Responsible Competitiveness: Reshaping Global Markets through Responsible Business Practices. *Corporate Governance*, *6*(4), 334–348. doi:10.1108/14720700610689469

Zaheeruddin, & Manas, M. (2015). Renewable energy management through microgrid central controller design: An approach to integrate solar, wind and biomass with battery. *Energy Reports, 1*, 156–163. doi:10.1016/j.egyr.2015.06.003

Zahra, S. A., Hayton, J. C., Neubaum, D. O., Dibrell, C., & Craig, J. (2008). Culture of family commitment and strategic flexibility: The moderating effect of stewardship. *Entrepreneurship Theory and Practice*, *32*(6), 1035–1054. doi:10.1111/j.1540-6520.2008.00271.x

Zeng, J. W., & Qiao, W. (2011). Support Vector Machine-Based Short-Term Wind Power Forecasting. *Proceedings of the IEEE/PES Power Systems Conference and Exposition*, 1-8. 10.1109/PSCE.2011.5772573

Zeng, J., & Qiao, W. (2013). Short-term Solar Power Prediction using a Support Vector Machine. *Renewable Energy*, *52*, 118–127. doi:10.1016/j.renene.2012.10.009

Zhang, X., & Ukil, A. (2019). Enhanced Hierarchical Control of Hybrid Energy Storage System in Microgrids. *IECON 2018 - 44th Annual Conference of the IEEE Industrial Electronics Society, 1*, 1801–1806. 10.1109/iecon.2018.8591343

Zhang, K., Wang, Q., Liang, Q., & Chen, H. (2016). A bibliometric analysis of research on carbon tax from 1989 to 2014. *Renewable & Sustainable Energy Reviews*, *58*, 297–310. doi:10.1016/j.rser.2015.12.089

Zhang, Y., Lundblad, A., Campana, P. E., Benavente, F., & Yan, J. (2017). Battery sizing and rule-based operation of grid-connected photovoltaic-battery system: A case study in Sweden. *Energy Conversion and Management*, *133*, 249–263. doi:10.1016/j.enconman.2016.11.060

Zhao, G., & Davison, M. (2009). Optimal control of hydroelectric facility incorporating pump storage. *Renewable Energy*, *34*(4), 1064–1077. doi:10.1016/j.renene.2008.07.005

Zhao, Y., & Zhu, K. (2014). CH3NH3Cl-assisted one-step solution growth of CH3NH3PbI3: Structure, charge-carrier dynamics, and photovoltaic properties of perovskite solar cells. *The Journal of Physical Chemistry C*, *118*(18), 9412–9418. doi:10.1021/jp502696w

Zhao, Z.-Y., Chen, Y.-L., & Chang, R.-D. (2016). How to stimulate renewable energy power generation effectively? - China's incentive approaches and lessons. *Renewable Energy*, *92*, 147–156. doi:10.1016/j.renene.2016.02.001

Zogo, B., Cedrick, E., & Long, W. (2017). International motivation in renewable energy: A PPP approach. *Energy Procedia*, 229–238.

Zurich, E. T. H. (2017). *The role of multilateral development banks as enabler for new power generation technologies*. Zurich: ETH Zurich.

About the Contributors

Peter Yang's main research area is renewable energy and energy efficiency as solutions to environmental and ecological challenges of the global economy. Research projects he has completed include environmental and ecological impact of carbon based energy production and consumption in major economics; investment, installation and consumption of renewable energy technologies; renewable energy promotion policies and regulations, such as feed-in tariffs, fuel and carbon taxes, renewable energy targets, and CO2 reduction targets. These projects resulted in two books, Rolling Back the Tide of Climate Change: Renewable Solutions and Policy Instruments in the USA and China (2015) and Renewables Are Getting Cheaper (2016), and a number of peer reviewed journal papers, book chapters, book reviews, and conference papers. His current energy related research interests include latest R&D of renewable energy technologies; challenges and solutions of renewable energy technologies, grid integration and energy storage; energy efficiency in transportation and buildings; as well as teaching, training, and public education of renewable energy transformation.

* * *

Maysam Abbod is attached to Brunel University London as Reader. He received BSc degree in Electrical Engineering from University of Technology in 1987. PhD in Control Engineering from University of Sheffield in 1992. From 1993 to 2006 he was with the Department of Automatic Control and Systems Engineering at the University of Sheffield as a research associate and senior research fellow.

Janice Ashworth has been managing Ottawa Renewable Energy Cooperative (OREC) since 2011 and has over 10 years of experience in wind and solar energy. At OREC, she is responsible for business development, project management, and financing. Janice is a member of the Sustainability Committee of the Ottawa Chamber of Commerce and sits on the Energy Evolution Steering Committee with the City of Ottawa. She coordinated the Nova Scotia Sustainable Electricity Alliance and worked in wind energy with the community-owned wind field. She has a background in community organizing with Ecology Ottawa and Ecology Action Centre. Janice has a Master's in Environmental Studies from Dalhousie University with a focus on community energy. In her time as General Manager of OREC, Janice has been named as a 40 Under 40 business leader by the Ottawa Business Journal and was awarded the 2017 Solar Woman of Distinction by Women in Renewable Energy.

Abderrahim Assab is an Analyst at the European Bank for Reconstruction and Development (EBRD) working on the financing of Energy Efficiency and Climate Change projects. Prior to joining the EBRD, Abderrahim was part of the Organization for Economic Cooperation and Development (OECD) Institutional Investors and Long-term investment project. Abderrahim is a graduate of Imperial College London.

Sana Badruddin received her M.Sc in Environmental Sustainability from the University of Ottawa in 2018 and presently works as an analyst assessing the environmental impacts of infrastructure projects at Infrastructure Canada. Sana's major research during her Master's focused on understanding the contribution of renewables on Ontario's rising electricity prices, after which she worked with David Wright to perform data analytics of the economic viability of commercial-scale solar microgrids in Ontario. Previously, she helped fund small-scale renewable energy projects in off-grid communities at Indigenous and Northern Affairs Canada.

gree in electrical engineering from Quaid-e-Awam University Nawabshah Pakistan in 2009 and the M.E degree in electrical power engineering from Mehran University Jamshoro Pakistan in 2015. He is a Lecturer at Quaid-e-Awam University Larkana campus and currently doing his PhD studies from Aston University Birmingham United Kingdom. His research interests include the design of electrical machines and drives and analyzing especial machines, Microgrids, Energy Storage and Electric Vehicles. He is a member of IEEE UK and Ireland section.

Umit Cali is a seasoned technology researcher conducting academic-based, peer-review quality research and industrial experience in machine learning, energy systems, blockchain technology and related areas. He completed his PhD degree in the University of Kassel in Germany in 2010. He has worked for IBM International (Istanbul/Turkey) as a network engineer and project manager, a senior researcher in energy informatics and economics (Fraunhofer Institute – Germany), an Energy Consultant (Lahmeyer International- Germany), a Manager in wind energy project development and energy markets/energy economics (One of the largest German utilities EnBW – Germany), Chief Technology Officer for an international service provider in energy efficiency (KREEN Renewables GmbH- Germany/Switzerland) and assistant professor in three universities (Gediz University- Turkey, the University of Wisconsin at Platteville and the University of North Carolina at Charlotte). Dr. Cali is also founder of a startup companies (Ennova Enerji AS and EnergyXchain LLC) which is established in Turkey and in US, respectively, to develop machine learning and blockchain technology solutions to the energy industry.

Wenping Cao received the B.Eng in electrical engineering from Beijing Jiaotong University, Beijing, China, in 1991, and the PhD degree in electrical machines and drives from the University of Nottingham, Nottingham, U.K., in 2004. He is currently Chair Professor of Electrical Power Engineering and Head of Power Electronics, Machines and Power System (PEMPS) Group at Aston University, Birmingham, U.K. Prof. Cao is presently the Chairman for the Industrial Electronics Society, IEEE UK and Ireland Section, and also a “Royal Society Wolfson Research Merit Award” holder, U.K. He was a semi-finalist at the “Annual MIT-CHIEF Business Plan Contest”, U.S.A., in 2015; the “Dragon’s Den Competition Award” winner from Queen’s University Belfast, U.K., in 2014, the “Innovator of the Year Award” winner from Newcastle University, U.K., in 2013. His research interests include fault analysis and condition monitoring of electrical machines, power electronics and batteries.

Xiangping Chen received her PhD degree in Electronic and Electrical Engineering in 2013, Newcastle University, Newcastle upon Tyne, UK. She is currently an associate professor in Guizhou University, China. Her research interests are in renewable energy with energy storage, energy management and energy efficiency improvement. She is also a Guest Editor-In-Chief, InTech book, “Advancements in Energy Storage Technologies”, 2018.

Dumisani Chirambo is a diligent and resourceful analyst, entrepreneur and researcher with over 9 years’ experience in the fields of climate change strategies (finance, mitigation and adaptation), environmental management, energy policy, environmental law and policy, and project management. Other competencies include international work experience and knowledge of various applications such as Geographic Information Systems. Some of his notable publications include Addressing the Renewable Energy Financing Gap in Africa: Integrated Renewable Energy Financing in Malawi. Journal of Renewable & Sustainable Energy Reviews, 62, 793-803. 2016.; and The Climate Finance and Energy Investment Dilemma in Africa: Lacking Amidst Plenty. Journal of Developing Societies, 30 (4), 415–440. 2014.

Kate Clark is the Director of the Undergraduate Environment & Sustainability program at Western Colorado University in Gunnison, Colorado. She also teaches in the Master of Environmental Management program and undergraduate Sociology program. She studies environmental justice as well as the politics and political economy of electricity.

Keriann F. Conroy received her Master of Environmental Management at Western Colorado University, conducting qualitative research on electricity democracy among rural electric cooperatives. She currently serves as a fact-checker for High Country News, and works as a research associate with Sustainable Development Strategies Group in Gunnison, Colorado.

Cinzia Dessì is an Assistant Professor of Management in the Department of Economics and Business, University of Cagliari, Italy. She received her PhD in Management and holds a master's degree in Economics from the University of Cagliari. In March 2018 she received the CFBA Certificate in Family Business Advising by the Family Firm Institute. Prior to join the academia, she worked as manager in the business of her family. She has published articles in leading journals (e.g., Family Business Review, Journal of Small Business and Enterprise Development) and received relevant grants for her studies (e.g., Academy of Management Best Paper - 2009 and 2015 -, Carolyn Dexter Award – 2015 – FFI 2017 Scholarship). Her current research interest is in the areas of entrepreneurship and family firms, and focuses on growth, innovation, governance, and succession.

Angela Dettori is a Research Fellow at the Department of Economics and Business, University of Cagliari, Italy. She received her PhD in Management and holds a master's degree in Economics from the University of Cagliari. She has been teaching Corporate Social Responsibility for several years. She has published articles in various journals (for example, Journal of Management and Sustainability, The International Journal of Environmental Sustainability) and has participated in several conferences (for example, Toulon Verona Conference, Academy of Management). Her main interest in research focuses on sustainable development and sustainability-oriented behavior, especially in family businesses.

Alexander Edwards completed his PhD in Environmental Policy from King's College London, with a focus on public policy, climate change, and energy. His interests include fuel poverty, sustainable development, and the politics of energy use. He worked as a research analyst on energy and climate policy analysis at the Global Green Growth Institute.

Michela Floris is an Assistant Professor of Management in the Department of Economics and Business, University of Cagliari, Italy. She received her PhD in Management and holds a master's degree in Economics from the University of Cagliari. She has been teaching Family Business Management for several years. She has published articles in leading journals (e.g., Family Business Review, Journal of Small Business and Enterprise Development) and received relevant grants for her studies (e.g., Academy of Management Best Paper - 2009 and 2015 -, Carolyn Dexter Award - 2015). Her main interest of research focuses on strategy, organizational goals and goal-driven behaviors in family firms, including entrepreneurship, marketing, innovation, and growth.

Jeremy Gibberd is an Architect and Researcher. His research interests are in sustainable built environments, education and community buildings and building performance. Further detail is available at www.jeremygibberd.com.

Debjani Goswami Received Master in Electronics and Microwave communication from Burdwan University India and Master of Physics from Jadavpur University, India and currently the teaching fellow and the part time PhD researcher at Aston University, Birmingham, UK. Her research interest is Electrical Vehicle, power train, machine design and energy system.

Fayyaz Jandan received his B.E degree in Electrical Engineering in 2012, from Quaid-e-Awam University, Nawabshah Sindh Pakistan. He is currently a Lecturer at Quaid-e-Awam University Larkana campus, Sindh-Pakistan. He has done his M.E in Electrical Power Engineering from Mehran University of Engineering & Technology Jamshoro Pakistan. His research interests are Power system analysis and power quality, signal processing and machine learning based power quality disturbances analysis.

Aleksandr Khokhlov was born in 1950 in Samarkand city, Uzbekistan. September, 1967 – June, 1972: study at Tashkent Polytechnic Institute, Tashkent, Uzbekistan. Major: Hydropower sets, Engineer in hydro-electric engineering. September, 1972 - September, 1980: work at the Ministry of water resources, Tashkent, Uzbekistan. September, 1980 - February, 1990; work and post-graduate study at "SANIIRI" Research Institute, Tashkent, Uzbekistan. Major: Melioration. May, 1990: received the degree of PhD. Dissertation subject "Improvement of operating modes of pumping stations with long pipelines". February, 1990 – present: work in Research-and-production enterprise "Vodopodyomnik" ("Water lifting") as a director. He totally has 75 publications: two book chapters IGI-Global, four monographs, seventeen patents for invention, forty tree articles in scientific technical journals and others.

Vladimir Khokhlov was born in 1976 in Tashkent city, Uzbekistan. September, 1993 – June, 1998: study at Tashkent Institute of Irrigation and Mechanization of Agriculture, Tashkent, Uzbekistan Major: Water economic building, Engineer in hydro-technics. June, 1994 – present: work in Research-and-production enterprise "Vodopodyomnik" ("Water lifting") in series as a technician, engineer, researcher and deputy director. September, 1998- September, 2001: post-graduate study at Tashkent State Technical University, Tashkent, Uzbekistan. Major: Hydro-technical and meliorative building. February, 2002: received the degree of PhD. Dissertation subject "Improvement of Drainage Water Pumping from the Buildings of Hydro-power Sets". January, 2004 – December, 2006: work and doctor of science degree study at Tashkent State Technical University. Major: Power systems and complexes. March, 2009: Received the degree of Doctor of science in technics. Dissertation subject "Power Saving Operating Modes of Pumps and Pumping Stations with Long Pipelines". Since September, 2011: Professor on Fluid mechanics and Hydraulics for the third academic level and for masters at Department of applied sciences of Turin Polytechnic University in Tashkent, Uzbekistan. Activities include lecturing, practice and laboratory training. He totally has 96 publications: two book chapter IGI-Global, two textbooks, six monographs, twelve patents for inventions, fifty two articles in scientific technical journals, twenty two reports in scientific conferences and others.

Ashish Kulkarni is currently holding the position of Assistant Professor of Economics at the institute. He completed his PhD studying the macro economics factors influencing business cycles in India. In his tenure with the industry, he has served in the chief economist office at a reputed renewable energy company, where he used to analyse economic variables that could impact the growth of the sector.

Claudio Lima is a thought leader in advanced energy (utilities, oil and gas) and telecom/IT, working with emerging technologies and digital transformation. He has a PhD in Electronic Engineering in 1995 at UKC (England, 1995), M.Sc. and B.Sc. in Electrical Engineering. He served as Vice Chair of the IEEE 2030 Smart Grid Standards. He led smart grid and advanced telecom/IT initiatives as Global Smart Grid CTO of Huawei Technologies and as Distinguished Member of Technical Staff (DMTS)/ Sr. Research Scientist at Sprint Advanced Technology Labs (Sprint ATL), in Silicon Valley-CA. He also served as co-chair of the Industrial Internet Consortium (IIC)/Energy, Utility and Oil & Gas. Currently, Dr. Lima is the chair of IEEE Blockchain in Energy Standards WG - P2418.5, vice-chair of IEEE Blockchain IoT Standards WG- P2418.1, member of the Blockchain/DLT Cybersecurity Industry Advisory Board (IAB) of PNNL/DOE, technical chair of the IEEE-PES Blockchain in Energy Workshop and IEEE-NIST Blockchain Global Summit, and former global standards director of EEA Ethereum Enterprise Alliance. He has more than 12+ USPTO patents and +150 conference and peer-review papers publications.

Jody Luna is the Founder of Conscious Designs, she has a Bachelor of Architecture from University of Miami, Coral Gables, FL and master's degree in architecture from Illinois Institute of Technology, Chicago, IL. Her master's thesis focused on development of a rural residential property in southwest Florida. Using sustainability as her guide, she wrote An Evolution of Architecture: A Hot, Humid Case Study focusing on ways to incorporate passive sustainable solutions into the standard architecture design process. She earned her Doctor of Education in instructional leadership, her dissertation researched sustainable education methodologies in an effort to understand how to increase sustainability efforts in an educational setting.

Maheswari M. is currently working as Professor, Department of Electrical and Electronics Engineering in Nalla Malla Reddy Engineering College, Hyderabad.She completed her B.E. Electrical and Electronics Engineering degree in K.S.Rangasamy College of Technology Thiruchengode with first class with distinction, M.E. Power Systems in Government College of Technology, Coimbatore and with first class with distinction and PhD at Anna University, Chennai in the year 2015. She has 18 years of teaching experience and having various administrative positions. She published her research work in 15 various international journals and 17 International Conferences. She is also a reviewer of leading journals.

Yazan Migdadi is an Associate Professor of Operations Management at Management and Marketing Department, Qatar University, Qatar. He was awarded PhD in Operations Strategy from Bradford University, UK, also awarded BA and MBA from Yarmouk University, Jordan. He worked as an Assistant and Associate Professor of Operations Management at Yarmouk university, Jordan. Also he worked as a Teaching Assistant and a Researcher of Operations and Information Management at Bradford University, School of Management, UK. His main research interest is reporting operations trategies in general and green operations strategy in particular: best practices, taxonomies, process, and typologies.

Maria Petrova is a researcher and teacher focused on understanding the human-energy-economy nexus by investigating the factors that influence acceptance of renewable energy technologies. She develops and uses case studies to inform state officials, policy makers, and managers about smoothing society's transition to new renewable energy technologies. She also examines the barriers for adopting renewable energy at the individual and state level by conducting surveys and interviews at the national and international level.

Padmavathi Rajangam is an Assistant Professor of Chemistry at Sree Sastha Institute of Engineering and Technology, Chennai-123. She has completed her B.Sc., Chemistry at Govt. Arts College, Ariyalur and M.Sc., Chemistry in Kundavai Nachiyar Govt. Women's College, Thanjavur. She got her Master of Technology in Polymer Science and Engineering at Anna University, Chennai and completed her PhD in Science and Humanities at Anna University, Chennai. Her area of Research experience is on Proton Exchange Membrane Fuel Cells. She has published 2 books and 15 papers in National and International journals.

Cameron Robertson-Gillis is in the process of completing his Bachelor of Commerce at the University of Ottawa, specializing in Finance. He is presently working on developing partnerships for a startup focused on manufacturing small-scale recycling facilities that turn discarded plastic into consumer ready products. Cameron has a professional background in policy, healthcare and the environment. Previously, he worked alongside David Wright to determine the economic viability of solar installations in Ontario and Southern California by performing data analytics and developing a model that optimizes the flow of energy throughout a solar microgrid. Additionally, he has experience supporting a research team focused on improving the delivery of care to people living with HIV, conducted statistical analyses for the Department of National Defence and has served as a Page in the House of Commons.

Tatyana Ruseva is an Associate Professor of Environmental Policy in the Department of Government and Justice Studies at Appalachian State University. Her research has appeared in the Journal of Environmental Management, Environmental Science and Policy, Ecological Economics, Society and Ecology, and edited volumes on human-environment interactions.

Gunasekharan S. is currently working as Professor in Mechanical Engineering Department in Lords Institute of Engineering & Technology, Hyderabad.He completed his B.E. Mechanical Engineering degree in K. S. Rangasamy College of Technology, Thiruchengode, M.E. Industrial Engineering in Kumaraguru College of Technology, Coimbatore and PhD at Karpagam University, Coimbatore in the area of Lean Manufacturing. He has 18 years of teaching experience and having various administrative positions. He published his research work in ten various international journals.

Ranjit Singh Sarban Singh is currently attached to Department of Computer Engineering, Faculty of Electronics and Computer Engineering, Universiti Teknikal Malaysia Melaka. He received his PhD in General Control System Engineering from

Brunel University London, United Kingdom in 2016. He has authored more than 50 research papers involving innovation on Consumer Electronics and Embedded System Engineering.

Vasundhara Sen is an adjunct faculty at Symbiosis International (Deemed University) and teaches subjects related to economics and renewable energy. Prior to joining academics, she gained industry experience of over 8 years in the area of wind energy policies and regulations.

Dereje Azemraw Senshaw is Principal Energy Specialist of GGGI's Office of Thought Leadership. In this role, he provides thought leadership to a sound 'equitable green growth' goal, tools and approaches. His priorities are focused on generating new knowledge and insights on green growth topics, and making this knowledge available and accessible to policy-makers and the public through various publications and outreach methods; Advocating green growth through analytical insights gained from country activities, leading climate diplomacy programs and activities, and findings from knowledge sharing, South-South cooperation, and country-level capacity building activities; And also, providing technical know-how and content-based services to operational/country teams to support their transition to a sustainable energy future and to implement their Nationally Determined Contributions(NDCs) under the Paris Agreement and achieve their Sustainable Development Goal (SDG) targets. Prior to the current post, he worked as Senior Energy Specialist at the Global Green Growth Institute (GGGI), and he was responsible for the development and execution of GGGI's energy projects/programs within GGGI's value chain. His work focused on the large portfolio from an advisory and technical assistance perspective, spanning all aspects of the energy system, from demand to supply, conducting (sub-) sectoral policy and technical analysis, preparing green energy plans, designing renewable energy deployment, developing bankable projects for off-grid/grid-connected renewable energy and energy efficiency, market assessments and capacity development and knowledge sharing in close interaction with operational/country teams.

Tatiana Shestopalova was born in 1954 in Samarkand. In 1978, she graduated from the Frunze Polytechnic Institute with a degree in Power Supply of Industrial Enterprises and Cities. She has been engaged in pedagogical activity since 1982, during which time lecture courses were prepared in six disciplines, consultations were held on students' research work, course and diploma projects. In 2000, she completed her postgraduate studies at the Kyrgyz-Russian Slavic University. In 2004 she successfully defended her thesis for the degree of PhD in technical sciences, in 2006 she was awarded the academic title of associate professor. In recent years, she has been intensively engaged in the problems of using renewable energy sources, taking part in a number of research projects with the Ministry of Education of the Russian Federation. Since 2017, she is the Head of the Department of Hydropower and Renewable Energy, since 2019 she is the Director of the Institute of Hydropower and Renewable Energy of the National Research University "MPEI" (Russia). She is the author of 101 published works.

Janna Titova was born in 1976 in Tashkent city, Uzbekistan. September, 1993 – June, 1998: study at Tashkent Institute of Irrigation and Mechanization of Agriculture, Tashkent, Uzbekistan. Major: Mechanization of water economic and meliorative, Engineer-mechanic. June 2006 – present: work in Research-and-production enterprise "Vodopodyomnik" ("Water lifting") as research assistant, main researcher. January, 2014 – December, 2016: doctor of science degree study at Tashkent State Technical University, Tashkent, Uzbekistan. Major: Power systems and complexes. December, 2016: received the degree of Doctor of science in technics. Dissertation subject "Algorithms and Methods of Regulation of the Pumping Stations Operating Modes". Professor's assistant, docent on Heating and cooling of electrical machines and transformers for the second academic level at Tashkent State Technical University, Department of electrical machines. Activities include lecturing and practice training. She totally has totally has 44 publications: two book chapter IGI-Global, two textbooks, five monographs, four patents for invention and software, sixteen articles in scientific technical journals, seventeen reports in scientific conferences and others.

David Wright combines an Engineering PhD from Cambridge University, UK, with his current position as Full Professor in The University of Ottawa, Telfer School of Management to provide a business perspective on emerging technologies, including their use to mitigate and adapt to climate change. He collaborates with researchers in the uOttawa SunLab and with industrial partners to provide a business analysis of solar power technologies. He is a member of the Standards Council of Canada committee working on development of new international ISO/IEC standards on "Sustainability for and by Information Technology". Dr Wright is cited in Who's Who in the World, Who's Who in Canadian Business and Who's Who in Science and Engineering. His recent publications are in the areas of Solar Power and Smart Grid.

Index

E

F

G

H

I

K

L

M

N

T

U

V

W

X

Z

Ensure Quality Research is Introduced to the Academic Community

Become an IGI Global Reviewer for Authored Book Projects

The overall success of an authored book project is dependent on quality and timely reviews.

In this competitive age of scholarly publishing, constructive and timely feedback significantly expedites the turnaround time of manuscripts from submission to acceptance, allowing the publication and discovery of forward-thinking research at a much more expeditious rate. Several IGI Global authored book projects are currently seeking highly qualified experts in the field to fill vacancies on their respective editorial review boards:

Applications may be sent to:
development@igi-global.com

Applicants must have a doctorate (or an equivalent degree) as well as publishing and reviewing experience. Reviewers are asked to write reviews in a timely, collegial, and constructive manner. All reviewers will begin their role on an ad-hoc basis for a period of one year, and upon successful completion of this term can be considered for full editorial review board status, with the potential for a subsequent promotion to Associate Editor.

If you have a colleague that may be interested in this opportunity, we encourage you to share this information with them.

www.ingramcontent.com/pod-product-compliance
Lightning Source LLC
LaVergne TN
LVHW080355011026
849161LV00009B/2308

* 9 7 8 1 5 2 2 5 8 5 5 9 6 *